MATERIALS UNDER EXTREME CONDITIONS AND NANOPHASE MATERIALS

STIC

EUROPEAN MATERIALS RESEARCH SOCIETY SYMPOSIA PROCEEDINGS

Volume 1: Ceramic Materials Research (ed. R.J. Brook)
Volume 2: Photon, Beam and Plasma Assisted Processing (eds. I.W. Boyd and E.F. Krimmel)
Volume 3: Deep Implants (eds. G.G. Bentini, A. Golanski and S. Kalbitzer)
Volume 4: Metastable Alloys: Preparation and Properties (eds. K. Samwer, M. von Allmen, J. Bøttinger and B. Stritzker)
Volume 5: Superconducting and Low-Temperature Particle Detectors (eds. G. Waysand and G. Chardin)
Volumes 6A, 6B: High T_c Superconductors (eds. P.F. Bongers, C. Schlenker and B. Stritzker)
Volume 7: Solid State Ionics (eds. M. Balkanski and C. Julien)
Volume 8: Rare-Earth Permanent Magnets (ed. I.R. Harris)
Volume 9: Defects in Silicon (eds. C.A.J. Ammerlaan, A. Chantre and P. Wagner)
Volume 10A, 10B: Silicon Molecular Beam Epitaxy (eds. E. Kasper and E.H.C. Parker)
Volume 11: Acoustic, Thermal Wave and Optical Characterization of Materials (eds. G.M. Crean, M. Locatelli and J. McGilp)
Volume 12: Beam Processing and Laser Chemistry (eds. I.W. Boyd and E. Rimini)
Volume 13: Interfaces in Biomaterials Sciences (eds. D. Muster and G. Hastings)
Volume 14A, 14B: High T_c Superconductor Materials (eds. H.-U. Habermeier, E. Kaldis and J. Schoenes)
Volume 15: Metal Matrix Composites (eds. G. Chadwick and L. Froyen)
Volume 16: Magnetic Thin Films, Multilayers and Superlattices (eds. A. Fert, G. Güntherodt, B. Heinrich, E.E. Marinero and M. Maurer)
Volume 17: Metallurgical Coatings and Materials Surface Modifications (eds. H. Hintermann and J. Spitz)
Volume 18: Surface Processing and Laser Assisted Chemistry (eds. I.W. Boyd, E. Fogarassy and M. Stuke)
Volume 19: Materials for Optoelectronic Devices, OEICs and Photonics (eds. H. Schlötterer, M. Quillec, P.D. Greene and M. Bertolotti)
Volume 20: SiC, Natural and Synthetic Diamond and Related Materials (eds. A.A. Gippius, R. Helbig and J.P.F. Sellschop)
Volume 21: Analytical Techniques for the Characterization of Compound Semiconductors (eds. G. Bastard and H. Oppolzer)
Volume 22: Micronic Integrated Sensors (ed. J.L. Robert)
Volume 23: High Energy and High Dose Ion Implantation (eds. S.U. Campisano, J. Gyulai, P.L.F. Hemment and J.A. Kilner)
Volume 24: Laser Surface Processing and Characterization (ed. I.W. Boyd)
Volume 25: Nuclear Methods in Semiconductor Physics (eds. G. Langouche, J.C. Soares and J.P. Stoquert)
Volume 26: Clays and Hydrosilicate Gels in Nuclear Fields (ed. A. Meunier)
Volume 27: Chemistry of Cements for Nuclear Applications (eds. P. Barret and F.P. Glasser)
Volume 28: Nuclear Materials for Fission Reactors (eds. Hj. Matzke and G. Schumacher)
Volume 29: Modifications Induced by Irradiation in Glasses (ed. P. Mazzoldi)
Volume 30: Polyconjugated Materials (ed. G. Zerbi)
Volume 31: SiGe Based Technologies (eds. E. Kaspar, Y. Shiraki and T.P. Pearsall)
Volume 32: Materials Surface Processing (eds. M. Stuke, E.E. Marinero and I. Nishiyama)
Volume 33: Chemistry for Electronic Materials (eds. Y. Pauleau, G. Wahl, K.F. Jensen and T. Hirai)
Volume 34: Semiconductor Materials Analysis and Fabrication Process Control (eds. G. Crean, R. Stuck and J. Woollam)
Volume 35: Synthetic Materials for Non-Linear Optics and Electronics (eds. C. Taliani, Z.V. Vardeny and Y. Maruyama)
Volume 36: CdTe and Related Cd Rich Alloys (eds. R. Triboulet, W.R. Wilcox and O. Oda)
Volume 37: Single Chamber Processing (eds. Y.I. Nissim and A. Katz)
Volume 38: High T_c Superconductors (eds. J. Dumas, H.W. Neumüller, H.F. Braun, V.P. Seminozhenko and Yu.D. Tretyakov)
Volume 39: Materials under Extreme Conditions and Nanophase Materials (eds. G. Kostorz, I. Gorynin, V. Trevilov, H.-D. Kunze, B. Skorokhod and V. Troitsky)

MATERIALS UNDER EXTREME CONDITIONS AND NANOPHASE MATERIALS

PROCEEDINGS OF SYMPOSIUM B ON
MATERIALS UNDER EXTREME CONDITIONS
AND SYMPOSIUM C ON NANOPHASE MATERIALS
OF THE 1992 E-MRS FALL CONFERENCE

STRASBOURG, FRANCE, NOVEMBER 3–5, 1992

Edited by

G. KOSTORZ
ETH, Zürich, Switzerland

I. GORYNIN
CRISM 'Prometey', St. Petersburg, Russia

V. TREFILOV
Frantsevich Institute for Problems of Materials Sciences, Kiev, Ukraine

H.-D. KUNZE
B. GÜNTHER
Fraunhofer IFAM, Bremen, Germany

B. SKOROKHOD
V. TROITSKY
Frantsevich Institute for Problems of Materials Sciences, Kiev, Ukraine

1993
NORTH-HOLLAND
AMSTERDAM-LONDON-NEW YORK-TOKYO

Printed on acid-free paper

ISBN: 0 444 81763 8

Published by:
North-Holland
Elsevier Science Publishers B.V.
Sara Burgerhartstraat 25
P.O. Box 211
1000 AE Amsterdam
The Netherlands

Reprinted from:
MATERIALS SCIENCE AND ENGINEERING A168 (1, 2)

The manuscripts for the Proceedings were received by the Publisher:
February–April 1993

Printed in The Netherlands

Preface

The programme of *Symposium B on Materials Under Extreme Conditions* covered a variety of extremes, in particular high rates of deformation, low temperatures, high temperatures, high pressure, thermal shock, high currents, corrosion and wear, surface processing, even high purity. Only a limited number of papers were accepted for this volume. They are arranged according to the following sequence; mechanical properties of metallic and ceramic materials under various extreme conditions, wear, surface properties and processing, internal stresses and high pressure, and high purity.

Symposium C on Nanophase Materials dealt primarily with the preparation methods and physical/chemical properties of ultrafine grained polycrystalline materials. These nanophase materials appear to be of increasing importance now with respect to both the scientific as well as the application point of view. At the moment we are far from understanding the effects of preparation conditions on grain boundary structure and particle size effects. Symposium C was intended to contribute especially to this subject, i.e. to the relationship between manufacturing conditions and properties of these nanophase materials. Out of a total of 59 contributions to the field 26 have been selected for publication within this proceedings volume. They are introduced by a short review which is followed by the topics of ultrafine particle synthesis, consolidation and properties of (a) nanophase metals, (b) nanoceramics and (c) nanocomposites.

This limited collection of papers for both symposia will not give a complete and well-balanced overview of the many complex facets of the broad subjects. This was neither possible nor intended – also in view of the short-term preparation of the symposia.

G. Kostorz
Zürich, Switzerland

H.-D. Kunze
Bremen, Germany

Sponsors

This conference was held under the auspices of:

– The Council of Europe
– The Commission of the European Communities

It is our pleasure to acknowledge with gratitude the financial assistance provided by:

– Banque Populaire	(France)
– Brasserie de Kronenbourg	(France)
– Centre de Recherches Nucléaires	(France)
– Centre National de la Recherche Scientifique	(France)
– Elsevier Science Publishers B.V.	(The Netherlands)
– EOARD "United States Air Force – European Office of Aerospace Research and Development"	(U.K.)
– La Maison du Café – Ergersheim	(France)
– Office du Tourisme de la Ville de Strasbourg	(France)
– The Commision of the European Communities DG I	
– The Commision of the European Communities DG XII	
– The Council of Europe	
– The European Parliament	

Contents

II. Symposium C on Nanophase Materials

PART I

Symposium B
on
Materials under Extreme Conditions

Materials Science and Engineering, *A168* (1993) 1–3

Flow stress behavior of cast Co–Cr–Mo alloy evaluated by high strain rate of compression tests

Jong-Cheng Tsai, Jing-Bang Duh and Shan-Chang Cheuh

Materials Research Laboratories, Industrial Technology Research Institute 195 Chung Hsing Rd., Section 4, Chutung, Hsinchu 310 (Taiwan)

Abstract

The effect of the temperature and strain rate on the flow stress behavior of cast Co–Cr–Mo alloy was evaluated by the hot compression test. Results show that the flow stress increases rapidly with increasing strain rate and decreasing temperature. The flow softening behavior is due to dynamic recrystallization during hot working. The strain to recrystallize increases as the strain rate increases from 10^{-3} s to 1.0 s^{-1}, and then decreases with increasing strain rate above 1.0 s^{-1}.

1. Introduction

The cast Co–Cr–Mo surgical implant alloy is a hard to work material, owing to its poor hot ductility and high deformation resistance [1]. Previous research has shown that the hot worked alloy may provide a more sound surgical implant than the as-cast condition [2]. The low strain rate (less than 0.1 s^{-1}) of isothermal forging was studied to examine the feasibility of fabricating wrought implants [3]. In this study, hot compression testing was carried out to investigate the effect of strain rates up to 100 s^{-1} on the flow stress behavior and microstructure.

2. Experimental procedure

The Co–28.5Cr–6.2Mo–0.24C alloy was prepared as a 10 mm diameter and 15 mm high cylinder from a 70 mm diameter ingot for the hot compression test. The flow stress behavior was evaluated using a Gleeble-2000 thermal–mechanical tester. The true stress–true strain curves were measured by a diametrical strain gage attached to the mid-length surface of the specimen. The compression jaws were cooled by circulated water to prevent an increase in temperature of the specimen during compression. The hot compression tests were conducted at the temperature of 700–1200 °C and strain rate of 10^{-3}–10^{2} s. Once the compression was finished, the compressed specimens were quenched at a cooling rate of 100 °C s^{-1}. The centers of the compressed specimens were selected to examine microstructural changes. The etchant for metallographic observation was 90% HCl and 10% H_2O_2.

3. Results and discussion

The true stress–true strain curves measured at 700–1200 °C at a high strain rate of 100 s^{-1} are presented in Fig. 1. The flow stress shows a strong dependence on temperature, which decreases with increasing temperature. It is found that at 800 °C or lower, a significant amount of strain hardening occurs, *i.e.* the flow stress increases rapidly with increasing strain until rupture. At temperatures higher than 900 °C, the stress curves show characteristics of dynamic recrystallization during hot working, exhibiting a peak at medium strains and flow softening at

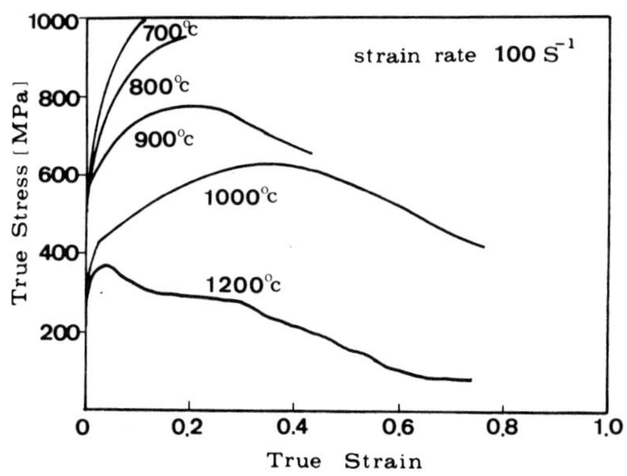

Fig. 1. Temperature dependence of true stress–true strain curves for compression at a strain rate of 100 s^{-1}.

0921-5093/93/$6.00

larger strains [4]. At 700 °C, the alloy shows a cored cobalt matrix with large grains and a dispersion of eutectic carbides located interdendritically and along the grain boundaries, as shown in Fig. 2(a). This indicates that the microstructure still retains some of the cast features after deformation at 700 °C. Nevertheless, after a strain of 0.2 at 900 °C, a recrystallized structure appeared within the carbide-free dendrite cores, shown in Fig. 2(b). It is also observed that break down and spheroidization of the eutectic carbides occur in the intedendritic regions. This microstructural change correlated with the flow softening behavior at temperatures above 900 °C shown in Fig. 1. In addition, the flow stress decreases rapidly with increasing strain at 1200 °C, almost half that at 1000 °C. This drop-off strength at 1200 °C is attributed to incipient melting of the interdendritic eutectic along grain boundaries [1]. When the alloy is deformed at a high strain rate, the temperature increase due to deformation and interface friction may raise the temperature of the alloy to near the melting point of the eutectic, 1235 °C [5], and drastically reduce the hot strength.

Figure 3 shows the effect of strain rate on the flow stress behavior at 1000 °C. The stress at fixed strain increases rapidly with increasing strain rate. The flow curves at all strain rates exhibit flow softening. Owing to the high work hardening effect at the high strain rates and low temperatures, the degree of flow softening is large (*cf.* Figs. 1 and 3). At the fastest strain rate of 100 s^{-1} the microstructure shows severe flow localization along intense shear bands (see Fig. 4(a)), whereas at a lower strain rate of 1.0 s^{-1} (Fig. 4(b)) the plastic flow lines appear more homogeneous. For all

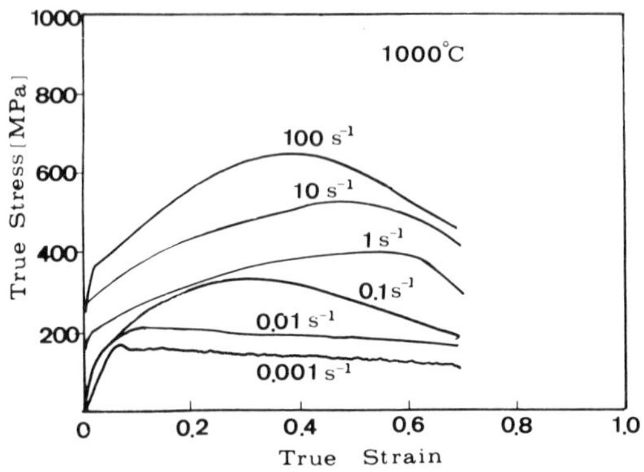

Fig. 3. The effect of strain rate on the flow stress at 1000 °C.

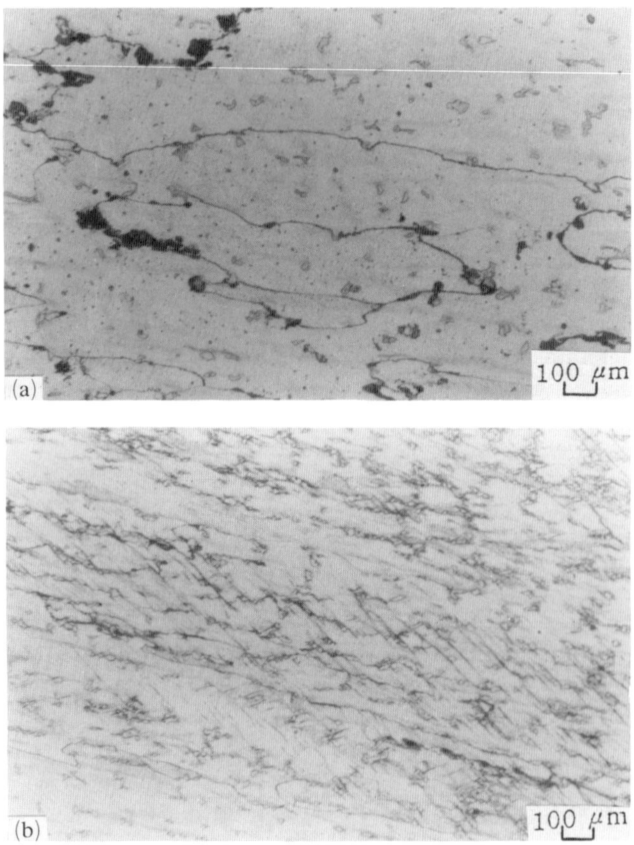

Fig. 2. Microstructure deformed at different temperatures at a strain rate of 100 s^{-1}: (a) 700 °C, (b) 900 °C.

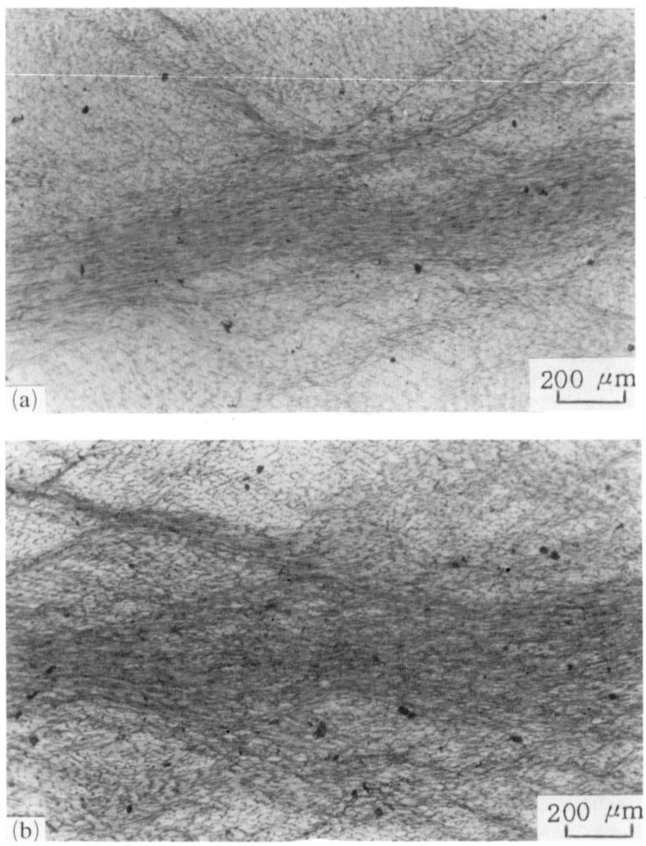

Fig. 4. Effect of strain rate on the microstructure at 1000 °C: (a) 100 s^{-1}, (b) 1 s^{-1}.

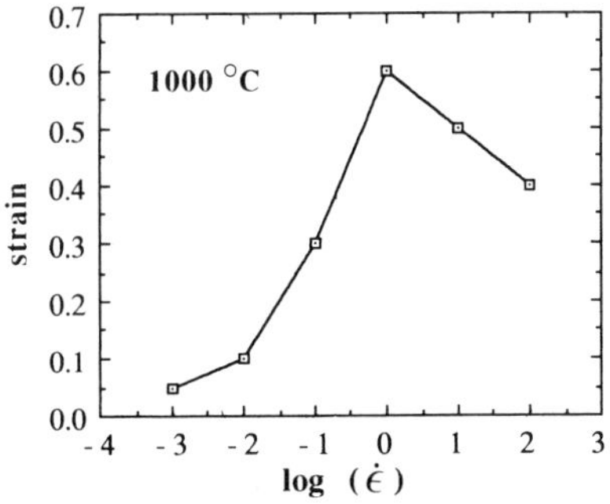

Fig. 5. Effect of strain rate on the strain to induce dynamic recrystallization at 1000 °C.

strain rates at 1000 °C, recrystallized grains exist at the center of the specimens. However, the area near the end surfaces of the pancake still remains uncrystallized in the dead zone owing to the friction effect.

It has been recognized that the strain for initiation of dynamic recrystallization during hot working is of the order of the strain at the peak strength [6, 7]. Based on the peak in Fig. 3, the effect of strain rate on the strain to recrystallize is shown in Fig. 5. The strain at peak strength increases as the strain rate increases from 10^{-3} s^{-1} to 1 s^{-1}. The trend of this result is similar to previous results for compression at strain rates below 0.1 s^{-1} [3]. It is suggested that deformation at decreasing strain rates from 1 s^{-1} to 10^{-3} s^{-1} will take longer at the elevated temperature and promote the restoration process, thus reducing the strain to recrystallize. The temperature increase during high strain rates of compression above 10 s^{-1} is measured to be below 20 °C. Therefore, the effect of deformation heating on the flow softening behavior may be minor. It shows that the minimum strain for the initiation of dynamic recrystallization at a faster strain rate (100 s^{-1}) is lower than that at the lower strain rate (1 s^{-1}). This would be explained by the considerable inhomogeneous deformation at the high strain rate of deformation, inducing the appearance of local dynamic recrystallization [3]. The maximum strain to recrystallize will occur at the strain rate of 1.0 s^{-1} for Co–Cr–Mo alloy at 1000 °C.

4. Conclusion

(1) The flow stress of cast Co–Cr–Mo alloy increases very rapidly with increasing strain rate and decreasing temperature.

(2) The main flow softening mechanism is dynamic recrystallization.

(3) The strain to recrystallize will increase with increasing strain rate from 10^{-3} s^{-1} to 1 s^{-1}, and decrease with increasing strain rate higher than 1.0 s^{-1}.

Acknowledgment

The authors express their appreciation to the Ministry of Economic Affairs of Taiwan for financial support.

References

1 J. C. Tsai and J. B. Duh, *Scripta Metall., 27* (1992) 561.
2 T. M. Devine and J. Wulff, *J. Biomed. Mater. Res., 9* (1975) 151.
3 J. P. Immarigeon, K. Rajan and W. Wallace, *Metall. Trans. A, 15* (1984) 339.
4 H. J. McQueen and J. J. Jonas, in R. J. Arsenault (ed.), *Plastic Deformation of Materials, Treatise on Materials and Technology*, Vol. 6, Academic Press, New York, 1975, p. 393.
5 T. Kilner, R. M. Pilliar and C. Allibert, *J. Biomed. Mater. Res., 16* (1982) 63.
6 J. R. Lane and N. J. Grant, *ASME Trans., 44* (1952) 113.
7 C. M. Sellar, *Philos. Trans. R. Soc. London, Ser. A, 288* (1978) 147.

Dynamic compaction of Al–Li–X powder obtained by a rotating electrode process

M. Vukčević
Faculty of Metallurgy, Podgorica (Montenegro)

S. Glišić
Institute of Physics, Belgrade (Serbia)

D. Uskoković*
Institute of Technical Sciences, Serbian Academy of Science and Arts, Belgrade (Serbia)

(Received November 3, 1992; in revised form January 15, 1993)

Abstract

The microstructures of dynamically compacted Al–Li–X powders obtained by the rotating electrode process were studied. Dynamic compaction was carried out using a LEXAN projectile whose velocity was 1520 m s^{-1} and the dynamic pressure obtained in the area of the (pseudo) planar shock wave was 7.1 GPa. Strong bonding between Al–Li–X particles with a thin melted contact zone was observed after dynamic compaction. Coarse particles had fewer interparticle contacts, but the contacts were stronger than for small particles. In the first layer of the sample in the area of (pseudo) planar shock wave the presence of 0.44 pores mm^{-2} and a total porosity of 4.6% were detected by quantitative metallography. Interparticle bonding was not detected in the first particle layer in the area affected by the release wave.

1. Introduction

The synthesis of new materials by powder processing under extreme conditions has been extensively studied over the last few years, with special emphasis on dynamic compaction owing to its many advantages compared with other techniques of powder consolidation [1, 2]. In the case of dynamic compaction, consolidation near to the theoretical density is achieved owing to the strong shock wave (up to 50 GPa) acting on the powder for a very short time, only a few microseconds. The temperatures generated in the samples are much lower than the temperatures of recrystallization of many materials. Hence, dynamic compaction is especially significant for processing of rapidly solidified crystalline and amorphous powders.

There are basically two methods of dynamic compaction: explosive detonation [3] and the impact of a high velocity projectile [4]. The advantage of the first method is lower capital investment, the possibility of producing larger compacts and designing various compact shapes. The advantage of the second method is better control of the shock parameters, thus offering greater feasibility for study of the physics of consolidation.

There are only a few publications about the dynamic compaction of Al–Li–X powders [5–7], describing the possibility of preparation of low porosity compacts (*ca.* 5%). Relatively weak interparticle bonding and cracks are the greatest problems for application of the products. However, there are no data for the compaction of these powders of defined geometry. From the standpoint of the study of dynamic compaction, the Al–Li–X system is especially interesting because of its high strength and high modulus of elasticity. Hence the aim of this paper was to report preliminary results of our study of phenomena occurring during the formation of contacts in Al–Li–X powders with defined geometry using shock waves with controlled properties.

2. Experiments

2.1. *Preparation of Al–Li–X powder by the rotating electrode process*

The alloy of nominal composition 95.3% Al, 2.5% Li, 1.3% Cu, 0.67% Mg, 0.12% Zr, 0.07% Ti, 0.04% Fe was prepared by melting in the laboratory vacuum-induction furnace. Electrodes for centrifugal atomiza-

*Correspondence address: Pro Ignis Consultancy, Mr. Miša Ignis, Postbus 82430, NL 2508 EK, Denhaag, Netherlands.

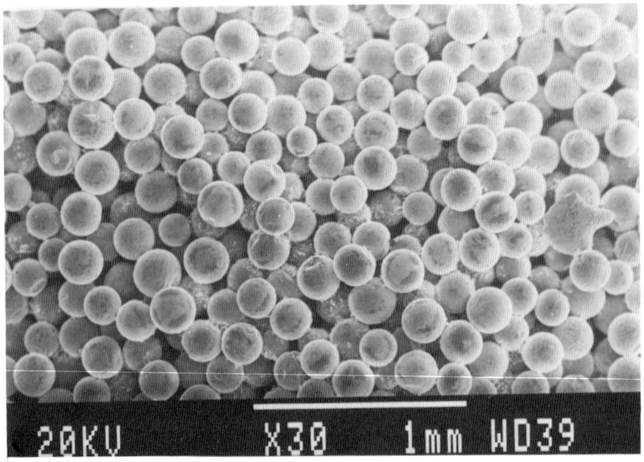

Fig. 1. Scanning electron micrograph of Al–Li–X powder obtained by centrifugal atomization, *i.e.* by REP.

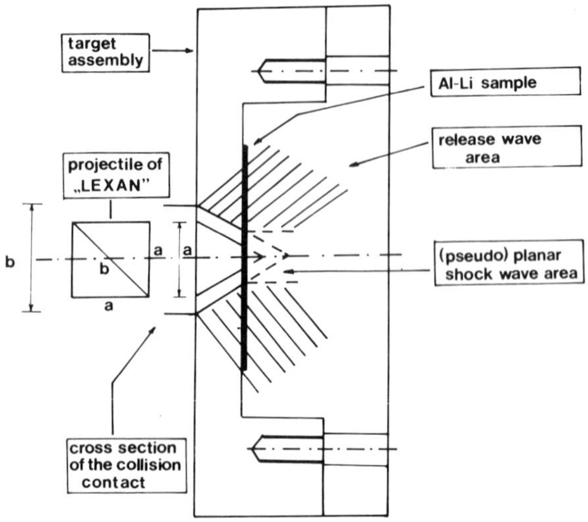

Fig. 2. Cross-section of the target in interaction with the projectile and the area of shock propagation.

tion were obtained from these ingots. Metal powder was processed in the laboratory REP equipment [9, 10]. The diameter of the consumed electrode was 16 mm. The angular velocity was 9000 rad min^{-1} and the current was 54 A. The experiments were done in helium with a small content of argon. Typical Al–Li–X powder obtained by the rotating electrode process is shown in Fig. 1. The particle structure shows "dendritic" characteristics in the case of smaller particles (50–150 μm) and a "cellular" structure in the case of coarse particles.

2.2. Dynamic compaction of Al–Li–X powders—method and technique

Dynamic compaction of Al–Li–X powders was carried out by the (pseudo)planar shock wave generated by the high velocity impact of a projectile. The projectile was launched by an electrodynamic macroparticle accelerator (EDA) [8], in English often called a rail gun. The EDA is a laboratory device which uses electromagnetic and plasma propulsion to accelerate a dielectric projectile (or sabot) up to the required velocity. The working impulse of the device is controlled by the electrical regime only, defined by an accumulated energy of up to 100 kJ, a peak current of up to 480 kA, and a total coefficient of efficiency of over 7%. The planar shock wave inside the target was produced by the impact of rectangular parallelepiped LEXAN projectile with cross-section 12 mm × 12 mm, with mass measured before launching and velocity measured just before impact. The target assembly consists of a copper container which contains the powder as a thin disc.

A cross-section of the target assembly containing a disc-shape charge of powder and the shock propagation inside it is illustrated in Fig. 2. The shaded area designates the release wave as the result of the finite dimensions of the projectile and of edge effects. This is an area of uncontrolled parameters of impact. The area of known properties of the shock wave is a cone in which the planar shock wave propagates through part of the powder sample disc. The shape of this cone is defined by target material properties and the shock velocity. In the target assembly, the area of the entire disc powder sample is greater than the area subjected to shock wave effects. The thickness of the porous powder disc sample (in the direction of an incoming shock) is just a few layers of particles of the powder.

The structure of shock wave propagation through a porous material is quite complex owing to the noncontinuous nature of the material. The first layer of the powder sample disc (in the direction of the incoming shock wave) is called the interface between the container (copper) and the powder sample. We wanted to determine the local state in this layer under the realized shock wave conditions. Moreover, the dynamic compaction of powder depends strongly on the process of sample preparation.

According to the well known shock-Hugoniots for isotropic homogeneous projectile materials (LEXAN) and target containers (copper), by measuring the projectile velocity just before impact, and the mass density of both materials, we can determine the pressure, particle velocity and shock wave velocity at the interface layer in the copper container. In deeper layers of the porous sample one can estimate the same quantities according to different constitutive models [11]. At this point we ignore the contribution to shock pressure from the reflection wave from the rear side of the sample (Fig. 2).

Using the experimental method known as the collision method or momentum transfer method [12], equations of continuity and measured velocity of the projectile W, one can equalize the pressures at the collision interface on both the projectile and target sides:

$$P_p = \rho_p U_{sp}(W - U_p) \quad \text{(projectile)} \tag{1}$$

$$P_t = \rho_t U_{st} U_p \quad \text{(target)} \tag{2}$$

where P_p and P_t are pressures, U_s is the shock velocity, U_p is the particle velocity, W is the measured projectile velocity and ρ_p and ρ_t are densities of projectile and target materials respectively. The dynamic properties of materials under a high pressure load are determined by Hugoniot curves

$$U_{sp} = 2.4 + U_p \quad \text{(LEXAN)} \tag{3}$$

$$U_{st} = 3.91 + 1.51 U_p \quad \text{(copper)} \tag{4}$$

which are known from the data library [13]. The values of U_p, U_s, and W are given in millimetres per microsecond, P is in gigapascals and the densities are grams per cubic centimetre.

Substituting eqn. (3) into eqn. (1) and eqn. (4) into eqn. (2), and equalizing eqns. (1) and (2) with known densities for the projectile and target materials, one can determine the particle velocity on the copper side of the collision interface. By inserting U_p into eqn. (2), the dynamic pressure can also be determined. Both calculated values are only valid for the planar shock wave on the copper side of the collision interface. The same pressure load only occurs on the first layer of the porous sample at the interface. Measurement of the projectile velocity gives sufficient data for calculation of the pressure on the collision interface inside the target.

2.3. Static and dynamic compaction

The powder sample consists of Al–Li–X alloy spheres (of diameter 315–500 μm) arranged in five to six layers in a disc shape (diameter 40 mm) (Fig. 3) inside the target. The original powder packing density before the shock is very important. We studied the dependence of the green density on pressing pressure and found the first sign of consolidation at over 1.5 GPa for a few particles of powder only. Good contact between the spheres (particles) of the powder sample is obtained by cold pressing of 0.44 ± 0.01 GPa. Neither evacuation nor vibration of the powder sample were done before or during cold pressing at room temperature. The same procedure was used for a number of experiments.

The copper target with encapsulated powder sample was at room temperature before impact of the projectile. The samples were compacted dynamically by projectile velocities of 1500–1520 m s^{-1}, producing interface pressures of 7.0–7.1 GPa. We carried out a number of impact experiments under the same conditions and we selected the (pseudo)planar impacts for analysis only. We neglected the experiments with an impact tilt between the projectile head plane and the target plane. This is a basic condition for the use of eqns. (1)–(4) for calculation of the dynamic pressure at the forward copper–sample interface inside the target. The criteria for the selection of (pseudo)planar impacts was the appearance, symmetry, and dimensions of the target impact crater. The dynamic tilt of the projectile head was not measured.

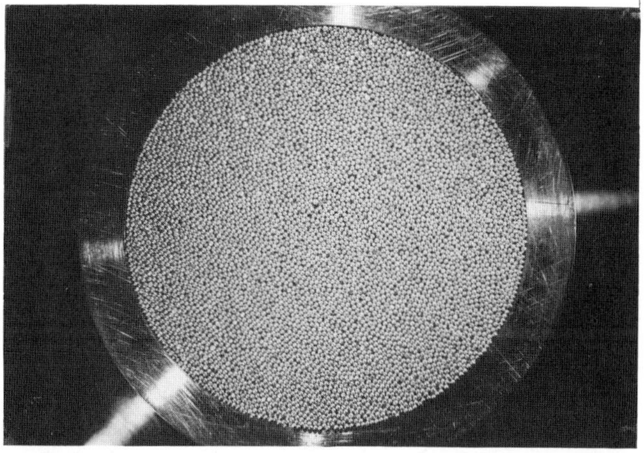

Fig. 3. Al–Li–X spheres in the target assembly before compaction.

3. Results and discussion

3.1. Microstructural analysis

Microstructures of Al–Li–X spheres compacted by a shock wave are illustrated in Fig. 4. It is obvious that strong bonds between Al–Li–X particles were formed in the zone of the (pseudo)planar impact wave, in the first layer of particles nearest to the projectile impact. The liquid phase between deformed Al–Li–X spheres is also obvious. It appears as the result of local melting caused by the release of heat due to friction between sliding particles and plastic deformation. The white areas in the grain boundaries or between a few grains are the result of melted and rapidly solidified material. Energy-dispersive analysis of X-rays indicates that there is no difference in chemical composition of analysed areas, although it is logical to expect some differences in the fine crystalline structure. It is also obvious that the character of the melted thin film at the interparticle contacts depends on the initial particle size distribution and on the type of deformation.

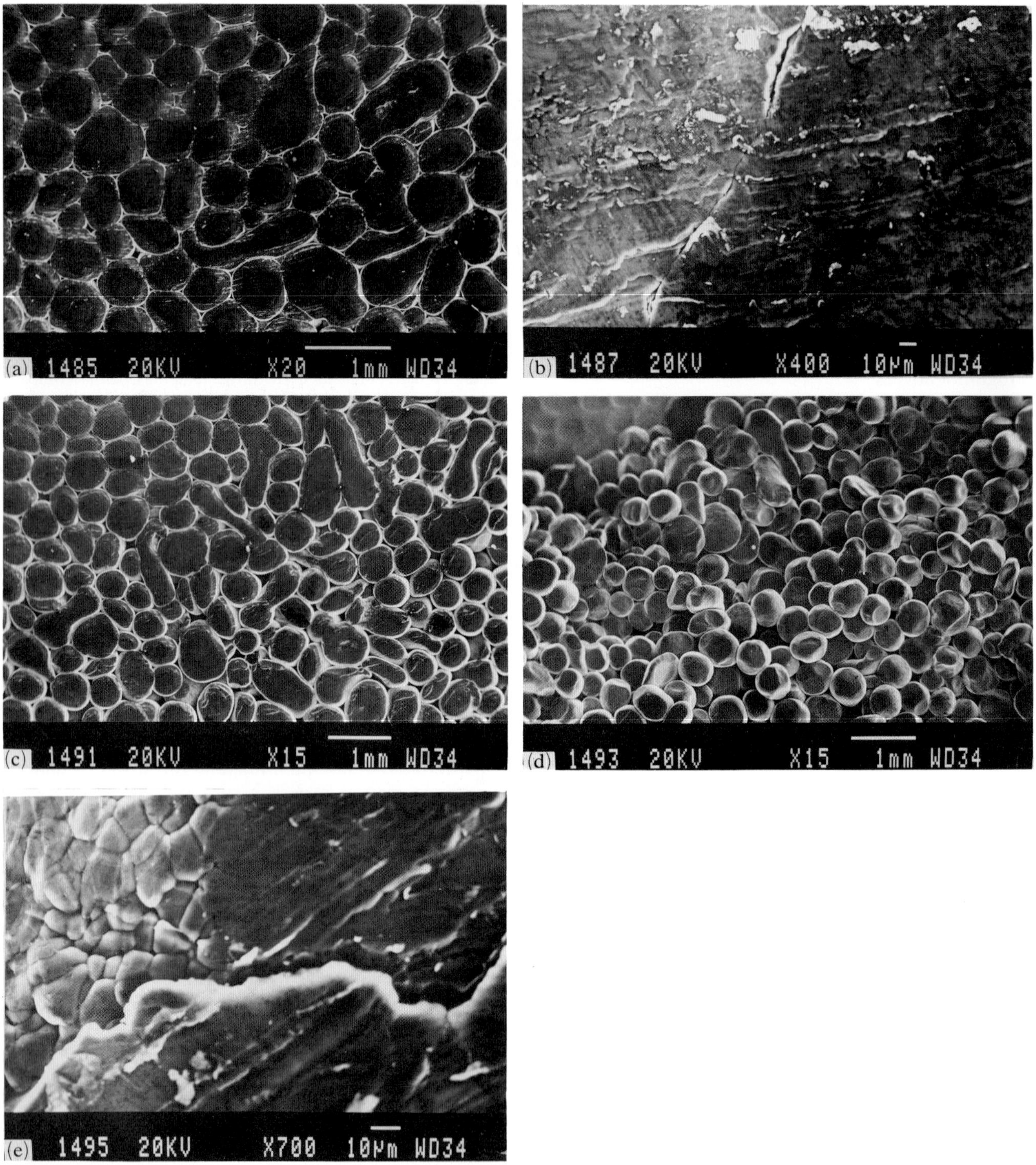

Fig. 4. Scanning electron micrographs of compacted Al–Li–X spheres at a dynamic impact of 7.1 GPa: (a) surface layer of Al–Li–X spheres in the zone of the (pseudo)planar shock wave; (b) magnified detail of a contact; (c) the effect of the shock wave at the boundary between the zones of the (pseudo)planar wave and released wave; (d), (e) the effect of compaction through the front of shock wave propagation.

It is obvious from Fig. 4(a) that the coarser particles have fewer interparticle contacts with stronger bonding than the smaller particles. The variation in particle size distribution over a wider range causes changes in packing, *i.e.* a decrease or increase in residual voids, with different effects on heat generation. The change in particle shape factor and deviations from an ideal sphere result in better opportunity for the formation of interparticle contacts. The mechanism of formation of contacts between particles of different sizes is also interesting. On the basis of residual deformations, most of the heat energy released owing to plastic deformation is accumulated in this case on the smaller particles. Hence, higher stresses during impact appear in the particles with smaller radius. This causes plastic flow, heating and softening, which result in further growth of the contact area between particles. As a consequence, the change in particle volume is realized mainly by changes in shape of the smaller particles.

The effect of shock wave propagation on the boundary between the zones of the (pseudo)planar and release waves on the development of the microstructure during dynamic compaction is illustrated in Fig. 4(c)–(e). In the zone of release, consolidation has not occurred in the first layer of particles, so that the boundary between the zones of (pseudo)planar and released waves is clearly visible. Deformation of spheres is much lower in this case and spherical particles with no changes in the initial "cellular" structure are often present (Fig. 4(e)). The liquid phase is probably only present in the zone of action of the (pseudo)planar impact wave (upper left part of Fig. 4(c)). The light parts on the right of the figure in the zone of action of the released wave are probably the consequence either of the quasiliquid (quasiviscous) state of deformed particles or of the geometry of the image-taking, *i.e.* the slope of the emitted or reflected electron beam from the brim of the spheres. The action of the shock wave, at the pressure used of 7.1 GPa, is only of interest for the first layer of sample.

The number of residual pores after compaction and their volume ratio in the bulk of compacted material was estimated by quantitative metallography [14]. This analysis was only possible in the first layer of the impact zone of the (pseudo)planar wave. Application of this method has shown the presence of 0.44 pores mm^{-2} and a total porosity of 4.6% in the impact zone.

3.2. Discussion of results

The processes occurring in the contact area in a particulate system during dynamic compaction depend on the particle size, density, thermodynamic characteristics of the material and the conditions of impact. For choice of the optimal regime of shock compaction it is necessary for the following common conditions to be fulfilled [15]:

$$e_0 > e_{min} \quad p > p_{min} \quad E_k > E_{kmin}$$

$$t > t_{min} \tag{5}$$

where e_0 is the specific internal energy, p is the pressure, E_k is the density of the energy across the contact surface unit, and t is the duration of impact.

The physical meaning of the first two conditions is that both a minimal pressure for plastic flow which will result in the disappearance of pores and a defined specific energy which depends on the pressure and initial particle distribution are needed. The condition $e_0 > e_{min}$ determines the transition to the dynamic regime of deformation of porous materials, necessary for the fracture of surface inclusions and shearing sufficient for the formation of strong interparticle bonding. This transition can be realized by a change in the initial temperature, particle size, porosity, etc. If we assume that most of the internal energy e_0 is generated in grain boundaries, this law defines the minimal quantity of melt formed at interparticle contacts necessary for the formation of strong contacts. The third condition is connected with the achievement of a certain level of specific surface energy needed for the activation of diffusion processes which lead to the formation of strong bonds between particles. The last condition ($t > t_{min}$) is usually connected with providing a sufficient duration of impact wave needed for the total disappearance of pores.

Few models have been proposed to describe densification and the mechanism of contact formation in aluminium alloys. Gourdin's [16, 17] model assumes a threshold stress of 2.25 GPa for the dynamic compaction of Al–6Si alloy and a specific energy input of 2×10^5 J kg^{-1} for melting at interparticle contacts and for the formation of strong contacts. The experiments done by Morris [18] illustrate intensive melting of interparticle contacts and densification in the case of 99% pure aluminium compacted at 3 GPa with a projectile velocity of 1500 m s^{-1}. Peng *et al.* [5] computed pressure profiles for different packing configurations of the explosive during dynamic compaction of Al–Cu–Li–Mg alloys obtained by vacuum atomization (particle size less than 45 μm). The compaction and formation of interparticle contacts are the result of the primary wave and rarefaction waves and the duration of impact. Compacts obtained at calculated impacts of 3.6 GPa have strong interparticle contacts and 60%–80% of the tensile strength of extruded samples.

The pressure of 7.1 GPa, applied in our experiments, is higher than the pressures used in other studies where lower porosities have been obtained after consolidation. Our Al–Li–X powder had larger particles than powders used in other studies. Hence, pores between Al–Li–X particles require much higher pressures for complete disappearance or hot dynamic compaction [19].

The shock impedances of the target copper container and the powder sample inside were very different, and there were manifold shock wave reflections inside the powder sample. It is very difficult to measure the shock velocity in a thin (about 1.5 mm) porous sample under shock wave reflections, finite dimension of plane shock, boundary effects etc. in order to obtain pressure–time profiles. From a theoretical point of view, it would be possible to calculate a superposition of reflected shocks on interfaces if the Hugoniot of the porous sample is known and idealized assumptions are used [20]. In principle, the Hugoniot of porous materials is strongly dependent on the preparation of the porous sample. We can only determine the shocked states of the porous material on the basis of changes in the porous states at the cross-section perpendicular to both sample interfaces. As for the preparation procedure, for our powder sample Al–Li–X, container system and container material, and shock conditions, we found the same changes in both interface sample layers (porosity and quantity of melting) and smaller changes in inner layers in the sample. That means that in our experiments, the dynamic pressures on both interfaces were similar, and were not greater on the back interface (due to the reflected shock), as would be concluded from idealized considerations [20]. In reality, there is dispersion of the shock front through inhomogeneous porous material, dependent on the type of porous material and its method of preparation. We assume that for an aluminium alloy container, the cross-sectional changes in the material, under the same experimental conditions, would be different because of the close shock impedances of the container material and the powder sample. Experimental work is in progress [19].

4. Conclusion

Microstructures of dynamically compacted Al–Li–X spheres obtained by the rotating electrode process were compacted using a projectile launched from an electrodynamic macroparticle accelerator with a controlled dynamic pressure of 7.1 GPa. Microstructural examinations show that there is a clear boundary between the zones of (pseudo)planar and release waves and a significant dependence of compaction on the particle shape, size and distribution. In the zone of the (pseudo)planar shock wave, quantitative metallographic examination showed that the first layer of Al–Li–X spheres had a porosity of 4.6%. In the zone of the release wave strong interparticle contacts were not formed, *i.e.* no signs of significant densification were found.

Acknowledgment

This study was financially supported by the Ministry of Science through the projects "Physical Chemical Processes in Homogeneous and Heterogenous Systems" and "Synthesis of New Materials in Disperse and Compact Forms by Combustion Synthesis and Dynamic Compaction". The authors wish to express their gratitude to Professor Vladimir Petrović, Ms Lidija Nelević, BSc, Mr. Predrag Živanović, BSc, and Mr. Dragan Tasić, for their contributions to the experimental part of this work.

References

1 L. E. Murr (ed.), *Shock Waves for Industrial Applications*, Noyes, Park Ridge, NJ, 1989, p. 532.
2 D. P. Uskoković, H. Palmour and R. M. Spriggs (eds.), *Science of Sintering: New Directions for Materials Processing and Microstructural Control*, Plenum, New York, 1989, p. 644.
3 G. S. Clyens and W. Johnson, *Mater. Sci. Eng., 30* (1977) 121.
4 D. G. Morris, *Met. Sci., 14* (1980) 215.
5 T. C. Peng, S. M. L. Sastry, J. E. O'Neal and D. Brasher, *Metall. Trans. A, 16* (1985) 1445.
6 L. H. Yu, *Mater. Sci. Eng., 132A* (1991) 257.
7 M. A. Meyers, N. N. Thadhani and L. H. Yu, in L. E. Murr (eds.), *Shock Waves for Industrial Applications*, Noyes, Park Ridge, NJ, 1989, p. 265.
8 S. Glišić, D. Šević, D. Tanaskoski and I. Belić, in N. Konjević *et al.* (eds.), *Summer School and Int. Symp. on Physics of Ionized Gases, XIV SPIG 1988, Sarajevo*, University of Sarajevo, 1988, p. 446.
9 M. Zdujić, M. Sokić, V. Petrović and D. Uskoković, *Powd. Met. Int., 18* (1986) 275, 325.
10 M. Šušić, M. Zdujić, Lj. Karanović and D. Uskoković, *Z. Metallkd., 80* (1989) 497.
11 K.-H. Oh and P.-A. Persson, *J. Appl. Phys., 66* (1989) 4736.
12 Y. B. Zeldovich and Y. P. Raizer, *Physics of Shock Waves and High Temperature Hydrodynamic Phenomena*, Vol. II, Academic Press, New York, 1967, p. 724.
13 S. P. Marsh (ed.), *LASL Shock Hugoniot Data*, University of California Press, 1980.
14 E. E. Underwood, *Quantitative Stereology*, Addison-Wesley, Reading, MA, 1968, p. 330.
15 O. V. Roman, V. F. Nesterenko and I. M. Pikus, *Phys. Gor. Vzriv, 15* (1979) 102.
16 W. H. Gourdin, *Lawrence Livermore Lab. Rep. UCRL-88902*, Part 3, May 1983 (Livermore, CA).
17 W. H. Gourdin, *J. Appl. Phys., 55* (1984) 172.
18 D. G. Morris, *Mater. Sci. Eng., 57* (1983) 187.
19 M. Vukčević, S. Glišić and D. Uskoković, *1993 Powder Metallurgy World Congress, Kyoto, July 1993* to be presented.
20 P. Kasiray, T. Vreeland, Jr., R. B. Schwarz and T. J. Ahrens, *Acta Metall., 32* (1984) 1235.

Materials Science and Engineering, A168 (1993) 11–16

Shear fracture at liquid nitrogen temperature

E. S. Dzidowski

Technical University of Wroclaw, I-24, Wybrzeze Wyspianskiego 27, 50-370 Wroclaw (Poland)

Abstract

Interpretational limitations of the commonly known models of shear fracture are indicated. Fundamental relationships between brittle and shear fracture occurring at the liquid nitrogen temperature of 77 K are discussed. In particular, the effect of grain size and the condition of grain boundaries on the propagation of shear fracture of 0.09% low-carbon steel at the temperature of 77 K has been determined. An experimental–analytical relation defining the influence of temperature (77–293 K) and grain size on the value of limiting displacements from the perspective of the nucleation and propagation of shear fracture in low- and medium-carbon steels has been formulated. Similarity has been found between the mechanisms of shear fracture at temperatures of 293 and 77 K.

1. Introduction

The mechanism of shear fracture is one of the least known mechanisms [1]. Interpretation of the mechanism and development of shear fracture is made difficult by the fact that the commonly known models of this mechanism do not take into account to a sufficient degree the crystalline structure of metals [2–4]. Classic fracture mechanics, in turn, treats shear fracture as less dangerous than brittle fracture and does not devote too much attention to it [5].

Figure 1 shows the development of a fracture during shearing of low-carbon steel at the temperature of liquid nitrogen (77 K). As can be seen, brittle fracture (1) appears first. This is revealed by the dramatic fall in shearing force (observed directly after plasticization of the sheared material) and the stabilization and gradual reduction in the value of shearing force P_c that follow it.

This means that the subsequent ductile fracture (2) takes place at a similar or lower load level. It follows from the above that ductile fracture (2) at low temperatures cannot be regarded as less dangerous than brittle fracture (1), from the viewpoint of the extreme values of the shearing force. It can also be seen that the brittle crack trajectories (1) do not coincide with the shear fracture trajectories (2). So one cannot say that we are dealing with the classic transition from one kind of fracture to another. Therefore, a closer look at and explanation of the development of shear fracture at low temperatures may be interesting not only from the point of view of the strength of materials but also from the standpoint of the theory of controlled decohesion.

The main purpose of this work was to determine and explain the influence of the structure of low-carbon steel on the development of shear fracture during shearing at the temperature of liquid nitrogen (77 K).

2. Materials and methods

Tests were carried out on round specimens of diameter $d_0 = 22$ mm and length $l_0 = 3d_0$ made of 0.09% C low-carbon steel. The chemical constitution of the tested steel is shown in Table 1. Three batches of specimens, with differing structures, were tested. The differences were produced by appropriate heat treatment. The kinds and conditions of the heat treatment and the size of grains before and after are given in Table 2. Toughening (atypical for this type of steel) was applied to change the shape of cementite precipitates and the volume of impurities at the grain boundaries. This treatment results in coagulation of the cementite precipitates and displacement of the impurities from the boundaries to the inside of the grains. The particular batches of specimens differed in the size of grains, the structure of cementite and the condition of grain boundaries (Fig. 2).

The specimens were quenched in a propyl alcohol bath cooled with liquid nitrogen. The lowest applied temperatures (77 K) were obtained by direct cooling in liquid nitrogen.

The specimens were sheared using a device with sleeve cutting tools. One of the cutting tools was fixed and the other slidable. A power press drove the

0921-5093/93/$6.00

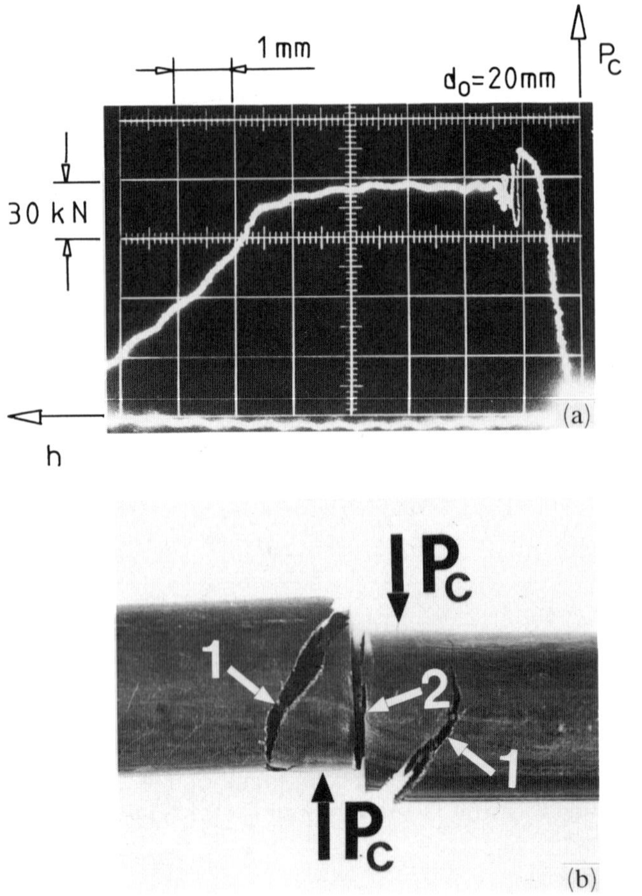

Fig. 1. (a) An oscillogram of shearing force P_c (a) and (b) an example of the cracking of steel sheared at the temperature of liquid nitrogen, 77 K. Simultaneous brittle fracture (1) and shear fracture (2) and the differences in trajectories of the fractures can be seen.

TABLE 1. The chemical constitution of 0.09% C steel

Element	C	Mn	Si	P	S	Cr	Ni	Cu
Content (%)	0.09	0.04	0.22	0.012	0.031	Trace	Trace	0.03

slidable tool. The specimens were sheared in the middle of their length.

Oscillograms of shearing forces $P_c = f(h)$ were used to determine the displacements: U_1 the limiting displacements due to loss of stability by force ($dP_c/dh = 0$); U_f the limiting displacements due to the initiation of shear fracture; U_{af} the displacements that accompany shear fracture.

The structure of the obtained fractures was examined by a scanning electron microscope.

3. Results of tests

3.1. Effect of heat treament of 0.09% C steel on the macroscopic course of cracking and microfractography at 77 K

The effect of heat treatment on the macroscopic course of cracking is shown in Fig. 3(a),(b),(c). Figure 3(d) is a microfractograph of the fracture from Fig. 3(a) whereas those of the fractures from Fig. 3(b),(c) are presented in Fig. 3(e). It follows from Fig. 3(a) and 3(d) that coarse-grained steel cracks in the brittle manner. This fracture is a typical intercrystalline fracture (Fig. 3(d)). The fractures of fine-grained steels, however, *i.e.* the standard steel (Fig. 3(b)) and the toughened steel, are typical ductile-shear fractures.

Figures 3 and 2 indicate that grain refinement rules out brittle fracture in the tested steel since fine-grained steel (Fig. 3(b),(c)) cracks exclusively in the shear fracture manner. The macroscopic course of shear fracture is determined by the condition of the grain boundaries. For similar grain sizes but different conditions of grain boundaries (steels 2 and 3, Fig. 3(b),(c)), far different shapes of fracture were obtained. The normalized steel fracture surface is concave–convex and resembles in shape the letter S (Fig. 3(b)). The toughened steel fracture surface (Fig. 3(c)) is almost ideally flat.

TABLE 2. The kind and conditions of heat treatment and the diameters of grains of 0.09% C steel before and after the heat treatment

Steel	Kind of heat treatment	Heat treatment conditions			Diameter d of grains (mm)
		Temperature T (K)	Time t of heating (min)	Manner of cooling	
1	Delivery condition	—		—	0.177–0.044
2	Normalizing	1163–1183	30	In air	0.022
3	Quench hardening	1163–1183	30	In water	0.015
	Tempering	953–973	60	With furnace	

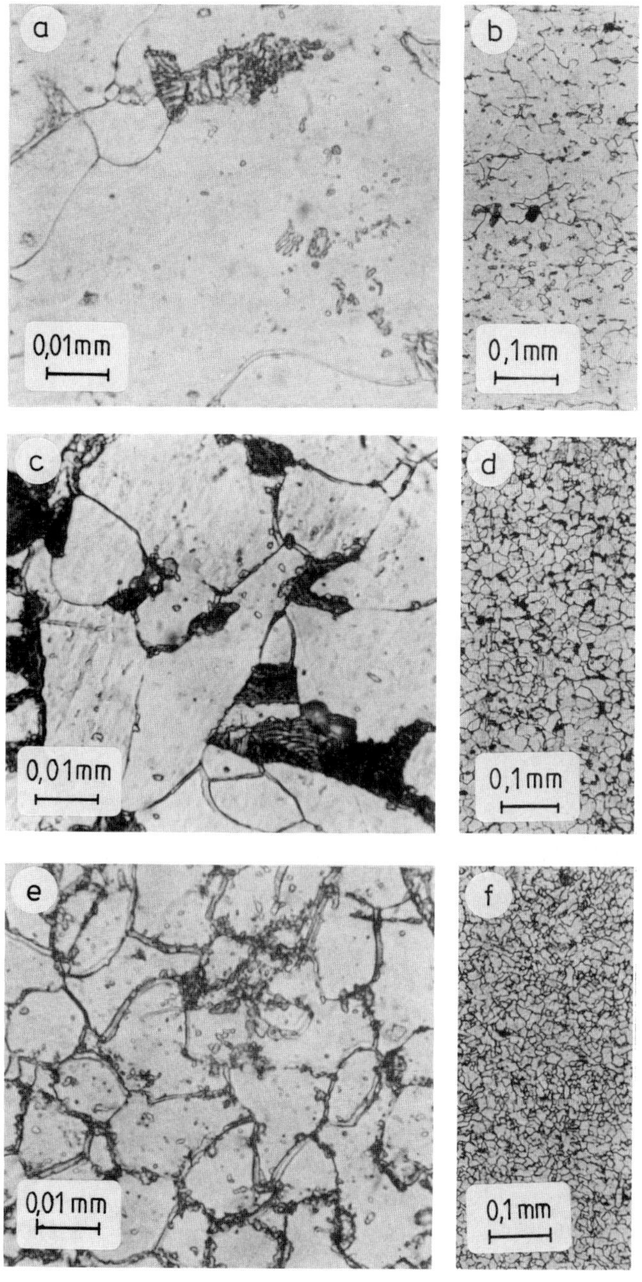

Fig. 2. The structure of 0.09% C steel before and after heat treatment (the heat treatment conditions are given in Table 2): (a), (b) steel in the delivery condition; (c), (d) steel after normalization; (e), (f) steel after toughening; (a), (c), (e) primary magnification 1000 ×; (b), (d), (f) primary magnification 100 ×.

3.2. Influence of heat treatment of 0.09% C steel on progress of the shearing force $P_c = f(h)$ at 77 K

The influence of heat treatment of the tested steel on the progress of the shearing force $P_c = f(h)$ is shown in Fig. 4(b),(c),(d),(e). Figure 4(a) is included for comparison with Fig. 4(b), to demonstrate the effect of lowering the temperature T from 293 to 77 K. Moreover, Fig. 4(f) is a schematic diagram of measurements of the limiting displacements U_l and U_f and bound-with-cracking displacements U_{af}. As one can see in Fig. 4, the course of shearing forces differs considerably. The effects of both temperature and the structure of the tested steel are pronounced.

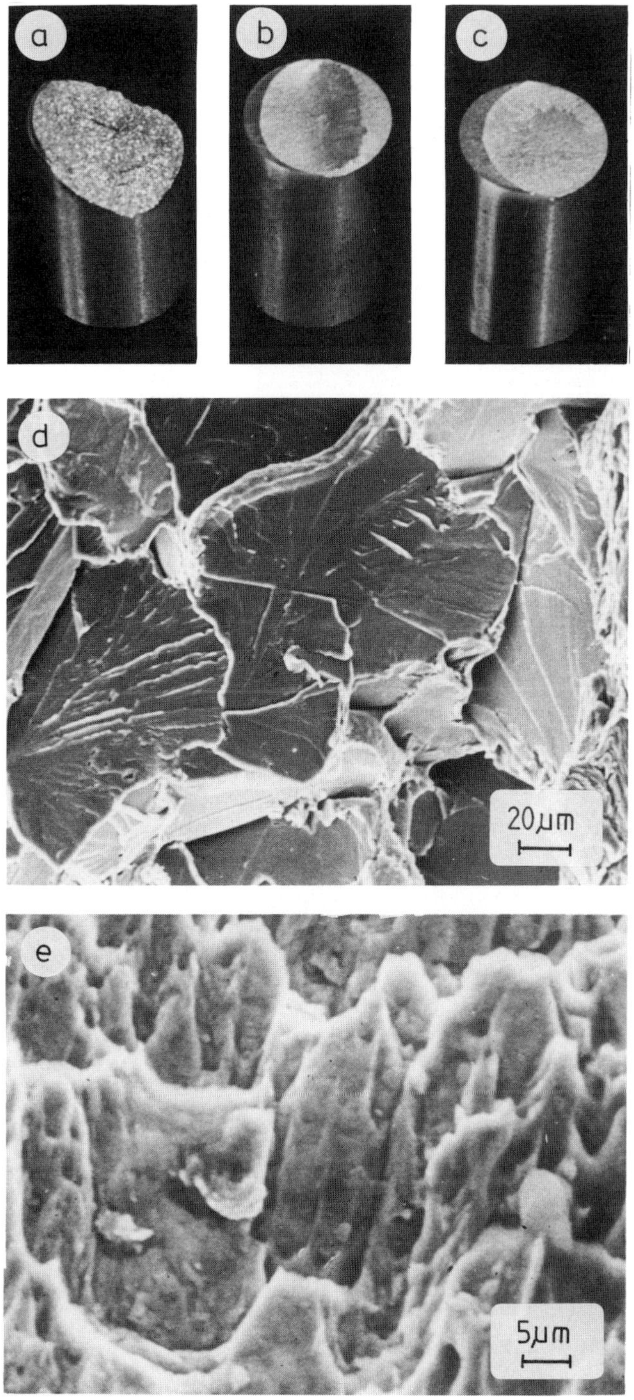

Fig. 3. The shapes of surfaces of fractures ((a), (b), (c)) and the morphology of fractures ((d), (e)) obtained after shearing of 0.09% C steel at the liquid nitrogen temperature 77 K: (a) fracture of steel in the delivery condition; (b) fracture of normalized steel; (c) fracture of toughened steel; (d) morphology of brittle fracture (a); (e) morphology of fractures (b), (c).

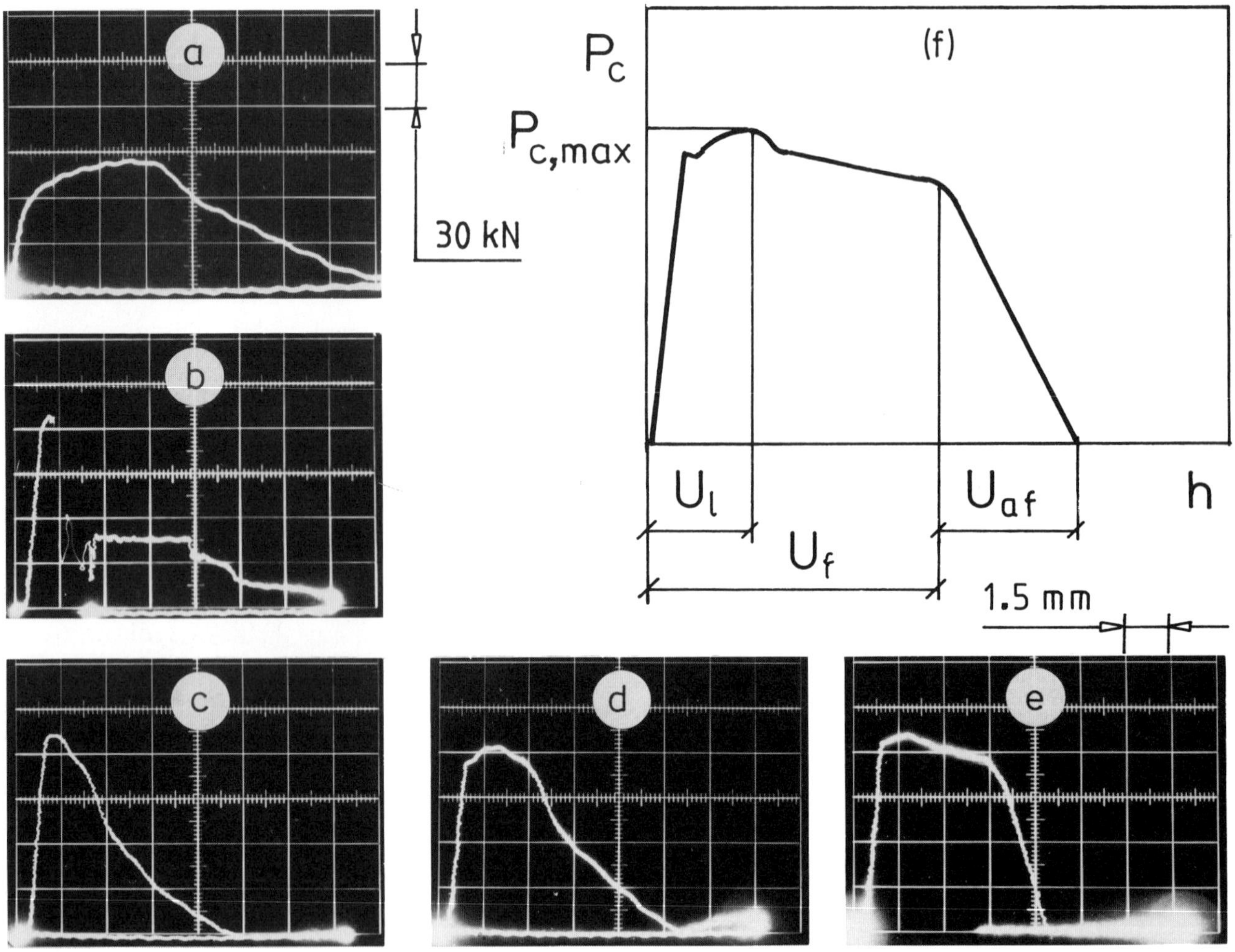

Fig. 4. The effect of heat treatment of 0.09% C steel on the course of shearing force $P_c = f(h)$ at (a) 293 K and (b), (c), (d), (e), 77 K and (f) a scheme of measuring displacements U_l, U_f, U_{af}: (a) steel in the delivery condition (shear fracture 293 K); (b) steel in the delivery condition (brittle fracture, 77 K); (c) steel in the delivery condition (shear fracture, 77 K); (d) steel after normalizing (shear fracture, 77 K); (e) steel after toughening (shear fracture, 77 K).

The effect of the temperature manifests itself mainly in lower values of displacements U_l and higher values of the yield point and maximum shearing force $P_{c,max}$ (Fig. 4(a),(b)).

The influence of the structure is reflected most clearly in the changes in values of displacements U_{af} which accompany fracturing (Figs. 4(c),(d),(e)). This is particularly visible for steel 3, *i.e.* the toughened steel. For steel 1 (steel in the delivery condition), surprisingly similar values of displacements U_{af} at the two temperatures of shearing (293 and 77 K) are observed (Fig. 4(a),(c)). In the toughened steel there is, moreover, a clear increment in the displacement U_f, critical for the initiation of shear fracture.

One should notice that coarse-grained steel (steel 1) can crack at the temperature of 77 K in brittle and shear fracture modes simultaneously (Fig. 1) or only in the brittle mode (Fig. 4(b)) or only in the shear mode (Fig. 4(c)). The shear fracture (Fig. 4(c)) occurs at the same load level as the brittle fracture (Fig. 4(a)). Thus the critical load for shear fracture is not much different from the load that plasticizes the sheared material and causes brittle fracture.

In the case of shear fracturing, it is more convenient to use the displacement U_l which is critical because of the loss of stability by the shearing force [6, 7]. The value of this displacement can be tied up with the strain-hardening ability of the material [7], which facilitates considerably the prediction of the nucleation and propagation of shear fracture.

3.3. Effect of low temperatures (77–293 K) and grain size d of 0.09% C steel on displacements U_l, limiting displacement due to loss of stability by force P_c

The effect of low temperatures (293–77 K) on the value of displacement $U_l = f(\log T)$, the limiting displacement due to loss of stability by the shearing force $P_c = f(h)$, is illustrated in Fig. 5. Curves 1 and 4 show the relationship $U_l = f(\log T)$ for 0.09% C and 0.44% C steel for comparison. The obtained experimentally straight lines 1, 2 and 3 (Fig. 5) illustrate relationship $U_l = f(\log T)$ for steel with the same carbon content (0.09% C) but a different grain size. The lines have similar angels of inclination whereby they run parallel to one another. As the carbon content increases (see straight line 4, steel 0.44% C), the angle of inclination of straight line $U_l = f(\log T)$ decreases. It follows from this that for low- and medium-carbon steel relationship $U_l = f(\log T)$ can be described by the equation of a straight line:

$$U_l = C_g + C_c \log T \tag{1}$$

where C_g is a constant dependent on grain size, C_c is a constant dependent on carbon content, T is the temperature in degrees kelvin.

In the case of the tested 0.09% C steel, the constant C_c has the same value for all the considered grain sizes, *i.e.* $C_c = 4.25$. However, the constant C_g assumes different negative values: −7.14 for steel 1, −6.32 for steel 2, and −6.21 for steel 3. It follows from Fig. 5 that lowering the temperature from 293 to 77 K results in a radical reduction in the value of limiting displacements U_l, according to eqn. (1). Refinement of the structure brings about an increase in the limiting displacement U_l, with the angle of inclination of the straight line $U_l = C_g + C_c \log T$ remaining unchanged.

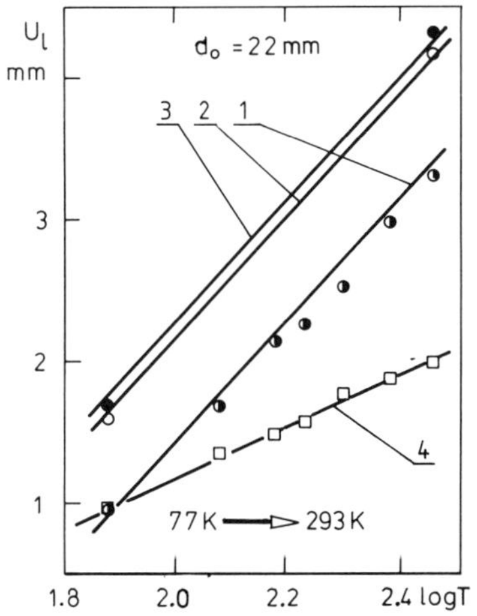

Fig. 5. The influence of log T on the values of limiting displacements U_l and the course of the relationship $U_l = C_g + C_c \log T$ for different grades of steel (0.09% and 0.44% C) and different grain sizes (from 0.015 to 0.177); curves 1, 2, 3 0.09% C (1 0.044–0.177 grain size, 2 0.022 grain size, 3 0.015 grain size); curve 4 0.44% C.

4. Discussion and conclusions

The tests have shown that the morphology of fractures in Fig. 3(e) does not differ radically from the

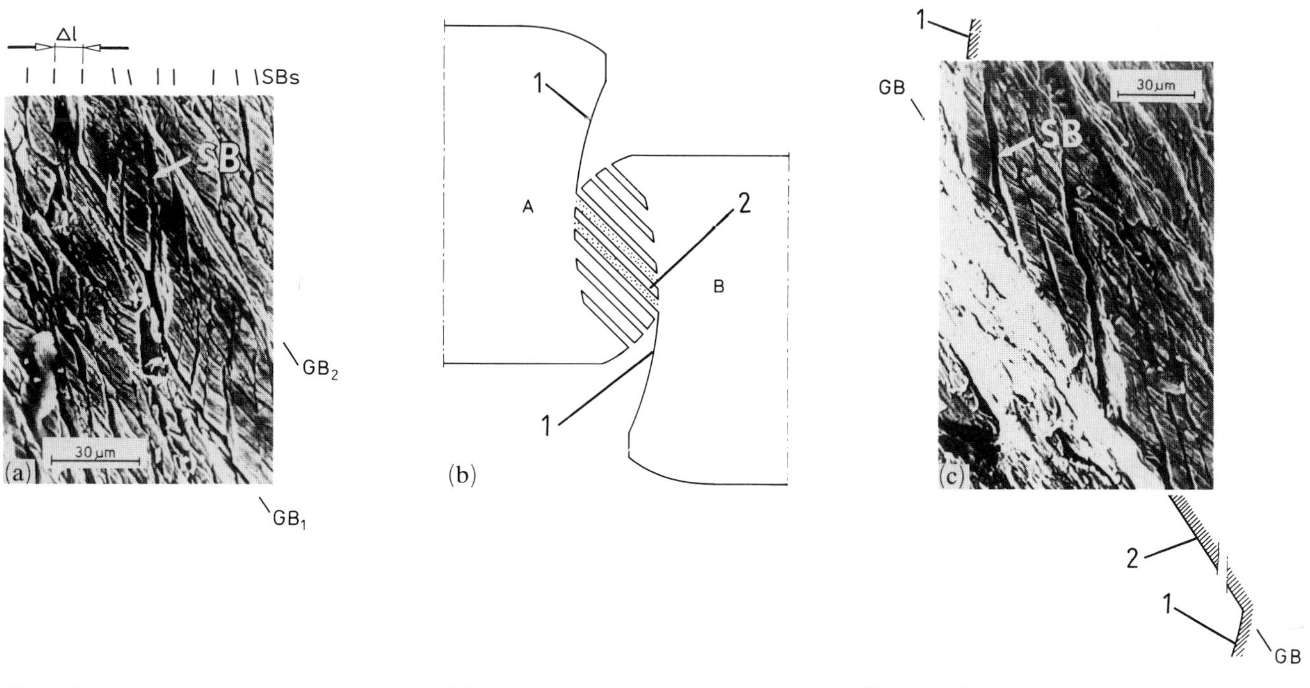

Fig. 6. (a) The mechanism of shear fracture, (b) the model and (c) the experimental verification of the propagation of shear fracture: 1 the propagation of fracture along shear bands (SB) and along the boundaries of macroscopic zone of strain localization, 2 the propagation of fracture along defective grain boundaries (GB) and across the macroscopic zone of strains localization.

morphology obtained at 293 K [8]. This indicates that there are no significant differences between the mechanism of shear fracture at the temperature of liquid nitrogen (77 K) and at room temperature (293 K). Lowering the temperature from 293 to 77 K only changes the values of critical loads $P_{c,max}$ (Fig. 4(a),(b)) and displacements U_l (Fig. 5) after which shear fracture starts. Refinement of the structure contributes to an increase in the limiting displacement U_l and an increase in the difference between the displacements U_f and U_l. The increase in displacement U_l can be ascribed to the increment in the strain hardening ability of the material which delays the loss of stability by the shearing force P_c [7]. The larger difference between displacements U_f and U_l (Fig. 4(c),(d),(e)) is due to the degree of purity of grain boundaries brought about by the applied heat treatment. This specific influence of the structure on the propagation of fracture and the shape of the fracture surface (Fig. 3(b),(c)) can be explained by referring to the model of shear fracture shown in Fig. 6 [9–11].

According to the model, the cracking of individual grains consists in separation of the material along shear bands SB. The necessary tensile stresses may be generated, among other things, by transverse displacement of the material along the defective boundaries of grains (Fig. 6(a)).

The mechanism of formation of the concave–convex fracture surface is illustrated by Fig. 6(b),(c). The S-shaped fracture surface is a result of a change in the manner and direction of cracking. The change consists in the transition from fracturing along the shear bands (1) to fracturing along the defective boundaries of grains (2). The convexity of a fracture surface (see Fig. 2(b)) is larger, the larger the displacements of the material along the defective grain boundaries are. The defects are a result of the interaction between shear bands and grain boundaries (Fig. 6(a)) and they are the more serious, the more impure the grain boundaries are. Hence there is considerable convexity of the fracture in the normalized steel (Fig. 3(b)) and very slight convexity in the toughened steel (Fig. 3(c)). The reduction in the propensity for forming boundary discontinuities facilitates the penetration of shear bands into the neighbouring grains and ensures the continuity of their direction. As a result, the initiation of fracture is delayed (the value of U_f increases, Fig. 4(e)) and the fracture itself propagates considerably faster (U_{af} decreases, Fig. 4(e)).

It follows from the above that incorporation of the interactions between shear bands and grain boundaries opens up new possibilities of interpretation (in the range of shear fracture) which are greater than those of the models based on the nucleation and development of oval voids. Models of the nucleation of brittle fracture, *i.e.* dislocation models, based on the flat pile-up of dislocations should not be applied to interpretation of the nucleation and propagation of shear fracture.

References

1 D. Teirlinck, T. Zok, J. D. Embury and M. F. Ashby, Fracture mechanism maps in stress space, *Acta Metall., 36* (1988) 1213.

2 F. A. McClintock and A. S. Argon, *Mechanical Behaviour of Materials*, Addison-Wesley, Reading, MA, 1966.

3 P. E. Thomason, A theory for ductile fracture by internal necking of cavities, *J. Inst. Met., 96* (1968) 360.

4 G. L. LeRoy, J. D. Embury, G. Edward and M. F. Ashby, A model of ductile fracture based on the nucleation and growth of voids, *Acta Metall., 29* (1981) 1509.

5 J. F. Knott, *Fundaments of Fracture Mechanics*, Butterworths, London, 1978.

6 E. S. Dzidowski, The occurrence of localized shear bands during transverse shearing, *COBEM 85, Anais do VIII Congresso Brasileiro de Engenharia Mecanica*, Instituto Technologico de Aeronautica, Sao Jose dos Campos, 1985, p. 907.

7 E. S. Dzidowski, A study of limiting displacements in the shearing of bars, *J. Mech. Work. Technol., 12* (1986) 297.

8 E. S. Dzidowski, Mechanism of shear fracture in the aspect of controlled decohesion of metals, *Scientific Papers of the Technical University of Wroclaw, Monographs No. 11* (1990) 93 (in Polish).

9 E. S. Dzidowski, The mechanism of shear fracture in the aspect of the micro- and macrolocalization of strains, *Proc. Int. Seminar on Large Plastic Deformation, MECAMAT 1991*, Paris, August 7–9, 1991, A. A. Balkema, Rotterdam, 1993, in press.

10 E. S. Dzidowski, The genesis of S-like fracture shapes during transverse shearing of metals. In J.-P. Boehler and A. S. Khan (eds.), *Anisotropy and Localization of Plastic Deformation*, Elsevier, London, 1991, p. 56.

11 E. S. Dzidowski, Mechanism of shear fracture during transverse shearing with uniaxial compression, in *Proc. 13th Risø Int. Symp. on Materials Science: Modeling of Plastic Deformation and Its Engineering Applications*, Risø National Laboratory, Roskilde, 1992, p. 253.

Materials Science and Engineering, A168 (1993) 17–22

Tensile test on *in situ* solidified notched specimens: effects of temperature history and strain rate on the hot ductility of Nb and Nb–V microalloyed steels

P. Deprez, J. P. Bricout and J. Oudin

Mechanical Engineering Laboratory, C.N.R.S. Research Unit D 1401, MECAMAT, Université de Valenciennes et du Hainaut Cambrésis, BP 311, 59304 Valenciennes Cedex (France)

Abstract

Hot steel ductility in continuous casting processes is determined by tensile tests of notched cylindrical specimens which are obtained by melting and solidification and then tested after rate-controlled cooling with a new specific device available for any standard test machine. Hot ductility is proposed to be related to ultimate area reduction in the notched zone of the specimens. Experiments in the range of 700–1000 °C are achieved on two types of niobium–vanadium microalloyed steels which only differ by their vanadium content. Ultimate area reductions are determinate by using *in situ* solidified specimens and reheated ones from room temperature, 10^{-3} and 0.5 s^{-1} mean strain rate being considered. The change of ultimate area reductions mainly depends on austenitic grain size, segregation's distribution, dynamic precipitation of niobium–vanadium carbonitrides and so far the recrystallization occurrence of austenite.

1. Introduction

Transverse cracking of C–Mn–Nb–N–V steel grade slabs from continuous casting is often observed. Such cracks are mainly intergranular, meandering along the prior austenite grain boundaries [1] and their occurrence is highly dependent on the composition and temperature history of the steel. Cracks arise as a result of strain and stress concentrations at the austenite boundaries caused either by the presence of thin films of the softer deformation-induced ferrite surrounding the austenite grains, or by grain boundary sliding in the austenite. Both failure mechanisms are promoted by the presence of grain boundary precipitates, coarser grain sizes, and lower strain rates [2]. The common assumptions state that transverse cracking is due to mechanical strains associated with the bending and straightening of the strain [3]; straightening is achieved in the range of 1000 down to 700 °C, which corresponds to the minimum ductility zone found by laboratory hot tensile tests, the strain rates being generally in the range 10^{-3}–10^{-4} s^{-1}.

The present investigation develops an *in situ* solidified notched specimen tensile test; the microstructure after solidification according to a given temperature history in a continuous casting process is reproduced [4]. In the following, the effect of temperature history and strain rate on the hot ductility of niobium–vanadium microalloyed steels is considered for temperatures ranging from 1100 to 700 °C.

2. Experimental conditions

The hot ductility of steels is usually defined from standard tensile tests performed under the conditions described in Fig. 1(a). Samples are heated to a temperature above the solution temperature of the microalloy precipitates, both to dissolve all these particles and to produce a coarse grain size typical of the continuously cast microstructure before the straightening operation. Unfortunately, the microstructure at the testing temperature obtained in this way may often differ from the as-cast structure, and more sophisticated simulations of the continuous casting operation involving melting are absolutely necessary. The experimental difficulty was to find a way of simulating the solidification procedure and at the same time perform tensile tests at the required temperatures as shown in Fig. 1(b). Such simulations have rarely been used because of difficulties in avoiding the detrimental effect induced by the solidification shrinkage cavity [5–7].

We have developed a new tensile test for an *in situ* solidified specimen able to reproduce the continuously cast microstructure before the straightening operation. The testing facility used is illustrated schematically in Fig. 2. The main components are the tension/compression testing machine and the induction heater. The starting set of specimens consists of two machined round-bar blanks and a calibrated notched silica crucible; the specimens and crucible are placed inside an inductor and then the two blanks are melted in a

0921-5093/93/$6.00

flowing argon atmosphere (Fig. 2(a)). Progressive raising of the testing machine lower ram allows the crucible to fill up (Fig. 2(b)); the crucible is easily destroyed after solidification of the specimen by two sliding rods (Fig. 2(c)) then the tensile test is carried out at the required temperature which is time controlled by a germanium photodiode pyrometer sighted between the turns of the induction coil (Fig. 2(d)). The choice of specimen and crucible geometry was determined so as to shift the shrinkage cavity out of the notched zone and avoid any problems during test measurements. This was realized using a cone-shaped crucible so as to shift the cooling isotherms and favour formation of the shrinkage cavity towards the upper part of the crucible about 10 mm above the notched zone (Fig. 2(b)). This gives a basic advantage for accurate measurements as well as clear observations of fracture types. Moreover tensile testing of notched specimens is of great interest because it favours decohesion and related defects [8].

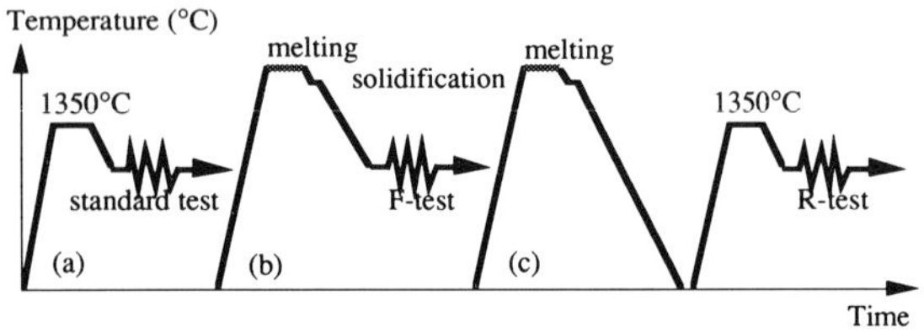

Fig. 1. Temperature history: (a) standard tensile test, (b) F-test, tensile test of *in situ* solidified specimens, (c) R-test, tensile test of *in situ* solidified, cooled and reheated specimens.

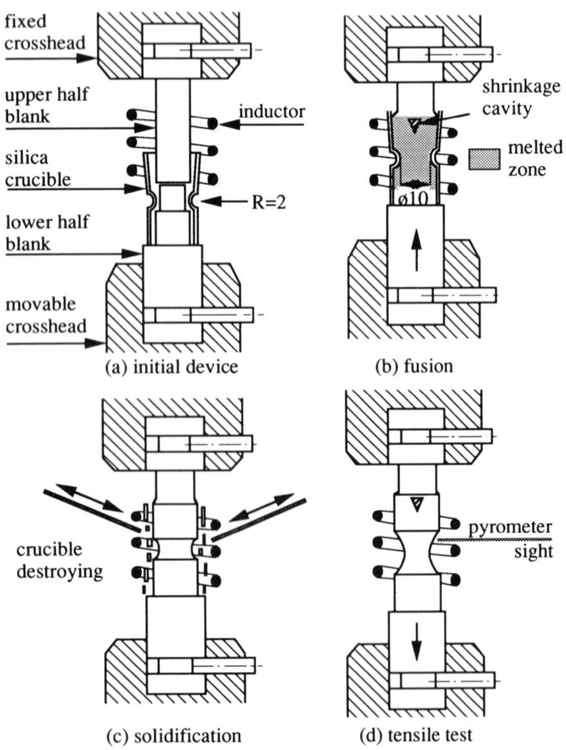

Fig. 2. Tensile test of *in situ* solidified notched specimens.

The composition of the two niobium-microalloyed steels examined is given in Table 1. Steels A and B are C–Mn–Nb–N grades with low sulphur content mainly differing in vanadium content: steel B is high in vanadium. It should be noted that sufficient manganese was present in these two steels to prevent classical hot shortness even under conditions far removed from equilibrium. The round-bar blanks were heated at 10 °C s^{-1} up to about 30 °C above the liquidus. After holding for 1 min, the specimen was solidified and cooled at 15 °C s^{-1} then strained until rupture at temperatures ranging from 1100 to 700 °C, the elongation rates being 0.25 and 125 mm min^{-1}. The corresponding mean strain rates defined as the ratio of the elongation rate to the initial notch width were respectively 10^{-3} and 0.5 s^{-1}. For each test, values of the ratio of the peak load to initial notch cross-section, total elongation and notch diameter were recorded as strength and hot ductility parameters. In order to evaluate the effect of the temperature history, a second procedure, the so-called R-test was implemented: the specimen is first solidified *in situ*, then cooled to room temperature, heated at 1350 °C for 1.5 min and tested at the required temperature (Fig. 1(c)). The specimen geometry is identical for the two procedures, so it is possible to compare accurately the F- and R-tests.

3. Experimental results

The peak stress, defined as the ratio of the maximum tensile force to initial notch cross-section, increases with decreasing temperature; the temperature history has no influence except at 700 °C where the R-test peak stress is always above the F-test peak stress. An

TABLE 1. Microalloyed steel composition (wt.%)

Steel	Type	C	Mn	P	S	Si	Al	Nb	V	N
A	C–Mn–Al–Nb	0.110	1.367	0.010	0.0023	0.30	0.027	0.034	0.003	0.0057
B	C–Mn–Al–Nb–V	0.106	1.334	0.010	0.0013	0.29	0.027	0.033	0.024	0.0039

Cr, Ni, Mo and Cu contents are less than 0.020, 0.024, 0.004 and less than 0.010 wt.% respectively for both steels.

increase in strain rate strongly increases the peak stress, especially for the lower temperatures. The peak stress *vs.* temperature curves of steels A and B are also quite similar for elongation rates of 0.25 and 125 mm min^{-1}.

Figure 3 shows the ultimate area reduction $RA=(S_0-S_r)/S_0$, S_0 being the initial notch cross-section and S_r the ultimate cross-section, *vs.* temperature. The curves are identified by the steel (A or B), elongation rate in millimetres per minute (0.25 or 125) and test type (F or R). At first sight, it appears that the ductility is strongly influenced by the strain rate and temperature history. An increase in the strain rate improves the hot ductility. Narrowing of the ductility trough by displacement of the upper temperature boundary to a lower temperature is also apparent. Nevertheless, the most significant feature is the dependence on strain rate of the effect of temperature history. Thus, for the lowest elongation rate, the R-test ductility curve is always above that of the F-test, whereas for the highest elongation rate the effect is reversed. It is interesting to notice that RA values are nearly zero for the 0.25 mm min^{-1} elongation rate and F-tests at 800 °C for steel A, and at 900 °C for steel B, the difference from the R-tests being over 20% in reduction in area. In addition, it may be noticed for steel B that the decay of ductility from higher temperatures is very sharp for both F- and R-tests. Steel B will be more sensitive to transverse cracking if the straightening operation takes place in the temperature range 900–1000 °C.

Figure 4 illustrates microstructures of the unstrained zone after F- and R-tests of steel B tested at 900 °C. The significant feature is that the austenite grain size is always much coarser for F-structures than R-structures for both steels A and B. Moreover, columnar grains perpendicular to the tensile direction are present at the surface of the *in situ* solidified samples. The structure is thus similarly stressed to a real strand shell. Figure 5 shows two examples of metallographic structure related to the broken notched zone. Intergranular cracks similar to transverse surface cracking of continuous casting slabs were only observed for the slow elongation rate and F-test (Fig. 5(a)), and dynamic recrystallization was observed for the higher elongation rate and F-test (Fig. 5(b)).

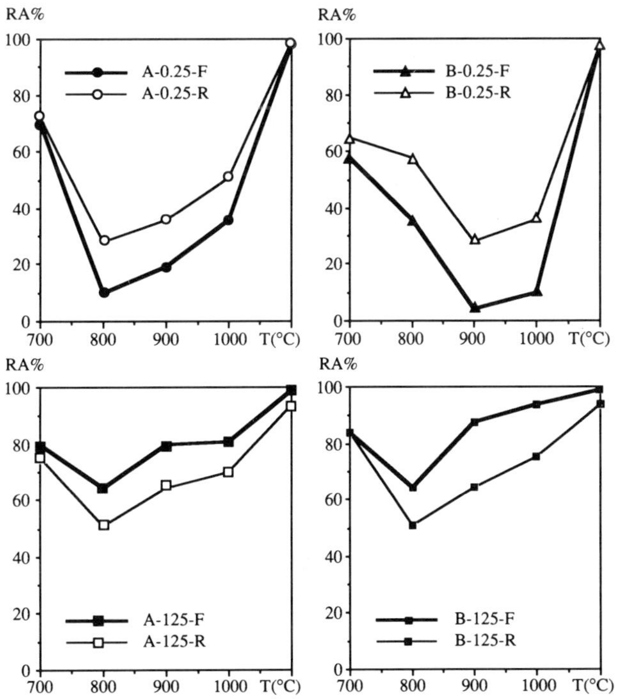

Fig. 3. Ultimate area reduction *vs.* temperature: F and R tests, 0.25 and 125 mm min^{-1} elongation rates for steels A and B.

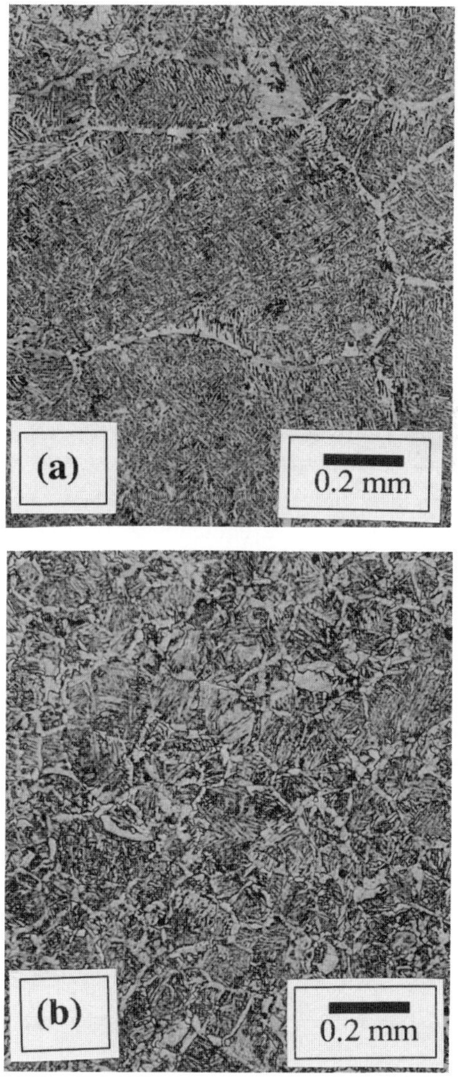

Fig. 4. Metallographic structures of the unstrained zone related to steel B tested at 900 °C: (a) after F-test, (b) after R-test.

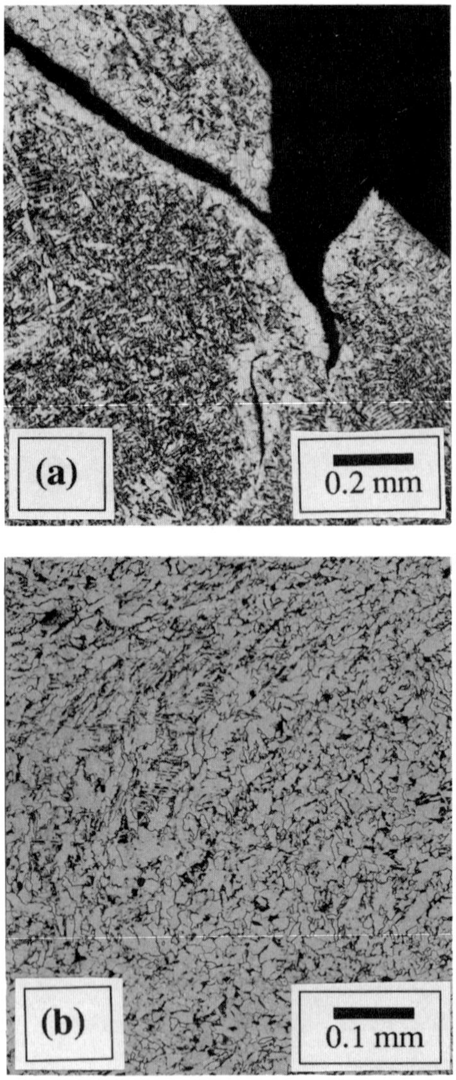

Fig. 5. Metallographic structures of the broken notched zone related to steel B tested at 900 °C by the F-test: (a) elongation rate 0.25 mm min^{-1}, (b) 125 mm min^{-1}.

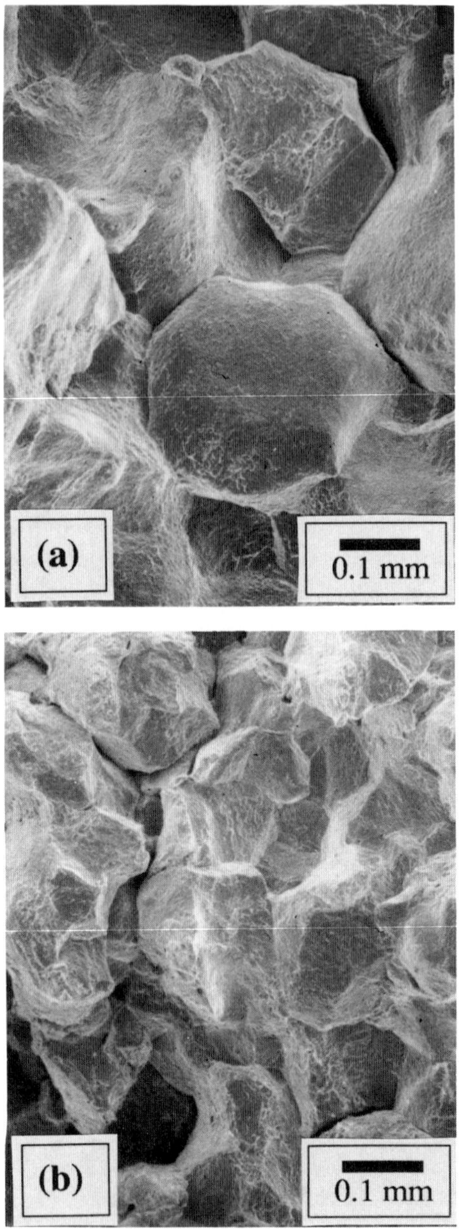

Fig. 6. Fracture surfaces related to steel B tested at 900 °C and 0.25 mm min^{-1}: (a) F-testing procedure, (b) R-testing procedure.

For the 0.25 mm min^{-1} elongation rate, the trough region is invariably associated with intergranular fracture, the fracture facets being covered with fine dimples or microvoids (Fig. 6). Preferential deformation in regions close to the grain boundary initiates voids at grain boundary inclusions or precipitates, which leads to intergranular failure by way of microvoid coalescence. For the 125 mm min^{-1} elongation rate, fully ductile dimple fractures are observed for F-fracture surfaces, whereas some intergranular zones are visible on R-fracture surfaces for both steels A and B (Fig. 7).

4. Discussion

The trough region of the ultimate area reduction *vs.* temperature curve is often associated with intergranular fracture, the fracture facets being either covered with fine dimples or smooth, suggesting two distinct mechanisms: embrittlement by strain concentration and microvoid coalescence at grain boundaries, and embrittlement by grain boundary sliding. A common feature of these embrittlement mechanisms is preferential deformation at or in the proximity of grain boundaries. So, to avoid embrittlement, it is essential to reduce the strain concentration at grain boundaries and therefore the austenitic grain size. As the straightening operation involves such low strains and strain rates, only dynamic recovery is possible and grain refinement

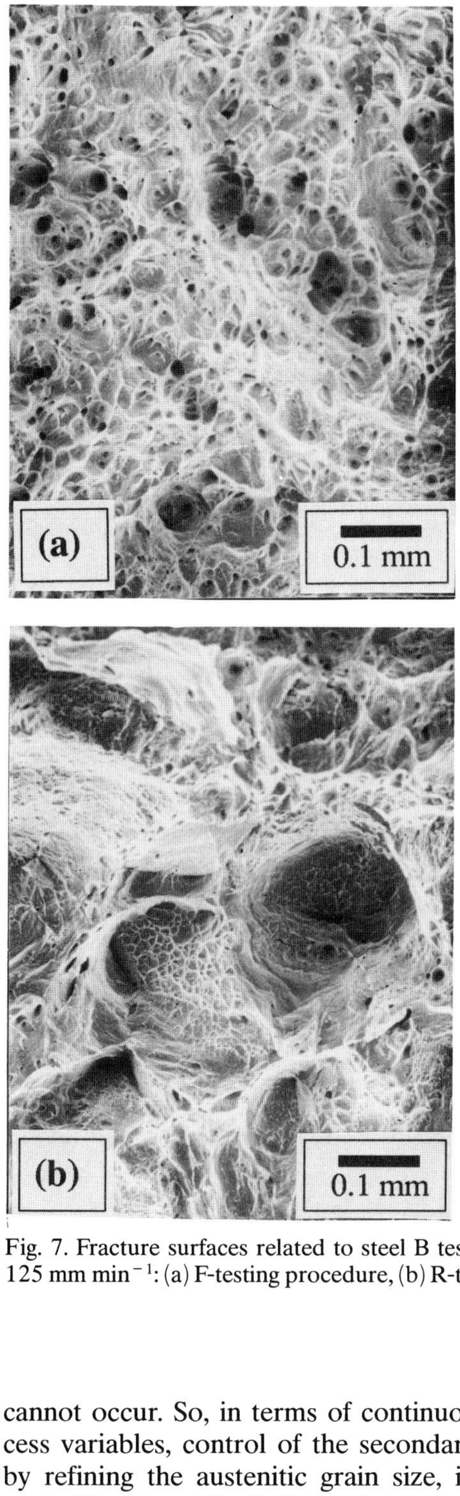

Fig. 7. Fracture surfaces related to steel B tested at 900 °C and 125 mm min^{-1}: (a) F-testing procedure, (b) R-testing procedure.

cannot occur. So, in terms of continuous casting process variables, control of the secondary cooling flow, by refining the austenitic grain size, is probably the most effective method of reducing transverse cracking.

In Nb–V-containing steels, the main mechanism of transverse surface cracking of continuous cast slabs is due to dynamic precipitation of carbide, nitride or carbonitride such as AlN, NbC(N) and VC(N) occurring at slow strain rates in the temperature range 700–1100 °C [9]. It is well known that the ductility trough is greatly widened and deepened by the addition of small amounts of niobium. Moreover, it can be seen that the ductility loss due to niobium addition is also enhanced by the addition of vanadium; the ductility trough of steel B is indeed deeper and strongly shifted toward higher temperatures. So, it seems that the dynamic precipitation of VC(N) which occurs in the temperature range 850–1000 °C [10] adds its detrimental effect to that due to NbC(N) precipitates.

The effect of strain rate is strongly marked for the F-test procedure: an increase in strain rate from 0.25 to 125 mm min^{-1} increases by about six times the ultimate area reduction values and changes the fracture appearance from intergranular to ductile. Higher strain rates improve the hot ductility because there is insufficient time for strain induced precipitation, the amount of grain boundary sliding is reduced and there is insufficient time for the formation and diffusion controlled growth of voids around the precipitates located at the grain boundaries. The difference between the F- and R-tests is very important and provides evidence of the effect of the temperature history on the hot ductility of steels. For the lowest strain rate (10^{-3} s^{-1}), F-tests give the lower ultimate area reduction values whereas for the highest strain rate (0.5 s^{-1}) the effect is reversed. Metallographic examinations of the unstrained zone of the tensile specimens have shown that the austenite grain size is always much coarser for F-structures than for R-structures for both steels A and B. This follows from a decrease in the specific grain boundary area which raises the precipitate density on the grain boundaries. Moreover, the F-test procedure produces a solidification structure with columnar grains perpendicular to the main direction of stressing in the tensile test. Coarse columnar grains and lower strain rates encourage grain boundary crack propagation due to carbonitride strain induced precipitation and thus increase the tendency to form transverse cracks. However, higher strain rates improve the hot ductility by making dynamic recrystallization possible and because there is insufficient time for strain induced carbonitride precipitation. Recrystallization was only observed for F-tests down to 900 °C for both steels A and B because the intergranular solidification segregations act as recrystallization nuclei. Moreover, intense segregation of the microalloying addition to the interdendritic boundaries probably occurs during solidification, and this will lead to precipitation at temperatures higher than those normally found after a solution treatment such as in the R-testing procedure. Less precipitation of fine NbC(N) at the austenite boundaries and within the matrix has been found, giving improved hot ductility over material solution treated and cooled to the test temperature [11].

5. Conclusion

Hot ductility and surface cracking of microalloyed continuously cast steels were analysed by tensile tests of *in situ* solidified notched specimens.

(i) The *in situ* solidification procedure produces a solidification structure with coarse columnar grains perpendicular to the tensile direction, similar to the continuously cast microstructure before strand straightening.

(ii) Coarse columnar grains and lower strain rates encourage grain boundary crack propagation due to carbonitride strain induced precipitation and thus increase the tendency to form transverse cracks.

(iii) The difference between the *in situ* solidification procedure and the reheating procedure is very important and provides evidence of the effect of temperature history on the hot ductility; the effect of precipitates on the hot ductility strongly depends on their size distribution and location and therefore the temperature history.

(iv) For low strain rates, the *in situ* solidification procedure gives very low ultimate area reduction values whereas the behaviour is reversed for high strain rates.

(v) Ductility loss due to the addition of niobium is enhanced by the addition of vanadium, the ductility trough being deeper and strongly shifted towards higher temperatures; the likelihood of transverse cracking will be enhanced when the temperature during straightening is raised into the temperature range 850–1000 °C.

Acknowledgments

The authors are very grateful to CNRS, Ministère de la Recherche et de l'Espace and to Sollac for support of the above developments.

References

1 B. Mintz and J. M. Arrowsmith, *Met. Technol., 6* (1979) 24.
2 B. Mintz, S. Yue and J. J. Jonas, *Int. Mater. Rev., 36* (1991) 221.
3 C. R. Taylor, *Metall. Trans. B, 6* (1975) 359.
4 P. Deprez, J. P. Bricout and J. Oudin, *J. Mater. Proc. Technol., 32* (1992) 325.
5 G. A. Wilber, R. Batra, W. F. Savage and W. J. Childs, *Metall. Trans. A, 6* (1975) 1727.
6 H. G. Suzuki, S. Nishimura, J. Imamura and Y. Nakamura, *Transactions ISIJ, 24* (1984) 169.
7 J. Hertel, H. Litterscheidt, U. Lotter and H. Pircher, *Rev. Métall. CIT*, 1 (1992) 73.
8 J. Oudin, Y. Ravalard, J. C., Gelin, G., Lacombe and T. Labarthe-Vacquier, *Matér. Tech., 11–12* (1988) 39.
9 Y. Maehara and Y. Ohmori, *Mater. Sci. Eng., 62* (1984) 109.
10 H. Su, W. Zhang and Z. Yan, Chin, *J. Met. Sci. Technol., 4* (1988) 220.
11 B. Mintz, J. R. Wilcox and D. N. Crowther, *Mater. Sci. Technol., 2* (1986) 589.

Materials Science and Engineering, A168 (1993) 23–28

Structure and mechanical properties of high-temperature titanium alloys after rapid heat treatment

O. M. Ivasishin
Institute for Metal Physics, Kiev (Ukraine)

G. Lütjering
Technical University Hamburg-Harburg, 2100 Hamburg 90 (Germany)

Abstract

In this study a new approach to optimizing the mechanical properties of high-temperature titanium alloys was developed. It is based on using rapid heating of equiaxed structures into the β-field to achieve a fine β grain size (less than or equal to 100 μm), transforming on subsequent cooling into a fully lamellar structure. This fine β grain size is an order of magnitude smaller than the grain sizes achieved by conventional furnace β-treatment. Structures and mechanical properties (tensile, fatigue and creep) of high temperature alloys after rapid and conventional furnace heat treatments were compared. The results are discussed in terms of structure–property relationships.

1. Introduction

Titanium alloys in hot sections of advanced jet engines offer significant weight reductions compared with nickel-base superalloys, resulting in an increase in engine fuel efficiency and thrust-to-weight ratio. A combination of high tensile and fatigue strength values with good creep resistance is important for elevated temperature applications. Structure requirements to maximize these properties are often contradictory. Creep resistance of alloys with lamellar morphology is higher, while globular structures offer better tensile and fatigue properties. All the properties can be balanced to some extent by processing bimodal structures consisting of equiaxed primary α-grains in a lamellar $(\alpha+\beta)$-matrix [1].

The aim of this study was to optimize the properties of high-temperature alloys using rapid heat treatment which included rapid heating above the β-transus temperature and subsequent cooling. It was shown previously [2, 3] that by starting from equiaxed structures it is possible to produce fully transformed lamellar structures with fine β-grains, changing the morphology and the size of the lamellar structure by controlling the cooling rate.

2. Materials and experimental procedure

Typical near-α type alloys developed in the USA (Ti 6242) and Russia (VT 18 Y) for high-temperature application were studied. The chemical compositions of the alloys are given in Table 1. The starting equiaxed microstructures for both alloys were produced by rolling in the $(\alpha+\beta)$-field and subsequent annealing for 6 h at 940 °C (Ti 6242) or 6 h at 900 °C (VT 18 Y). Specimen blanks ($7\times7\times60$ mm^3) were taken from these materials and β-heat treated either conventionally in a furnace at 1050 °C for 1 h (hereafter designated FH), or rapidly by a direct resistance method (hereafter designated RH). The parameters of rapid heating were chosen from the influence of the heating rate on the β-transus temperature (Fig. 1) and β-grain size. To achieve the same β-grain sizes (approximately 100 μm) the alloys were heated at a rate of 50 °C s^{-1} to 1150 °C (Ti 6242) or to 1130 °C (VT 18 Y), overheating above the corresponding β-transus temperatures by 25 °C and 70 °C respectively. After β-treatment the blanks of Ti 6242 were either water quenched (hereafter designated WQ) or air cooled (hereafter designated AC). Water spray quenching was used after rapid heating while samples were immersed in water after conventional furnace heating. Only AC was done for the VT 18 Y blanks. Final heat treatments were 8 h at 590 °C (Ti 6242) or 6 h at 600 °C (VT 18 Y).

After the final heat treatment specimens for mechanical testing were machined. Tensile tests were performed at room temperature and 450 °C with an initial strain rate of 8×10^{-4} s^{-1}. Fatigue tests were done in air on electrolytically polished hourglass shaped specimens by rotating beam loading for high

Elsevier Sequoia

TABLE 1. Chemical compositions of Ti 6242 and VT 18 Y

Alloy	Al	Mo	Zr	Nb	Sn	Si	C	Fe	O_2	H_2
Ti 6242	6.1	2.0	4.1	—	2.0	0.08	0.02	0.04	0.10	0.003
VT 18 Y	6.9	0.9	4.1	1.0	2.7	0.15	0.08	0.06	0.08	0.002

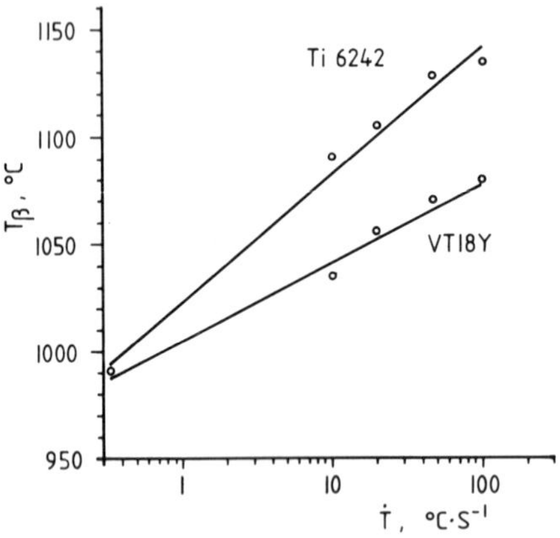

Fig. 1. The influence of heating rate on the β-transus temperature of Ti 6242 and VT 18 Y.

cycle fatigue (room temperature, $R = -1$, $f = 50$ Hz) or by axial loading for low cycle fatigue (450 °C, $R = 0.1$, $f = 0.3$ Hz). Fatigue crack propagation behavior was observed by optical microscopy. Creep resistance was evaluated by plastic strain after 100 h (210 MPa, 510 °C).

3. Experimental results

The microstructures resulting from furnace and rapid heat treatments are shown in Figs. 2 and 3. For both alloys rapid heating resulted in β-grain sizes which were much smaller than those after furnace treatment. WQ produced martensitic structures in the Ti 6242 alloy. Long primary martensite plates were formed under the FH-WQ condition, whereas these long plates were not observed to this extent after RH-WQ treatment. As can be seen from transmission electron microscopy (TEM) images, there was also some tendency for the secondary martensite plates to be finer in the RH-WQ material, compared with the FH-WQ material, probably owing to a higher cooling rate for spray quenching compared with conventional water quenching.

The AC treatment of the Ti 6242 blanks resulted in a mixture of fine martensite plates and coarser diffusionally grown plates. Under the FH-AC condition the fine martensite structure dominated whereas under the RH-AC condition the opposite was found. High chemical inhomogeneities in the β-solid solution on the scale of the starting equiaxed grain size might be the reason for the low martensite fraction in the RH-AC microstructure.

In the VT 18 Y alloy only diffusionally grown plates were observed after AC, presumably owing to the higher temperature in the β field used in the rapid heating experiments. The width of the α-lamellae was somewhat smaller after the RH-AC treatment compared with the FH-AC treatment, probably owing to a slightly faster cooling rate for RH-AC.

The results of tensile and creep tests of both alloys are listed in Table 2. It can be seen that the yield stress and ultimate tensile strength were only slightly affected by the β-grain size at both room and elevated temperature. The higher $\sigma_{0.2}$ and UTS values for the RH-WQ condition compared with those observed for the FH-WQ condition (Ti 6242) resulted from the finer lamellar structure after spray quenching. Reducing the cooling rate from WQ to AC decreased the $\sigma_{0.2}$ and UTS of Ti 6242 by about 200 MPa for the FH condition and 300 MPa for the RH condition. There was opposite ranking of strength values between the FH-AC and RH-AC conditions for the two alloys but this is in good agreement with observations of the microstructural features mentioned above. A significant improvement in ductlity was observed with β-grain size refinement, resulting in much higher true fracture stress and true fracture strain values for the fine-grained microstructures. This can be attributed to the absence of long α-lamellae in the fine-grained conditions in which cracks can nucleate and propagate in the first stage without changing direction. Comparing the ductility of the WQ and AC conditions for the Ti 6242 alloy, it can be seen that the ductility was also affected by the cooling rate.

In many respects, the influence of the heating rate on tensile properties of high-temperature alloys found in the present investigation is similar to that observed earlier for Ti–6Al–4V and VT 6 alloys [3]. Some similarity with the data obtained in ref. 3 can also be seen in the fatigue properties of Ti 6242 (Fig. 4). For the WQ conditions, rapid heating (small β grain size) increased both the HCF strength and the lifetime at higher stress amplitudes. Examples of the crack nucleation site and initial crack propagation path are shown in Fig. 5 for the two WQ conditions. It can be seen that

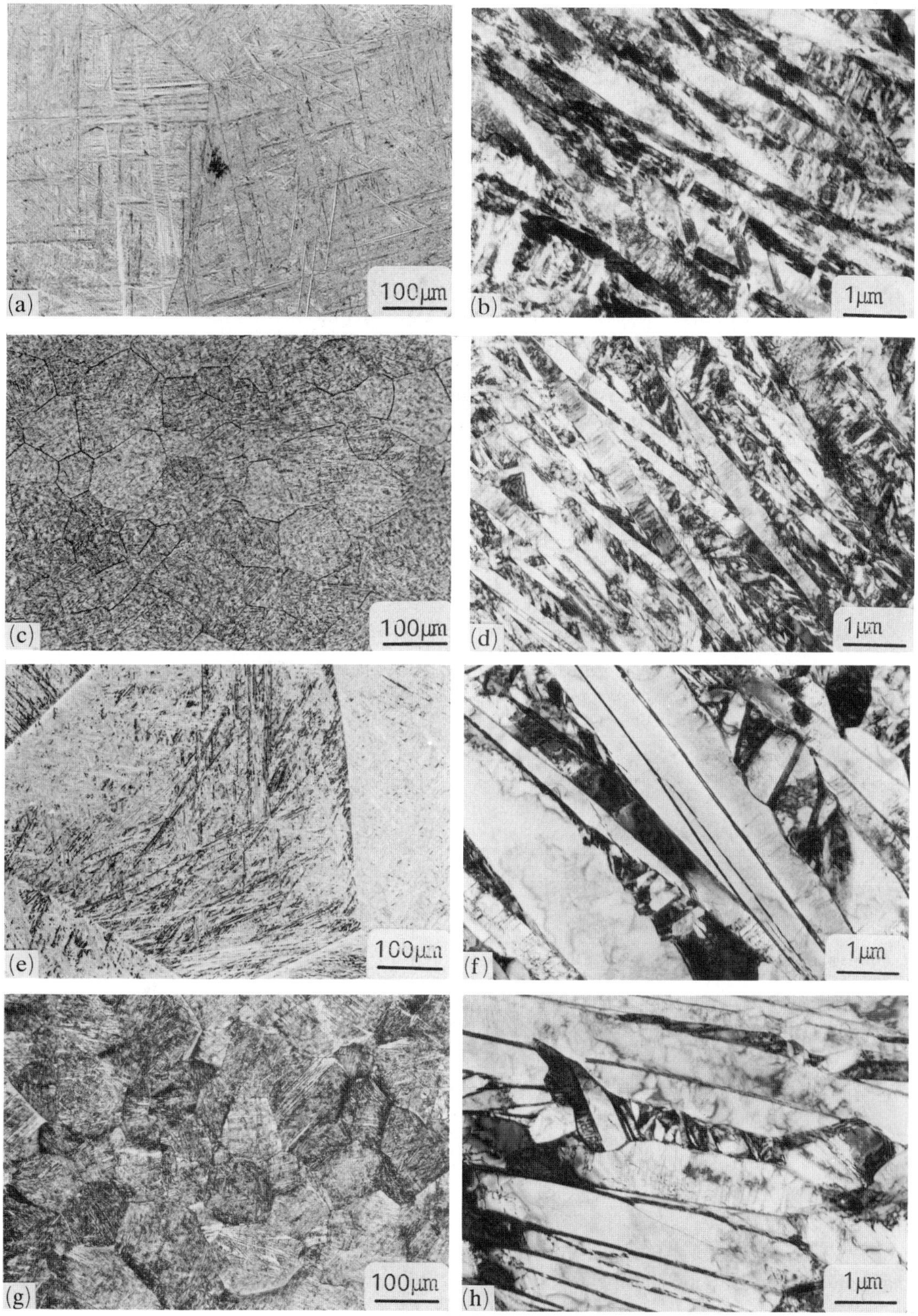

Fig. 2. Microstructure of Ti 6242: (a) FH-WQ (LM), (b) FH-WQ (TEM), (c) RH-WQ (LM), (d) RH-WQ (TEM), (e) FH-AC (LM), (f) FH-AC (TEM), (g) RH-AC (LM), (h) RH-AC (TEM).

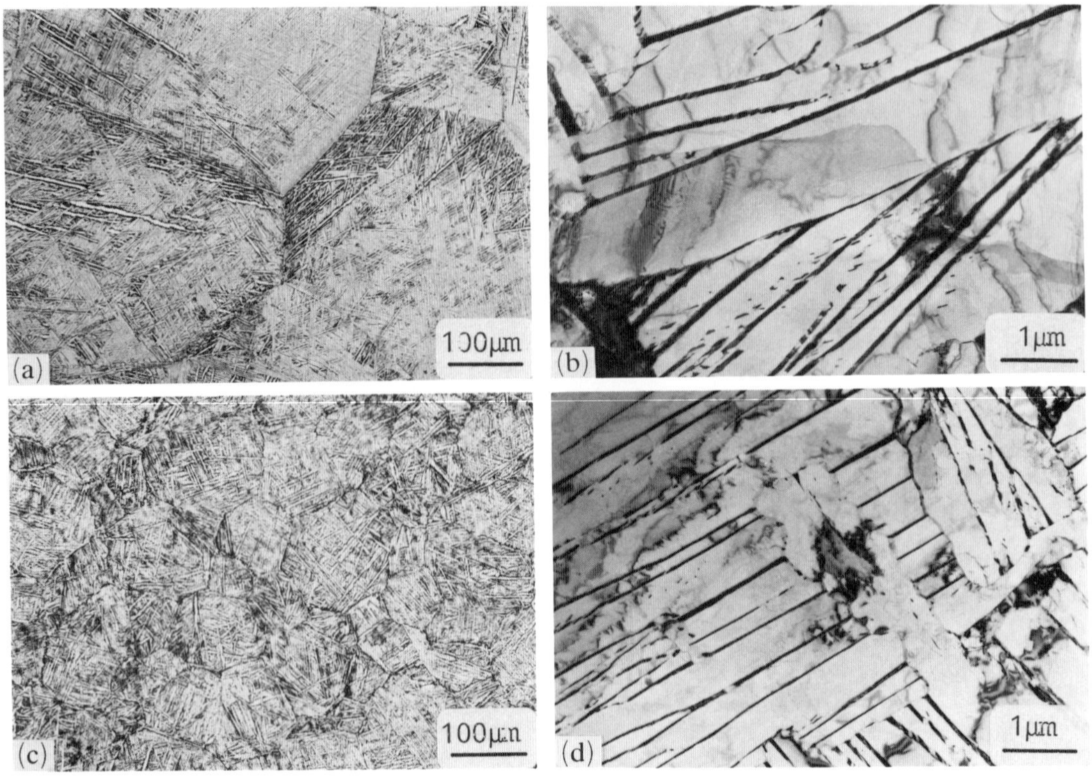

Fig. 3. Microstructure of VT 18 Y: (a) FH-AC (LM), (b) FH-AC (TEM), (c) RH-AC (LM), (d) RH-AC (TEM).

TABLE 2. Mechanical properties of Ti 6242 and VT 18 Y

Condition	Tensile properties						Creep resistance
	Temperature (°C)	$\sigma_{0.2}$ (MPa)	UTS (MPa)	σ_F (MPa)	A_5 (%)	ε_F	ε_{pl} (%)
Ti 6242,	RT	1225	1270	1335	0.7	0.01	0.24
FH-WQ	450	840	975	1120	3.7	0.18	
Ti 6242,	RT	1255	1335	1475	4.2	0.12	0.30
RH-WQ	450	865	1015	1320	7.8	0.45	
Ti 6242,	RT	1000	1105	1180	4.6	0.12	0.11
FH-AC	450	645	810	1040	6.8	0.30	
Ti 6242,	RT	930	1065	1405	11.2	0.40	0.12
RH-AC	450	615	805	1290	14.0	0.58	
VT 18 Y,	RT	930	1040	1250	11.2	0.20	0.12
FH-AC	450	630	805	1070	11.7	0.33	
VT 18 Y,	RT	970	1075	1435	14.0	0.34	0.11
RH-AC	450	645	805	1325	16.9	0.65	

the fatigue cracks nucleated within the long primary martensite plates, mentioned earlier, which were much less pronounced in the material with small β grain size (RH-WQ).

For the AC conditions, only the lifetime at high stress amplitudes was improved, resulting in a cross-over of the S–N curves and higher HCF strength for the coarse-grained mircrostructure. The higher HCF strength for the FH-AC condition was obviously due to the difference in volume fraction of martensitically transformed and diffusionally grown plates, as discussed above, resulting in a higher yield stress for the FH-AC condition.

Preliminary results of LCF tests performed at 450 °C confirmed the tendency of longer lifetimes for fine-grained structures but this effect was difficult to

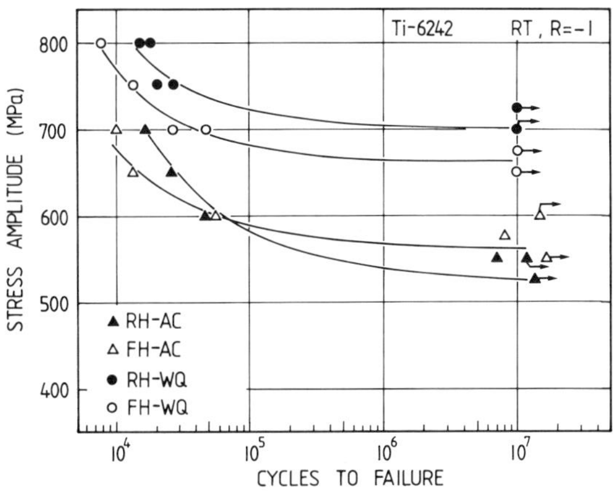

Fig. 4. S–N curves of Ti 6242.

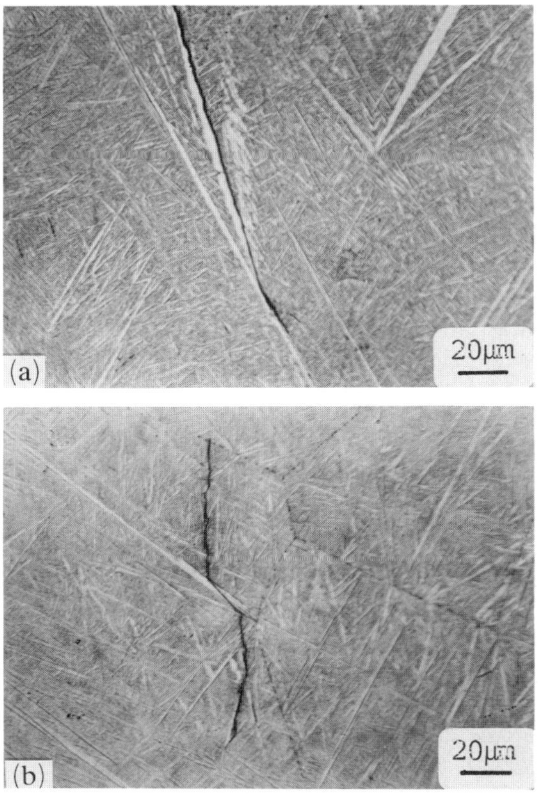

Fig. 5. Fatigue crack nucleation sites: (a) FH-WQ, (b) RH-WQ.

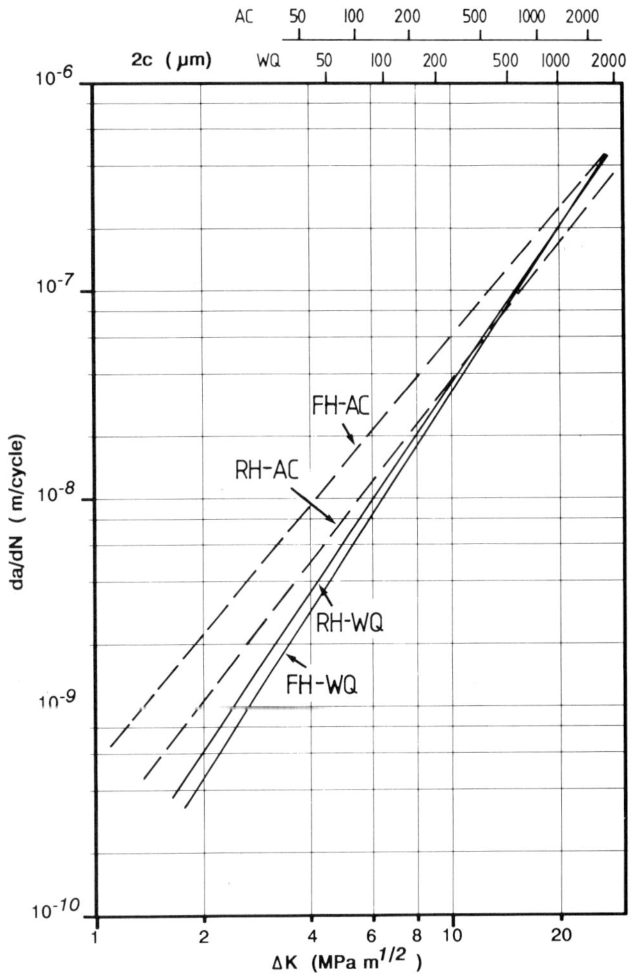

Fig. 6. Fatigue crack propagation in Ti 6242.

evaluate quantitatively because of the high fatigue strength values at 10^4 cycles being very close to the respective UTS values.

Results of the investigation of the propagation of small self-initiated fatigue cracks in Ti 6242 are summarized in Fig. 6. For the AC conditions, the material with the finer β grain size (RH-AC) exhibited a slower propagation rate of small fatigue cracks than the material with coarse β grain size (FH-AC). This behavior can be explained by the fact that β grain boundaries and α colony boundaries are strong obstacles to crack propagation. For the two WQ conditions with different β grain sizes no differences in propagation rates were found. Furthermore, the small cracks propagated more slowly under the WQ conditions compared with the AC conditions. Both results can be explained by assuming that the individual martensite plate boundaries already serve as strong obstacles for crack propagation on a smaller scale compared with β grain boundaries and α colony boundaries.

Creep tests (see Table 2) showed a much lower creep resistance of the WQ microstructures compared with the AC microstructures. This was not unexpected, taking into account that at temperatures above 500 °C creep deformation also depends on the perfection of plate boundaries [4] which is lower in martensite microstructures. The similarity of the creep resistances of the AC conditions indicates that the β grain size has a negligible effect on creep resistance.

TABLE 3. Comparison of mechanical properties of Ti 6242 in RH-AC and bimodal [1] conditions

Condition	Tensile properties (RT)				Creep (510 °C), 210 MPa,	HCF (RT), $R=-1$,	LCF (450 °C), $R=0.1$,
	$\sigma_{0.2}$ (MPa)	UTS (MPa)	A_5 (%)	ε_F	100 h, ε_{pl} (%)	$\sigma_{max\ 10^7}$ (MPa)	$\sigma_{max\ 10^4}$ (MPa)
RH-AC	930	1065	11	0.40	0.12	525	750
Bimodal	945	1055	12	0.47	0.15	500	675

4. Discussion

The results of the mechanical properties of furnace and rapidly β-heat treated high-temperature titanium alloys show that the tensile ductility of fine-grained materials is superior to that of their coarse-grained counterparts, mainly owing to a much shorter length of α-lamellae. Long individual α-colonies, being oriented favorably to the loading direction (slip planes), can serve as crack nucleation sites thereby drastically deteriorating the tensile ductility. For the WQ conditions, these long individual α-lamellae also serve as fatigue crack nucleation sites decreasing the HCF strength of the coarse-grained material compared with the rapidly β-heated material, although at least part of the improvement in fatigue strength due to the RH-WQ condition is a result of the faster cooling rate obtained by spray quenching and is not a result of the finer β grain size.

The difference in creep resistance between the fine-grained RH-WQ and the coarse-grained FH-WQ conditions should also be attributed more to the different cooling rates than to different grain sizes. Some specific features of martensite transformation in chemically inhomogeneous β-phase solid solutions [5] should also be mentioned as a possible explanation of the low creep resistance of the RH-WQ material. Comparing the creep properties of the WQ and AC conditions, the AC conditions appear much more attractive, having a similar low value of $\varepsilon_{pl} \approx 0.1\%$ for both alloys and, we should point out, for both fine-grained and coarse-grained conditions. A possible explanation for this behavior was published recently [4].

Considering the potential of β-heat treatment for high-temperature alloys it should be pointed out that the rapid heat treatment combined with air cooling (RH-AC) provides the best compromise of tensile, fatigue, and creep properties. It is interesting to compare the combination of properties for RH-AC of Ti 6242 with the properties of optimized bimodal structures, described earlier [1] (Table 3). From this comparison, the positive contribution of the RH-AC treatment can be concluded. One more point to mention is the isotropic character of properties of RH-AC treated material, while bimodal structures are sensitive to the crystallographic texture of the hexagonal α-phase and should be processed carefully to provide the best combination of properties. High-strength RH-WQ conditions can only be considered for short-term application because of the obvious drawback of low creep resistance.

There is no full set of data so far to compare the effectiveness of rapid heat treatment between Ti 6242 and VT 18 Y. Considering the available data on tensile and creep behavior, it would seem that similar performances can be achieved, despite some differences in composition. Additional fatigue tests should be performed to confirm this assumption.

Acknowledgments

This work was supported by the Academy of Sciences of Ukraine, Kiev, Ukraine, and the Deutsche Forschungsgemeinschaft, Bonn, FRG.

References

1 S. Saal, L. Wagner, G. Lütjering, H. Pillhöfer and M. Daeubler, *Z. Metallkd., 81* (1990) 535–539.
2 V. N. Grydnev, O. M. Ivasishin and S. P. Oshkaderov, *Physical Basis of Rapid Heat Treatment of Titanium Alloys*, Naukova dumka, Kiev, 1986 (in Russian).
3 O. M. Ivasishin, P. E. Markovsky, L. Wagner and G. Lütjering, *Titanium 1990: Products and Application*, TDA, 1990, pp. 99–110.
4 C. Andres, A. Gysler and G. Lütjering, *7th World Conf. on Titanium, San Diego, CA, 1992*, TMS, 1993.
5 V. N. Grydnev, O. M. Ivasishin and S. P. Oshkaderov, *5th World Conf. on Titanium, Munich, 1984*, DGM, 1985, pp. 1635–1641.

Materials Science and Engineering, *A168* (1993) 29–33

Structure-stable weldable KhN55MVC (ChS57) alloy for large-scale equipment operating at temperatures of up to 950 °C during a service life of up to 50 000 h

Yu. A. Dushin, A. V. Ivanov, V. V. Kunilovskij and N. A. Medvedev

Central Research Institute of Structural Materials "Prometey", Monastyrka River Emb. 1, 193167 Saint-Petersburg (Russian Federation)

(Received November 3, 1992; in revised form March 12, 1993)

Abstract

The mechanical, corrosion and radiation properties and the processing characteristics of KhN55MVC (ChS57) superalloy used in the chemical industry and in the energetics of large power installations are considered. A model useful for the prediction of long-term mechanical properties is proposed.

1. Characteristic properties of KhN55MVC alloy

The selection of materials for large power installations operating at high temperatures during a long service life is determined by the requirements of processing, structure stability and heat and corrosion resistance. For example, a material used for the production of heat exchanger components for the nuclear reactor system VGM-200 must serve for 5×10^4 h at temperatures of up to 950 °C.

In order to develop materials to cope with these adverse conditions, we should concentrate our research on strained Ni-based alloys with strengthening elements such as Mo and W. In such alloys, however, possible close-packed phase precipitation under operating conditions is a cause for concern: the solid solution becomes impoverished in refractory elements and loses strength, and brittle particles of σ^-, μ^- and ε phases initiate premature material fracture.

In KhN55MVC alloy the main alloying elements (Cr, Mo, W) are in such proportions as to prevent this phase precipitation [1]. (The chemical compositions of KhN55MVC alloy, both nominal and measured for the investigated melt 5V8859, are given in Table 1.) After 3000 h exposure at 900 °C only finely dispersed carbide particles are found (Fig. 1). In order to prevent environmental effects, the whole rod was exposed. Then specimens made from this rod were fractured by shock loading at room temperature. "Cup" fracture confirms a ductile intragranular fracture mode; the fracture energy even after exposure at an unfavourable temperature of 750 °C for 3000 h exceeds 120 J cm^{-2}. For comparison, Hastelloy XR structures are also shown in Fig. 1. Intermetallic phase precipitation and

TABLE 1. Nominal and measured chemical compositions of KhN55MVC (ChS57) alloy (mass%)

Alloy	C	Si	Ni	Cr	Mo	W	Fe	Al	Zr	Mn	Other
KhN55MVC (ChS57) according to TU 302.02.083-90	<0.05	<0.3	53–56	18.0–20.0	5.0–7.0	2.0–3.0	Bal.	<0.15	0.05–0.15	1.3–1.7	Y=0.05 B=0.005 S<0.01 P<0.015 N<0.03
KhN55MVC (ChS57) melt 5V8859	0.03	0.16	54.5	19.18	6.0	2.85	Bal.	0.07	0.15	1.48	Y=0.05 B=0.005 S=0.004 P=0.008

Elsevier Sequoia

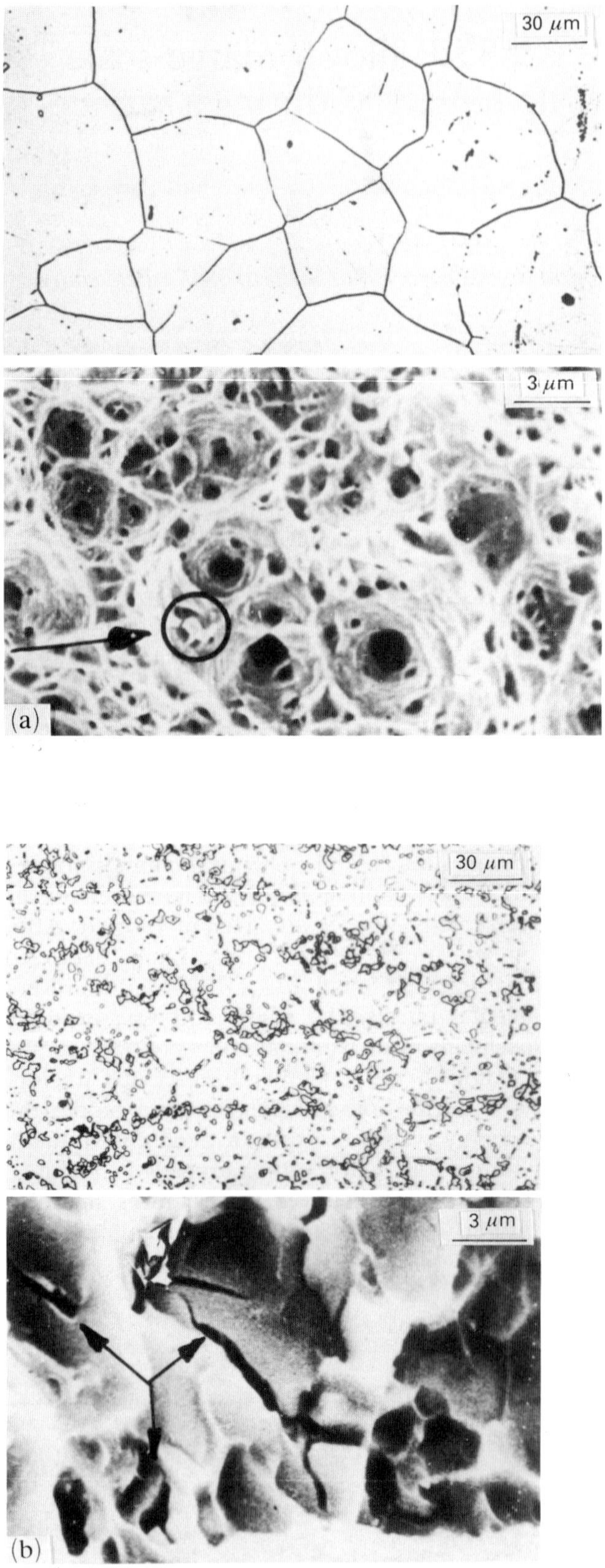

Fig. 1. Fractographs of (a) KhN55MVC (carbides M_6C, ZrC) and (b) Hastelloy XR (μ phase: 20 mass % Ni, 71 mass % Mo, 8 mass % W) alloys after exposure at 900 °C for 3000 h.

fracture in the region of brittle particles and cracks are observed; the fracture energy is reduced to 30 J cm^{-2} after high temperature exposure (900 °C, 3000 h).

A moderate Mo and W concentration in KhN55MVC alloy allows us to deform ingots of 6 t weight and to manufacture all necessary semifinished products, but limits the high temperature strength. The latter problem can be solved by Zr alloying. In this case it should be remembered that the operating stresses at maximum temperature are one order lower than the yield strength. Thus, in creep, shear along grain boundaries plays an important role and is accompanied by void formation and metal density reduction (plastic loosening). The zirconium concentrates on grain boundaries and inhibits intergranular slip (the displacement along grain boundaries is reduced by 5–10 times), voids and crack formation. When the Zr concentration is increased from 0.04% to 0.09%, the relative density loosening $\Delta\rho/\rho_0$ at 950 °C decreases by a factor of 10 and the creep life increases threefold. Thus it is possible to gain a harmonious combination of processing and service characteristics in this alloy.

Welding of the alloy in thicknesses up to 172 mm can be achieved by automatic methods: electron beam, submerged arc and electroslag welding. The mechanical properties of the weld metal allow the forge rolling of welded sheets and plates. The alloy is adaptable to forming, cutting off, expansion, reduction, coiling and flattering. With respect to cutting, the alloy corresponds to the third group and is among the austenitic steels. In terms of strength, creep life, radiation properties and hydrogen permeability the material is quite similar to alloys of the Hastelloy XR type. The advantage of KhN55MVC alloy is its high ductility (both short term and long term) and hence good fatigue resistance. This characteristic is especially important under conditions of reactor cooling.

2. Prediction of long-term mechanical properties

The prolonged service life (up to 10^5 h) and high operating temperature (up to 950 °C) of heat exchangers in nuclear power plants make it necessary to limit the value of applied stress to 5–8 MPa. Such a stress–temperature combination has only existed recently in power engineering, so material structure variations and deformation mechanisms are as yet insufficiently studied.

Under such unusual operating conditions, traditional calculation methods for material long-term strength and ductility prediction appear to be useless, since with stress reduction a change in deformation and fracture mechanisms takes place, while high temperatures lead to deep diffusive penetration of coolant elements into materials and consequently the material mechanical properties become dependent on coolant composition and heat exchanger tube wall thickness.

In developing the calculation model, the following physical preconditions were taken into account.

(1) At high temperatures and low stresses creep occurs by a diffusion mechanism associated with intergranular slip [2]. An electron microscopy study of the dislocation structure evolution during the creep of KhN55MVC alloy at 950 °C indicated that a transition from a dislocation creep mechanism to a diffusion one resulted on decreasing the applied stress from 13 to 5 MPa, *i.e.* to the level of operating stresses.

In order to verify this, special creep tests were performed on KhN55MVC alloy at 950 °C and two stress levels, 5 and 10 MPa. It was found that a twofold increase in applied stress caused the steady state creep rate to increase twofold. Such a directly proportional dependence of creep rate on applied stress is characteristic of diffusional creep.

Thus the creep strain contains both dislocation (index "s") and diffusion (index "f") constituents and the rate of steady state creep should be defined as

$$\dot{\varepsilon} = A_s(\sigma - \sigma_b)^{n_s} + A_f(\sigma - \sigma_b)^{n_f} \qquad (1)$$

where A_s, n_s, A_f and n_f are material constants at a given temperature and σ_b is the back stress.

Experimental values of the steady state creep rate of KhN55MVC alloy at 950 °C (Fig. 2) are in satisfactory agreement with eqn. (1) if the constants have the following values: $A_s = 3.5 \times 10^{-11}$ MPa$^{-n_s}$ h^{-1}, $n_s = 4.92$, $A_f = 2.18 \times 10^{-7}$ MPa^{-1} h^{-1}, $n_f = 1.0$, $\sigma_b = 0$.

(2) Grain boundary slip at high temperatures and low stresses is followed by nucleation and growth of grain boundary cavities. Grain boundary cavitation is

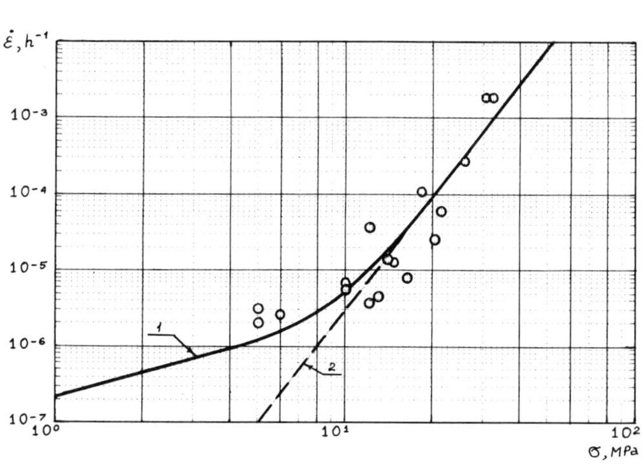

Fig. 2. Dependence of steady state creep rate of KhN55MVC alloy on real stress at 950 °C: 1, calculated curve; 2, dislocation component of creep rate; ○, experimental points.

usually described by some scalar damage parameter ω (*e.g.* the relative variation in material density) for which a kinetic equation and fracture criterion are postulated. In deriving the fracture criterion, one usually considers the critical damage level reached [3] and, recently, the phenomenon of plastic instability [4].

The criterion of plastic instability is based on investigation of the close interrelation between material deformation and damage mechanisms and allows correlation of the long- and short-term properties.

The basis of the plastic instability model is the consideration of effective net section reduction at the expense of nucleation and growth of grain boundary cavities. According to ref. 4, net section stability loss occurs when a sudden accidental force increase in a net section as a result of its reduction cannot be compensated by a load increase due to material hardening. In this case the variation in force acting in the net section of a material structural element is equal to zero, $\delta F = 0$. (A structural element is defined as an element of material containing a grain boundary in which the stress–strain conditions are assumed to be uniform.) Taking into account that $F = S_n \sigma_n$, where S_n is the net section area normalized by the section of the structural element, $\sigma_n = \sigma / S_n$ is the stress acting in the structural element and $\omega = 1 - S_n$ is the fraction of section area occupied by cavities (a damage parameter), we have

$$\delta S_n \sigma_n + \delta \sigma_n S_n = 0 \qquad (2)$$

In order to describe the process of cavity accumulation "on average", we apply the well-known parametric equation used for a wide range of materials [5],

$$\omega = B \varepsilon \tau \sigma^m \qquad (3)$$

where τ is time and B and m are material constants at a given temperature. Then, using the law of hardening in the form $\sigma_n = \sigma_1 + \varphi \varepsilon^p$, where σ_1 is the physical yield stress and φ and p are hardening parameters, to determine $\delta \sigma_n$ and the equation of deformation cavity growth for δS_n, we obtain

$$\varphi p \varepsilon_*^{p-1}(1 - B\varepsilon_* \tau_* \sigma^m)^2 - 2.5 B \varepsilon_* \tau_* \sigma^{m+1} = 0 \qquad (4)$$

Here ε_* is the real breaking deformation and τ_* is the time to fracture, which is expressed as

$$\tau_* = \int_0^{\varepsilon_*} \frac{d\varepsilon}{A_s(\sigma - \sigma_b)^{n_s} + A_f(\sigma - \sigma_b)^{n_f}} \qquad (5)$$

in accordance with eqn. (1).

The numerical solution of the system of equations (4) and (5) gives the fracture characteristics ε_* and τ_* as functions of the real stress σ.

In creep fracture tests the specimen elongation δ_* and applied stress σ_0 are usually measured. These values were calculated using the well-known relations $\delta_* = \exp\varepsilon_* - 1$ and $\sigma_0 = \sigma/\exp\varepsilon$, which are true in the case of uniform deformation. Then the damage parameters B and m were obtained from experimental data on fracture elongation $\delta_*(\sigma_0)$ in such a way as to achieve the best agreement with eqns. (4) and (5). It was found that $B = 2.8\times10^{-8}$ MPa^{-m} h^{-1} and $m = 2.8$ for KhN55MVC alloy at 950 °C.

The hardening parameters φ and p were obtained from measured tensile diagrams, the localization of deformation being taken into acount. The following values were obtained for KhN55MVC alloy: $\varphi = 116$ MPa, $p = 0.11$.

(3) The prolonged effect of coolant on a material at high temperatures results in the diffusive penetration of environmental elements into it. In Kh55MVC alloy tests in the helium (5 MPa) environment of the intermediate circuit of the VGM-200 power plant (partial pressures of impurities: CO, 250 Pa; CO_2, 25 Pa; CH_4, 25 Pa; H_2, 250 Pa; H_2O, 10 Pa) at 900 °C, distinct particles of carbide phase are visible on grain boundaries.

Thus corrosion attack by helium coolant is mainly defined by a carburizing process and carbide particle nucleation. The effect of these particles on the mechanical properties is not simple.

On one hand, the presence of carbide particles leads to an increase in the back stress σ_b and a reduction in the creep rate. The maximum effect of the back stress on the creep rate is determined by the low stresses applied, *i.e.* in the range of diffusion creep. In this case, in accordance with Burton's model [6] which has been confirmed experimentally, the back stress is directly proportional to the volume fraction of the second phase, f:

$$\sigma_b = \beta f \tag{6}$$

where β is a material constant at a given temperature. To determine this coefficient of the effect of carbides, plane specimen tests were performed on specimens of thickness 0.5 and 1.0 mm in helium coolant at $T = 950$ °C and $\sigma_0 = 8$ MPa. The creep curves were treated with the KVARTET program to define the steady state creep rate. Then the carbide phase content in the middle of the steady state creep stage was found with the use of the OKA program and the coefficient β from eqns. (1) and (6) was calculated to be about 164 MPa.

On the other hand, the preferable nucleation of cavities on carbide particles gives rise to intensive cavitation on grain boundaries and to decreasing long-term ductility. Then the constant B in eqn. (3) in a first approximation is modified as

$$B = B_0(1 + \gamma f) \tag{7}$$

where γ is a constant. It was defined on the basis of test results obtained with plane specimens 1 mm thick in helium coolant ($\sigma_0 = 16$ MPa). To calculate γ, the equation of plastic instability (4) was applied together with the lifetime equation (5). This procedure gave the value $\gamma = 1.03$.

The character of the creep rupture strength variation is determined by the interaction of the above-mentioned competing factors, *i.e.* creep rate reduction and cavity nucleation intensification.

Calculation of the long-term mechanical properties of KhN55MVC alloy at 950 °C was carried out using the developed programme complex for the IBM PC computer. To verify the model reliability, the calculation data were related to the specimen state during the creep process (dislocation creep strain fraction, density variation) and to the actual characteristics of fracture (ultimate values of stress and elongation).

The calculation curves of damage $\omega(\tau_*)$, long-term ductility $\delta_*(\tau_*)$ and strength $\sigma_0(\tau_*)$ are given in Fig. 3. Experimental points were obtained on standard round (5 mm) specimens made from various alloy melts and semifinished products.

Damage was controlled by measuring the metal density change $\Delta\rho$ during the creep process. Determination of $\Delta\rho$ was carried out by means of precision hydrostatic weighing of elements cut from the working part of specimens after creep tests of various durations. Then the relative density change $\Delta\rho/\rho_0$ was identified with the volume fraction of cavities, which is equal to the fraction of cross-section occupied by cavities, ω, according to metallography principles [3]. It is seen that quite satisfactory results were obtained for both the damage kinetics and its response to the applied stress level change.

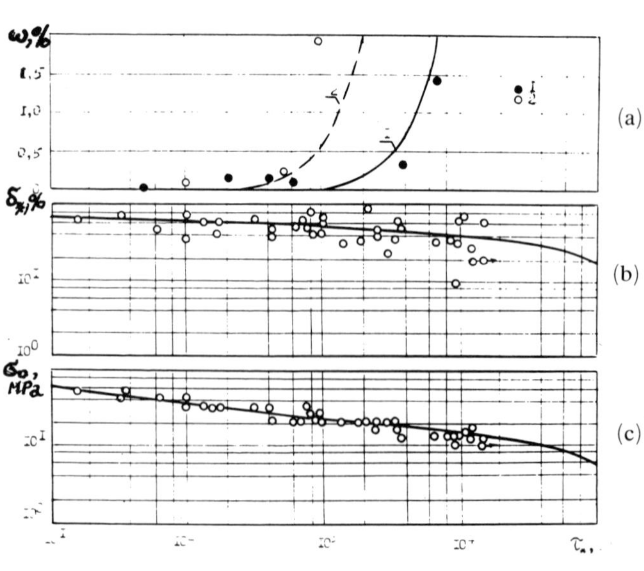

Fig. 3. (a) Volume fraction of cavities for two applied stresses (1, $\sigma_0 = 13$ MPa; 2, $\sigma_0 = 18$ MPa), (b) long-term ductility and (c) strength of KhN55MVC alloy at 950 °C.

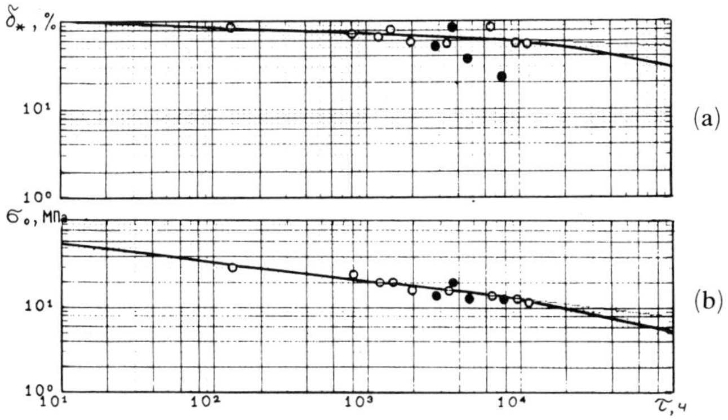

Fig. 4. (a) Long-term ductility and (b) strength of KhN55MVC alloy at 950 °C in (1) air and (2) helium coolant.

TABLE 2. Calculated long-term mechanical properties of KhN55MVC alloy at $T = 950$ °C and $\sigma_0 = 8$ MPa

	$\dot{\varepsilon}$ (h^{-1})	δ_* (%)	τ_* (h)
Dislocation creep	1.62×10^{-6}	23.4	135030
Dislocation + diffusion creep	3.56×10^{-6}	24.3	60848
Dislocation + diffusion creep + coolant effect	1.74×10^{-6}	10.5	65084

Concerning the graphs $\delta_*(\tau_*)$ and $\sigma_0(\tau_*)$, one can see a scatter of experimental points; nevertheless, the points are grouped around the calculation curves. The increasing slope of these curves is visible, especially at $\tau_* > 10^4$ h. This is explained both by the change in creep rate relation due to the change in creep mechanism and by premature fracture because of steady plastic deformation breakdown.

Figure 4 illustrates the results of long-term mechanical tests at 950 °C in air and in helium circuits with controlled coolant composition. For better comparison of the test results obtained in coolant and in air, specimens of the same melt were used. It is evident that the results obtained in air are grouped round the calculation curves of long-term ductility and strength. The long-term ductility δ_* of specimens tested in helium is basically lower than the calculated value, while the long-term strength σ_0 lies in the range of the experimental point scatter for tests in air.

As regards the operating conditions of the intermediate heat exchanger of the power plant VGM-200, calculation of the carburization characteristics and determination of the long-term mechanical properties of KhN55MVC alloy at 950 °C and 8 MPa stress (*i.e.* the required creep strength for 50 000 h) were performed with a heat exchanger tube wall thickness of 2.5 mm. The calculation results are given in Table 2.

References

1 Y. A. Dushin, N. N. Gribov, V. A. Jgnatov and N. A. Medvedev, Structure-stable alloy for structural components of the high-temperature gas-cooled reactor with coolant temperature of 950 °C, *Energy*, **16** (1–2) (1991) 317–326.
2 J. Cadek, *Creep Kovovych Materialu*, Academia, Prague, 1984.
3 V. I. Kumanin, L. A. Kovaleva and S. V. Alekseev, *Lifetime of Metals under Creep Conditions*, Metallurgia, Moscow, 1988.
4 O. V. Kuklina and B. Z. Margolin, Physical and mechanical model of cavitation fracture by creep, *Problemy Prochnosti (Kiev)*, (9) (1990) 3–10.
5 G. W. Greenwood, Cavity nucleation in the early stages of creep, *Philos. Mag.*, **19** (1969) 423–427.
6 B. Burton, On the mechanism of the inhibition of diffusional creep by second phase particles, *Mater. Sci. Eng.*, **11** (6) (1973) 337–343.

Materials Science and Engineering, A168 (1993) 35–40

High strength conductor for pulsed magnets

D. Dew-Hughes
Oxford University, Department of Engineering Science, Oxford, OX1 3PJ (UK)

Abstract

Magnetic inductions in excess of 30 T are produced in pulsed magnet systems. The coils for such systems call for a conductor with a combination of high conductivity, about 50% that of pure copper, and high strength, in excess of 1 GPa. High conductivity is required to limit the temperature rise during the pulse. High strength is required to resist the enormous Lorentz forces exerted on the conductor. These properties must be developed at 77 K, the temperature to which coils are precooled. In addition, the conductor must be ductile at room temperature to allow fabrication and coil winding. The desired combination of properties can only be achieved in a composite conductor in which the major component is copper. Macrocomposite conductors consist of copper and reinforcement arranged in such a way that the resultant properties are described by the rule of mixtures. Examples are copper with a stainless steel jacket or copper with a titanium core. In microcomposites the reinforcement is distributed on such a fine scale that strengths exceed the rule of mixtures. Examples are copper–niobium *in situ* composites and silver–copper two-phase alloys. The development of both types of composite is outlined in this paper.

1. Introduction

High magnetic fields are required for much research in condensed matter physics. Inductions in the range 30–70 T are provided by non-destructive pulsed magnet systems. Stored electrical energy is discharged into resistive coils which then generate pulsed magnetic fields with durations from 10 ms to 1 s. While a peak induction of 100 T and a duration of 1 s may be regarded as the ultimate goal for a non-destructive system (inductions in excess of 100 T are generated in ultra-short pulses by explosive compression which results in the destruction of the coil and the experiment), the immediate aims are for long (quasi-stationary) 100 ms–1 s pulses up to 60 T, and standard 10–100 ms pulses up to 70 T in peak magnitude. Several high magnetic field laboratories throughout the world are developing facilities aimed at these goals.

One of the most severe limitations on pulsed-coil performance which must be overcome is that of the mechanical integrity of the coil. The interaction of the electrical current flowing within the coil with the magnetic induction generated by it results in a Lorentz force which, acting like an internal pressure, tries to burst the coil. The resultant stresses in the coil must be contained either by the conductor from which the coil is constructed, or by reinforcement, internal or external, or a combination of these.

Coils wound from pure copper, with interlayer reinforcement of glass fibre, are capable of producing inductions of 69 T [1]. The failure mode of these coils is dramatic. With its combination of high strength and low modulus, glass fibre has a high density of stored elastic energy at its breaking stress. Once the breaking stress of the fibre is exceeded the release of this energy is sufficient to cause total disintegration of the coil. The failure of coils in which the strength is mainly derived from that of the conductor is much more gentle, and usually electrical in origin. Plastic deformation of the coil at some stage produces a local breakdown in insulation, sparking and local melting. An example of this is shown in Fig. 1, which is a cross-section of a coil which has failed in this manner. Although the coil has failed, it retains its overall integrity without catastrophic destruction of the experiment. This paper is concerned with the development of such a high strength conductor for pulsed magnet coils.

2. Properties of conductor for pulsed magnets

As indicated in the introduction, the most important property of a pulsed magnet conductor is mechanical strength. The maximum, tensile, hoop stress is developed in the centre turn of the innermost layer of a solenoid when generating a magnetic field. This maximum stress, for a solenoid of infinite length generating an induction of magnitude B tesla at its axis, is given by [2]:

0921-5093/93/$6.00

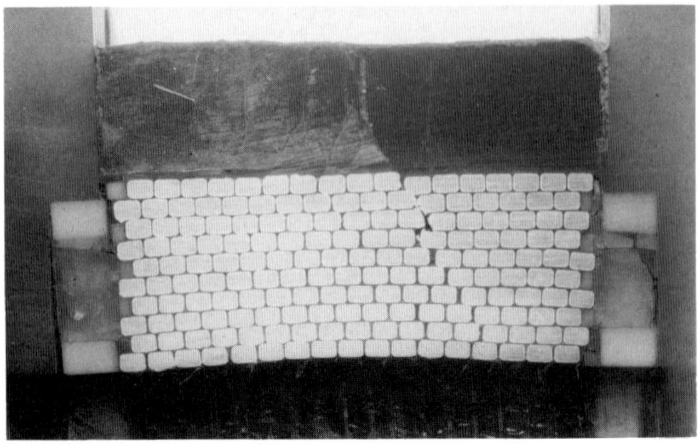

Fig. 1. Cross-section of failed coil, showing "barrelling" of the inner layer, and propagation of disruption to the point of insulation failure between the eighth and ninth layers. The conductor is a 2 mm × 3 mm CuNb microcomposite. (Courtesy of H. Jones, Clarendon Laboratory, Oxford University.)

$$\sigma_{max} = \frac{B^2}{6\mu_0} \frac{(\alpha+2)}{(\alpha-1)} \quad (1)$$

where μ_0 is the magnetic permeability of free space and α is the ratio of outer to inner diameter of the solenoid. For small coils α lies typically in the range 3–5; taking α as 4 the maximum tensile stresses are 0.95 GPa and 1.30 GPa respectively at 60 T and 70 T. This indicates the level of strength which must be attained in the conductor, though this can be reduced somewhat by the use of external support, such as a high strength steel or titanium alloy jacket or a fibre wrap, to the windings.

At the position of maximum stress, *i.e.* in the central innermost turn, the conductor will be subjected to a tensile stress of the above magnitude. Coils experience plastic deformation during high field pulses without necessarily undergoing failure. This is confirmed by measurements of the inductance of the coil which increases with the number and intensity of the pulses. An increase in inductance of about 2%–3% can be observed before the coil fails [3]. Clearly the yield stress (YS) of the conductor may be exceeded without failure. Examination of failed coils never reveals tensile failure of the conductor, and thus the tensile strength (TS) of the conductor is presumably not reached. However, each individual turn of the conductor is reinforced and constrained by the surrounding turns, and tensile failure can only occur at a constrained tensile strength in excess of the TS measured in a uniaxial test. It is found in practice that the measured TS of the conductor is a very good indicator of coil performance.

Thus the prime requirement of the pulsed coil conductor is that it has a tensile strength in the range 0.95–1.30 GPa. Were this the only requirement there would be little difficulty in producing conductor which readily exceeded it. However, this level of strength must be combined with high electrical conductivity, necessary to limit the temperature rise in the resistive coil during the pulse. It is usual to precool the coil in liquid nitrogen in order to take advantage of its large thermal capacity between 77 K and room temperature. The specified high values of strength and conductivity are those for the conductor at 77 K; fortunately these are generally larger than the values at room temperature. Acceptable values of electrical conductivity are about 50% that of pure copper at 77 K, 0.23 GS m^{-1} or 380% IACs. High values of thermal conductivity are also desirable, and in particular for long pulse coils, high values of transverse thermal conductivity, to ensure a more uniform temperature distribution during and after the pulse. The conductor should also have reasonable ductility at room temperature, both to allow it to be processed into wire, and to allow it to be wound into a coil. At the inner turns and at the terminations the maximum outer fibre strain can be as high as 10%. Thus a material which undergoes a considerable increase in strength as the temperature is lowered is highly desirable.

In order to satisfy the electrical and thermal conductivity requirements, the conductor must be based on copper. A number of commercial copper alloys such as Cu–Cd and CuNiZr have been developed to combine conductivity with strength, for applications such as overhead conductors in electric traction systems. Their tensile strengths range from 0.5 GPa at room temperature to 0.66 GPa at 77 K, and conductivities are about one tenth of that of pure copper at 77 K. These alloys are therefore unsuitable for pulsed coil applications. Age-hardening alloys, such as Cu–Be, Cu–Co–Be, and Cu–Ni–Si, can achieve the required levels of strength on peak ageing, but retain too much of the alloying elements in solid solution to give adequate conductivity. For example a TS (77 K) of 1.5 GPa can be achieved in a CuBe alloy, but its conductivity is only about one quarter that of pure copper at 77 K.

A satisfactory combination of strength and conductivity can only be achieved by a composite conductor in which pure copper is combined with a suitable reinforcement. This may be either a macrocomposite, in which the reinforcement is distributed on a relatively coarse scale and the properties are described by the rule of mixtures (ROM), or microcomposites where the reinforcement is in the form of fine filaments and the properties exceed those calculated from the ROM. The development of both types of conductor under the EUROMAGTECH collaboration will be described in turn.

3. Macrocomposite conductors

The simplest form of macrocomposite consists of a copper core surrounded by an external reinforcing jacket. The 300-series austenitic stainless steels (SS) are particularly suitable as reinforcement as they can show dramatic increases in TS from room temperature to 77 K [4]. This increase is due to the formation of deformation-induced martensite and is more pronounced the more unstable the steel. The increase may be as high as 100% in the steels of lowest stability, such as 304 [4]. Tensile strengths are raised by prior cold work. The combination of cold work and low temperature can raise the TS (77 K) of 304 to a value in excess of 2 GPa [5].

The development of a conductor consisting of 50%–60% copper, jacketed with 300-series stainless steel, has resulted from an initial collaboration between the Clarendon Laboratory of the University of Oxford and AEA Technology, Harwell [6], and has continued under the auspices of EUROMAGTECH. The conductivity depends almost entirely on the cross-section area fraction of copper; the strength developed depends on the ratio of copper to stainless steel, the grade and composition of the stainless steel and the degree of cold deformation after the final anneal. In practice the latter is limited to a maximum of about 75% reduction in area (RA). Figure 2 shows the cross-section of a coil wound from conductor of this type.

The conductor is fabricated by inserting an OFHC copper rod inside a thick-walled stainless steel tube. This is then swaged and wire drawn down to final dimensions (2×3 mm^2 rectangle with radiused corners) with intermediate anneals of 2 h at 800 °C. The drawing and annealing schedules are chosen to give a final RA of 65%–75%. (An initial attempt was made to reinforce copper internally with stainless steel rod, but during the drawing stage the copper pulled away from the steel.) The conductor is then insulated with a wrapped Kapton(TM) film which is cured at 290 °C for 2 h. This treatment effectively anneals the copper core, raising its conductivity back to close to that of the unworked state.

Fig. 2. Cross-section of coil wound from 2 mm × 3 mm stainless steel clad copper macrocomposite conductor. (Courtesy H. Jones, Clarendon Laboratory, Oxford University.)

The results of this programme to date are summarized in Table 1. A conductor based on 304 stainless steel can attain, reproducibly, TS (77 K) values in excess of 1.2 GPa, and a value of 1.29 GPa was seen in one instance. The relatively less stable 316 (304 plus 2% Mo) generates insufficient strength. 321 (titanium-stabilized 304), because the titanium removes carbon from solid solution, is expected to be less stable than 304, but any martensite formed will also be deficient in carbon and have lower strength. The latter effect appears, just, to win out, as conductor based on 321 is consistently marginally less strong than that based on 304. Its low temperature mechanical properties match those of the low carbon 304L [4]. Its reduced stability is demonstrated by the fact that it becomes magnetic at 77 K, indicating some transformation to martensite without further deformation. There is an interesting inconsistency between the two samples, HEF1 and HEF6, based on 321 stainless steel. HEF6 with 65% RA, is stronger than HEF1 with 75% RA. The conductor is drawn down in the round, and only converted to a rectangular cross-section in the final stages of reduction. If this is done in one pass, the corners of the cross-section undergo a much greater than average degree of cold work, and if this average is the desired 75% RA then the degree of cold work at the corners is so extreme as to cause local longitudinal cracking, resulting in degraded properties in a tensile test. Seventy-five per cent RA can be achieved satisfactorily if round to rectangular is carried out in two or more passes.

The conductivity of all of these conductors is above the minimum requirement, and a further gain in strength can be achieved by some reduction in the copper cross-section. A multifilamentary composite was fabricated by bundling a mixture of pure copper and stainless-jacketed copper hexagonal section wires into a stainless outer tube and subsequent drawing to final dimensions (sample HEF3). The ratio of copper to stainless steel was 1/1. This exercise resulted in a loss of conductivity with no increase in strength.

Stainless steel clad copper conductor should prove to be the simplest and cheapest conductor for 70 T, standard pulse magnets. Unfortunately, this conductor is unsuitable for long-pulse applications. The stainless steel sheath confers poor transverse thermal conductivity. This problem would be overcome if it were possible to develop a processing route for the internal reinforcement of copper with stainless steel. Heavily

TABLE 1. Strength and conductivity of pulsed magnet conductors

Material		Condition	290 K				77 K			
			YS (GPa)	TS (GPa)	Elongation (%)	Conductivity (% IACS)	YS (GPa)	TS (GPa)	Elongation (%)	Conductivity (% IACS)
TP copper		Annealed	0.11	0.25	45	94.3	0.13	0.40	55	694
		Hard drawn	0.33	0.40	25	82	0.40	0.49	23	570
Macrocomposites										
Cu/SS										
DCF 95	316	52% Cu	0.54	0.60	1.1	73	0.64	0.83	8.4	449
DCF 101	304	60% Cu	0.67	0.73	1.1	63	0.75	1.00	2.5	423
DCF 400	304	60% Cu	0.77	0.86	5.2	—	1.05	1.23	1.5	—
DCF 401	304	60% Cu	0.72	0.85	5.3	58	1.00	1.29	1.9	389
HEF 4	304	60% Cu, 75% RA	0.70	0.87	1.8	64	0.88	1.24	2.1	444
HEF 3	304	50% Cu m/f 70% RA	0.71	0.90	2.0	51	0.88	1.20	2.0	293
HEF 1	321	60% Cu, 75% RA	0.43	0.80	2.0	66	0.82	1.12	1.8	452
HEF 6	321	60% Cu, 65% RA	0.83	0.89	2.0	—	1.07	1.15	7.2	—
Cu/Ti										
CP125		Pure Ti, 98% RA	0.89	1.14	7.6	—	1.44	1.56	4.4	—
TCC		30% Cu	0.59	0.68	6.3	45	0.86	0.94	6.2	288
IO		50% Cu, 95% RA	0.47	0.77	6.5	60		1.03		384
Microcomposites										
CuNb										
Supercon			0.65	0.99	3.2	76	0.94	1.19	2.5	331
Bochvar			0.77	1.07	2.0	71	1.17	1.33	2.4	261
CuAg										
10% Ag		99% RA	0.51	0.55	4.2	—	0.61	0.68	8.0	—
		1 int. anneal	0.66	0.74	3.5	—	0.82	0.90	2.8	—
		2 int. anneals	0.66	0.74	3.0	—	0.86	0.91	2.3	—
16% Ag			0.85	0.95	7.2	80	1.03	1.17	7.7	306
18% Ag			0.95	1.02	2.9	80	1.15	1.25	2.5	226

Data in this table are, in most cases, the average of several tests. (Measurements were mainly made at Oxford by the author, I. Johnston, P. Frings and R. Schaap.)

cold-worked commercial purity titanium (CP125) shows a 50% increase in tensile strength between room temperature and 77 K. IMI Titanium market a titanium clad copper conductor (TCC) with only 30% copper in the cross-section. In the as-supplied (annealed) condition this has a TS (77 K) of 0.94 GPa. An inside out (IO) version of this conductor, with 50% copper surrounding a titanium core has been investigated; the TS (77 K) is greater than 1 GPa. These results are given in Table 1 together with the results for a 33%Cu–67%CuBe jacketed conductor developed for the 60 T long pulse coil at the University of Amsterdam [7]. This conductor has adequate strength, but the conductivity is low. The tensile properties of all of these macrocomposites are consistent with the application of the ROM to the properties of the two constituents.

It can be seen from the foregoing that the macrocomposite approach, which is relatively simple and straightforward, is able to produce a variety of conductors with adequate properties for the proposed applications in pulsed magnet coils.

4. Microcomposite conductors

A microcomposite is a matrix reinforced with filaments or whiskers of another material. When the spacing between filaments is of the order of micrometres or less, deformation of the matrix is inhibited to such an extent that the overall mechanical properties exceed those predicted from the ROM [8–10]. The mechanism of matrix strengthening is believed to be the same as that in eutectoid steels, in which plates of non-deforming iron carbide restrict dislocation generation and motion in the ferrite lamellae. The flow stress at large strains is given by [11]:

$$\sigma = \frac{AGb}{\pi d} \ln\left(\frac{d}{2b}\right) + \nu_f \sigma_r + \sigma_0 \quad (2)$$

where A is a constant, G and b are the shear modulus and Burgers vector of the matrix, d is the filament spacing, ν_f and σ_r the volume fraction and strength of the filaments, and σ_0 the Peierls stress for the matrix. The only terms in eqn. (2) which are likely to show any strong temperature dependence are σ_r and σ_0. Some further increase in strength as the temperature is lowered may be gained from the suppression of dynamic recovery in the matrix.

High strength requires a high volume fraction of high strength filaments at very small spacings. Unfortunately, the filament spacings at which significant increases in strength begin to be observed correspond to the electron mean free path in pure copper at 77 K, approximately 0.25 μm. Thus the increases in strength are accompanied by a decrease in conductivity, which also is no longer described by the ROM.

Microcomposites are produced from two-phase ingots, prepared either by powder metallurgy or by casting, followed by extreme cold work. Consumable arc-melted, *in situ*, composite ingots are cast in alloy systems which show complete solubility in the liquid state and almost complete immiscibility in the liquid state. Systems based on copper include Cu–Fe, Cu–V, Cu–Ta and Cu–Nb. Of these, Cu–Nb, initially investigated as a precursor for multifilamentary Nb_3Sn superconductors, has received the most attention [12]. A pulsed coil wound from conductor developed from this material has produced a peak induction of 68.4 T [13].

Microcomposites are capable of remarkable strengths, for example a TS (RT) of greater than 2 GPa, at submicrometre spacings [14]. Processing reductions in diameter of over a thousandfold are required from the initial ingot to produce microstructures which can develop this level of strength. Such reductions involve several rebundlings in order to end up with a conductor with a cross-section of a few millimetres. Experience suggests that control of the microstructure during the casting process is vital, and reproducibility is difficult. Cu–Nb microcomposites have been prepared with up to 40% niobium, but the preferred level seems to be 15%–20%, presumably because of difficulties in processing the high volume fractions. Cu–Nb microcomposites processed into wire form with dimensions suitable for pulsed coil applications have a TS (77 K) in the range 1.0–1.2 GPa. The product is rather variable, and absolute process control has not been achieved.

A recent interesting development is that of copper–silver alloys. The copper–silver system is a simple eutectic with terminal solubility at the eutectic temperature (779 °C) of 4.9 at.% Ag in copper and 14.1 at.% Cu in silver. The solubility limits for both terminal solutions decrease to very low values at room temperature [15]. Two-phase alloys in this system, when subject to cold deformation, result in an ultrafine mixture of copper-rich and silver-rich filaments. The material is a microcomposite whose mechanical properties exceed the ROM [16]. Additionally, the electrical conductivity remains high. Both strength and conductivity can be enhanced by heat treatments which are believed to cause precipitation within the already fine, ribbon-like, microstructure [17]. Room temperature tensile strengths in excess of 1 GPa, coupled with an electrical conductivity of 80% IACS, can be obtained.

The copper–silver alloys have the advantage of being extremely ductile and easy to process, although care must be taken to prevent melting during the initial extrusion. A low extrusion temperature (around 500 °C) and fairly low extrusion ratio (9:1) is sufficient to break down the cast structure of an induction-melted billet. A subsequent cold deformation of 99% RA to a 2 mm × 3 mm cross-section can be carried out without the necessity for intermediate anneals, although the properties can be improved by the appropriate choice of intermediate anneals. This material has great development potential and a full metallurgical and metallographic investigation of the Cu–Ag system would be highly desirable.

Results for microcomposite materials are also included in Table 1.

5. Conclusions

Results for the best examples of various systems proposed for pulsed magnet conductors are plotted as TS (77 K) *vs.* conductivity at 77 K, in Fig. 3.

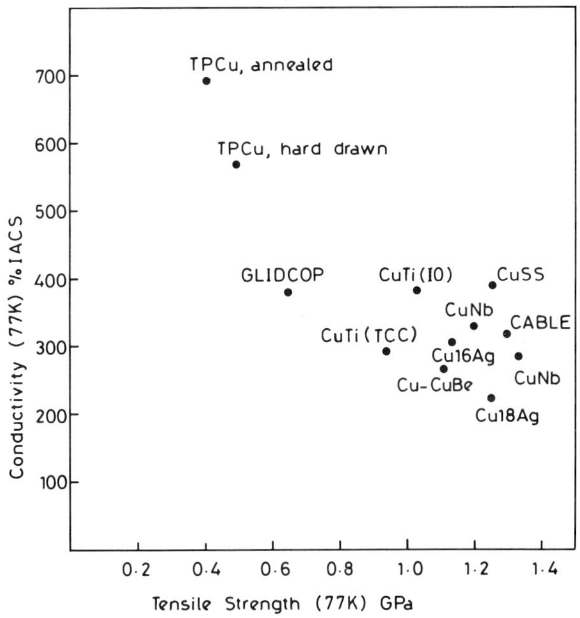

Fig. 3. Conductivity *vs.* tensile strength, both at 77 K, for candidate pulsed conductor materials. Glidcop is internally oxidized CuAl alloy prepared by powder metallurgy, data of Maryann Hills, Los Alamos Scientific Laboratory. Cable is a cabled reinforcement of ultrahigh strength pearlitic steel filaments woven around a copper core, fabricated by the Bekaert Company.

Serious candidate materials, all still capable of further development, would seem to be Cu–stainless steel and Cu–Ti macrocomposites, and Cu–Nb and Cu–Ag microcomposites.

The stainless steel jacketed copper has values of strength which should enable it to produce 70 T in a standard pulse, with a reserve of conductivity. Its poor transverse thermal conductivity rules it out for long pulse applications. However, the IO Cu–Ti could be developed for a long pulse conductor.

The highest strengths have been observed in CuNb microcomposites, but at the expense of some conductivity, difficulty of processing and lack of reproducibility. The CuAg system has not yet matched the performance of CuNb, but is a very recent material and has undergone very little development to date. Its relative ease of processing over CuNb is strongly in its favour. Microcomposite materials are suitable for both standard and long pulse coils.

Acknowledgments

The EUROMAGTECH consortium for the development of pulsed magnet conductor comprises the magnet laboratories of the Universities of Amsterdam, Leuven, Oxford and Toulouse together with that of the Max-Planck Institute, Grenoble, and is supported by DGXII of the Commission of the European Communities, under the Science Programme, grant number SCI-CT88-0066. Processing of the Cu–stainless-steel macrocomposites was carried out by J. A. Lee and H. Feyrit at AEA Technology, Harwell. Some of the tensile and conductivity data were supplied by R. Schaap (Amsterdam) and L. van Bockstal (Leuven). I wish to thank the Head of the Department of Engineering Science, Professor J. M. Brady, for the provision of experimental facilities, and my colleagues in the EUROMAGTECH consortium for many valuable discussions.

References

1 F. Herlach, Catholic University of Leuven, 1992, private communication.
2 F. Herlach, G. de Vos and J. Witters, *J. Phys. (Paris), 45* (1984) C1-915–921.
3 H. Jones and L. Van Bokstal, 1992, private communication.
4 G. P. Sanderson and D. T. Llewellyn, *J. Iron Steel Inst., 207* (1969) 1129–1146.
5 *Mechanical Properties of Austenitic Chromium–Nickel Stainless Steels at Subzero Temperatures*, International Nickel Co., Inc., New York, 1963.
6 H. Jones, F. Herlach, J. A. Lee, H. M. Whitworth, A. G. Day, D. J. Jeffrey, D. Dew-Hughes and G. Sherratt, *IEEE Trans. Magn., 24* (1968) 1055–1058.
7 R. Schaap, 1992, private communication.
8 J. Bevk, J. P. Harbison and J. L. Bell, *J. Appl. Phys., 49* (1978) 6031–6038.
9 P. D. Funkenbusch and T. H. Courtney, *Acta Metall., 33* (1985) 913–922.
10 W. A. Spitzig, A. R. Pelton and F. C. Laabs, *Acta Metall., 35* (1987) 2427–2442.
11 J. G. Sevillano, in P. Haasen, V. Gerold and G. Kostorz (eds.), *Strength of Metals and Alloys*, Pergamon Press, Oxford, 1980, pp. 819–824.
12 C. V. Renaud, E. Gregory and J. Wong, *Adv. Cryog. Eng. Mater., 34* (1988) 435–442.
13 S. Foner, E. Bobrov, C. Renaud, E. Gregory and J. Wong, *IEEE Trans. Magn., 24* (1980) 1059–1062.
14 L. I. Martynyuk, M. M. Oleksienko, V. T. Petrenko, A. I. Pikalov, M. A. Tikhonovsky and V. M. Tretyakov, *Cryogenics, 32*, (1993) 145–148 (ICMC Supplement).
15 M. Hansen, *Constitution of Binary Alloys*, McGraw Hill, New York, 1985, p. 18.
16 G. Frommeyer and G. Wasserman, *Acta Metall., 23* (1975) 1353–1360.
17 Y. Sakai, K. Inoue, T. Asano, H. Wada and H. Maeda, *Appl. Phys. Lett., 59* (1991) 2965–2967.

Contact damage submitted to thermal shock: a method to evaluate and simulate thermal shock resistance of brittle materials

F. Osterstock

LERMAT, URA CNRS No. 1317, ISMRA, 6 Bd du Maréchal Juin, 14050 Caen Cedex (France)

Abstract

A criterion for study of the thermal shock resistance of brittle materials is derived. It makes use of the stable extension stage of Vickers indentation cracks under the action of the thermal transient stress which appears during quenching. During this stage, the relative increase in crack length can be measured. If plotted as a function of indentation load, it can be used as an indicator of thermal shock resistance. On the basis of these well defined conditions, it is possible to reflect the empirical ranking of various materials (float glass, glass-ceramics and polycrystalline α-SiC). Furthermore, the anisotropy of the thermal transient stresses, as deduced from the shape of the quenched part, can be revealed. Finally a critical point, separating stable and unstable crack extension, is defined. It may thus become possible to evaluate quantitatively the maximum value of thermal transient stress, without knowledge of the heat transfer process during quenching.

1. Introduction

Since the publications of Buessem [1] and Kingery [2], much effort has been devoted to the thermal shock resistance of brittle materials. The aim was a comprehensive physical approach with a view to improving reliability. Buessem reviewed the physical parameters of the material involved in thermal heat transfer. Kingery derived a critical stress level in the skin of the quenched part. This was a first proposal to explain the sudden drop in strength after a critical quenching temperature had been attained. Theoretical approaches to describe the empirically well known strength *vs.* quenching temperature difference behavior have been proposed by Hasselman [3–5]. However, the transient thermal stresses are still poorly known, because the heat transfer processes are also poorly defined, as well as the distribution in size of surface defects.

In this respect, Faber *et al.* [6] proposed to make a Knoop indentation in the center of a disc to be quenched, to determine the origin of rupture at a critical quenching temperature. At the same time, studies of Vickers indentation techniques were published [7–10]. They showed that the as-indented crack system describes an equilibrium between a central residual opening force and the toughness of the material. If an external stress is then applied, this equilibrium will be disrupted and the cracks will propagate in a stable way before reaching certain critical states which define the onset of catastrophic failure.

In the present work, we made use of the stable (and measurable) crack extension which can occur under the action of a thermal transient stress, in order to define a criterion for thermal shock resistance. This derivation is confirmed by experimental measurements of materials whose empirical ranking is already well established.

2. The median–radial crack system

Lawn, Marshall and coworkers have produced a critical analysis of the symmetric crack pattern which appears at the corners of a Vickers indentation on brittle materials [7–10]. A median–radial crack system is defined, which results in a semi-elliptical surface crack, the traces of which (the radial cracks) can only be seen on a metallographically polished surface. These cracks, of length c_0, are in equilibrium between the toughness of the material K_c and a central opening force. For a given indentation load P and in the as-indented stage, this gives

$$K_c = \chi_r P c_0^{-3.2} \qquad (1)$$

The proportionality factor χ_r is proposed as

$$\chi_r = \varphi_r (E/H)^{1/2} \qquad (2)$$

The term (E/H) is the ratio of the Young modulus to the hardness. It indicates the wedge effect of a plastically deformed zone within an otherwise elastic bulk.

φ_r depends on the shape and effectiveness of the indenter and is considered a constant.

If an external increasing stress σ_a is now applied, the median–radial cracks will increase their size c under the superimposition of the decreasing residual and increasing applied stress intensity fields. This is written as

$$K_c = \chi_r P c^{-3/2} + \sigma_a(\Omega \pi c)^{1/2} \tag{3}$$

and indicates the possible existence of stable crack extension, in equilibrium with the toughness of the material. The term $\sigma_a(\Omega \pi c)^{1/2}$ describes the stress intensity field of a semi-circular surface crack. Differentiation of σ_a with respect to the actual crack length c shows that stable crack growth occurs between the initial crack length c_0 and a given crack length c_m. Both depend on the indentation load and on the toughness of the materials. The value of c_m is obtained as

$$c_m = (4\chi_r P/K_c)^{2/3} \tag{4}$$

It follows that the maximum value attained by the applied stress during stable crack extension is

$$\sigma_m = 3K_c/4(\pi\Omega c_m)^{1/2} \tag{5}$$

This behaviour is depicted by the increasing (stable crack extension) and then decreasing (unstable crack extension) σ_a *vs.* c curves in Fig. 1. The value $\chi_r = 1$ was used for simplicity.

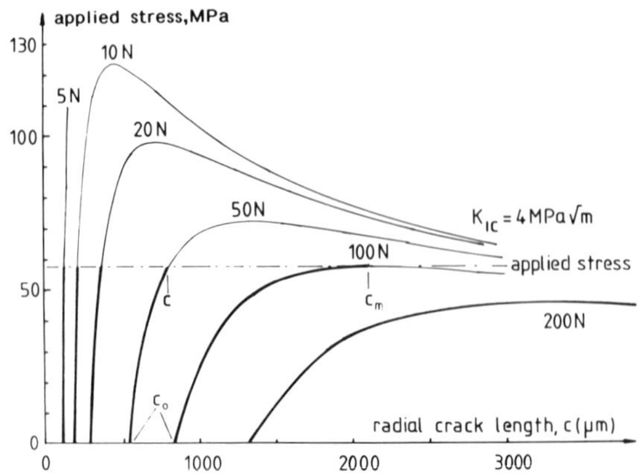

Fig. 1. Stable (increasing) and unstable (decreasing) branches of the extension of Vickers indentation radial cracks under the action of an applied stress. The value of the indentation load is given on the top of each curve. The extension for a given level of applied stress is the thick part of each line ($\chi_r = 1$).

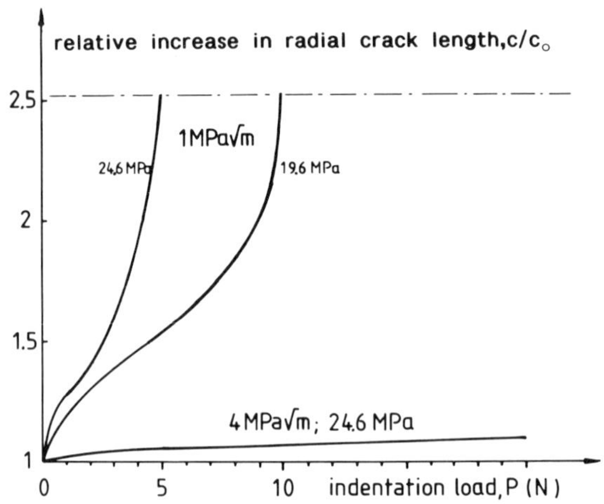

Fig. 2. Comparative effects of toughness and given applied stress on the relative increase in length of the radial cracks from Vickers indentations, as a function of indentation load.

3. Influence of transient thermal stresses

Transient thermal stresses act like other applied mechanical stresses. Thus, if an indented brittle part is quenched, the median–radial cracks will extend. This extension is, however, limited by the maximum value of the thermal transient stress. As depicted in Fig. 1, a level of constant stress is thus considered (58 MPa). The cuts with stable propagation stages thus define relative increases in radial crack length, as a function of indentation load. Such curves are plotted in Fig. 2. The influence of both the material toughness and the stress level is considered.

It can be seen from Fig. 2 that the variation in relative increase in radial crack length is strongly dependent on the level of the applied stress and on the toughness of the material. These curves can thus be used to investigate the relative resistance of various materials to given quenching conditions, or the increasing level of thermal transient stress as the quenching temperature difference ΔT is increased.

A further point should be underlined. In Fig. 1, the line of constant applied stress ($\sigma_a = 58$ MPa) is drawn. It intersects the stages of stable crack growth for indentation loads less than 100 N. For these loads, the lines touch in only one point which defines the onset of catastrophic failure beyond the critical crack length c_m (eqn. (4)) for the given critical applied stress, $\sigma_m = \sigma_a = 58$ MPa (eqn. (5)). Independent of the indentation load, the ratio (c_m/c_0) is constant and equal to 2.52 [9]. Therefore, the limiting value of 2.52 on the ordinates of Fig. 2 defines the onset of catastrophic crack extension, as well as critical values for thermal shock damage.

4. Experiments and results

The experimental aims were to verify whether the predicted shape of the curves in Fig. 2 can be reproduced, and whether an empirical ranking can be reflected.

The quenching medium was liquid nitrogen. Several brittle materials were used: float glass, a precursor of a glass-ceramic (VM-ST3), a class-ceramic (ST3) and finally a polycrystalline α-silicon carbide. The samples were parallelepipeds with dimensions of approximately $8 \times 4 \times 16$ mm^3, indented on the 8×16 mm^2 face. The toughness of the materials is depicted in Fig. 3, as derived from ref. 9. A plot of $P^{2/3}$ *vs.* c was used, because (from $K_c = \chi_r P c^{-3/2}$, see eqn. (1)) the slope of these straight lines is a direct indication of their relative toughness. In Fig. 4, the relative increases in radial crack length, as a function of indentation load, are compared for the float glass and the precursor VM-ST3. The samples were indented at room temperature and then quenched. The difference in response is striking. The relative increase in radial crack length is larger for float glass than for the glass-ceramic precursor. This behavior can be compared with practical experience. For the same quenching conditions the glass-ceramic did not exhibit any increase in crack length.

The relative resistance to thermal shock of brittle materials, judged according to the proposed method, is even more apparent in Fig. 5. After indentation at room temperature, the samples were first heated to 160 °C and then quenched in liquid nitrogen. The extension of the radial cracks was again measured at room temperature in air. The experimental results

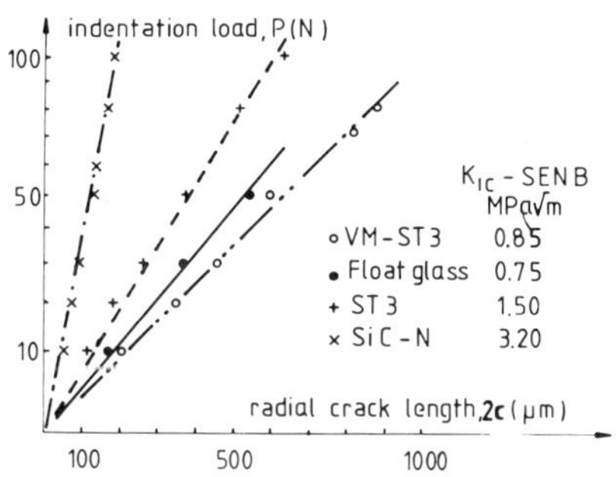

Fig. 3. Fracture mechanical properties of the investigated samples in a $P^{2/3}$ *vs.* c plot. The slope of the lines gives an indication of the relative toughness.

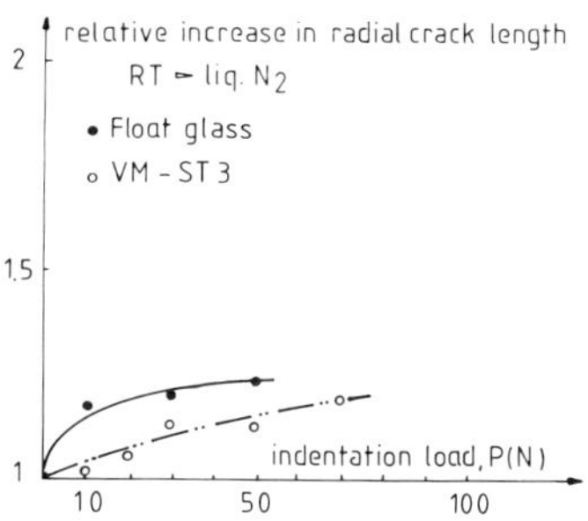

Fig. 4. Relative increases in radial crack length of the float glass and VM-ST3 precursor when indented with various loads at room temperature and then quenched into liquid N_2.

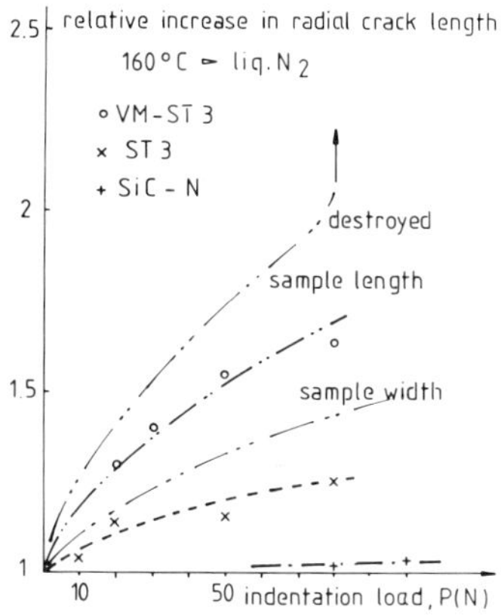

Fig. 5. Relative increases in radial crack length of the VM-ST3-precursor, the ST3 glass-ceramic and SiC when quenched from 160 °C into liquid N_2. In the case of the VM-ST3-precursor (which was destroyed) the borders (thin lines) indicate the effect of thermal stress anisotropy.

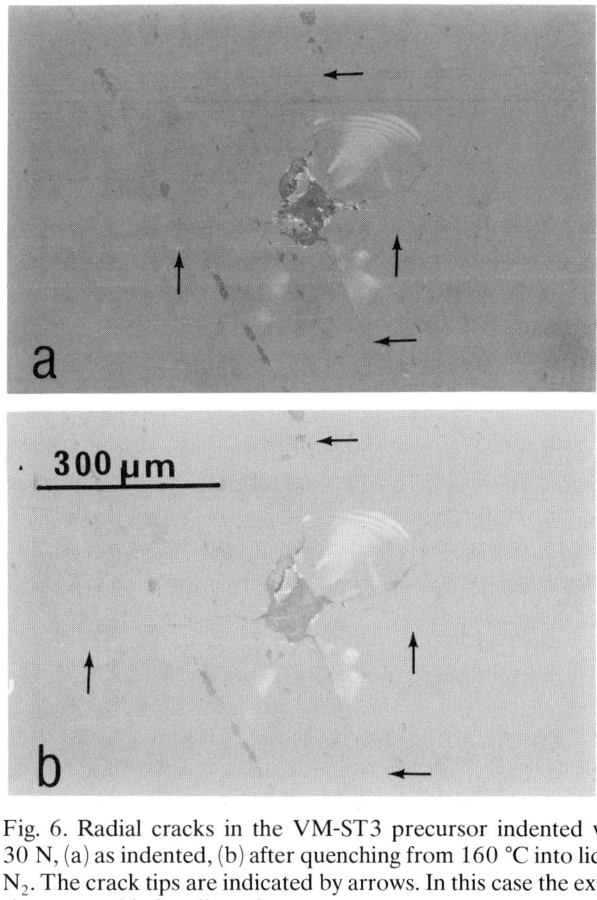

Fig. 6. Radial cracks in the VM-ST3 precursor indented with 30 N, (a) as indented, (b) after quenching from 160 °C into liquid N_2. The crack tips are indicated by arrows. In this case the extension was stable for all cracks.

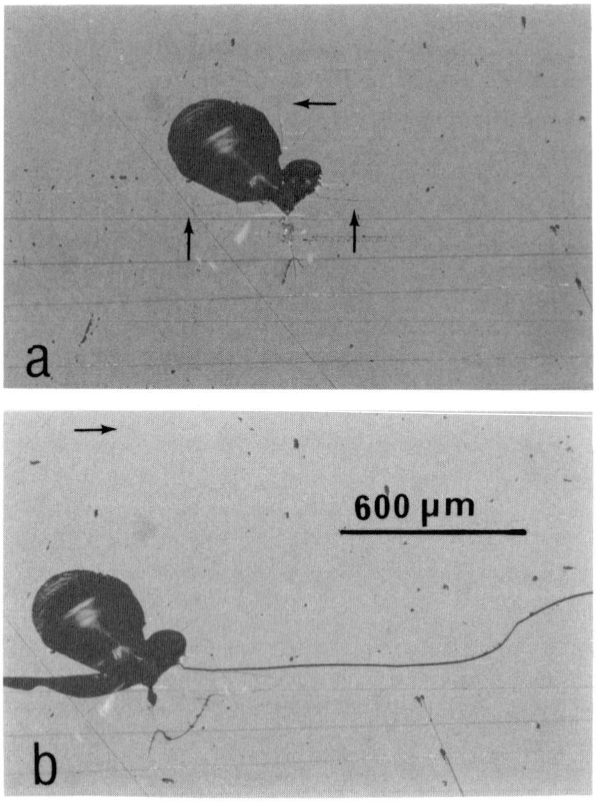

Fig. 7. Radial cracks in the VM-ST3 precursor indented with 50 N, (a) as indented, (b) after quenching from 160 °C into liquid N_2. The crack tips are indicated by arrows. Unstable extension, as a result of thermal stress anisotropy, only occurred for the "horizontal" cracks (without arrow in case (b)).

reflect the empirical ranking of these materials. In another investigation, such curves allowed a ranking of three grades of polycrystalline "YBaCuO" [11].

For the case of the parallelepiped samples investigated, the thermal stresses which appear during quenching are not isotropic. This is shown for material VM-ST3 in Fig. 5. Limits have been drawn, the lower corresponding to the width of the parallepiped and the upper to its length. Since the sample was destroyed in the latter direction a critical point with respect to quenching parameters, toughness of the material and the indentation load, is defined (Figs. 6 and 7).

5. Concluding remarks and perspectives

Within the framework of the Vickers indentation median-radial crack system, a criterion has been derived which allows the thermal shock resistance of brittle materials to be defined. For simplicity and ease of use, in a first step liquid nitrogen was used as the quenching medium. The experimental results show a rather good fit with the shape of the predicted curves, and reflect an empirical ranking. It was also shown that the anisotropy of the thermal stresses can be revealed. These results need and should induce further work on this subject.

Finally the detection of a critical indentation load for given quenching conditions should give rise to further developments. The maximum value of transient thermal stress is often poorly known. However, mechanical and thermal stresses have the same effect on the extension of median-radial cracks. Now, for a given thermal shock and a given indentation load, the critical ratio $c_m/c_0 = 2.52$ can be detected. Thus it would be sufficient first to indent a sample with the same load, and then to load it mechanically to rupture (unstable crack extension). The mechanical stress at rupture may then be put equal to the maximum value of the thermal transient stress, without knowledge of the heat transfer process and of the real value of χ_r.

References

1 W. R. Buessem, Thermal shock testing, *J. Am. Ceram. Soc., 38*(1)(1955) 15–17.
2 W. D. Kingery, Factors affecting the thermal shock resistance of ceramic materials, *J. Am. Ceram. Soc., 38*(1)(1955) 3–15.
3 D. P. H. Hasselman, Unified theory of thermal shock fracture initiation and crack propagation of brittle ceramics, *J. Am. Ceram. Soc., 52*(1)(1965) 600–604.
4 D. P. H. Hasselman, Thermal stress resistance parameters of brittle refractory ceramics: a compendium, *Am. Ceram. Soc. Bull., 49*(12)(1970) 1933–1937.
5 D. P. H. Hasselman, Thermal stress resistance of engineering ceramics, *Mater. Sci. Eng., 71* (1985) 261–264.
6 K. T. Faber, M. D. Huang and A. G. Evans, Quantitative studies of thermal shock in ceramics based on a novel technique, *J. Am. Ceram. Soc., 64*(5)(1981) 296–301.
7 B. R. Lawn and M. V. Swain, Microfracture beneath point indentation in brittle solids, *J. Mater. Sci., 10* (1975) 113–122.
8 B. R. Lawn and E. R. Fuller, Equilibrium penny-like cracks in indentation fracture, *J. Mater. Sci., 10*(1975) 2061–2024.
9 G. R. Anstis, P. Chantikul, B. R. Lawn and D. B. Marshall, A critical evaluation of indentation techniques for measuring fracture toughness: I, direct crack measurement, *J. Am. Ceram. Soc., 64*(9)(1981) 533–538.
10 P. Chantikul, G. R. Anstis, B. R. Lawn and D. B. Marshall, A critical evaluation of indentation techniques for measuring fracture toughness: II, strength method, *J. Am. Ceram. Soc., 64*(9)(1981) 539–544.
11 F. Osterstock, S. Strauss, B. L. Mordike, I. Monnot and G. Desgardin, Toughness and thermal shock resistance of polycristalline YBaCuO, in *J. Alloy Comp., 195* (1993) 679–682.

Materials Science and Engineering, A168 (1993) 45–47

Superplasticity of CuO-doped $ZrO_2(Y_2O_3)$

W. Hendrix, S. Kuypers, J. Vangrunderbeek, J. Luyten and W. Vandermeulen
Materials Division, V.I.T.O., Vlaamse Instelling voor Technologisch Onderzoek, B-2400 Mol (Belgium)

Abstract

Low temperature superplastic behaviour (1100 °C) was demonstrated in creep compression tests of three different CuO-doped yttria stabilized tetragonal zirconia polycrystalline (YTZP) materials, synthesized using commercially available powders. The microstructures obtained were analysed by different characterization techniques such as ceramography, impedance spectroscopy, transmission electron microscopy and X-ray diffraction. Although some observations point to the formation of a thin liquid phase which could explain the low temperature superplasticity of these CuO-doped YTZP materials, such a phase could not be detected.

1. Introduction

The recent discovery of superplasticity in ceramics by Wakai *et al.* [1, 2] opens the way to net-shape superplastic forming of these materials. The practical application of such a process requires a high deformation rate at a relatively low temperature.

The aim of the present study was the development of a material which would satisfy these requirements. Previous experiments [3] have shown that ZrO_2-based materials are the most suitable for modification for low temperature superplasticity. Their grain boundaries have good cohesive strength and grain growth at the superplastic deformation temperature is rather limited.

In analogy with the work of Chen *et al.* [4] in a previous investigation, yttria stabilized tetragonal zirconia polycrystalline (YTZP) powders doped with 0.3 mol% CuO were sintered and tested for superplastic behaviour. Figure 1 shows that, as a result of the addition of CuO, the compression creep rate under a load of 40 MPa in the temperature range 1100–1200 °C is increased by two orders of magnitude [5]. In the present contribution we attempt to clarify the mechanism of this low temperature superplasticity.

2. Experimental procedure

Fine ZrO_2 (Y_2O_3) powders with 2.5–3 mol% Y_2O_3 were supplied by the companies Tioxide, Lonza and Tosoh. These base powders were mixed with 0.3 mol% CuO, uniaxially cold pressed and sintered at 1250 °C. After sintering, the ceramics were characterized using scanning and transmission electron microscopies (SEM, TEM), impedance spectroscopy (IS), X-ray diffraction (XRD) and ceramography. Table 1 summarizes the properties of the ceramic materials used in this study. The grain size was estimated from SEM observations of fracture surfaces.

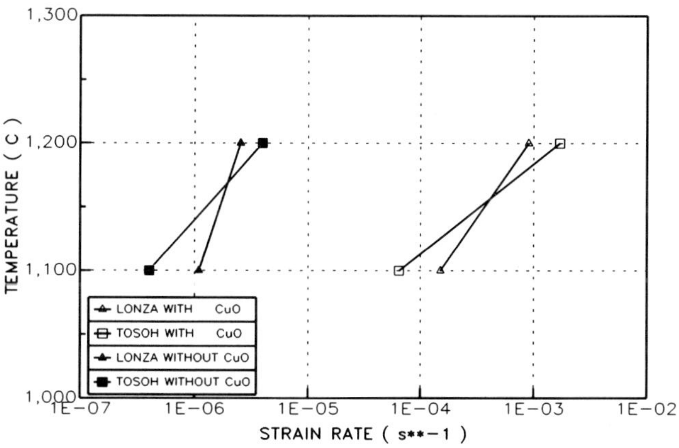

Fig. 1. Temperature *vs.* strain rate durig compression creep on 3YTZP (Lonza and Tosoh) with and without CuO.

TABLE 1. Characteristics of the ceramics

Composition	Density (% TD)	Grain size (μm)	Crystal structure (XRD)
3Y–TZP, Lonza, +0.3 mol% CuO	96.3	0.3	Tetragonal (f.c.c., Mon)
3Y–TZP, Tosoh, +0.3 mol% CuO	97.7	0.3	Tetragonal (f.c.c., Mon)
2.5Y–TZP, Tioxide, +0.3 mol% CuO	94.9	0.3	Tetragonal (f.c.c., Mon)

0921-5093/93/$6.00

The superplastic behaviour was tested by compression creep tests under constant load on pellets (7 mm in diameter and 7 or 13.7 mm in height) at temperatures ranging from 1000 °C to 1400 °C. The activation energy of deformation was determined from the strain rate between 3% and 5% strain. The sintering behaviour at 1250 °C of materials with up to 1 mol% CuO was also studied in order to improve the superplastic behaviour. However, no characterization tests have as yet been performed.

3. Results and discussion

Figure 2 shows the creep rate *vs.* stress for different materials and test temperatures. It can be seen that the effect of different starting powders on the creep rate is generally within one order of magnitude.

The effect of temperature on the creep rate can be described by an activation energy. At a stress of 40 MPa, values of 480, 500 and 490 kJ mol^{-1} are obtained for the three starting materials (Lonza, Tosoh and Tioxide). These values are of the same order as those found by Hwang and Chen [6] for their 2YTZP+CuO ceramic (445 kJ mol^{-1}). These authors proposed Coble creep and grain-boundary-phase enhanced diffusional creep, and liquid-enhanced interface creep, as the dominant deformation mechanisms below and above the eutectic temperature (1130 °C). From the present creep test results, however, no conclusions on the existence of two mechanisms can be drawn since the activation energy appears to be the same below and above the eutectic temperature.

Until now, our TEM observations have not revealed a thin layer at the grain boundaries. This is contradictory to some observations reported in the literature [6]. However, IS did indicate grain boundary modifications due to the addition of CuO. The grain boundary ion conductivity measured in CuO-containing specimens was strongly increased [7], whereas the lattice ion conductivity of the grain interior remained unchanged (Fig. 3).

The sintering experiments performed on two powders with different CuO contents show that for normal heating rates, the final density decreases for CuO contents above 0.5 mol% (Fig. 4). Specimens sintered with a higher heating rate did not show such a remarkable drop in density. The lack of densification after relatively slow heating has been explained by the loss of a liquid phase during heating. At high heating rates the sintering temperature is supposed to be reached with less loss, and increased densification can then be expected. The CuO content of the present material was only 0.3 mol%, so if a liquid phase were formed it would only be present in very small amounts. It may also be the case that only CuO-rich

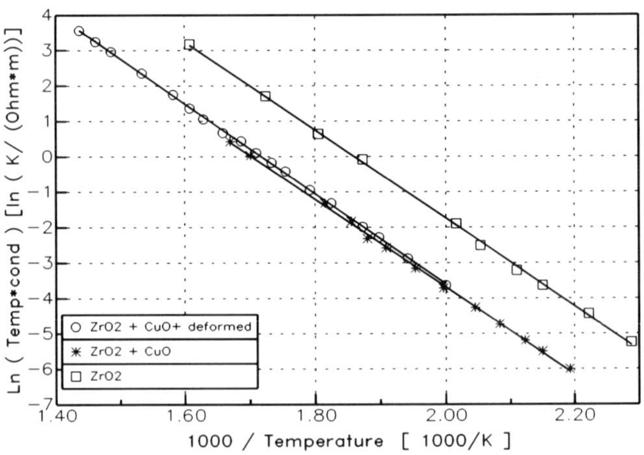

Fig. 3. Arrhenius plot of the ionic grain boundary conductivity. The conductivity is drastically decreased for the samples doped with 0.3 mol% CuO.

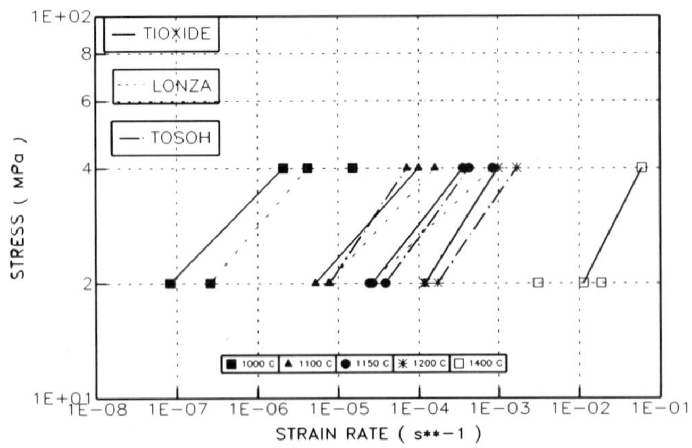

Fig. 2. The effect of temperature on the strain rate for the three 3YTZP materials.

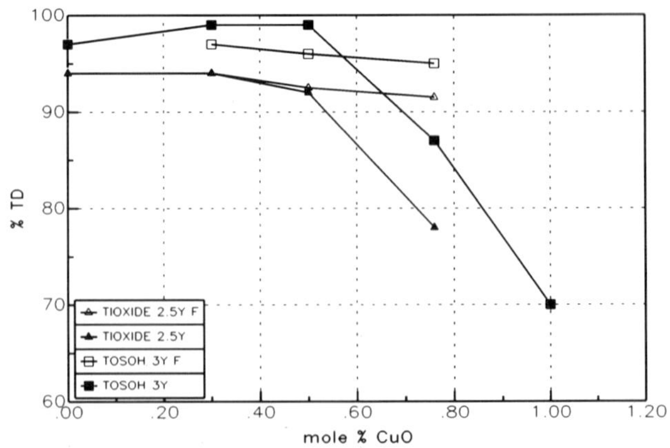

Fig. 4. Density (TD theoretical density) *vs.* mol% CuO for two starting powders, sintered at 1250 °C, after normal heating and after accelerated heating (F).

grain boundaries are formed, which would explain the IS and TEM results.

4. Conclusions

(1) Although some optimization of the synthesis is needed, the low temperature superplastic behaviour of 3 YTZP ceramics doped with 0.3 mol% CuO can be demonstrated.

(2) There are only indirect indications that a liquid film is formed around the grains at about 1130 °C, enhancing the diffusion and enabling low temperature superplastic deformation.

Acknowledgments

The authors wish to thank the group headed by P. Diels for careful characterization of the materials. They are also indebted to J. Cooymans and G. Verreyt for technical assistance.

References

1 F. Wakai, S. Sakaguchi and Y. Matsuno, *Adv. Ceram. Mater., 3* (1986) 259–263.
2 F. Wakai, *Br. Ceram. Trans. J., 88* (1989) 205–208.
3 J. Luyten, W. Hendrix, J. Sleurs and W. Vandermeulen, *Proc. Mechanics of Creep in Brittle Materials, Leicester, September 2–4, 1991.*
4 I.-W. Chen and L. A. Xue, *J. Am. Ceram. Soc., 73* (9) (1990) 2585–2609.
5 J. Luyten, W. Hendrix, J. Sleurs, W. Vandermeulen and A. Stalios, *Proc. 1st Pacific Rim Int. Conf. on Advanced Materials and Processes, Hangzhou, June 23–27, 1992.*
6 C.-M. J. Hwang and I.-W. Chen, *J. Am. Ceram. Soc., 73* (6) (1990) 1626–1632.
7 J. Vangrunderbeek, J. Luyten, S. Kuypers, W. Hendrix and F. De Schutter, *Proc. 2nd Int. Symp. on Electrochemical Impedance Spectroscopy, Santa Barbara, CA, July 12–17, 1992.*

Wear properties of steel plasma nitrided at high temperatures

Mehmet Baki Karamış
Department of Mechanical Engineering, Erciyes University, 38090 Kayseri (Turkey)

Abstract

The wear behaviour of 722M24 (En40B) steel plasma nitrided at high temperatures for long durations was investigated. Under dry sliding and abrasive-containing lubricated conditions, the wear characteristics of the steel were studied as a function of load, time and white layer thickness. The white layer and the wear debris were characterized by X-ray diffraction. The topography of the wear surface was examined in a scanning electron microscope. The plasma nitriding process was carried out using a 20 kW plasma processing unit in cracked ammonia atmosphere. Treatment times of up to 100 h were employed with temperatures of 550, 570 and 590 °C. Wear tests were performed with plasma nitrided AMS 6472 as the counterface material. The tests under abrasive-containing lubricated conditions were carried out with 10% carborundum powder 10 μm in size in SAE-30 machine oil. The wear resistance of the steel was improved by plasma nitriding and the abrasive wear resistance of nitrided steel was higher than the wear resistance under dry sliding conditions. However, the surface hardness and white layer thickness are very important factors in the wear behaviour of steel.

1. Introduction

It has been common practice for many years to improve the wear resistance of ferrous material by some form of surface treatment which may consist of thermal hardening or a chemical–thermal process such as conversion and/or diffusion. Nitriding, a ferritic thermochemical treatment, has been particularly effective in improving the wear resistance of components manufactured from chromium–molybdenum steels [1]. This includes engineering components such as large heavy gears, pinions, pinion shafts etc.; the small, fast-running gears manufactured from AISI 722M24 (En40B) material which are used in engine and vehicle construction are plasma nitrided [2].

The application of nitriding in engineering is usually limited by the relatively shallow depth of case which can be imparted (up to 1 mm) compared with other heat treatments. However, this does not necessarily entail a reduction in the rolling fatigue strength of gears and other properties such as tooth root fatigue strength, slip and wear behaviour, resistance to seizure and abrasion. Nitrided components are equal to or even better than case hardened gears [3].

Since the abrasive wear resistance of a material is closely related to its hardness, plasma nitriding which increases the surface hardness can greatly decrease the abrasive wear. High surface hardnesses are also desirable for reduction in weight loss under dry sliding conditions. Although the process can produce a minimum surface hardness level of 700 $HV_{0.3}$ with about 1 mm case depth for 100 h at 590 °C, this hardness level is quite adequate for most engineering requirements [4]. However, the wear rate is not only a function of the surface hardness but is also a function of the case depth [5, 6]. With plasma nitriding it is possible to form a monophase compound layer. A thin tenacious monophase layer of γ' provides good lubricity to a surface and should not be removed from the component. It is a product of the plasma nitriding process and accounts for the long service life of parts in plastic processing, gears etc. [2, 7]. The process has many advantages such as negligible distortion, controllable layer structures, complete suppression of compound layers etc. [2, 8–10].

2. Experimental details

The material used in this work was a low alloy steel (*i.e.* AISI 722M24, equivalent to En40B). This contains, in per cent, 0.258 C, 0.219 Si, 0.630 Mn, 0.208 Ni, 3.32 Cr, and 0.677 Mo. Cylindrical samples 50 mm long and 25 mm in diameter were machined from hardened and tempered material and subjected to plasma nitriding in cracked ammonia at temperatures of 550, 570 and 590 °C for times up to 100 h. The pressure of the treatment gas was kept at 2.5 mbar. The plasma nitriding conditions and the surface properties of the test steel have been published elsewhere [4].

0921-5093/93/$6.00

After the treatment the samples were prepared for metallographic examination. The hardness distribution of the cases, and thickness and phase of the white layer were determined.

Wear tests were carried out under dry sliding and abrasive-containing lubricated conditions. Wear test discs 5 mm thick were cut from the plasma nitrided cylindrical samples. To facilitate fastening to the wear tester, the discs were drilled at the centre, making a hole 4 mm in diameter. The counterface disc of 20 mm inside diameter, 41 mm outside diameter and 10 mm thick was made from AMS 6472 material which was plasma nitrided at 530 °C for 36 h. The counterface material contains, in per cent, 0.382 C, 0.243 Si, 0.653 Mn, 0.007 P, 0.014 S, 1.61 Cr, 0.345 Mo and 0.965 Al.

Wear testing was carried out on a general purpose wear tester using the disc–disc test configuration. Before testing, the wear couple was cleaned and weighed on a microbalance with 0.1 mg accuracy. The top (722 M24) and bottom (AMS 6472) discs were inserted into the wear machine. The test loads applied to the top disc were varied between 6 and 15 kg in 3 kg increments. After loading, rotation of the bottom disc was started. The top disc was held stationary while the bottom disc was rotated at 385 rev min^{-1}, equivalent to a linear velocity of 0.826 m s^{-1}.

During the dry wear test, the wear debris was collected carefully and analysed by X-ray diffraction. The abrasive wear test was carried out with 10% carborundum powder 10 μm in size in SAE-30 oil.

The abrasive lubricant was introduced into the friction zone by a circulation pump. The lubricant was stirred continuously so that the abrasive particles did not remain in the bottom of the bath. During the experiment, the frictional forces were recorded.

The experiment was interrupted after 5 min intervals. The samples were removed and cleaned with methanol and reweighed. The sliding distance corresponding to 5 min is 247.8 m. Tests were conducted for 20 min which corresponds to a sliding distance of 991.2 m. Since the test discs were capable of being indexed in several positions, several results could be obtained from each test piece.

3. Results

3.1. Properties of the surface

It is well known that higher surface hardnesses are obtained at lower treatment temperatures for the same treatment duration. The higher treatment temperatures produce a lower surface hardness owing to the production of coarser nitride precipitates within the case. Therefore, the sample treated at 590 + 550 °C for 2 × 50 h had the lowest surface hardness (*i.e.* 710 $HV_{0.3}$) with the deepest nitrided case (*i.e.* 0.95 mm). Parallel situations are observed for treatment at temperatures of 550 and 570 °C: nitriding at a temperature of 550 °C produces a higher hardness and shallower case depth than nitriding at 570 °C. However, the chemical composition of the material to be treated also has a major effect on the surface hardness. It is clear that aluminium-containing material, the counterface material, has a higher hardness than the test material. This is evidence for the better nitridability of Cr–Al nitriding steels [5, 6].

The microstructure of nitrided steels consists of a compound layer (white layer) and diffusion zone (Fig. 1). There are some precipitations suppressed at grain boundaries. The white layer thickness of the steel nitrided at 570 °C for 49 h is 9.40 μm. X-ray diffraction analysis revealed that the white layer consists of iron nitride (γ-Fe_4N), which has the best ductility of all nitrides. The growth of the white layer at the treatment temperatures is shown in Fig. 2. The thickness of the white layer at the treatment temperature does not exceed 10.1 μm, even for very long treatment times. According to the diffusion laws, the diffusion zone produced for short treatment times at low temperatures is comparatively thin.

Fig. 1. Microstructure of the steel nitrided at 570 °C for 49 h.

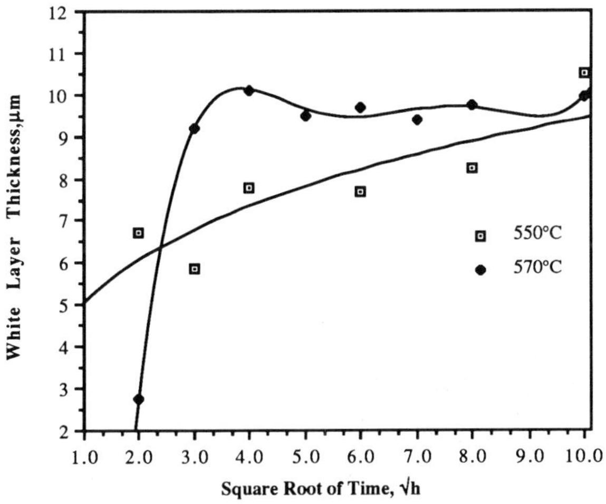

Fig. 2. The relationship between the white layer thickness and treatment time at 550 and 570 °C [4].

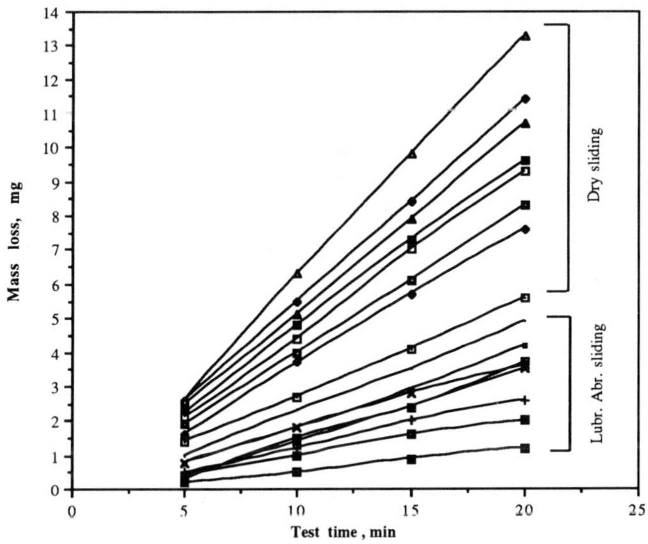

Fig. 3. Mass loss with test time under dry sliding and lubricated abrasive conditions, load 88.29 N. Dry sliding treatment conditions: ⊡ 550 °C for 9 h; ◆ 550 °C for 64 h; ◘ 570 °C for 9 h; ◇ 570 °C for 64 h; ■ 550 °C for 50 + 50 h; □ 570 + 550 °C for 50 + 50 h; ▲ 590 + 550 °C for 50 + 50 h; △ untreated. Lubricated-abrasive sliding conditions: ■ 550 °C for 9 h; + 550 °C for 64 h; ⊞ 570 °C for 9 h; × 570 °C for 64 h; ✱ 550 °C for 50 + 50 h; ◘ 570 + 550 °C for 50 + 50 h; ▬ 590 + 550 °C for 50 + 50 h; ● untreated.

3.2. Wear tests

The results of the abrasive-containing lubricated and dry sliding wear tests are shown in Figs. 3 and 4. In Fig. 3, it can be seen that the weight loss increased with test time. The nitrided case depth increased with the square root of the treatment time [4]. The hardness of the surface is the most effective factor with respect to wear rate. Under the same conditions, it is evident that the weight loss of the sample treated at 550 °C for 9 h

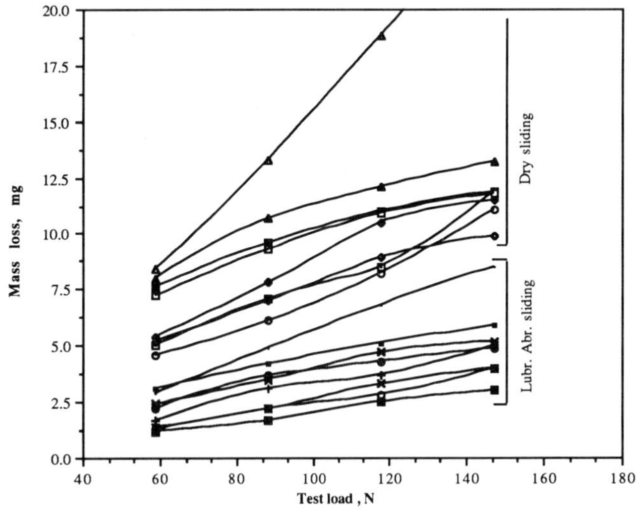

Fig. 4. Variation in mass loss with test load under dry sliding and lubricated abrasive conditions, test time 20 min. Dry sliding treatment conditions: ⊡ 570 °C for 4 h; ◆ 570 °C for 36 h; ○ 550 °C for 4 h; ◇ 550 °C for 36 h; ■ 550 °C for 50 + 50 h; □ 570 + 550 °C for 50 + 50 h; ▲ 590 + 550 °C for 50 + 50 h; △ untreated. Lubricated-abrasive sliding conditions: ○ 570 °C for 4 h; + 570 °C for 36 h; ⊞ 550 °C for 4 h; × 550 °C for 36 h; ✱ 550 °C for 50 + 50 h; ● 570 + 550 °C for 50 + 50 h; ▬ 590 + 550 °C for 50 + 50 h; ▪ untreated.

is the least, as it has a high hardness value, *i.e.* 920 $HV_{0.3}$ (Fig. 3).

When the material loss is compared on the basis of the test results, it is found that the combination of high hardness and shallow case is better in reducing the extent of material loss under both test conditions. Under both test conditions, the weight loss of the plasma nitrided samples increases as the applied load increases (Fig. 4). It can be seen that the wear rates for samples treated at 550 and 570 °C for 36 h increase more slowly than those of the other samples after a test load of 120 N under dry sliding conditions. The same tendencies are observed from the beginning of the tests for the samples double treated at higher temperatures (*i.e.* 590 °C).

It is clear that the weight loss generated under lubricated conditions is less than that under dry sliding conditions, owing to the lubricant medium.

Figure 5 shows the total weight loss corresponding to the treatment durations; the samples treated for 9 h at 550 °C and 16 h at 570 °C have the least weight loss under both test conditions. These nitriding times reduce the wear at each plasma nitriding temperature. The total weight loss against white layer thickness can be seen in Fig. 6. Under both test conditions, the total weight loss is increased by increasing the white layer thickness. The weight loss under dry conditions is more rapid than that under abrasive lubricated conditions. X-ray diffraction indicated that the wear debris

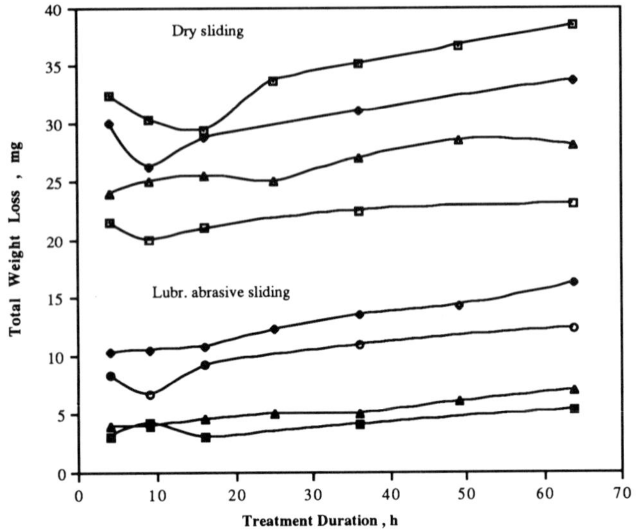

Fig. 5. Total weight loss with treatment duration: ⊡ 570 °C; ◆ 550 °C; △ 570 °C counterface; □ 550 °C counterface; ◇ 570 °C; ○ 550 °C; ▲ 570 °C counterface; ■ 550 °C counterface.

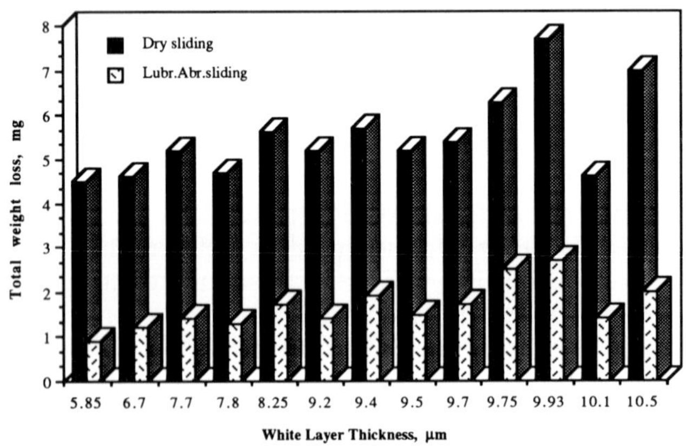

Fig. 6. Variation in total weight loss with white layer thickness.

produced during dry wear test was predominantly Fe_2O_3.

4. Discussion

The wear tests undertaken during this study indicate that plasma nitriding gives good wear resistance to the material surface. The production of relatively deep cases on the steels can be achieved by plasma nitriding at temperatures of 550, 570 and 590 °C. The Cr–Al nitriding steels nitride more rapidly than the Cr–Mo grades and develop higher hardness. Higher nitriding temperatures produce lower surface hardness owing to the production of coarser nitride precipitates within

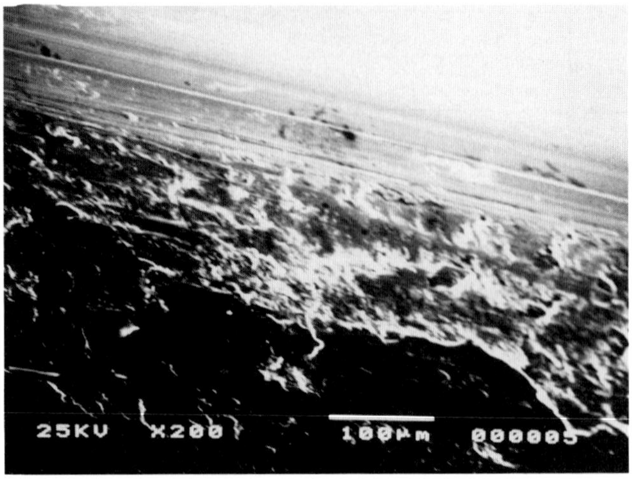

Fig. 7. Scanning electron micrograph observed after dry wear tests: 590 + 550 °C for 2 × 50 h, test load 117.72 N.

the case. However, the abrasive wear resistance is generally governed by the surface hardness, while the microstructure also plays a major role. The hardness is also an important factor in dry sliding wear. Therefore, the combination of high hardness and shallow case is better for reducing the extent of material loss under both wear test conditions. The nitride produced by plasma nitriding not only has a high hardness but also a low dynamic coefficient of friction. The friction coefficients range from 0.4 to 0.6 and from 0.15 to 0.3 under dry and lubricated abrasive conditions respectively [6].

The results illustrated in Fig. 6 show that the white layer thickness is a very effective factor for reducing the amount of wear under the test conditions presented in this study. It can be seen that the total weight loss is increased with a thicker white layer. This is because the particles generated from breakdown of the thicker white layers change the wear mechanism from adhesive to abrasive. Thus, the wear rate is increased rapidly under dry sliding conditions. Under lubricated abrasive conditions, the outer porous part of the white layer is not only able to absorb a quantity of oil but also traps abrasive particles. Therefore, a lower wear rate is achieved under these conditions.

It is obvious that the main factor for reducing the wear rate is the lubricant under the test conditions. Figure 7 shows the topography of the worn surface under dry sliding conditions. It can be seen that the wear traces are parallel to the sliding direction. These grooves were caused by abrasive debris generated during the early stage of the test. If the thicker white layer is present, its breakdown may also change the wear mode from adhesive to abrasive, because the debris of the white layer acts as an abrasive in the friction zone. There are also some oxide particles and wear debris attached loosely to the surface.

5. Conclusions

The following conclusions may be summarized from the results presented in this study.

(1) The dry and abrasive wear resistance of 722 M24 steel is enhanced by plasma nitriding.

(2) Longer plasma nitriding times at higher treatment temperatures produce deep hardened cases which have industrial benefit. The nitrided cases also have a good hardness transition between the bulk and the surface.

(3) The surface hardness plays a very important role in material loss as a result of microstructure and depth of the hardened case.

(4) The thinner white layer produced by plasma nitriding improves the wear resistance. The thicker compound layer is more easily flaked than the thinner layer.

Acknowledgments

The author wishes to thank Professor T. Bell (Wolfson Institute for Surface Engineering, School of Metallurgy and Materials, University of Birmingham) for his comments on the work. Thanks also go to Dr. A. M. Staines (Nitriding Services Ltd., Pleasant Street, West Bromwich, West Midlands B70 7DP).

References

1 T. Bell and N. L. Loh, *J. Heat Treat., 2* (3) (1982) 232–237.
2 B. Edenhofer, *Metall. Mater. Technol., 8* (8) (1976) 421–426.
3 K. Keller, *Antriebstechnik, 10* (3) (1971) 1–7.
4 M. B. Karamiş and A. M. Staines, *Heat Treat. Met., 3* (1989) 79–82.
5 M. B. Karamiş, *Wear, 147* (1991) 385–399.
6 M. B. Karamiş, *Thin Solid Films, 203* (1991) 49–60.
7 T. Bell, *Plastics Rubber Process., 1* (4) (1976) 161–166.
8 B. Edenhofer, *Trait. Therm., 80* (1973) 31–43.
9 B. Edenhofer, *Fachber. Oberflaechentech., 12* (4) (1974) 97–102.
10 B. Edenhofer, *Werkstatt Betr., 109* (5) (1976) 289–293.

The role of nitrogen in the intergranular glass phase of Si_3N_4 on high temperature applications and wear

R. F. Silva and A. P. Moreira
Departamento de Engenharia Cerâmica e do Vidro, Universidade de Aveiro, 3800 Aveiro (Portugal)

J. M. Gomes and A. S. Miranda
Departamento de Engenharia Mecânica, Universidade do Minho, 4800 Guimarães (Portugal)

J. M. Vieira
Departamento de Engenharia Cerâmica e do Vidro, Universidade de Aveiro, 3800 Aveiro (Portugal)

Abstract

The hot hardness of Si_3N_4 tool materials of the CeO_2 and CeO_2–AlN doping systems, oxidation in air from 900 to 1450 °C, and turning of tool steel with ceramic inserts of the same compositions were investigated to determine the effect of the N content of the intergranular phase. Si/N grain boundary diffusion was found to be the main controlling mechanism at high temperature for densification, mechanical behavior and chemical degradation of these materials in air, the passivation oxidation rate of the CeO_2–AlN–Si_3N_4 system being dependent on the atomic fraction of N of the intergranular phase, with lower values for the N-rich compositions.

1. Introduction

The densification of Si_3N_4 occurs by liquid phase sintering with the addition of different oxides or nitrides or mixtures of both, of which the CeO_2 and CeO_2–AlN doping systems are examples [1–4]. The low cost rare earth oxide CeO_2 is reported to give superior chemical resistance [5] and high fracture toughness [6]. Intergranular diffusion of Si and N is the controlling mechanism of densification, $\alpha \rightarrow \beta$-Si_3N_4 phase transformation and creep behavior [1, 7–11]. The chemical nature of the liquid phase is important for sintering, and for any property related to the liquid viscosity. The viscosity of the oxynitride glass is known to be enhanced by the N content and to be dependent on the glass-modifying cations [12–14].

In the present study the effect of N content on the sintering kinetics is correlated with two properties which are directly relevant to the use of Si_3N_4 as cutting tool inserts: the hot hardness [15–17] and oxidation resistance [18, 19].

2. Materials and experiments

Dense Si_3N_4 samples were uniaxially hot pressed at 20–30 MPa applied pressure, for 300–5400 s at 1500–1750 °C. Two samples were post-HIPed at 1750 °C for 1 h at 100 MPa. The phase composition, intergranular phase volume and N content of each batch are given in Table 1. The volume fraction of intergranular phase and N content were calculated from the amount and composition of sintering aids and the density of glasses of similar compositions [4].

TABLE 1. Composition of Si_3N_4-based materials, volume content and nitrogen content of corresponding intergranular phases

Batch	Composition (mol.%)				Intergranular phase	
	Si_3N_4	SiO_2	Ce_2O_3	AlN	Volume fraction V_L (%)	N-content[a] (at.%)
1Ce	88.8	10.2	1.0	—	7.5	—
2Ce	87.4	10.6	2.0	—	9.2	—
3Ce	84.6	11.4	4.0	—	12.8	—
1CeAl	76.2	7.3	0.9	15.6	10.1	27.1
2CeAl	73.2	7.5	1.8	17.5	12.6	26.3
3CeAl	69.0	8.0	3.5	19.5	16.9	24.2

Raw materials: Si_3N_4, HcStarck LC10 for batches 1–3Ce and HcStarck LC12 for batches 1–3CeAl; CeO_2, Fluka puriss; AlN, HcStarck grade C.
[a]From added AlN.

0921-5093/93/$6.00

The experimental procedures for hot-pressing dilatometry, hot-hardness testing and turning tests are reported elsewhere [15, 20, 21]. Oxidation tests were done in air [22] at 900–1450 °C for 0.5–25 h.

3. Results and discussion

3.1. Hot-pressing kinetics

Shrinkage data of pressure dilatometry in the range 1300–1800 °C were analyzed by non-linear curve fitting to a general densification equation which is based on Coble's equation [23] for the intermediate stage of sintering by a grain boundary diffusion mechanism, with an additional contribution of particle rearrangement in the early stages of the process [20, 24]. The effects of initial particle size and purity of the powders, of the sintering aid composition (either Ce^{3+}, Al^{3+}, N^{3-} or Mg^{2+}), and of the liquid volume fraction V_L, on the densification rate and α–β Si_3N_4 phase transformation and grain coarsening during sintering were analyzed and discussed separately [20, 24]. The full set of densification parameters for each of the batches in Table 1 is published elsewhere [20]. The values of effective diffusion coefficient, $D = D_0 \exp(-Q/RT)$, were calculated after correcting the densification rate for dependences on porosity, grain size and intergranular layer thickness. The values of D, V_L and relative N content are retained for comparison with the passive oxidation rate constants and the changes in hot hardness.

The values of D_0 and Q are dependent on the sintering aid, CeO_2, CeO_2–AlN or MgO–AlN, but are independent of the amount of each compound, Table 2. The differences in D_0 and Q between the three sets of batches of this study have been explained on the basis of the compensation law [24, 25]: $\ln D_0 = Q/RT^* + \ln B$, where $B = 4.9 \times 10^{-17}$ m^2 s^{-1}, $T^* = 1740$ K is a value of T in the center of the temperature range of the hot-pressing experiments, and the correlation coefficient is $c = 0.992$. The diffusion coefficients of N and Si in Si_3N and the effective diffusion coefficients of hot-pressing of Si_3N_4 have been reviewed [11]. The straight line for the compensation law joins the experimental points (D_0, Q) of hot-pressing of MgO-doped Si_3N_4 [8] and Y_2O_3–SiO_2-doped Si_3N_4 [9], and the diffusion coefficient of N in β-Si_3N_4, $D = 1.4 \times 10^7 \exp(-778$ kJ mol$^{-1}/RT)$ m^2 s^{-1} [11]. Since the values of D fit a single straight line, the same diffusion species must be controlling the solution–precipitation mechanism of densification of several Si_3N_4 materials, so the species responsible for the effective values of the Si/N diffusion coefficient is the N ion. Differences in the values of Q in Table 2 are explained as the effect of sintering aids on reactions that bring the solution–preparation of Si_3N_4 in the liquid to equilibrium. Dissolved N from the Si_3N_4 powder, or intentionally added as AlN, lowers the eutectic point of the system [4, 24, 26, 27]. A transient liquid, which recrystallizes as $CeAlO_3$ on cooling, is already observed at 1300 °C during hot pressing of the CeAl systems [24]. Transient liquids of analogous compositions were reported in Y–Si–Al–O–N [28, 29]. The batches 1–3 CeAl have lower values of Q, owing to the early contribution of the transient liquid to the hot pressing rate.

3.2. Hot hardness

The hot hardness results show that the N-rich compositions (batches 1-3CeAl), have high H_V values at room temperature which are retained up to 900 °C,

TABLE 2. Diffusion coefficient for densification and passive oxidation rate constant

Batch	Effective diffusion coefficient hot-pressing		Passive oxidation (1050–1350 °C in air)	
	D_0 (m^2 s^{-1})	Q (kJ mol^{-1})	K_0 (mg^2 cm^{-4} h^{-1})	Q_{ox} (kJ mol^{-1})
1Ce	1.2×10^{-2}	504	1.1×10^{11}	422
2Ce	1.8×10^{-2}	471	2.4×10^{11}	422
3Ce	1.6×10^{-2}	481	4.4×10^{11}	422
Average	1.5×10^{-2}	485		
1CeAl	1.6×10^{-6}	337	1.8×10^{13}	455
2CeAl	1.8×10^{-6}	385	1.1×10^{14}	455
3CeAl	1.4×10^{-6}	337	1.3×10^{15}	455
Average	1.6×10^{-6}	353		
MgAlG[a]	2.8×10^{-4}	412	—	—
MgAlF[b]	3.1×10^{-4}	425	—	—
Average	3.0×10^{-4}	418.5		

[a]97.5% Si_3N_4(HcStrack LC10) + 1% MgO(Merck puriss.) + 1.5% (weight per cent) AlN.
[b]97.5% Si_3N_4(HcStrack LC12) + 1% MgO + 1.5% AlN (weight per cent).

while H_V for batches 1–3Ce has a more pronounced dependence on T [15]. Two temperature ranges of H_V are observed. The values of the activation energy of H_V, in the range from room temperature (RT) to 800–900 °C, amount to 2.5–5.1 kJ mol^{-1} and 0.6–3.4 kJ mol^{-1} for 1–3Ce and 2–3CeAl respectively [24, 30]. The corresponding values of the activation energy of H_V in the high temperature range Q_H are given in Table 3. The values of Q_H of the 2CeAl samples are proportional to the logarithm of their room temperature hardness, Table 3. This proportionality is understood as resulting from competition, in the range 900–1200 °C, between the low and high temperature yielding mechanisms that relate to hardness [24, 30].

When the compensation law is applied to the reciprocals of H_V of the 2CeAl samples, to separate tentatively the contributions of the deformation mechanisms of Si_3N_4, a single value of activation energy, $Q_e = 49.5$ kJ mol^{-1}, is calculated for the high temperature mechanism of hardness of Si_3N_4, Table 3. This has been identified as grain boundary viscous flow [17], the hardness being related to the flow rate equations as $H^{-n} = [K'\exp(-Q/R)]t$, where K' is a constant which includes the pre-exponential factor of D, t is the indentation time and $n \geqslant 8$ [17]. By multiplying Q_e by $n = 8$, a value that is almost identical with Q, the activation energy of hot-pressing for the same batch (Table 2), is obtained ($nQ_e = 396$ kJ mol^{-1}). A relative increase in K^* from batch 3CeAl (K_3^*) to batch 2CeAl (K_2^*), $n \ln(K_2^*/K_3^*) = 2.5$, at $Q_H \approx 20$ kJ mol^{-1}, is observed. The intergranular phase of batch 2CeAl has a higher N-content than that of batch 3CeAl, Table 1.

The wear resistance of ceramics in dry friction is a function of their hardness H, toughness K_{Ic}, and the chemical stability of the ceramic to the second material of the sliding couple. Abrasion and chemical dissolution are competing mechanisms of wear for Si_3N_4 ceramics under high speed sliding contact with metals and alloys [21]. The abrasion wear resistance becomes proportional to $K_{Ic}{}^a H^b$, where $a \approx 2/3$ and $b \approx 1/2$ are real and positive constants [15]. The exact values of a and b are dependent on the mechanical properties, the microstructure of the two materials of the sliding couple, the test conditions and the kinetics of the tribological contact. The room temperature indentation toughness K_{Ic} is $K_{Ic} = 4.2 \pm 0.3$ MPa m$^{1/2}$ for the 2Ce cutting insert material and $K_{Ic} = 4.3 \pm 1.0$ MPa m$^{1/2}$ for the 2CeAl inserts [21]. For the dense Si_3N_4 materials, K_{Ic} is almost independent of T up to 1000 °C [31]. A net increase in K_{Ic} can be observed in the range 1000–1200 °C for those materials of high liquid volume fractions [31]. As discussed above, the effects of T on H_V are large and dependent on material composition. The specific wear resistance of the Ce-doped Si_3N_4 inserts to abrasion in cutting of iron alloys are improved by N enrichment of the intergranular layer [15, 21]. The rake face of the CeO_2–AlN–Si_3N_4 inserts (2CeAl) remains stable at cutting speeds of 2.5–4 m s^{-1}, whereas an extensive abraded crater is already observed in the CeO_2–Si_3N_4 material (2Ce) [21]. The relationship between wear resistance, cutting speed and hot hardness $H_V(T)$, was amenable to quantification [15]. The dissolution rate of Si_3N_4 in the chips of different iron alloys also shows that the N-rich compositions can survive at high cutting speeds for longer times [21].

3.3. Oxidation in air

The oxidation of CeO_2 and CeO_2–AlN doped Si_3N_4 in air reveals three temperature ranges [22]: (i) nonprotective oxidation for $T < 1050$ °C, (ii) passivation at 1050–1350 °C and (iii) oxidation with extensive dissolution of the Si_3N_4 grains above 1350 °C. The oxidation rate K in range (i) is limited by the porosity of the reaction products and it decreases as T approaches 1050 °C. Range (iii) was only observed for the 2CeAl and 3CeAl HIP samples. The passive oxidation, range (ii), follows the parabolic law $W^2 = K(T)t$, where W is the weight gain and $K(T) = K_0 \exp(-Q_{ox}/RT)$, is the oxidation rate constant, Table 2. Within the experimental error of the present results (± 30 kJ mol^{-1}) Q_{ox} is constant for all batches, being almost identical with the value Q for densification of the CeO_2 doped batches, Table 2. Within ± 50 kJ mol^{-1}, the value of

TABLE 3. Room temperature hardness and hot hardness, $H = K^* \exp(Q_H/RT)$, of the Si_3N_4–CeO_2–AlN systems

Batch	Sample	H_V-RT (kgf mm^{-2})	900–1200 °C		
			K^* (kgf mm^{-2})	Q_H (kJ mol^{-1})	Q_e (kJ mol^{-1})
2CeAl	1500 °C, 60 min	2014	131	24.7	51.6
	1550 °C, 60 min	1958	160	21.8	52.9
	1600 °C, 60 min	1864	238	17.6	46.0
	1750 °C, 15 min	1700	316	13.1	47.5
3CeAl	1650 °C, 5 min	1598	134	20.5	—
	HIP	1526	56	28.7	—

Q_{ox} is also close to the values of activation energy of passive oxidation of several other Si_3N_4 materials with dopants of MgO [32, 33], CeO_2 [32] and Y_2O_3–Al_2O_3 [34]. For batches 1–3Ce, the increase in K_0 is directly proportional to V_L, Tables 1 and 2. The same proportionality was reported for the oxidation of MgO-doped hot-pressed Si_3N_4 [35]. The oxidation kinetics of Si_3N_4 and related nitrogen ceramics has been reviewed [36]. The mechanism of passive oxidation of Si_3N_4 and sialon in range (ii) has been described [35] as rate controlled by the outward diffusion of N and cation impurities and the inward diffusion of oxygen, along the grain boundaries in the subscale region. The linear relationship between K_0 and V_L is also proportional to the fractional area of intergranular phase of the surface, which is the area for diffusion of N and cations from the bulk of the Si_3N_4 to the reaction interface, the effective diffusivity being independent of the CeO_2 concentration in the intergranular liquid [37], Tables 1 and 2. For batches 1–3CeAl, K_0 has an exponential dependence on the atomic fraction of N of the intergranular phase, the correlation coefficient being 0.986 for $\ln(K_0) \propto N$(at.%). The quality of the linear fit is further improved if $\ln(K_0)$ is plotted as a function of the molar ratio of AlN to (AlN + Ce_2O_3) in the sintering aids, being $c = 0.999$, Fig. 1.

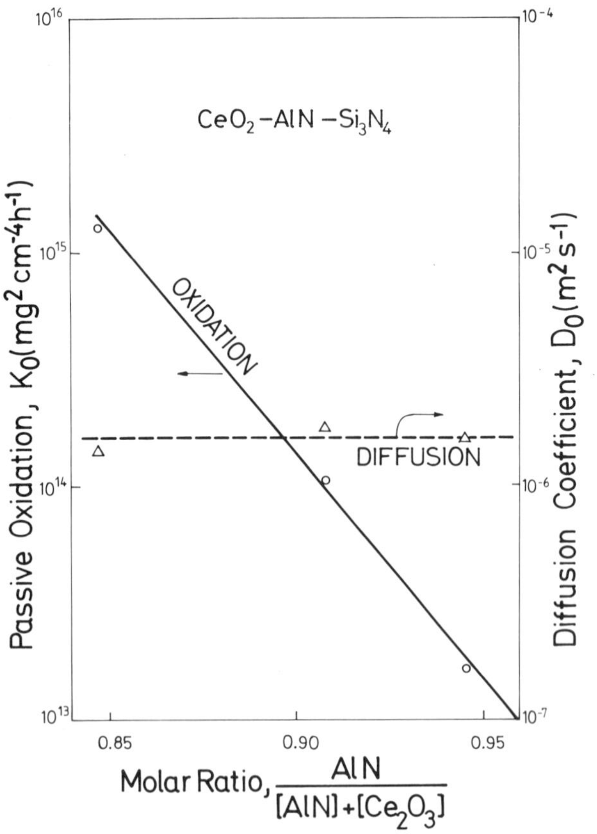

Fig. 1. Effect of AlN content on the pre-exponential factors of the diffusion coefficient in hot pressing and the passive oxidation rate constant.

Intergranular film thicknesses of 1.8–3.3 nm were calculated [20] for the samples in Table 2. The low atomic number of N, the limiting energy resolution and spatial resolution of existing equipment for surface analysis (Auger electron spectroscopy) and of analytical electron microscopy preclude the accurate determination of the N content in the intergranular films of Si_3N_4 [38]. An N content of 15 at.% was determined by electron energy loss spectroscopy in the grain boundary liquid phase of MgO-doped Si_3N_4 hot pressed at 28 MPa at 1500 °C for 2 h [38]. The calculated values of N content in Table 1 are above the reported saturation values of N concentration in N-containing glasses [39, 40], which increases from 15 at.% at 1650 °C and 0.1 MPa of N_2 pressure [39] to about 25 at.% at 1900 °C, 210 MPa of N_2 pressure [40]. The N content of the glasses is dependent on composition, glass melting temperature and time [14]. When reactive N sources, such as NH_3, are used as the atmosphere for melting the glass, the N content is also increased (20 at.%) [14]. In uniaxial hot pressing, the Si_3N_4 samples are highly confined by the dye walls, N losses for the short hot-pressing cycles at 1500–1750 °C being minimal. Except for the transient $CeAlO_3$, no secondary crystalline phases were detected by X-ray diffraction in the CeAl hot pressed samples, Table 2. Most of the N, added as AlN, remained in the sintering liquid in amounts that must be proportional to the dissolved AlN.

The substitution degree of β'-Si_3N_4 in the CeAl samples is below the detection limit of X-ray diffraction [24]. Al from AlN enters the liquid in the same amount as the N. The oxidation rate of sialons either decreases or increases when the Al content increases [36]. K decreases one order of magnitude at 1370 °C, from 2.36×10^{-12} g^2 cm^{-4} s^{-1} to 2.22×10^{-13} g^2 cm^{-4} s^{-1}, in sialons when the weight fraction of Al_2O_3 is increased from 20 wt.% to 50 wt.% [41]. The relative change in Al content of the CeAl samples in Tables 1 and 2 is much lower than these values.

The increase in K_0 from batch 2CeAl to batch 3CeAl, Table 2, is given as $\ln(K_0^{(3)})/K_0^{(2)}) = 2.45$, which is the same as the relative change in $n \ln K^*$ above. The substitution of N for O increases the average anion coordination number which leads to glass network cross-linking, high viscosities, lower diffusion coefficients and better high temperature resistance of the material [12–14]. The exponential dependence of viscosity of the oxynitride glasses on e/oN = 3N(at.%)/[3N(at.%) + 2O(at.%)], the equivalent concentration of N in the glasses [12–14], can explain the exponential relationship observed for K_0, Fig. 1.

4. Conclusions

The activation energy for densification Q, grain boundary viscous flow of hot hardness nQ_c and passive oxidation rate in air Q_{ox} of CeO_2–AlN doped Si_3N_4 are virtually the same, these phenomena all being controlled by the same diffusional mechanism. The effective diffusion for hot pressing is coupled to N diffusion in Si_3N_4.

In overlapping temperature ranges, below the hot pressing temperatures, the yielding rate of hot hardness and the passive oxidation rate of the CeO_2–AlN doped Si_3N_4 have identical relative increments with increasing N content, the oxidation rate being exponentially dependent on the AlN molecular fraction of the sintering aids.

The hot hardness and oxidation, being also controlled by diffusion in the intergranular phase, show similar trends to the wear resistance of Si_3N_4 cutting inserts as a function of N content.

Acknowledgments

The financial support from JNICT and Minas e Metalurgia under the research contract 87/80 MATR is gratefully acknowledged.

References

1 G. N. Babini, A. Bellosi and P. Vincenzini, *Ceram. Int., 6* (3) (1980) 91–98.
2 T. Mah, K. S. Mazdiyasni and R. Ruh, *J. Am. Ceram. Soc., 62* (1–2) (1979) 12–16.
3 F. F. Lange, *Am. Ceram. Soc. Bull., 59* (2) (1980) 239–240, 249.
4 J. Mukerji, P. K. Das, P. Greil and G. Petzow, *Ceram. Int., 13* (1987) 215–221.
5 S. T. Buljan and S. F. Wayne, *Adv. Ceram. Mater., 2* (4) (1987) 813–816.
6 E. Tani, S. Umebayashi, K. Kishi, K. Kobayashi and M. Nishijima, *Am. Ceram. Soc. Bull., 65* (9) (1986) 1311–1315.
7 R. J. Brook, T. G. Carruthers, L. J. Bowen and R. J. Weston, in F. Riley (ed.), *Nitrogen Ceramics*, Noordhoff, Leyden, 1977, pp. 383–390.
8 L. J. Bowen, R. J. Weston, T. G. Carruthers and R. J. Brook, *J. Mater. Sci., 13* (1978) 341–350.
9 L. J. Bowen, T. G. Carruthers and R. J. Brook, *J. Am. Ceram. Soc., 61* (7–8) (1978) 335–339.
10 O. Abe, *J. Mater. Sci., 25* (1990) 4018–4026.
11 G. Ziegler, J. Heinrich and G. Wotting, *J. Mater. Sci., 22* (1987) 3041–3086.
12 S. Hampshire, R. A. Drew and K. H. Jack, *Commun. Am. Ceram. Soc., 67* (1984) C-46–C-47.
13 R. A. Drew, S. Hampshire and K. H. Jack, *Proc. Br. Ceram. Soc., 31* (1981) 119–132.
14 R. E. Loehman, *J. Mater. Educ., 9* (5) (1987) 483–498.
15 R. Silva and J. Vieira, *J. Hard Mater., 3* (1) (1992) 63–72.
16 T. Ekstrom and J. Persson, *J. Am. Ceram. Soc., 73* (10) (1990) 2834–2838.
17 M. G. Naylor and T. F. Page, *Tech. Rep., DA-ERO-78-G-010*, University of Cambridge, 1981.
18 E. M. Trent, *Metal Cutting*, Butterworths, Guildford, 1984.
19 J. A. Schey, *Tribology in Metalworking*, American Society for Metals, Metals Park, OH, 1983.
20 R. Silva and J. Vieira, *Proc. European Ceramic Society Second Conf., Augsburg, 1991*, Deutsche Keramische Gessellshaft, Koln, in press.
21 R. Silva, J. Gomes, A. Miranda and J. Vieira, *Wear, 148* (1991) 69–89.
22 A. P. Moreira, Estabilidade química e tempo de vida no sistema Si–Ce–Al–O–N a temperaturas elevadas, *MSc Thesis*, University of Aveiro, Portugal, 1992.
23 R. L. Coble, *J. Appl. Phys., 41* (12) (1970) 4798–4807.
24 R. Silva, Cinética de sinterização e desgaste de pastilhas de corte do sistema Si_3N_4–CeO_2–AlN, *PhD Thesis*, University of Aveiro, Portugal, 1992.
25 J. M. Vieira and R. J. Brook, in W. D. Kingery (ed.), *Advances in Ceramics*, Vol. 10, American Ceramic Society, Columbus, OH, 1983, pp. 438–463.
26 K. H. Jack, in P. Vincenzini (ed.), *Energy and Ceramics, Materials Science Monographs*, Vol. 6, Elsevier, Amsterdam, 1980, pp. 534–549.
27 P. K. Das and J. Mukerji, *Key Engineering Materials*, Vol. 29–31, Trans Tech Publications, Aedermannsdorf, 1989, pp. 109–120.
28 P. K. Das and J. Mukerji, *J. Eur. Ceram. Soc., 5* (1989) 105–112.
29 J. Mukerji, P. Greil and G. Petzow, *Sci. Sintering, 15* (1) (1983) 43–53.
30 R. Silva and J. Vieira, Hot-hardness of Si_3N_4-based cutting tools, to be published.
31 B. S. Karunaratne and M. H. Lewis, *J. Mater. Sci., 15* (1980) 449–462.
32 D. Cubiccioti and K. H. Lau, *J. Am. Ceram. Soc., 61* (11–12) (1978) 512–517.
33 S. C. Singhal, *J. Am. Ceram. Soc., 59* (1–2) (1976) 81–82.
34 T. Chartier, J. L. Besson and P. Goursat, *Int. J. High-Technol. Ceram., 2* (1986) 33–45.
35 D. R. Clarke and F. F. Lange, *J. Am. Ceram. Soc., 63* (9–10) (1980) 586–593.
36 C. C. Sorrell, *J. Aust. Ceram. Soc., 19* (2) (1983) 48–67.
37 T. Ekstrom and M. Nygren, *J. Am. Ceram. Soc., 75* (2) (1992) 259–276.
38 D. R. Clarke and N. J. Zaluzec, *Commun. Am. Ceram. Soc., 65* (1982) C-132–C-133.
39 D. R. Messier and A. Broz, *Commun. Am. Ceram. Soc., 65* (1982) C-122–C-123.
40 J. C. Mittl, R. L. Tallman, P. V. Kelsey, Jr. and J. G. Jolley, *J. Non-Cryst. Solids, 71* (1985) 287–294.
41 S. C. Singhal and F. F. Lange, *J. Am. Ceram. Soc., 60* (3–4) (1977) 190–191.

Processing and investigation of tauboride-containing NiCrBTi–TiC plasma coatings

P. Vityaz, A. Verstak, T. Talako and S. Sobolevskiy
Byelorussian Powder Metallurgy Association, Minsk (Belarus)

E. Lugscheider, P. Jokiel and G. Beckschulte
Material Science Institute of Aachen Technical University, Aachen (Germany)

K. Yuschenko
Paton Welding Institute, Kiev (Ukraine)

G. Pursche
Surface Technologies Institute of Chemnitz Technical University, Chemnitz (Germany)

Abstract

Results of plasma-coating processing from tauboride-containing Ni–Cr–B–Ti alloy with 5–35 wt.% TiC grains added are presented. The coatings were sprayed in air, vacuum or underwater. The changes in the material phase content during spraying and the influence of the spraying parameters on the coating structure and properties are analysed.

1. Introduction

The tauboride nickel-based alloys were created as a result of the development of new hard alloys for wear-resistant coatings [1, 2]. Besides their hardening with precipitates of simple and double borides and carbides, which is useful for self-fluxing alloys, the mechanical properties of these alloys benefit from the presence of a triple complex boride, called tauboride, with $Cr_{23}C_6$ crystal type and $Ni_{20-21}Me_{3-2}B_6$ chemical composition, where "Me" is a refractory metal from the group of V, Nb, Ta, Ti, etc. stabilizing the tauboride. It is expected that the hardness and some tribological properties of tauboride alloys can be substantially improved by adding carbide grains to the material. The problems of phase transformations and interactions of the above-mentioned materials with each other and with the surrounding gas during thermal spraying and the possibilities of controlling the coating properties have to be investigated.

In this paper we present the results of an investigation of tauboride Ni–Cr–B–Ti coatings processed by different methods of plasma spraying. The potential for increasing their abrasive wear resistance by adding TiC particles is shown.

2. Materials and experimental procedures

For the investigations the following powders were used: (i) gas-atomized powders, Amperit 9930 (Ni–3.5B–10Cr–5Ti, wt.%), fraction 40–63 μm; (ii) mechanically alloyed powders of the same composition, fraction 63–100 μm; (iii) titanium carbide powders, Amperit 570.4, fraction 16–40 μm.

Composite powders with 5–35 wt.% TiC were processed by powder agglomeration with a binder or by mechanical alloying. Some TiC particles were plated with Ni–20%Cr alloy using a high frequency (HF) plasma (without changing the TiC stoichiometry). Mechanical mixtures of TiC with 20% NiCr powders were also used. In the latter case the NiCr particles were in the form of a scale 20–60 μm in size and 1–3 μm thick. They were produced by air stream grinding.

Air plasma spraying (APS), low pressure plasma spraying (LPPS or VPS) and underwater plasma spraying (UPS) were used for coating processing. A carbon steel was used as substrate material. For process optimization the "splat test" method and deposition efficiency measurements as a function of spraying distance, traverse speed and powder rate were used.

Optical metallography, X-ray diffraction and scan-

0921-5093/93/$6.00

ning and transmission electron microscopy were used to determine the material structure.

The mass losses during pin-on-disk wear tests and the scratching energy density W_r were used as the characteristics of coating abrasive wear resistance. In the latter case a Vickers pyramid indenter was applied with a normal force F_n to the coating surface. Then the coating was moved in a direction perpendicular to F_n and the tangential force F_t was measured. W_r was calculated as $W_r = F_t/A_r$, where A_r is the scratch cross-section.

3. Results and discussion

(1) The investigations of the form and structure of Ni–Cr–B–Ti atomized and mechanically alloyed powders and plasma coatings showed the following.

The atomized powders are found to consist mainly of primary (white colour) and secondary (grey colour) tauboride crystals with layers of eutectic and precipitates of simple and double borides (Fig. 1). The hardness of such material varies within the range 700–850 $HV_{0.5}$. After plasma spraying, the tauboride grain size decreases and the hardness increases to 800–1000 $HV_{0.5}$.

The mechanically alloyed powders do not contain any tauboride crystals initially. The latter form as a result of remelting during spraying to the extent of 10%–15% only. The coating hardness does not exceed 500 HV 0.5.

In both types of powders metal particles are surrounded by carbides (Fig. 2). Because of this, TiC oxidation and thermal decomposition take place during spraying.

As a result of the above characteristic features, the resistance to abrasive wear of the coatings from atomized powders is substantially higher than that of the coatings from mechanically alloyed powders (Fig. 3).

(2) Titanium carbide particles added to the sprayed powders in amounts from 5 to 35 wt.% increase the LPPS coating resistance to abrasive wear (Fig. 4). For the APS coatings, 10 wt.% TiC addition gives the best results. This is due to the oxidation of TiC particles during spraying in air and an enhanced coating porosity resulting from the addition of TiC particles to the powder.

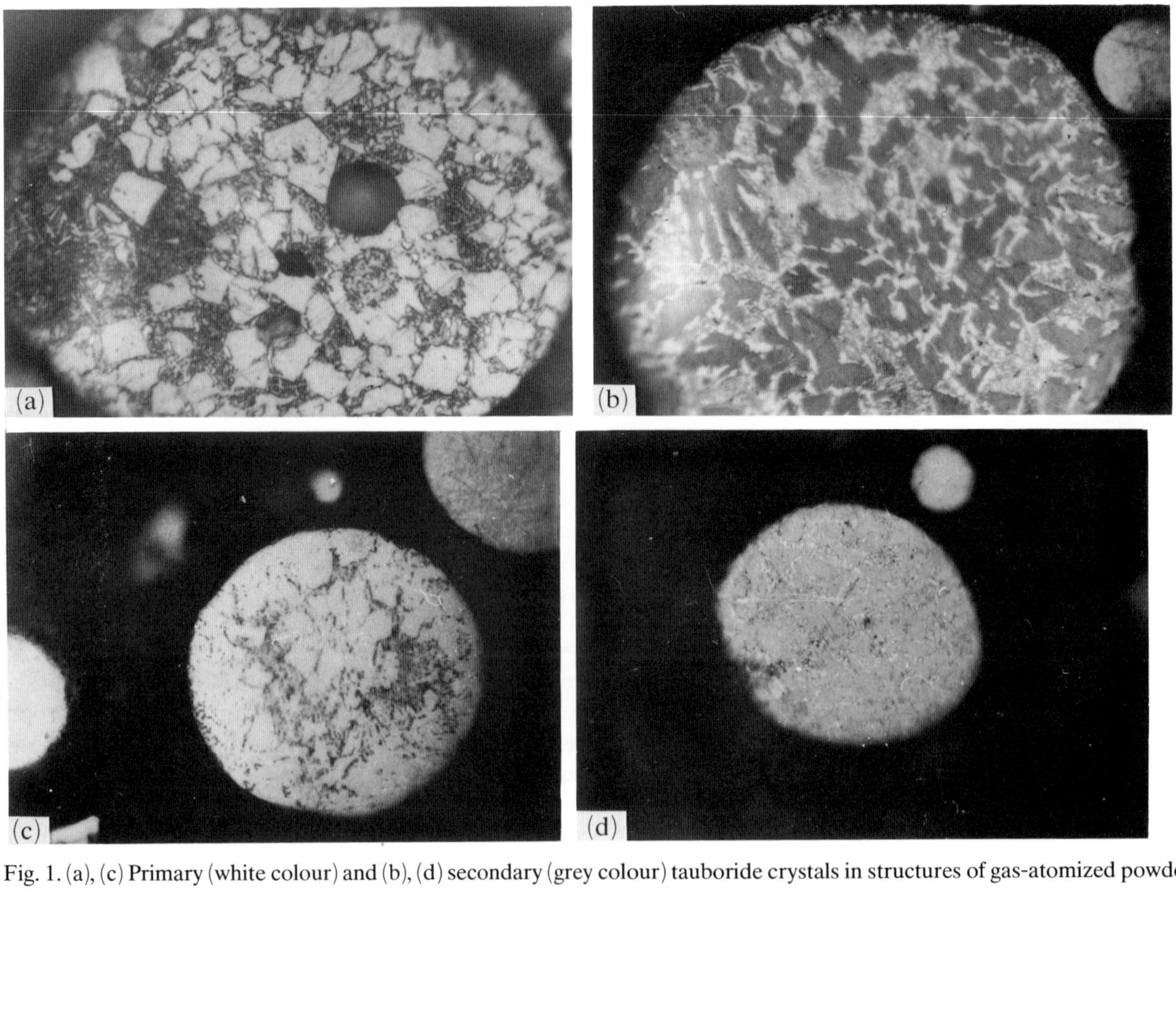

Fig. 1. (a), (c) Primary (white colour) and (b), (d) secondary (grey colour) tauboride crystals in structures of gas-atomized powders.

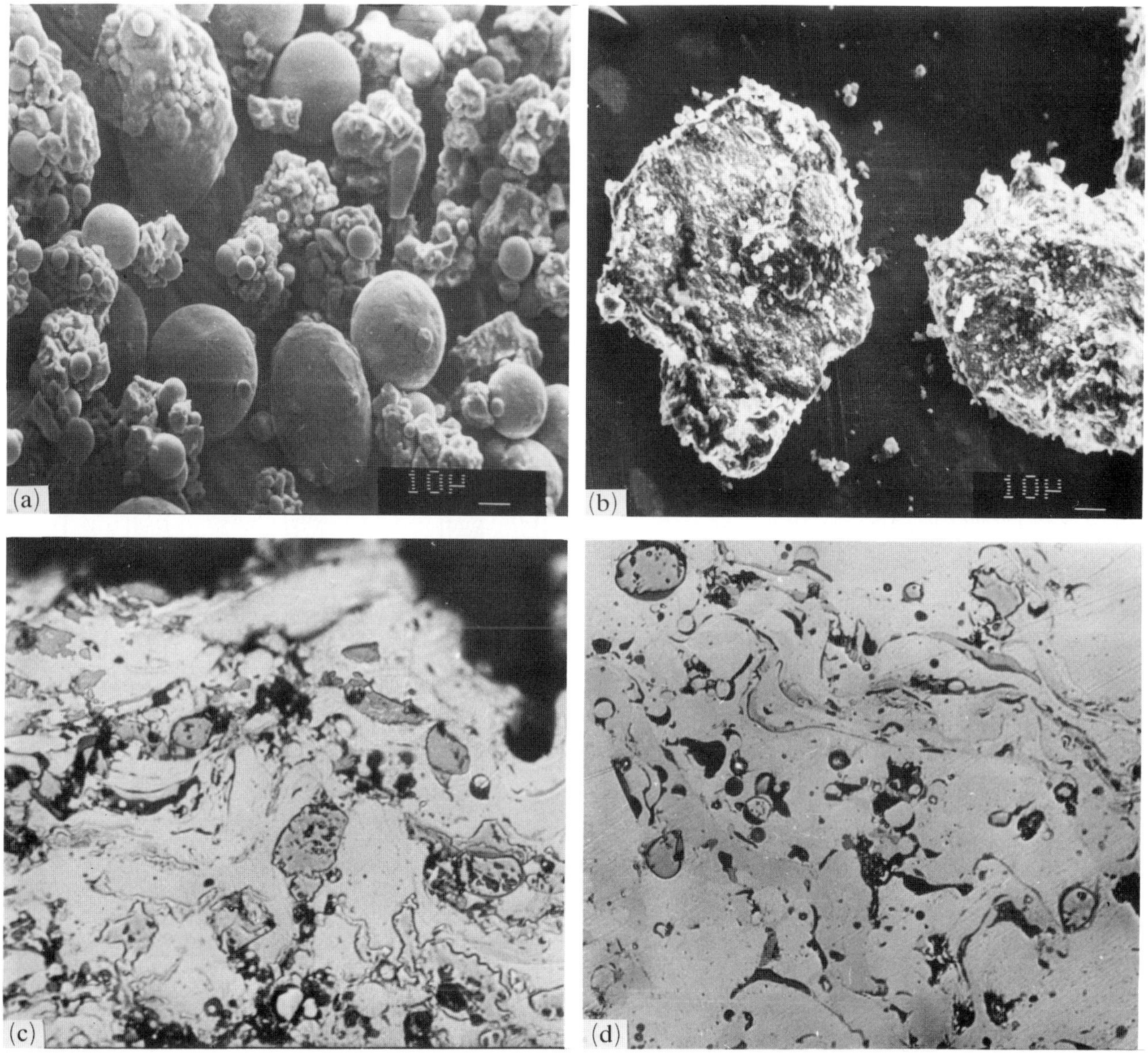

Fig. 2. General view of (a) agglomerated and (b) mechanically alloyed powders and (c), (d) structures of corresponding APS coatings.

(3) With increasing plasma jet power the LPPS coating deposition parameters of density and bond strength also increase. However, since some TiC thermal decomposition takes place, which is accompanied by carbon loss, the carbide grain hardness and the coating resistance to abrasive wear decrease (Fig. 5).

The addition of NiCr scale to the sprayed powder prevents TiC decomposition during spraying under "hard" parameters (Fig. 6), while the latter allow the processing of very dense coatings (Fig. 7) whose resistance to abrasive wear is substantially improved (Fig. 6).

(4) The underwater plasma-spraying method allows the processing of practically non-porous coatings from the tauboride alloy under consideration (Fig. 8(a)). Owing to the rapid quenching of the sprayed layers by the water, a very fine material structure forms: the tauboride crystal grain size is 0.01–0.10 μm. As a result, the sprayed material hardness exceeds 1100 $HV_{0.5}$.

The significant temperature increase of the spraying spot and the reduction in carbide oxidation due to process isolation underwater allow the use of UPS for TiC coating production even without metallic addition (Fig. 8(b)). It is possible to reduce the porosity of the TiC coatings to 7%–8% by using different kinds of plasma gun nozzles and optimizing the spraying parameters. However, in the latter case some carbon loss occurs in the carbide grains (up to $TiC_{0.92}$ stoichiometry). As a

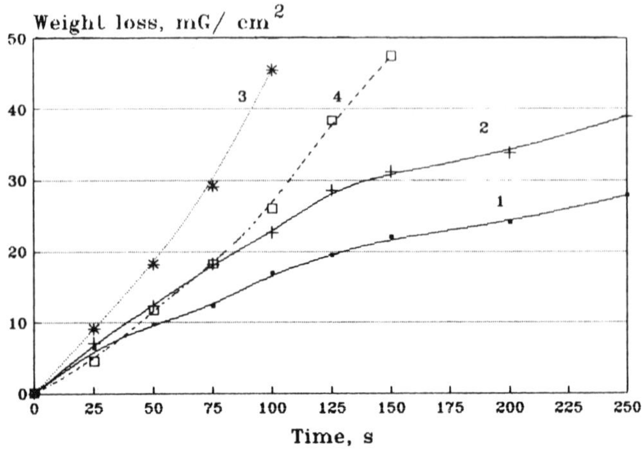

Fig. 3. Abrasive wear (weight loss) during pin-on-disk testing of APS NICrBTi–TiC coatings. The following powders were used for spraying: 1, agglomerated powder, 10 wt.% TiC; 2, agglomerated powder, 30 wt.% TiC; 3, mechanically alloyed powder, 5 wt.% TiC; 4, mechanically alloyed powder, 10 wt.% TiC.

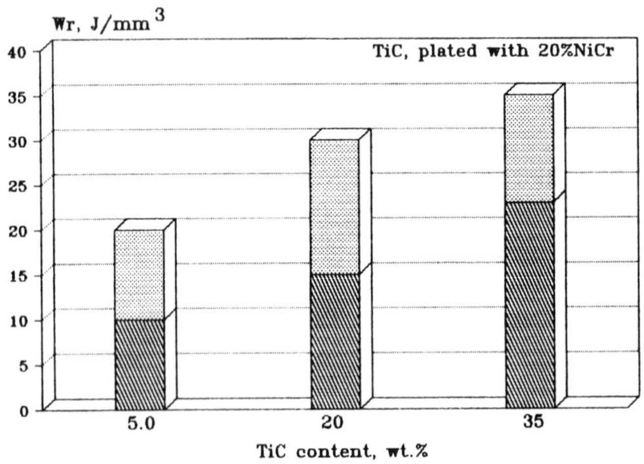

Fig. 4. Influence of TiC content in spraying powder on LPPS NiCrBTi–TiC coating resistance to abrasive wear, W_r.

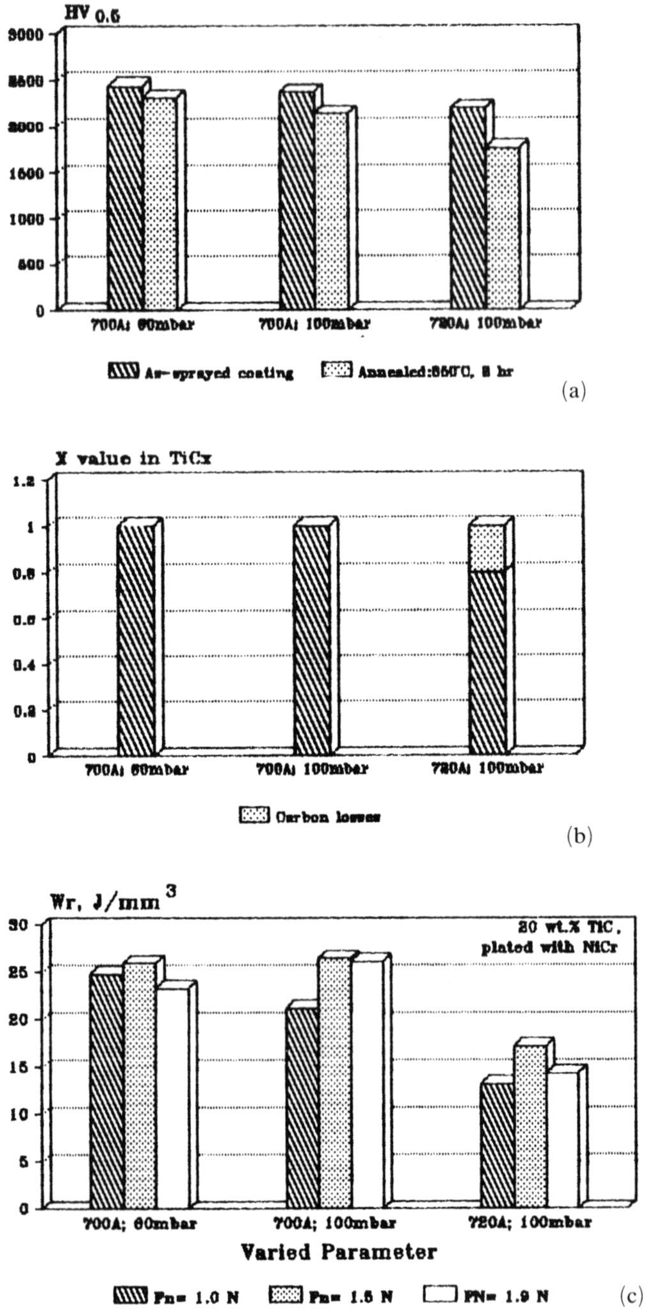

Fig. 5. Dependence of (a) hardness of TiC grains, (b) carbon loss in TiC and (c) resistance to abrasive wear, W_r, on spraying parameters of LPPS NiCrBTi–TiC coatings. Variable spraying parameters: arc current (A), total chamber pressure (mbar). Constant spraying parameters: argon rate 50 l min^{-1}, hydrogen rate 9 l min^{-1}, powder rate 0.35 l min^{-1}, spraying distance 270 mm, traverse speed, 50 mm s^{-1}.

result, the hardness of the densest coatings is only 1400–1700 $HV_{0.5}$. A hardness of 2000–2200 $HV_{0.5}$ is achieved for coatings with TiC_1 stoichiometry, porosity of 10%–12% and deposition efficiency of about 0.6.

4. Conclusions

The processing of coatings by various plasma-spraying methods from tauboride Ni–Cr–B–Ti alloy powders with titanium carbide added has been investigated. Gas-atomized metallic powders are preferable for spraying to achieve a high content of tauboride crystals in the coatings. The addition of NiCr scale to the sprayed powder prevents thermal decomposition of the carbides during spraying. The coating resistance to abrasive wear increases with increasing TiC grain content in the sprayed powder within the investigated interval from 5 to 35 wt.% in the case of low pressure

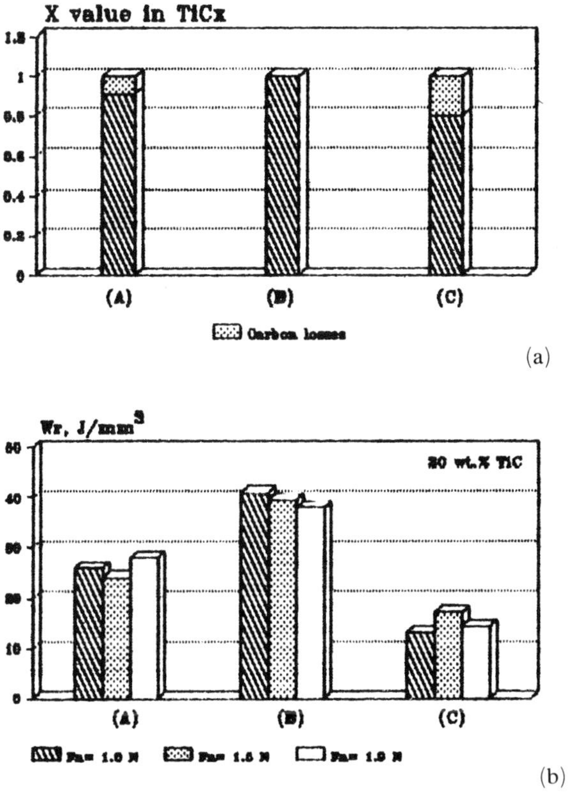

Fig. 6. Dependence of (a) carbon loss in TiC and (b) resistance to abrasive wear, W_r, of LPPS NiCrBTi–TiC coatings on type of spraying powder: (A) agglomerated powder; (B) mixture of Ni–Cr–B–Ti and TiC powders with 4 wt.% of NiCr scale; (C) mixture of Ni–Cr–B–Ti powder and TiC powder plated with NiCr by HF plasma. The spraying parameters correspond to group 3 of Fig. 5.

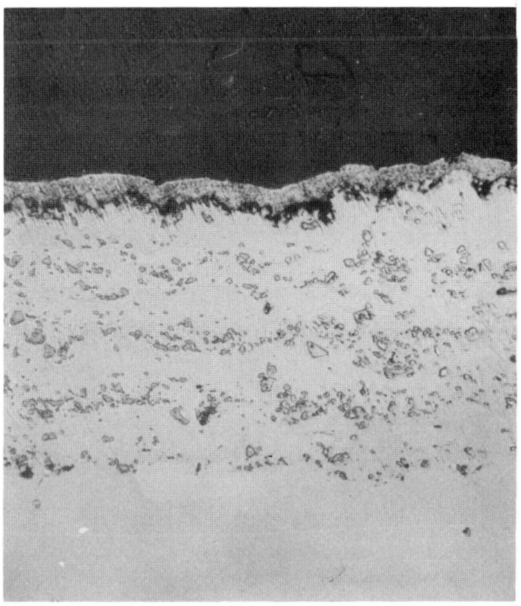

Fig. 7. Structure of LPPS NiCrBTi–20wt.%TiC coating sprayed with NiCr scale.

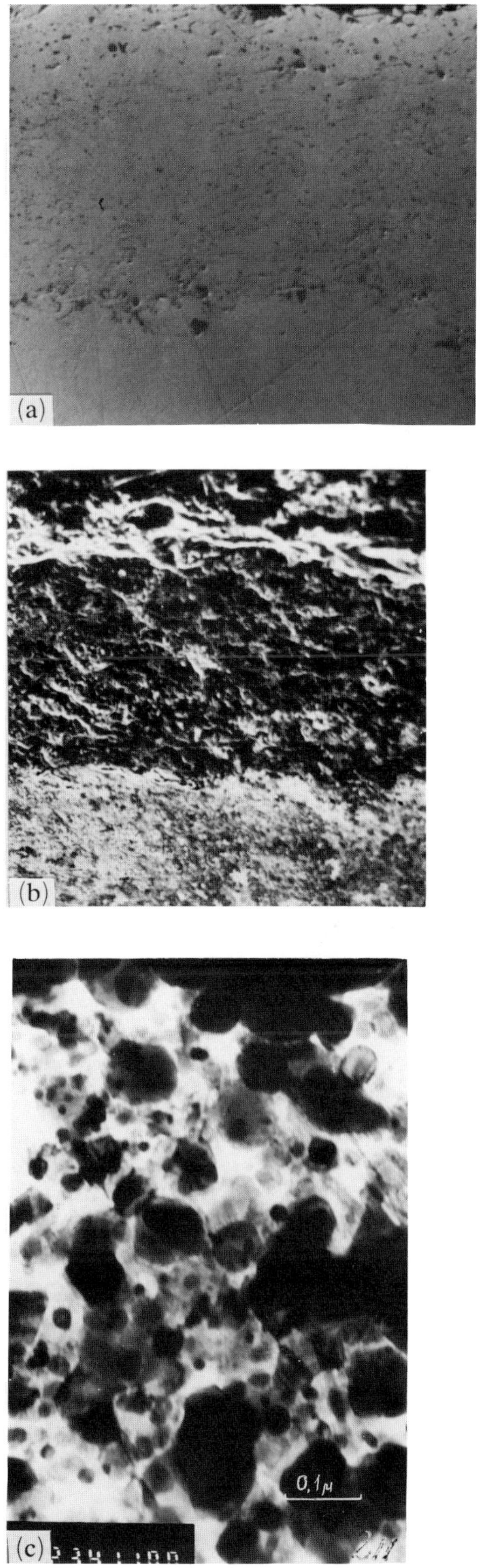

Fig. 8. Typical structures of UPS coatings from (a), (c) Ni–Cr–B–Ti and (b) TiC powders.

plasma spraying. For air-plasma-sprayed coatings the optimum is near 10 wt.% TiC. Underwater plasma spraying is conducive to the formation of tauboride coatings with a dense and fine structure and to the processing of coatings with a very high content of carbides.

References

1 E. Ganglberger, N. Nowotny and F. Benesowski, *Monatsh. Chem., 96* (1965) 1144–1146.

2 E. Lugscheider, A. Krautwald, H. Eschnauer, J. Wilden and H. Meinhardt, *Surf. Coat. Technol., 32* (1987) 273–284.

Selected properties of surface layers of metals alloyed with a pulsed laser

Ryszard Pawlak and Tomasz Raczynski

Technical University of Lodz, Department of Material Science, Institute of Principles of Electrical Power Engineering, Stefanowskiego 18/22, 90-924 Lodz (Poland)

Abstract

The use of laser radiation in alloying the surface layer of metals is a new and fast developing method for the laser treatment of surfaces. Lasers of a different concentration of alloying material can be obtained by alloying with a pulsed laser. The paper presents the results of studies on the effect of various factors (pulse duration τ_i, pulse energy E_i, coverage coefficient) on basic parameters of alloyed layers in tungsten, molybdenum and nickel (electrical conductivity, resistance to erosion in an electric arc).

1. Introduction

Technologies making use of laser processing of surfaces have already gained recognition in the field of surface layer engineering. Processes for producing alloyed surface layers, in which the laser constitutes a basic energy source of unconventional parameters, have been developing rapidly since the 1970s. This technique most frequently employs continuous wave lasers and was first used for constructional materials (different grades of steel). Now, however, other metals are becoming subjects of study in the field of laser engineering of the surface layer [1–3]. We are using a pulsed laser to study different base metals employed in electrical engineering (base metals tungsten, molybdenum, nickel, silver, alloying metals, gold, nickel, tin) [4–6].

2. Obtaining alloyed lasers

To produce alloyed lasers, an Nd-YAG (Kwant-15) laser, generating pulses of duration 1–4 ms and a regulated value of pulse energy E_i was used. The treatment used was "pulse after pulse" with overlapping of successive areas defined by the coefficient k_p. An alloying element was deposited onto the surface processed in the form of a thin foil. It was assumed that the layers had a uniform thickness $h = 50–250$ μm, a surface deformation $f < 30$ μm, and were free from macrodefects. A theoretical analysis of the fusion process and experimental studies have allowed optimization of the technological parameters necessary to satisfy these conditions [6], see Table 1.

3. Structure and mechanical properties

Metallographic images of the layer sections are shown in Fig. 1. In most cases, a rotational arrangement of phases in the layer can be observed. This arrangement results from the mechanism of thermocapillary convection and recurs regularly with the action of successive laser pulses [7]. Investigations using a microanalyser on W–Ni, Ni–Au and Mo–Au have shown considerable uniformity of distribution of the alloying element in the layer.

The microhardness of the layers was also measured. The results presented in Fig. 2 show a field of scatter $n > 30$ for measurements performed in different zones of the layer. An increase in microhardness (even by 230% for Ag–Ni) is characteristic, independent of the relation between the alloying metal microhardness and the base metal microhardness.

TABLE 1. Parameters for laser alloying

Base metal	Doping metal	E_i (J)	τ_i (ms)	k_p
W	Ni	5.0–7.2	4	≤0.5
Mo	Au	4.0–6.2	4	≤0.5
Ni	Au	3.8–5.7	4	0.3–0.4
Ag	Ni	5.5–9	4	0.3–0.5
Ag	Sn	3.6–5.5	4	≤0.5

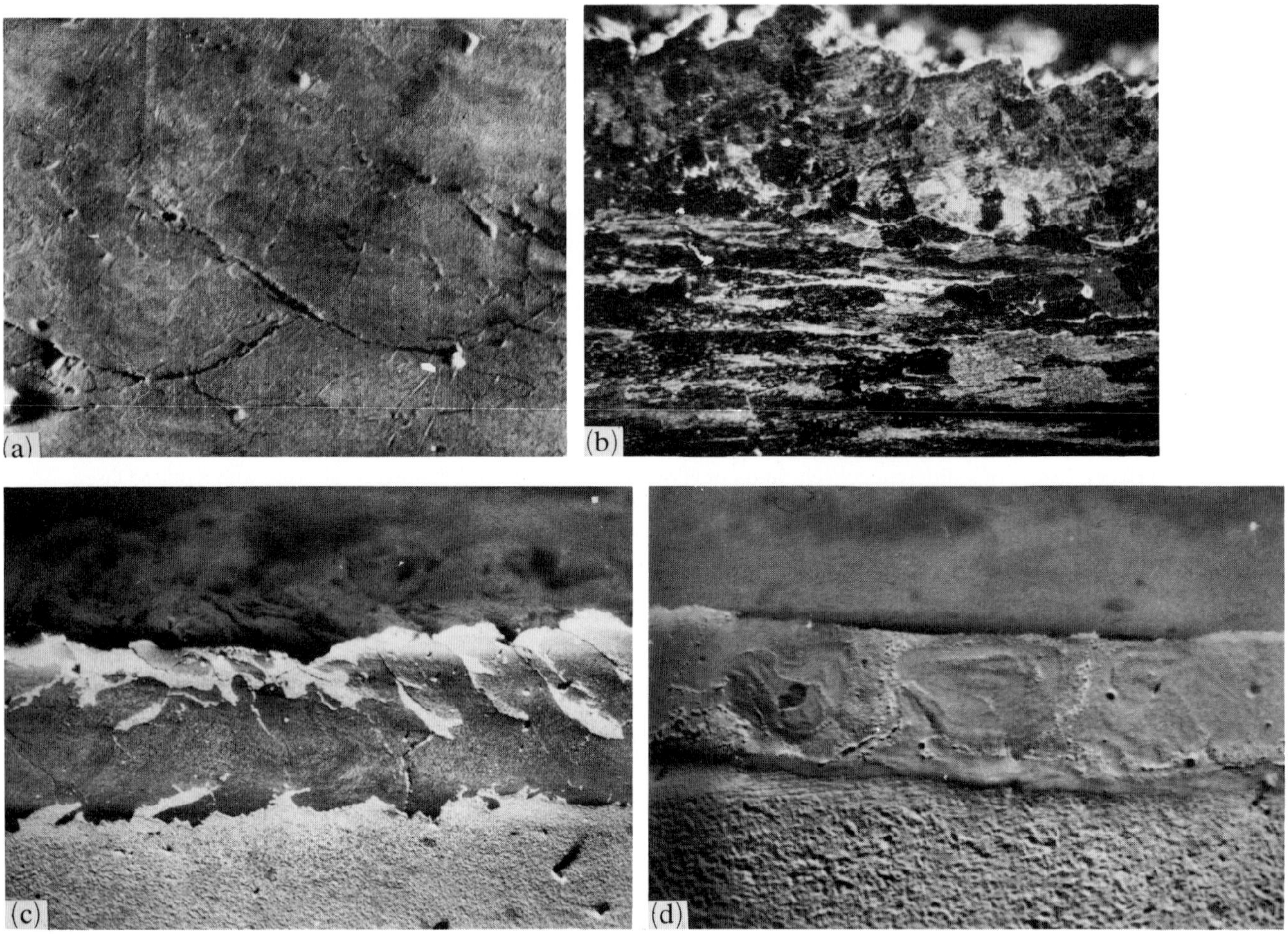

Fig. 1. View of metallurgical section of alloyed layers: (a) W–Ni, (b) Mo–Au, (c) Ag–Ni, (d) Ag–Sn.

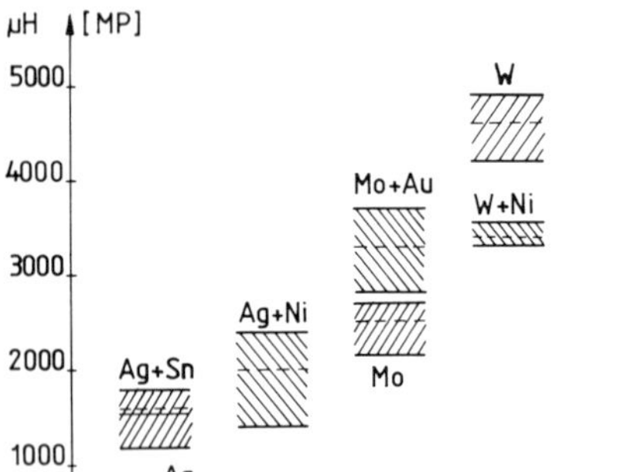

Fig. 2. Changes in microhardness of layers in comparison with the base metal.

4. Layer resistivity

Determination of the resistivity of an alloyed system such as these by theoretical analysis would require precise knowledge of the phase image, the proportion of particular phases, their dispersion and stresses formed during production [8]. A four-probe method was adapted for comparative measurements, measuring the resistivity of a specimen before and after alloying. The absolute resistivity values obtained contain an error resulting from the fact that the layer base is conductive and it is impossible to eliminate its contribution to current conduction. However, the aim of the measurements was to determine a change in the alloyed layer resistivity in relation to the base metal $\Delta\rho/\rho$. In order to investigate the effect of deformation of the electrical field in the specimen between the current electrodes on the value of resistance measured, measurements of ρ were taken as a function of distance from the voltage electrodes.

The reproducible results point to correct selection of the gap between the electrodes. Figure 3 shows, as an example, the results of measurements of ρ for Ni–Au layers, from which a reduction in the resistivity of the layer $\Delta\rho/\rho \approx 20\%$–$50\%$ is found, depending on the alloying process (variation of E_i in the range 3.8–5.7 J and a k_p in the range 0.35–0.5). In other cases an increase in resistivity was found, for example

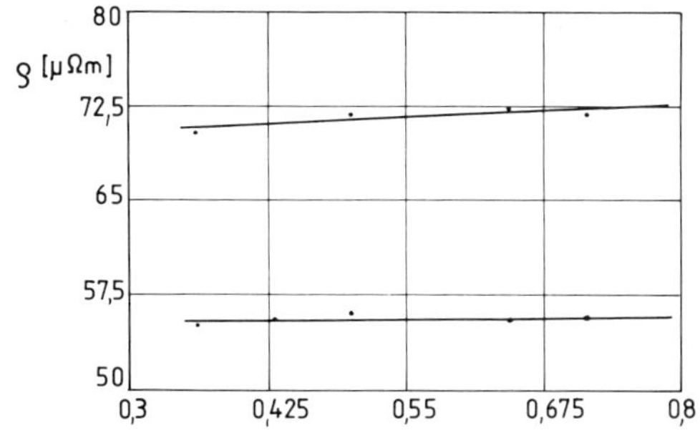

Fig. 3. Resistivity of Ni–Au layers as a function of l, the distance from the voltage electrodes.

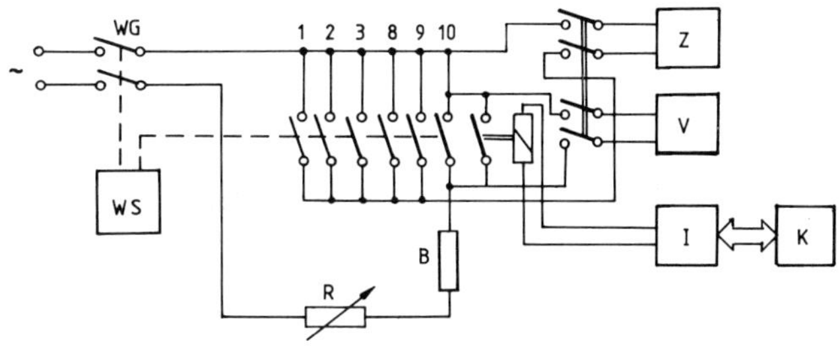

Fig. 4. Block diagram of the system for testing contact erosion: 1–10 switching contacts, Z d.c. supply, V voltmeter, R load resistor, I interface, K computer.

for Ag–Ni layers $\Delta\rho/\rho \approx 30\%$–$60\%$, while for Mo–Ni layers $\Delta\rho/\rho \approx 25\%$–$45\%$.

5. Investigation of resistance to erosion in an electric arc

Investigations were aimed at finding a change in the resistance to erosion in an electric arc of silver specimens with surface layers of Ag–Ni, in relation to silver specimens not subjected to laser treatment.

An electric arc, ignited between the opening contacts, reaches a temperature of greater than 10 000 K (even 50 000 K) and results in wear of the contact in the form of mass loss Δm and a change in shape of the contact surface. No general theory of contact wear due to arc erosion has been developed hitherto, although it is generally assumed that the most important role in this process is played by the following [9]: electrolytic phenomena in the arc plasma: bombardment of the cathode by ions and of the anode by electrons; detachment of ions from the contact surface by the electric field. In view of the complexity of the phenomena accompanying the burning of an arc, investigations of resistance of the material to erosion in an arc are always carried out comparatively, by investigating a material of known properties under the same conditions [9, 10].

The investigations consisted in determining the mass loss as a function of the number of switchings. This was carried out as shown in Fig. 4, under the following conditions: contact gap 5 mm, contact pressure 10 N, reflexive force 30 N, test voltage 220 V, test current 60 A, switching frequency 1/3 Hz, number of switchings 4×10^4.

The results of measurements of the erosion indicate clearly a slower loss of mass (threefold after 10^4 switchings and twofold after 4×10^4 switchings) of the alloyed Ag–Ni layers compared with silver specimens, particularly at the initial stage of investigations when the alloyed layer is of a considerable thickness. The

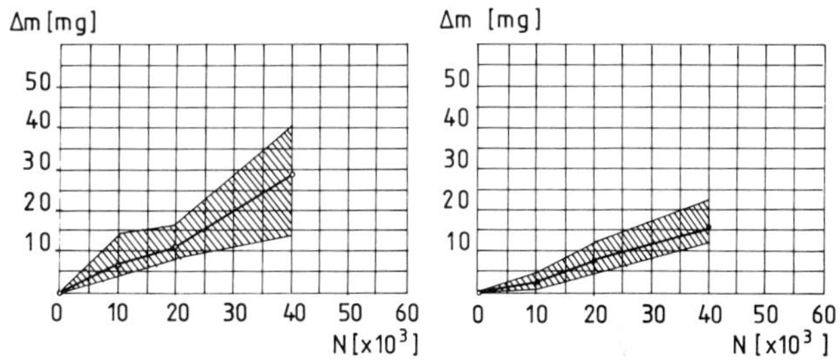

Fig. 5. The mass loss Δm as a function of N the number of switchings: (a) silver contacts, (b) Ag–Ni contacts.

results in the form of band characteristic curves are presented in Fig. 5.

6. Summary

The investigations show that alloyed surface layers of different depths and different concentration of the alloying material can be obtained by changing the process parameters (E_i, τ_i, k_p), while using a pulsed laser. In this way, the properties of the alloyed layers, for example microhardness, can be modified. There is usually an increase in microhardness relative to that of the base metal. The resistivity of the layers can also be changed; in this case both an increase in ρ (Ag–Ni) and a decrease in ρ (Ni–Au) were observed. Changes in microhardness and resistivity depend on the alloying process parameters. Alloyed surface layers can manifest greater resistance to the very destructive action of an electric arc. The above speaks in favour of use of the laser method to produce alloyed surface layers and to modify the properties of materials used in electrical engineering, for example for contact materials.

References

1 J. H. Abboud and R. F. West, *Mater. Sci. Technol.*, 7 (1991) 353–356.

2 E. S. Vlasova, N. T. Travina, R. E. Krizanovski, A. V. Kriklenko and N. I. Kasakova, *Phys. Chem. Mater. Treat., 24* (1990) 48–52.
3 H. Linder, H. W. Bergman and Th. Endres, *Proc. 3rd Euro. Conf. on Laser Treatment of Materials, ECLAT'90*, University Erlangen-Nurnberg, 1990, pp. 661–670.
4 R. Pawlak and T. Raczynski, *Proc. Conf. on Surface Physics 1989*, University Lodz, 1991, Vol. VI, pp. 91–94.
5 F. Kostrubiec and R. Pawlak, *Proc. Conf. on Surface Physics 1990*, University Lodz, 1991, Vol. IX, pp. 219–222.
6 R. Pawlak and T. Raczynski, *Laser Technology III, Proc. SPIE, 1391*, 170–173.
7 G. G. Gladush, *Kvantova Elektron., 9* (4) (1982) 660–667.
8 M. Keil, W. A. Merl and E. Vinaricky, *Elektrische Kontakte und ihre Werkstoffe*, Springer, Berlin, 1984.
9 O. A. Szmatko and J. V. Usov, *Elektriceskie i Magnitne Svojstva Metallov i Splavov*, Nauk. Dumka, Kiev, 1987.
10 S. Stolarz and E. Walczuk, *Metall.* (1) (1982) 40–43.

Specific features of copper and nickel oxides obtained under extreme conditions by CO_2 laser irradiation of inorganic salts

C. Popescu
LACECA Research Centre, Str. Siret 95, Sector 1, 78308, Bucharest (Romania)

R. Alexandrescu, I. Morjan and I. Voicu
Institute for Atomic Physics, P.O. Box MG-6, Bucharest-Măgurele (Romania)

Abstract

Several experiments showed that nickel and copper salts are decomposed to oxides by irradiation with a continuous wave CO_2 laser. The processes are characterized by an extremely high heating rate which develops within the impact region of the laser beam with the substrate. The salts were also decomposed under controlled heating conditions. A difference was noticed concerning the size and isotropy of final crystallites obtained by laser irradiation as compared with thermal heating. The magnitude of this difference seems to be related to the laser quanta absorptivity of the salts.

1. Introduction

The interaction of high-power laser radiation with solid systems is often investigated using a thermochemical approach, in which the strong localized action of the laser beam imposes the characteristics of the processes [1].

In earlier works [2, 3] it was shown that the decomposition of inorganic salts exposed to IR laser radiation takes place under unusual heating conditions, characterized by high heating rates developing within the impact region of the laser beam. Compared with the reaction products obtained in classical thermal decomposition, the oxides produced using laser processes exhibited peculiarities of structure and morphology. A correlation was found between these features and the absorptivity of the initial salts to the wavelength of laser radiation.

In this paper we present new results on the laser decomposition of some copper salts. The similarities of their behaviour and that of nickel salts under the same conditions are revealed.

The salts were also decomposed under controlled heating. Crystallographic investigations of the oxides produced (CuO and NiO) revealed differences in the size and isotropy of crystallites produced from laser experiments compared with those obtained by thermal heating. The high heating rates which characterize the laser processes together with the absorptivity of the substance to the laser wavelength should both be involved in the kinetics of the reactions.

2. Experimental details

Different nickel and copper salts of analytical purity grade were used. Their IR spectral characteristics within the range of CO_2 laser emission (10.6 μm) were investigated and the results are given in Table 1.

Laser irradiation experiments were performed using a continuous wave CO_2 laser of beam intensity 400–600 W cm^{-2}. The salts, in powdered form, were evenly distributed as a thin layer in a metal reaction cell, which was moved perpendicular to the beam at various speeds. After each passage through the laser beam the samples were weighed (accuracy $\pm 5 \times 10^{-5}$ g) until constant weight was achieved.

A Q 1500 D MOM derivatograph was used for the thermal analysis experiments. Heating rates of 5 and 10 K min^{-1} were used.

TABLE 1. Absorptivity values of nickel and copper salts at 10.6 μm

Substance	Code	Absorptivity (%)
$NiCO_3 \cdot Ni(OH)_2 \cdot H_2O$	K	42
$Ni(NO_3)_2 \cdot 6H_2O$	A	2
$NiCl_2 \cdot 6H_2O$	C	0
$NiSO_4 \cdot 7H_2O$	S	8
$Cu_2CO_3(OH)_2$	K′	15
$Cu(CH_3COO)_2 \cdot H_2O$	A′	5
$CuSO_{4 \cdot 5H2}O$	S′	13
$CuCl_2 \cdot 2H_2O$	C′	1–2

0921-5093/93/$6.00

X-ray diffraction measurements were taken using a Philips diffractometer, with Cu Kα radiation.

3. Results and discussion

The salts were processed in the laser beam, until a constant weight was observed. The weight losses *vs.* time can be used to study the kinetics of the laser-induced reactions [3].

The same salts were thermally decomposed in the derivatograph and the recorded thermogravimetric data were processed to give a kinetic analysis of the thermal reactions [4].

Table 2 presents the mass losses for the copper salts, decomposed thermally and by laser radiation. Thermal experiments all give CuO as the final product (in the limits of our experimental method). From laser experiments, either a non-stoichiometric oxide CuO_x, $x<1$, is obtained, or, more likely, a mixture of oxides Cu_2O and CuO.

Crystallographic investigations of both nickel and copper oxides by X-ray diffraction showed that the crystallites of laser-treated samples were larger than those obtained thermally and that this difference increased for salts with higher absorptivity to the wavelength of laser radiation. The mean size of the oxide crystallites was determined along the [111] and [200] crystallographic directions using the Debye–Sherrer formula with corrections calculated as indicated by Kaelbe [5]. The sizes of CuO crystallites obtained in laser and thermal experiments are listed in Table 3.

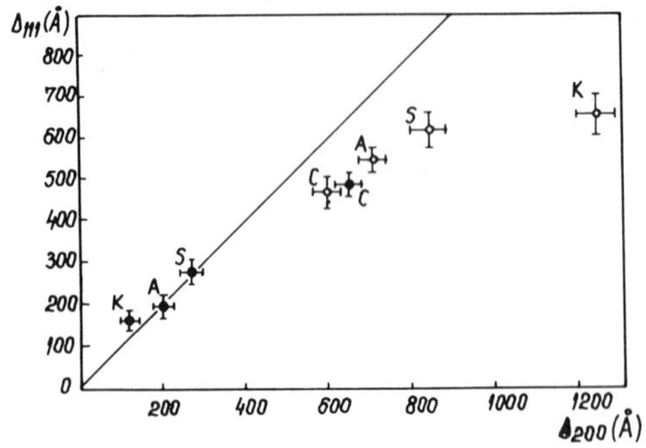

Fig. 1. Mean sizes of NiO_x crystallites, thermally treated (●) and laser treated (○).

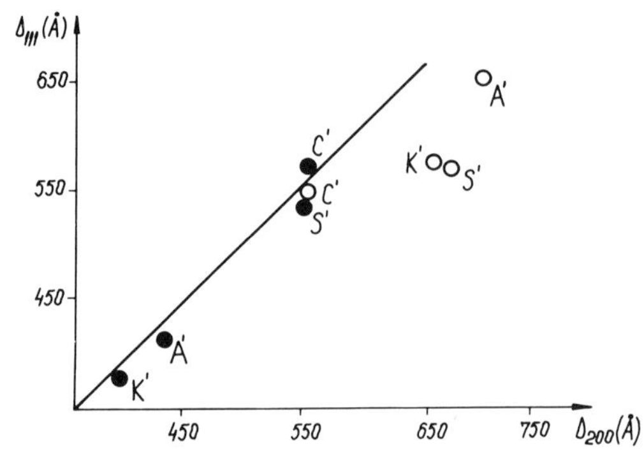

Fig. 2. Mean sizes of CuO_x crystallites obtained thermally (●) and by laser irradiation (○).

TABLE 2. Mass losses in laser and thermal experiments, compared with stoichiometric losses

Substance	Thermal mass loss (%)	Laser mass loss (%)	Theoretical mass loss (%)	
			To Cu_2O	To CuO
$CuSO_4 \cdot 5H_2O$	68.5	69.6	71.3	68.1
$Cu_2CO_3(OH)_2$	29.3	33.2	35.3	28.1
$CuCl_2 \cdot 2H_2O$	84.6	61.4	58.1	53.4
$Cu(CH_3COO)_2 \cdot H_2O$	60.2	—[a]	64.2	60.15

[a]The salt was scattered by the laser beam.

TABLE 3. Size of CuO crystallites obtained from thermal and laser processes, from different inorganic salts

Substance	Thermal process		Laser process	
	D_{111} (Å)	D_{200} (Å)	D_{111} (Å)	D_{200} (Å)
$Cu_2CO_3(OH)_2$	383	392	580	680
$CuSO_4 \cdot 5H_2O$	540	561	572	690
$Cu(CH_3COO)_2H_2O$	416	434	656	723
$CuCl_2 \cdot 2H_2O$	573	563	556	561

From the plots of Fig. 1 and Fig. 2 one may note that the crystallites produced by laser experiments have different sizes along the [111] and [200] crystallographic directions. For NiO_x crystallites (Fig. 1) [3], the largest anisotropy of shape is exhibited sample K (nickel hydroxycarbonate) which was the most absorbant to laser radiation. Very similar behaviour is observed for CuO_x crystallites in Fig. 2, obtained from copper salts with increased absorptivity in the order A′ < S′ < K′ (see Table 1). The crystallites obtained from non-absorbing salts (such as copper chloride) are situated towards the line of isotropy in both figures (Figs. 1 and 2).

We may conclude that the more the initial substance absorbs the laser radiation, the larger are the size and anisotropy of the oxide crystallites obtained (different lengths along the directions D_{200} and D_{111}).

Further support for the influence of salt absorptivity on the laser effect is given in Fig. 3, which shows the dependence of the mean crystallite size on the speed of passage through the laser beam. For the salt with high absorptivity (nickel hydroxycarbonate) the size of

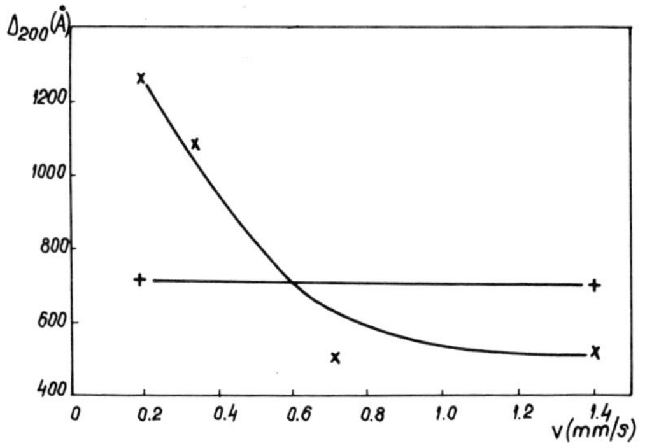

Fig. 3. Dependence of the mean size of NiO_x crystallites obtained on the speed of passage through the laser beam: × nickel hydroxycarbonate; + nickel nitrate.

crystallites in the direction D_{200} decreases as the speed increases [3], but remains unchanged for nickel nitrate.

A qualitative explanation of the effect of the laser on oxide morphology could be found by analysing the dynamics of the laser heating process. The laser irradiation area is characterized by extremely high heating rates (1000–1500 K is easily reached in a short time interval with a medium-power CO_2 laser). The heating rate of the laser-induced reactions is strongly related to the quanta absorbed by the substance, *i.e.* to the absorptivity of the salt at 10.6 μm, the power of the source, the speed of the cell through the laser beam, etc. Obviously the heating rate for the laser experiments is a non-linear function of temperature, and different heating rate functions (*i.e.* different powers, speeds, etc.) will lead to different physical forms of the final products. Thus, it is expected that substances with higher absorptivities will experience higher changes in the heating rates (as shown in Fig. 3) with the consequence that the crystallites grow larger and are more anisotropic.

For the case of salts with lower absorptivity, the laser heating is more uniform, with increased dissipative losses through the metal walls of the cell; the reaction products look like those which have undergone pure thermal decomposition.

References

1 D. Bauerle (ed.), *Laser Processing and Diagnostics*, Springer, Berlin, 1985.

2 C. Popescu, V. Jianu, R. Alexandrescu, I. N. Mihăilescu and I. Morjan, *Thermochim. Acta, 129* (1988) 269.

3 C. Popescu, I. Ursu, M. Popescu, R. Alexandrescu, I. Morjan and I. N. Mihăilescu, *Thermochim. Acta, 164* (1990) 79.

4 C. Popescu, R. Alexandrescu, I. Morjan and M. Popescu, *Thermochim. Acta, 184* (1991) 73.

5 D. Negoiu, *Treatise of Inorganic Chemistry*, Tehnica, Bucharest, 1972 (in Romanian).

Study of the effects of CO_2 laser irradiation on $Cd_{0.204}Hg_{0.796}Te$

Thierry Vogt

Centre de Recherches Nucléaires, IN2P3-CNRS, Université Louis Pasteur, Groupe Recherches Physiques et Matériaux, BP20, F-67037 Strasbourg Cedex 2 (France)
Institut Franco-Allemand de Recherches (ISL), BP34, F-68301 Saint-Louis Cedex (France)

René Joecklé

Institut Franco-Allemand de Recherches (ISL), BP34, F-68301 Saint-Louis Cedex (France)

Claude Schwab

Centre de Recherches Nucléaires, IN2P3-CNRS, Université Louis Pasteur, Groupe Recherches Physiques et Matériaux, BP20, F-67037 Strasbourg Cedex 2 (France)

Abstract

The behaviour of p-type $Cd_{0.204}Hg_{0.796}Te$ used in IR detectors was studied under high power CO_2 laser irradiation at room temperature. Its melting threshold was determined for two irradiation conditions. (1) For a pulsed focused laser (pulse duration $\tau = 0.3$ ms, spot diameter 225 μm), the melting threshold fluence ranged from 13 to 38 J cm^{-2}. (2) For a continuous wave CO_2 focused laser beam (spot diameter 240 μm, $I = 360$ kW cm^{-2}) swept on the target using a rotating mirror, the threshold fluence was $F = 11$ J cm^{-2}, corresponding to a scanning speed of 8 m s^{-1}. These experimental results were compared with calculated values assuming either a monodimensional model or a two-dimensional combined absorption and conduction model. The threshold calculated with the monodimensional model was $F = 15$ J cm^{-2}, whereas the threshold calculated with the two-dimensional model was $F = 19$ J cm^{-2} for the pulsed irradiation and $F = 17.5$ J cm^{-2} for the continuous irradiation. A decrease in mercury concentration from 39.8 at.%) down to 13.1 at.%, depending on the fluence, was observed using X-ray microanalysis.

1. Introduction

$Cd_xHg_{(1-x)}Te$ (CMT) is the main detector material used for night vision. Several investigations of the interaction of laser light with CMT were aimed at the removal of lattice damage associated with doping by ion implantation of this material. However, the first attempts in this direction were not as successful as for silicon. This is related to the different chemical nature of these two materials, *i.e.* an elemental as opposed to a compound semiconductor. There is indeed the possibility of a compositional change of CMT during heating, owing to volatile element losses even before melting [1].

The purpose of the present work was to determine the minimum fluence to obtain permanent damage, as revealed by surface melting, under high power laser irradiation conditions, corresponding to typical exposures for a single pixel in an IR camera. It was measured experimentally and compared with theoretical estimations. The rests were done on bulk $Cd_{0.204}Hg_{0.796}Te$ suitable for the fabrication of IR devices for the 8–14 μm range [2], to understand the fundamental laser–matter interaction prior to investigation of real devices. The illumination conditions with CO_2 lasers (10.6 μm) were selected to cover the two present types of IR camera. The possible modifications of composition were measured by X-ray microanalysis.

2. Experiments

The samples used for the tests were p-type $Cd_{0.204}Hg_{0.796}Te$ single crystals supplied by the Colorado Research Laboratory, Walsenburg, CO. All tests were done at room temperature. The minimum energy necessary to obtain a morphological modification on the surface of the sample was determined by scanning electron microscopy. The laser beams used for the experiments were assumed to be gaussian.

2.1. Pulsed irradiation

For a camera with a linear array of detectors and one-dimensional sweeping, the mean irradiation time of a pixel by a laser beam focused by the camera optics is approximately 0.3 ms. To simulate this irradiation time, a CO_2 laser was used for which the high voltage supply of the electrical discharge could be modulated to give an adjustable laser pulse duration. For the tests, this pulse duration was $\tau = 0.3$ ms. The laser beam was

0921-5093/93/$6.00

focused onto the target with a diameter of 225 μm. Part of the incoming laser beam was reflected by a potassium chloride prism and measured by an energy meter.

Several tests were done with energies ranging from 2.15 to 20 mJ. Figures 1(a) and 1(b) show two craters resulting from irradiation with 7.6 and 14 mJ energies. The melting threshold corresponded to an energy value between 2.6 and 7.6 mJ. (Unfortunately, it was not possible to obtain intermediate values between these two limts.) The fluence threshold derived from the relation $F=2E/S$ (where E is the energy and S the laser spot surface) is then within the range 13–38 J cm^{-2}.

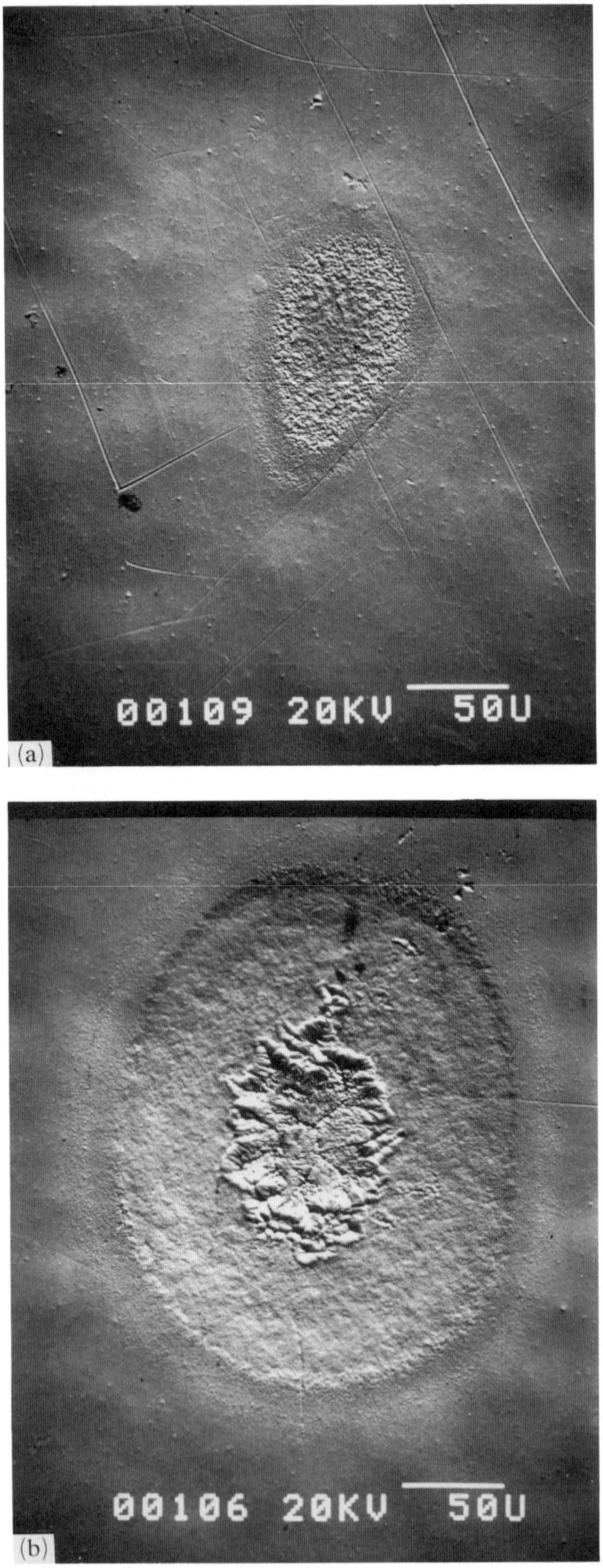

Fig. 1. Craters obtained with pulsed irradiation: fluence at the centre of the spot (a) $F_{max}=38$ J cm^{-2}, (b) $F_{max}=70$ J cm^{-2} (spot diameter 225 μm, $\tau=0.3$ ms).

2.2. Continuous irradiation

The optical source used in these experiments was a 76 W continuous wave CO_2 laser. In a single detector camera, the irradiation time of a pixel is several microseconds. To obtain such short times, the focused laser beam was swept onto the sample by a rotating mirror. The adjustable revolution speed of the mirror allowed the sweeping speed of the laser beam over the sample to be varied. Figure 2 shows the sweeping system. The laser spot diameter was 240 μm, the maximum power density was $I_{max}=360$ kW cm^{-2}. The spot diameter is divided by the scanning speed to obtain the irradiation time, the fluence is given by $F=I\tau$.

Several sweeps were done on the sample with scanning speeds ranging from 4.0 to 8.5 m s^{-1}, the corresponding fluences varying from 10 to 22 J cm^{-2}. The resulting damage was a groove whose width increases with fluence. Figure 3(a) shows grooves obtained under these conditions; Fig. 3(b) shows the groove resulting from a scanning speed of 5.5 m s^{-1} with a maximum fluence $F_{max}=16$ J cm^{-2}.

The melting threshold was obtained for a scanning speed of 8 m s^{-1}. The corresponding irradiation time and fluence were respectively $\tau=30$ μs and $F=11$ J cm^{-2}.

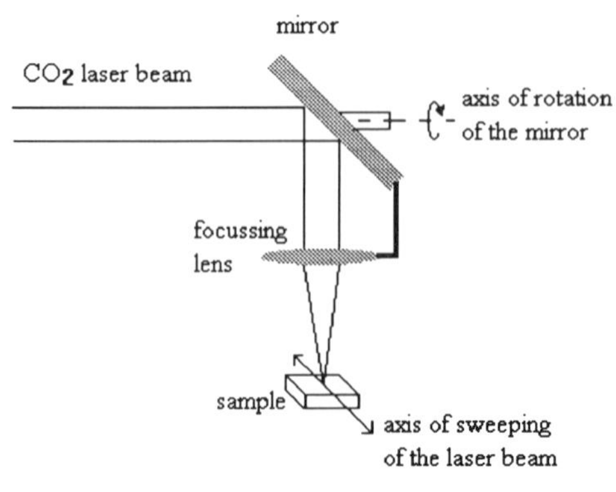

Fig. 2. Experimental set-up for the sweeping tests.

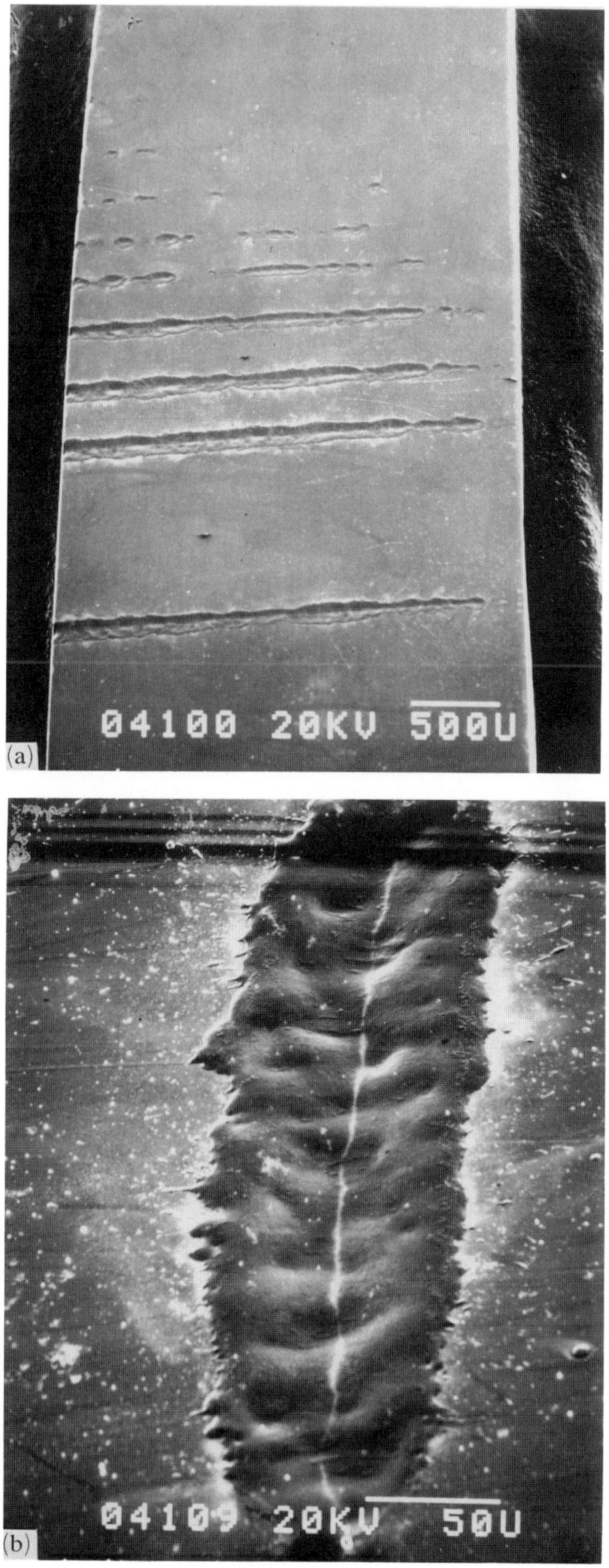

Fig. 3. (a) Grooves obtained under scanned irradiation; (b) groove obtained with scanned irradiation, fluence at the centre of the spot $F_{max}=16$ J cm^{-2} (spot diameter 240 μm, $\tau=44$ μs).

3. Theoretical melting threshold

The melting thresholds were estimated for both conditions of irradiation, using a conventional model and a more sophisticated model developed by Gautier at ISL [3].

3.1. Monodimensional absorption model

For a first approximation, monodimensional behaviour is assumed, *i.e.* we assume no radial heat conduction. Since at room temperature the absorption coefficient of CMT for a wavelength of $\lambda=10.6$ μm is $\alpha=100$ cm^{-1} [4], the optical absorption depth is about 100 μm. For an irradiation time of $\tau=0.3$ ms, the thermal conduction depth given by the relation $D=2(\kappa\tau)^{0.5}$ [5] is $D=35$ μm; for $\tau=30$ μs, we have $D=10$ μm (κ is the thermal diffusivity, *i.e.* the thermal conductivity divided by the specific weight and the specific heat). Since for both durations the absorption depth is larger than the thermal depth, we infer that an absorption model can be used and the conduction can be neglected. Further, since the thickness of the sample is ten times larger than the absorption depth, it can be assumed as an infinite half-space. In this case, the increase in temperature at the surface of the sample is given by the relation [5]:

$$\Delta T(t)=(1-R)I\alpha\tau/\rho C_p$$

where I is the laser power density, R is the reflection factor, α is the absorption coefficient, τ is the irradiation time, and C_p is the specific heat of the material. Our aim was to calculate the laser power density needed to reach the melting temperature at the surface of the material. The thermal and optical parameters are collected in Table 1.

For an irradiation time of 0.3 ms, the calculated power density is $I=49$ kW cm^{-2} with a corresponding fluence $F=15$ J cm^{-2}. For a scanning speed of 8 m s^{-1} ($\tau=30$ μs), the power density is $I=490$ kw cm^{-2} and the fluence is then $F=15$ J cm^{-2}.

TABLE 1. Thermal and optical parameters of CMT

α	100 cm^{-1} at room temperature for $\lambda=10.6$ μm [4]
κ	0.01 cm^2 s^{-1} at room temperature [6]
$T_{melting}$	707 °C for $x=0.204$ [6]
ρ	7.95 g cm^{-3} at room temperature[a]
k	0.015 W cm^{-1} K^{-1} at room temperature[a]
C_p	0.18 J g^{-1} K^{-1} [7]
R	0.30 at room temperature and 10.6 μm[b]

[a]Colorado Research Laboratory, PO Box 692, Walsenburgh, CO 81089, USA.
[b]Measured in our laboratory.

TABLE 2. Comparison of experimental and theoretical thresholds according to the models

Irradiation time and spot diameter	Threshold power density (kW cm^{-2}) and fluence (J cm^{-2})		
	Experimental value	Monodimensional model	Bidimensional model
0.3 ms, 225 μm	43–127	49	62
	13–38	15	19
30 μs, 240 μm	360	490	584
	11	15	17.5

3.2. Two-dimensional combined absorption and conduction model

For an irradiation time $\tau = 0.3$ ms, the thermal depth is only three times smaller than the absorption depth. Therefore, it is likely that the thermal conduction has some effect on the results. Furthermore, the radius of the spot is comparable with the thermal depth, hence the thermal radial conduction may also interfere. Therefore, a two-dimensional model developed at ISL [3] was used to calculate the thresholds for the two conditions.

The laser spot diameters are 225 μm for $\tau = 0.3$ ms, and 240 μm for $\tau = 30$ μs. Further we assume that the optical and thermal parameters remain constant during the process. The model then calculates the power density for which the surface temperature reaches the melting temperature for a predetermined duration.

For $\tau = 0.3$ ms, the calculated power density is $I = 62$ kW cm^{-2} corresponding to a fluence $F = 19$ J cm^{-2}, whereas for $\tau = 30$ μs the power density is $I = 584$ kW cm^{-2} and the fluence $F = 17.5$ J cm^{-2}.

The experimental and theoretical thresholds are listed in Table 2 and compared in Section 5.

4. X-ray microanalysis

It has been reported previously that some mercury losses can occur under laser irradiation of CMT [1, 8]. This is the reason why the damage resulting from irradiation was also investigated by X-ray microanalysis. Our samples had an initial value $x = 0.204 \pm 0.01$. From this value, we derive the following atomic proportions of the constituent elements: Hg 39.8%, Cd 10.2%, Te 50.0%. Our preliminary analysis indeed shows that significant mercury losses occur under laser irradiation.

4.1. Pulsed irradiation

We analysed a crater obtained with a maximum fluence at the centre of the spot $F_{max} = 38$ J cm^{-2}, which is slightly above the threshold value (crater shown in Fig. 1(a)). At the centre of the crater, the

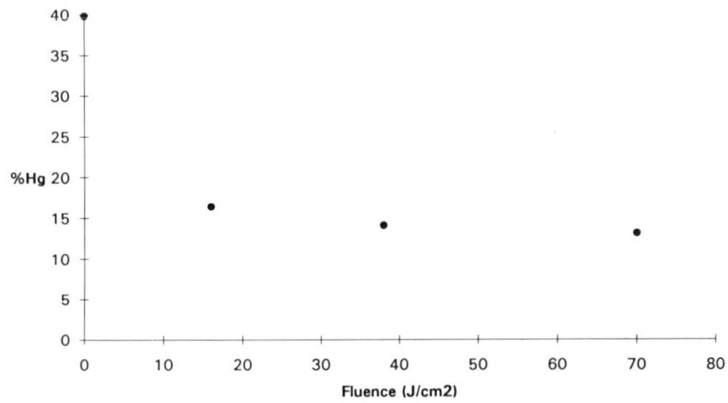

Fig. 4. Variation of mercury content (at.%) as a function of fluence.

mercury content decreased to 14.1% while at the edge of the crater, it was only reduced to 29.3%. On a crater obtained with a maximum fluence of 70 J cm^{-2}, (Fig. 1(b)) the mercury content was 13.1% at the centre and 20.5% at the edge.

4.2. Continuous irradiation

The best shaped groove obtained with a maximum fluence of 16 J cm^{-2} (slightly above the threshold value) was analysed (Fig. 3(b)). The mercury content decreased to 16.4% at the centre of the groove, and to 20.7% at its edge.

This first series of experiments shows that a very large amount of mercury is already outgased for a fluence of 16 J cm^{-2}. Moreover, the mercury losses do not seem to increase further for higher fluences (Fig. 4). In future, it would be interesting to determine the mercury losses at fluences smaller than the melting threshold values.

5. Discussion

The experiments reveal that the melting threshold fluence is smaller than both theoretical estimations. Perhaps other parameters should be taken into account in the assumptions, such as possible variation of the absorption coefficient of the material during heating or possible modification of its thermal and optical characteristics due to the mercury losses during irradiation.

6. Conclusion

The minimum fluence required to obtain surface melting of CMT under high power laser irradiation was determined for two experimental conditions. The experimental values are lower than the theoretical estimations. The reasons for these differences could be possible variations of the properties of the material during laser illumination. X-ray microanalysis revealed the expected mercury losses induced by laser irradiation. They are very important and further work will analyse the composition of the material submitted to fluences lower than the melting threshold value.

References

1 G. Bahir and R. Kalish, *Appl. Phys. Lett., 39* (1981) 730–732.
2 J. L. Schmitt and F. L. Stelzer, *J. Appl. Phys., 40* (1969) 4865–4869.
3 B. Gautier, *Rapport ISL S-R 908/84* (1984) (ISL, BP34 F-68301 Saint-Louis, France).
4 J. A. Mroczkowski and D. A. Nelson, *J. Appl. Phys., 54* (1983) 2041–2051.
5 H. S. Carlslaw and J. C. Jaeger, *Conduction of Heat in Solids*, Oxford University Press, London, 1959.
6 L. R. Holland, in J. Brice and P. Capper (eds.), *EMIS Datareview RN 15633*, EMIS Datareviews series 3, October 1986, pp. 18–19.
7 J. C. Brice, in J. Brice and P. Capper (eds.), *EMIS Datareview RN 15620*, EMIS Datareviews series 3, January 1986, pp. 13–15.
8 A. A. Zaginei, B. K. Kotlyarchuk, G. V. Plyatsko and V. G. Savitskii, *Inorganic Mater., 25* (1989) 932–935.

Morphological and electrical modifications in silicon submitted to high-intensity laser irradiation

Olivier Muller* and René Joecklé

Institut Franco-Allemand de Recherches de Saint-Louis (ISL), BP 34, F68301 Saint-Louis Cedex (France)

Abstract

In an effort to study the nature of the damage processes which occur when detectors for visible light are irradiated with high-power laser beams, bulk (111) n-type (P doped) silicon samples were submitted to high intensity 580 nm dye laser pulses for microsecond duration and the induced morphological (cracks, melting) as well as electrical (sheet resistance) modifications were investigated. The irradiation was either in a variable diameter focused mode, or a uniform power density beam. The uniform illumination mode allows one to obtain more precise values for the morphological degradation occurring with increasing fluence: from the first appearance of a white coating (over 1.7 J cm^{-2}) to the progressive formation of cracks (from 5.5 J cm^{-2} to melting) before melting occurs (at 10.7 J cm^{-2}). A detailed study of the resistivity as a function of fluence is reported, using the latter optical regime, and it is shown that melting is accompanied by a strong increase in sheet resistance (from 50 to 180 kΩ).

1. Introduction

During the past two decades, a great many works have been devoted to study of the interaction between intense laser beams and semiconducting materials. Most of these studies were stimulated by the fact that such an interaction paved the way for a new and rapidly growing technology called laser processing, with important applications in the field of microelectronics. Examples are the annealing of implantation-induced damage [1–3], or the processing of surfaces in terms of laser heating and melting of thin films on substrates [4] or crystallization of amorphous deposited layers [5–7].

Another important issue is study of the damage which is produced by the laser impact itself on semiconducting devices (*e.g.* flat-panels, CCD imaging sensors or photosensor arrays) and the possible resulting modifications of their properties.

Zhang *et al.* observed laser-induced changes in silicon MOS structures [8] as well as laser-induced damage of silicon photosensor arrays [9]; laser-induced damage of silicon CCD imaging sensors was reported by Becker *et al.* [10]; laser-induced degradation and morphological damage have been observed in p–i–n photodiodes by Watkins *et al.* [11] and by Huffaker *et al.* [12].

Laser-induced damage was also studied in crystalline silicon, which is the basic material in detectors for visible light.

Jhee *et al.* showed that first visible damage of silicon occurred at 0.45 J cm^{-2} ($\lambda = 1.06$ μm, $\tau_p = 60$ ps) [13], and it has been reported that the threshold for single-pulse damage of silicon is 1.6 J cm^{-2} ($\lambda = 1.06$ μm, $\tau_p = 5$ ns) [14].

In an effort to investigate the damage processes which occur when detectors for visible light are submitted to a high-power laser beam, we report in this paper the results of experiments on laser-induced morphological and electrical changes in (111) silicon.

The objective of this study was to contribute to our comprehension of the damage mechanism in silicon, going from the first visible surface modifications to the threshold of surface melting.

The paper is organized as follows. In the next section, details are given of the laser source and the optical set-up used to generate uniform or focused illumination. In Section 3, the morphological modifications of the silicon surface are described as a function of the laser fluence. Section 4 reports on preliminary resistivity measurements obtained by the four-point probe method.

2. Experiments

During these experiments, n-type, single-crystal silicon material was examined, with a nominal electri-

*Also with Centre de Recherches Nucléaires, IN2P3-CNRS/ULP, Groupe de Recherches Physiques et Matériaux, BP20, F67037 Strasbourg Cedex 2, France.

0921-5093/93/$6.00

cal resistivity value of 900 Ω cm. ⟨111⟩ oriented samples were cut into rectangular (5×10 mm^2 or 10×10 mm^2) pieces with thicknesses of 200 μm. They were laser irradiated without any specific surface treatment before illumination.

The laser source was a home-designed dye laser (R6G) delivering at 580 nm single pulses (2 μs (FWHM) in duration) of tunable energy up to 0.8 J per pulse. Compared with irradiation used elsewhere [13, 14], it should be noted that the pulse duration is much longer than that of the Q-switched YAG lasers, whereas the shorter wavelength is located inside the sensitivity range of the silicon visible light detector. The beam divergence was about 3 mrad. The spatial profile of the laser beam was measured using a CCD video camera and a laser-beam profiler, for the different types of irradiation used.

The irradiations were performed using two different modes: the uniform mode in which the optical beam is homogenized as much as possible using a suitable optical set-up, and the focused mode which simulates a more realistic situation.

In principle, *the uniform regime* of illumination allows well defined uniformly irradiated areas to be obtained on the semiconducting target, on which further measurements using for example electrical or optical probes can be taken appropriately. The experimental set-up used to obtain such a uniform regime is shown in Fig. 1. It consists of a BK7 glass rod with a square section ($40 \times 3 \times 3$ mm^3) onto which the laser beam is focused, and an output projection lens. The sample environment is arranged in such a manner that an energy meter or a video CCD camera can be used to control both the total energy of the pulse and its spatial profile. Fine optical adjustments are realized with the help of an auxiliary He–Ne laser beam. The position of the output projection lens F2 was chosen to obtain on the target square impact areas with edge 1.27 mm. Given the optical losses through the whole system (*i.e.* 65%) this set-up can provide energy fluences of up to about 12 J cm^{-2} on the target plane.

In the focused regime the optical set-up simply consists of a lens with a suitable focal length to achieve melted zones with diameter in the range 300–800 μm. *In the tightly focused regime* (*i.e.* for melted zone diameters of a few tens of micrometres), the main difficulty occurs when the position of the spot as well as its profile have to be controlled with a high accuracy. As shown in Fig. 2, this control was realized by lighting the target with an auxiliary incoherent source which gives, under specific conditions, the image of the incident laser spot on the CCD camera.

Detailed morphological examinations were subsequently performed using optical (50×200 Nomarski) and scanning electron (SEM) microscopes. Electrical characterizations were exclusively performed on samples irradiated in the uniform regime. We used the standard four-point probe (f.p.p.) method in the in-line equidistant configuration [15].

3. Morphological modifications

Figure 3 shows SEM (a) and Nomarski (b) images of quasi-uniform laser irradiation on silicon. These types of damage are typical of two distinct modifications which appear under increasing energy fluences. The SEM view was observed after irradiation at 8.2 J cm^{-2} and reveals a density of cracks, whereas the optical image (Fig. 3(b)) shows a well defined area where the molten material (10.7 J cm^{-2}) has again solidified.

As a function of the laser fluence, a complete chronology of the illumination-induced damage on

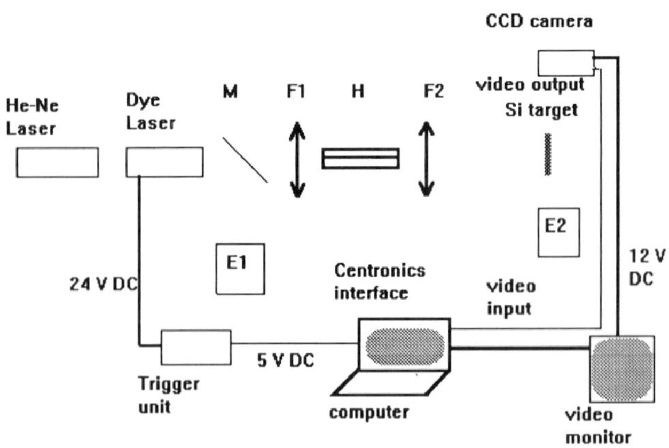

Fig. 1. Schematic diagram of the optical set-up used for uniform illumination of the sample (F1, F2 lenses; H homogenizer; M semi-transparent mirror ($RF/T = 8/92$); E1, E2 energy meters).

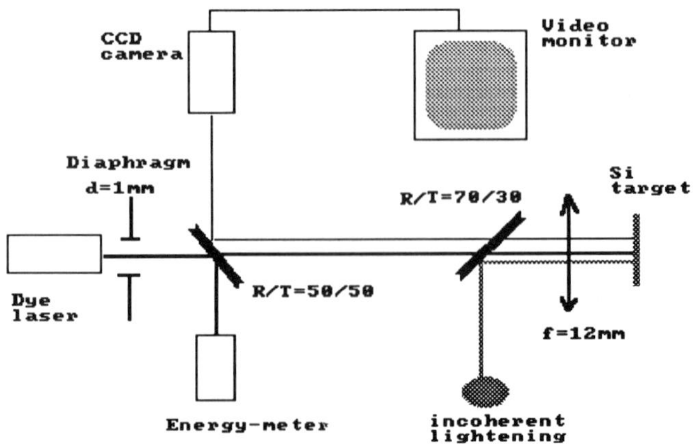

Fig. 2. Schematic diagram of the optical set-up used in the focused regime showing the auxiliary equipment allowing direct visualization of the irradiated area on the silicon sample surface.

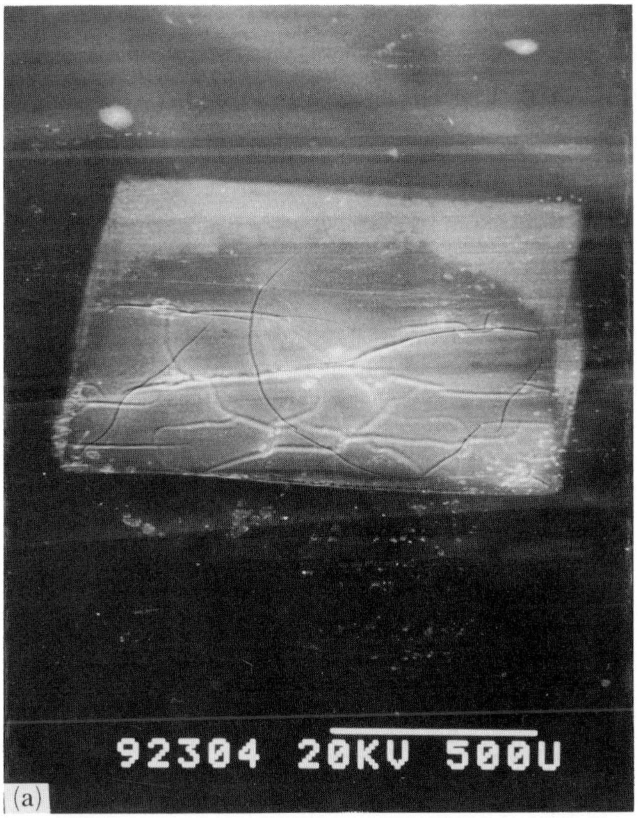

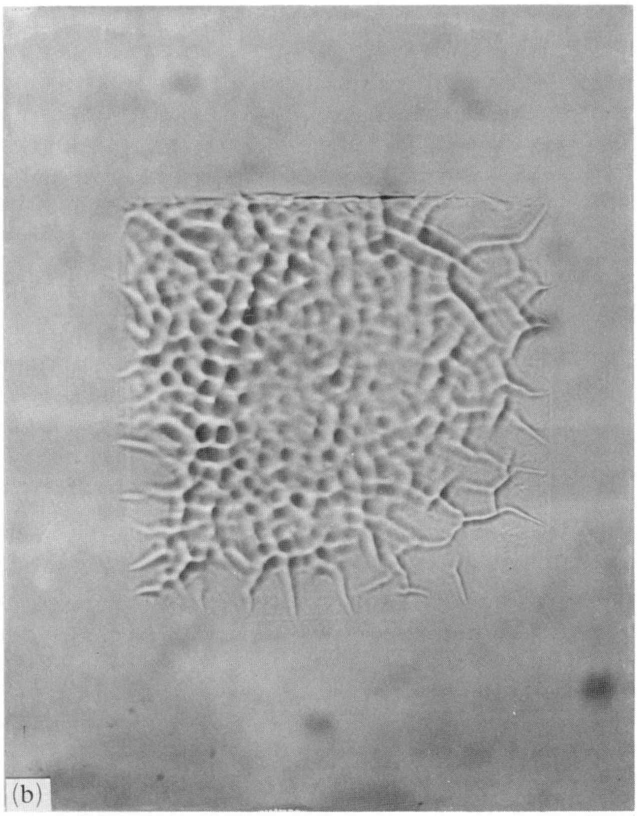

Fig. 3. (a) SEM and (b) Nomarski images showing quasi-uniform laser irradiations realized on n-Si: (a) high density of cracks at 8.2 J cm^{-2}; (b) uniform melting and resolidifying at 10.7 J cm^{-2}.

TABLE 1. Chronology of illumination-induced damage on silicon

Fluence (J cm^{-2})	Damage behaviour and comments
10.7	Uniform melting
8.7–10.4	Local melting
7.4–8.4	High density of cracks
5.5–6.4	Low density of cracks
1.7–4.1	Change in surface colour
<1.5	No visible damage

silicon was established for the uniform regime, where the fluences are more accurately determined. This chronology is reported in Table 1. As shown, it includes meaningful values, for example the threshold of no visible damage or the melting threshold.

Figure 4 shows similar results after illumination of a silicon sample in the focused mode, *i.e.* for melted zone diameters ranging from 300 to 800 μm. In this regime, the diameter of the optical impact at the FWHM of the laser pulse was found to vary with the total energy of the pulse: from about 400 μm at the lowest energy value to about 1600 μm for the highest value. Figure 4(a) shows an optical micrograph of a superficial crack that appeared at 8 J cm^{-2}; the angle between the linear traces is 120°. This characterizes the behaviour of (111) silicon surfaces and suggests that linear damage develops preferentially along dislocations, as reported by Birnbaum [16]. These cracks occur well below the melting threshold; moreover, the irradiation geometry is unidimensional, meaning that the temperature gradients and the related stresses are only important in the thickness direction. In that case, compressive stresses develop at the surface but are unable to produce cracks because the compressive yield strength of silicon is fairly high. The observed increase in density of cracks suggests that these cracks are initiated on particular sites, corresponding either to defects or dislocations at the surface, where small melted pits occur for low fluences, or to local non-uniformities of the laser beam, yielding local overheated zones, where melting can occur. During the cooling phase, strong tensile stresses develop around these sites and the observed cracks can develop.

The SEM image of Fig. 4(b) shows details of a crater obtained on the sme material after a more energetic (16 J cm^{-2}) irradiation. At the periphery of the melted zone, a ripple pattern is clearly visible. Such periodic ripples have also been observed by other authors [13, 14, 17–21], who explain that they are due to a constructive interference effect occurring between surface waves and the incident laser beam; thus the distance between two adjacent ripples should mirror the wave-

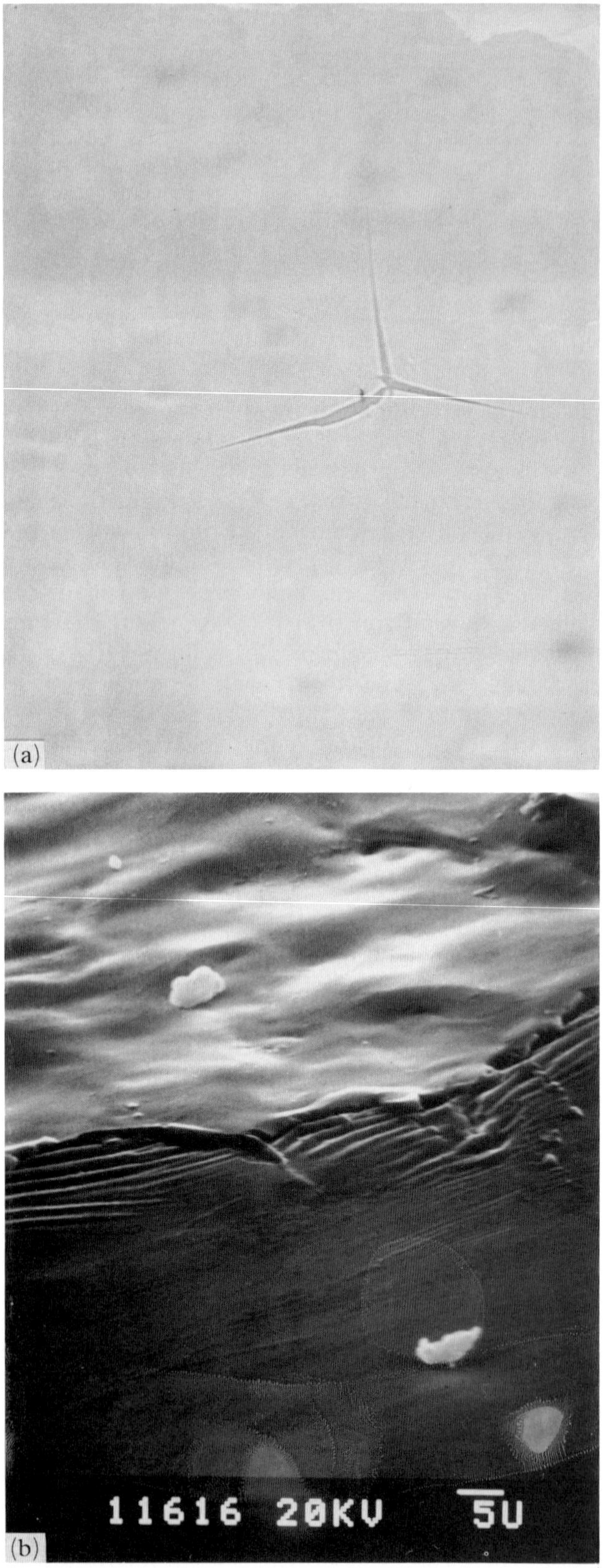

Fig. 4. (a) Nomarski and (b) SEM images of single-pulse damage observed with increasing energy fluence on n-Si targets: (a) superficial crack occurring at 8 J cm^{-2}; (b) ripple pattern at the periphery of the melted zone at 16 J cm^{-2}.

length of the laser beam, *i.e.* 0.58 μm. Measurements were taken from different SEM images, giving for this distance a value slightly lower than 1 μm, which could be considered in good agreement with theory [20].

In the tightly focused regime, *i.e.* for melted zone diameters of the order of 20 μm, experiments were performed at low deposited energies limited typically to 1 or 2 mJ per pulse. The SEM images shown in Fig. 5 are representative of what is observed for a single pulse of 0.65 mJ. As shown, the figure looks quite symmetric with a mean diameter of 20 μm and a characteristic cone-shaped protuberance in its centre. In this case the deposited energy fluence was of the order of 210 J cm^{-2}. Protuberances located in the centre of the laser beam impact have also been observed by other authors [22, 23] and their origin is to be found in the positive difference between the density of silicon in its liquid and solid phases [22]. The particular shape observed in Fig. 5 is believed to occur as a consequence of the very small dimension of the actual laser spot combined with the high deposited fluence.

4. Electrical characterization

4.1. Electrical measurements

As mentioned above, the four-point probe method was applied to samples processed under the uniform irradiation regime. Furthermore, it was assumed that the film being characterized reduces to the thin processed layer, the thickness of which is not known. Given these assumptions, the sheet resistance of the film is given by the following relation [15] (in-line equidistant configuration),

$$R_{\square}=(\pi/\ln 2)U/I$$

where U is the open-voltage measured between the inner points, for a current I flowing through the outer points.

Our results are reported in Fig. 6 as a function of the laser fluence. In the investigated range of fluence, the sheet resistance of the sample has a rather constant value of the order of 50 kΩ at low fluence values, and shows a sudden increase for irradiation fluences higher than 11 J cm^{-2} which, according to Table 1, corresponds to the melting threshold of this material.

4.2. Discussion

Such an increase in the sheet resistance value could result from a decrease in the residual free-carrier concentration and/or in their mobility in the processed layer. This was the hypothesis of Kamins [24] and Seto [25] for the analysis of their Hall-effect measurements on laser-irradiated silicon. These authors assumed that

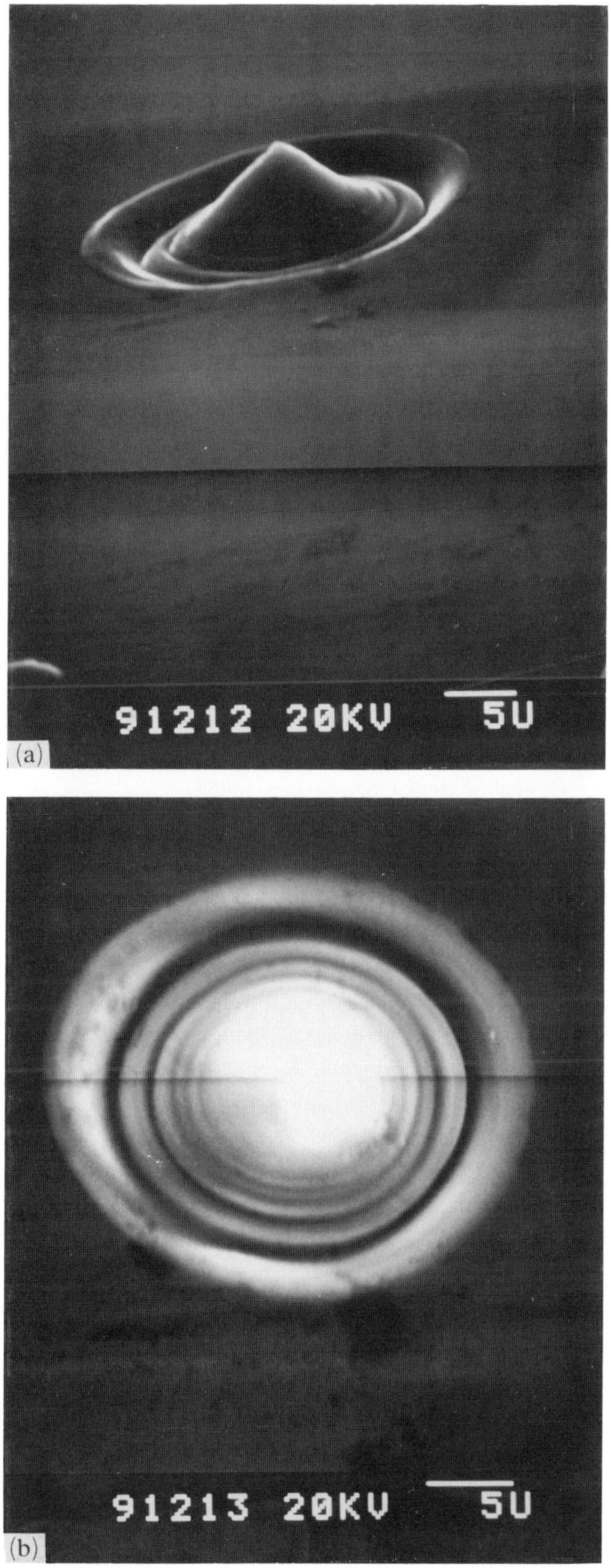

Fig. 5. SEM views of single-pulse damage on the surface of n-Si in the tightly focused case: (a) cone-shaped protuberance rising at 210 J cm^{-2}; (b) top view of the same impact.

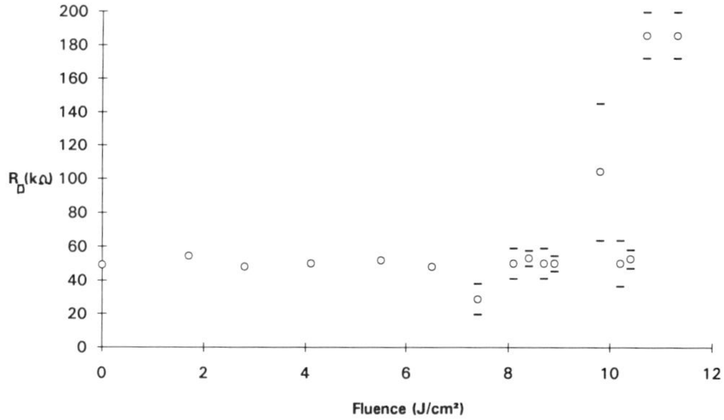

Fig. 6. Sheet resistance *vs.* incident laser energy fluence for n-Si.

under irradiation the free-carriers are trapped by voids or dangling bonds, thus giving rise to a barrier potential which reduces strongly both their population and their mobility.

However, taking into account that in the present case the irradiation was performed in ambient atmosphere, it is also possible that the increase in resistivity observed on n-Si for fluences higher than the melting threshold, is due to an oxidation process which occurs during resolidification of the molten area.

Further experiments, such as resistivity measurements of silicon samples preliminarily etched and illuminated in a neutral atmosphere, and other measurements using more doped silicon will be undertaken to understand the reason for such an increase in sheet resistance value.

5. Summary

Bulk (weakly) n-type silicon samples were submitted to high-power visible (580 nm) laser pulses of 2 μs in duration with a view to describing the induced morphological and, possibly electrical, modifications. By working in the uniform illumination regime, in which the deposited energy fluence is determined with good accuracy, a typical test-list of the observed damage was established as a function of the irradiation fluence. The test-list includes the following: a change in surface colour, the appearance of an increasing density of cracks, local and finally uniform melting with subsequent resolidification.

In these experiments, the sheet resistance was also measured as a function of the laser fluence. No noticeable variation was observed, excepted for fluences higher than the melting threshold where an oxide layer is probably grown at the surface of the resolidified volume.

Acknowledgment

We would like to acknowledge B. Prevot for helpful discussion.

References

1 J. C. Muller, A. Grob, J. J. Grob, R. Stuck and P. Siffert, *Appl. Phys. Lett., 33* (1978) 287.
2 A. Gat, J. F. Gibbons, T. J. Magee, J. Peng, V. R. Deline, P. Williams and C. A. Evans, Jr., *Appl. Phys. Lett., 32* (1978) 276.
3 W. A. Porter, D. L. Parker, T. Wm. Richardson and E. J. Swenson, *Appl. Phys. Lett., 33* (1978) 886.
4 R. A. Ghez and R. A. Laff, *J. Appl. Phys., 46* (1975) 2103.
5 J. C. Bean, H. J. Leamy, J. M. Poate, G. A. Rozgonyi, T. T. Sheng, J. S. Williams and G. K. Celler, *Appl. Phys. Lett., 33* (1978) 227.
6 J. Feinleib, J. de Neufville, S. C. Moss and S. R. Ovshinsky, *Appl. Phys. Lett., 18* (1971) 254.
7 E. Fujii, K. Senda, F. Emoto, A. Yamamoto, A. Nakamura, Y. Uemoto and G. Kano, *IEEE Trans. Electron. Devices, 37* (1990) 121.
8 C. Zhang, S. E. Watkins, R. M. Walser and M. F. Becker, Laser-induced changes in the electrical performance of silicon MOS device structures, *Proc. 20th Annu. Symp. on Optical Materials for Higher Power Lasers, NIST (NBS) Special Publication 775*, NIST, Boulder, CO, 1989, p. 105.
9 C. Zhang, T. Benchetrit, S. E. Watkins, R. M. Walser and M. F. Becker, Laser-induced damage to silicon photosensor arrays, *Proc. 20th Annu. Symp. on Optical Materials for Higher Power Lasers, NIST (NBS) Special Publication 775*, NIST, Boulder, CO, 1989.
10 M. F. Becker, C. Zhang, S. E. Watkins and R. M. Walser, *SPIE, 1105* (1989) 68.
11 S. E. Watkins, C. Zhang, R. M. Walser and M. F. Becker, *Appl. Opt., 29* (1990) 827.
12 D. L. Huffaker, R. M. Walser and M. F. Becker, Correlation of surface topography and coating damage with changes in the responsivity of silicon PIN photodiodes, *Proc. 21st Annu. Symp. on Optical Materials for Higher Power Lasers, NIST (NBS) Special Publication*, NIST, Boulder, CO, 1990.
13 Y. K. Jhee, M. F. Becker and R. M. Walser, *J. Opt. Soc. Am. B, 2* (1985) 1626.
14 M. F. Becker, F. E. Domann, A. F. Stewart and A. H. Guenther, Charge emission and related precursor events associated with laser damage, *15th Annu. Symp. on Optical Materials for High Power Lasers, NIST (NBS) Special Publication, 668*, NIST, Boulder, CO, 1984.
15 L. B. Valdes, *Proc. IRE, 42* (1954) 420.
16 M. Birnbaum, *J. Appl. Phys., 36* (1965) 3688.
17 Y. Jee, M. F. Becker and R. M. Walser, *J. Opt. Soc. Am. B, 5* (1988) 648.
18 A. L. Huang, M. F. Becker and R. M. Walser, *Appl. Opt., 25* (1986) 3864.
19 D. C. Emmony, S. E. Clark, N. C. Kerr and B. A. Omar, *SPIE, 1397* (1990) 651.
20 J. F. Young, J. S. Preston, H. M. van Driel and J. E. Sipe, *Phys. Rev. B, 27* (1983) 1155.
21 H. J. Leamy, G. A. Rozgonyi and T. T. Sheng, *Appl. Phys. Lett., 32* (1978) 535.
22 D. C. Emmony, N. J. Phillips, J. H. Toyer and L. J. Willis, *J. Phys. D, 8* (1975) 1472.
23 M. Bertolotti, F. de Pasquale, P. Marietti, D. Sette and G. Vitali, *J. Appl. Phys., 38* (1967) 4088.
24 T. I. Kamins, *J. Appl. Phys., 42* (1971) 4357.
25 J. Y. W. Seto, *J. Appl. Phys., 46* (1975) 5247.

Materials Science and Engineering, *A168* (1993) 87–92

Materials characterization at high temperatures using laser ultrasound

M. Paul, B. Haberer and W. Arnold

Fraunhofer Institute for Nondestructive Testing (IzfP), Bldg. 37, University, W-6600 Saarbrücken 11 (Germany)

Abstract

A new technique is presented allowing measurement of ultrasonic absorption up to very high temperatures in polycrystalline and other materials exhibiting strong scattering, by generating reverberation fields by short laser pulses, and by observing their decay using optical interferometers for detection. Furthermore, the elastic constants in the material examined can be evaluated by measuring the time-of-flight of the ultrasonic pulses. Ultrasonic absorption data and elastic constants in aluminum, SiC ceramic, and white cast iron are presented as a function of temperature up to 1400 K.

1. Introduction

Ultrasonic methods have been successfully applied for many years in non-destructive testing [1], physical acuostics [2], and solid state physics [3]. It is not only possible to detect flaws such as cracks, pores, and inclusions by direct or indirect reflection of ultrasound, but also to obtain information on the elastic and the inelastic properties, microstructure, and other properties describing the state of the material under investigation. A standard technique for such measurements in the megahertz range is the pulse-echo method [2]. The quantities of interest are then the time-of-flight of the pulse from which group and phase velocity can be derived and the decrease in its amplitude as it travels through the material, *i.e.* the ultrasonic attenuation as described by the coefficient α in Beer's law. From the time-of-flight of transverse and longitudinal waves, the sound velocities can be easily calculated, and hence the elastic moduli [4] and their changes due to, for example, internal stresses, inhomogeneities, second phases and temperature.

The ultrasonic attenuation contains mainly two types of losses, elastic losses which conserve the mechanical energy in the sample volume, and inelastic losses where the mechanical energy is dissipated, *i.e.* transferred into heat. For single crystals and isotropic media, the elastic losses are mainly due to diffraction because of the finite aperture of the transducer, and due to non-parallelism of the sample faces [3]. Inelastic losses result from the anelastic behavior of a material caused by dislocations, point defects, conduction electrons, thermoelasticity, magnetoelasticity, and other mechanisms [3, 5, 6]. These losses are all associated with a change in velocity and hence in elastic constants. Most of these mechanisms can be described as relaxations of a standard linear, viscoelastic solid [7]. Hence, ultrasonic measurements give information on the elastic and anelastic interaction mechanisms, and on the relationship between elastic moduli and absorption.

In polycrystalline materials an important elastic loss occurs when the ultrasound is scattered by grain boundaries owing to the change in acoustic impedance from grain to grain. Ultrasonic scattering has been studied intensively [8], and if the wavelength is much larger than the grain size, the attenuation coefficient α_S due to scattering can be described by Rayleigh's law:

$$\alpha_S = Sd^3f^4 \qquad (1)$$

Here, S is the scattering parameter, d is the grain size, and f is the ultrasonic frequency. The total attenuation coefficient α can then be expressed as

$$\alpha = \alpha_G + \alpha_A + \alpha_S \qquad (2)$$

Here, α_G stands for the attenuation coefficient representing geometric losses and α_A for the attenuation coefficient for inelastic losses. Using ultrasonic backscattering measurements and exploiting eqn. (1), one can determine the grain size non-destructively and characterize the microstructure [8], but such measurements rely on accurate knowledge of α_A.

Recently, several methods for absorption measurements in scattering media have been suggested by observing the diffusion of scattered ultrasound [9], the IR radiation caused by the absorption [10], and damping of the resonance of small cylindrical samples [11]. Another method of direct measurement of absorption is the reverberation technique. In this technique, the property that an incident ultrasonic

0921-5093/93/$6.00

beam loses its coherence by scattering is exploited [12]. After a certain time, the energy of the beam will be distributed over the volume of the sample, and the temporal decay of this "homogeneous diffuse field" will only be determined by absorption. The distribution of ultrasound and its decay correspond to the reverberation of sound in rooms. Furthermore, it has been demonstrated that by oblique insonification of an aluminum plate, the generation of a reverberant field is possible in a weak scattering medium solely owing to sidewall reflections and mode conversions in analogy to room acoustics [13]. In addition, the energy distribution of the different wave types in a diffuse field has been calculated showing that the shear modes are dominant [14]. The ratio of the energy content of the transverse and the longitudinal waves in the reverberation field is constant and does not depend on the volume of the sample whereas the energy content of the surface waves varies with the sample volume and with frequency. Because all modes contribute to the reverberation field, the absorption coefficient α_A can only be given in units per time and not per distance. If the reverberation fields are generated by contact transducers, their coupling causes an additional damping which becomes increasingly important with increasing ratio of transducer area to sample volume. Hence, an absolute measurement of the absorption in small samples is no longer reliable.

A non-contact technique for such measurements would avoid the above mentioned difficulties allowing one to carry out measurements not only at room temperature but also at very high temperatures. In this paper, we report on a technique allowing such measurements, and our intentions are twofold. Firstly, we want to show the capability of laser ultrasonics not only for accurate velocity measurements as demonstrated earlier [15] but also to obtain reliable data on the absorption coefficient α_A in technical materials. For example, such data are necessary for grain-size measurements (see eqns. (1) and (2)) [8, 16]. Secondly, we want to use α_A as a parameter for material characterization. We measured the absorption and the elastic moduli of fine grained aluminum, ceramic silicon carbide, white cast iron exhibiting a phase transition near 480 K, and several other materials.

2. Generation of ultrasound by laser pulses and interferometric detection

When a solid is irradiated by short laser pulses, broadband ultrasonic pulses are launched [17] allowing ultrasound to be remotely excited. Depending on the optical power of the laser pulses, one can distingush two different generation mechanisms. In the thermoelastic regime where the laser power density is below the material ablation threshold, pressure, shear and surface waves are generated simultaneously as a result of stresses induced by thermal expansion of the irradiated volume. At higher power densities, material ablation changes the mechanism. In this case, a plasma is formed close to the sample surface, resulting in an increased pressure wave amplitude as a result of momentum transfer due to the plasma pressure. As discussed earlier, the generated waveform depends on the laser intensity [18].

The interferometric detection of ultrasound can also be peformed over large distances. Therefore, in conjunction with interferometric detection, laser ultrasonics is suitable for application at very high temperatures. Interferometers either measure the phase modulation directly or measure the Doppler shift of the beam reflected off the surface which is displaced by ultrasound. Low frequency modulation due to instabilities such as ambient vibrations or thermal fluctuations can be suppressed by several techniques. Laser interferometers are less sensitive than piezoelectric, capacitive or electromagnetic transducers but are suitable for non-contact and remote detection. There are different types of interferometers with typical sensitivies of 10^{-5} nm Hz$^{-1/2}$ at 1 mW collected light power. A review of inteferometric measurement techniques can be found in ref. 19.

3. Experimental arrangement

Figure 1 shows the experimental set-up used. The laser pulse for generation is directed through a hole of the furnace and focused onto the sample. The beam of

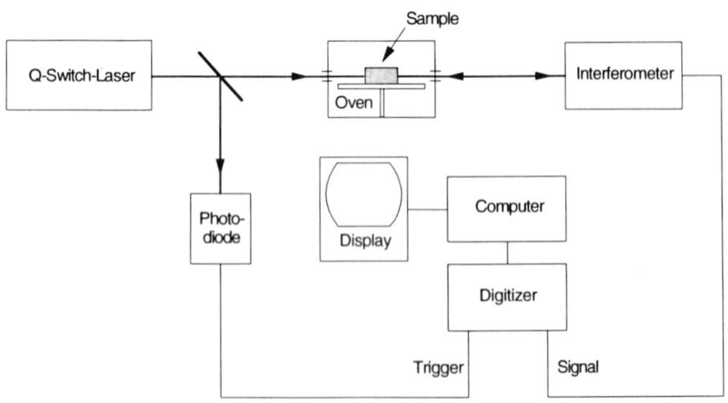

Fig. 1. Principle of the experimental arrangement. A laser pulse impinging on the surface of the sample generates broadband ultrasonic pulses containing both P- and S-type signals. After traveling through the sample, they are detected on the other side of the sample by an interferometer of suitable design and bandwidth. The sample is placed in an oven to obtain data up to a temperature of 1400 K.

the interferometer impinges onto the opposite and polished surface of the sample through another hole in the front door. The ultrasonic signals are detected by the interferometer, demodulated, stored in a digital transient recorder, and then processed by a computer.

For generation of the ultrasonic signals, an Nd-YAG laser at a wavelength of 1064 nm with maximal energy of 450 mJ and about 10 ns pulse duration was used. The beam diameter of 7 mm was focused to 3 mm and 1 mm on the sample surface for absorption and velocity measurements respectively. For detection we used a heterodyne Michelson interferometer which is stabilized against vibrations [20].

The velocity measurements were carried out at power densities between 40 and 60 MW cm^{-2} and the absorption measurements at power densities from 400 to 600 MW cm^{-2}. In the first case, the thermoelastic effect prevailed resulting in an ultrasonic waveform containing P- and S-components of comparable amplitude (see Fig. 2). Here, P and S stand for signals traveling with the velocity c_L of longitudinal waves and c_T of transverse waves respectively. In the second case, a reverberation field of large amplitude was generated (Fig. 3). For absorption measurements at a set frequency, the demodulated signal passes a variable narrow-band logarithmic amplifier. Depending on the signal-to-noise ratio of the reverberation field in a given sample, absorption data can be taken from 200 kHz to about 15 MHz. In order to improve the signal-to-noise ratio, the signals were averaged several times.

Three samples of different material, size and shape were chosen for measurements: (i) a disc of aluminum AlMg3 5 cm in diameter and 1 cm thick with a grain size of about 20 μm and containing 3 wt.% Mg inclusions (size 3 μm); (ii) a ceramic SiC plate of about 5 μm grain size with dimensions $4.7 \times 4.8 \times 0.5$ cm^3; (iii) a piece of white cast iron containing 57.6 wt.% cementite with dimensions $2.0 \times 1.4 \times 0.5$ cm^3. The samples were supported in the furnace on ceramic rings. The maximal temperature of the furnace was 1400 K. The temperature was measured with a thermocouple near the sample. During a given absorption measurement, temperature variations of ± 2 K were possible. For the absorption measurements, signals excited at different locations in one sample (usually along a line) were averaged. This results in a decay pattern which is more easily evaluated by linear regression or manually. Additionally, the signals were sometimes digitally filtered in order to smooth them out.

For the velocity measurements, the exciting beam was directed at the center of the sample surface in order to avoid side-wall reflections and the ensuing mode conversions. The interferometer beam was adjusted on the acoustical axis by minimizing the arrival time of the first pressure pulse. From the time delay of two pressure pulses, the longitudinal velocity as well as the starting point of the measurement could be determined with an accuracy better than 0.5%. From this starting point, the arrival time of the shear component was then measured. The minimal time-step was equal to the sampling time of 2.5 ns of the digitizing oscilloscope. The error of the measurements was

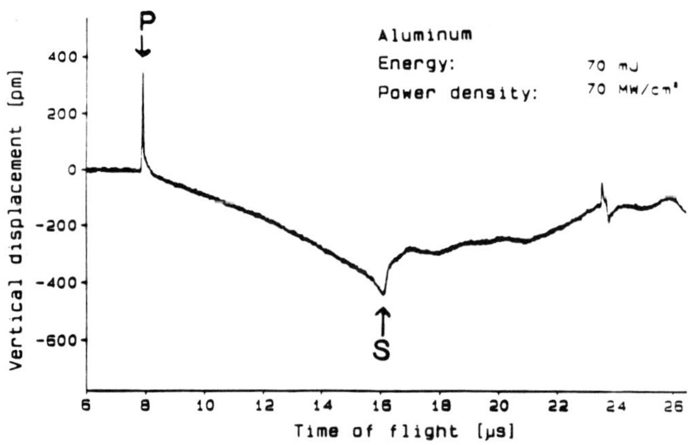

Fig. 2. Ultrasonic waveform obtained by laser-ultrasonics in an aluminum sample (optical intensity 70 MW cm^{-2}). Signals traveling with longitudinal velocity c_L and signals traveling with transverse velocity c_T are generated. From the time-of-flight of the P- and S-signals, the shear modulus G, Young's modulus, E, and Poisson's ratio μ are determined using the following formulae:

$$G = \rho c_T^2$$

$$E = \rho c_T^2 \frac{3(c_L/c_T)^2 - 4}{(c_L/c_T)^2 - 1}$$

$$\mu = \frac{(c_L/c_T)^2}{2(c_L/c_T)^{2-2}}$$

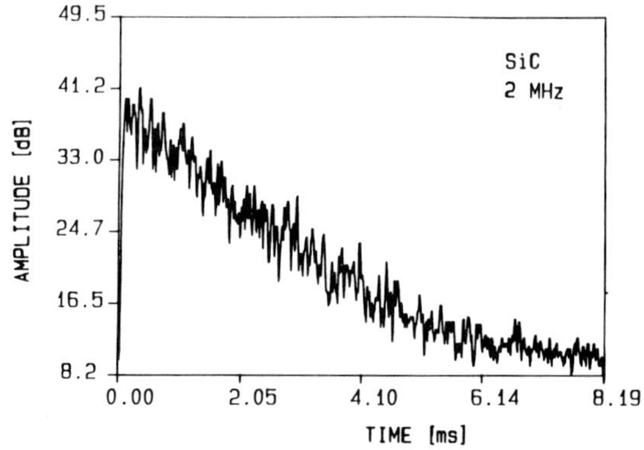

Fig. 3. Decay of a reverberation field obtained in SiC at 2 MHz at room temperature. As can be seen, the decay time is approximately 4 ms. Assuming that transverse waves are dominant in the reverberation field, this corresponds to a path-length of 30 m.

not determined by the sampling interval but mainly by the inaccuracies in determining the point of arrival of the S-pulse leading to a maximal error of about 3% in the moduli.

4. Results

Figure 4 displays the elastic moduli and the Poisson ratio μ of an aluminum sample as a function of temperature up to 830 K. The shear modulus G and the Young modulus E decrease with increasing temperature whereas the Poisson ratio increases. Note that the relative change $\Delta G/G$ is larger than the relative change $\Delta E/E$. In the same material we measured the absorption at frequencies of 0.5 MHz, 3 MHz, and 5 MHz respectively. Figure 5 displays the logarithmic decrement δ ($\delta = \alpha/f$) as a function of temperature up to 730 K. As can be seen, δ increases dramatically above 600 K. Figure 6 shows a logarithmic plot of δ *vs.* inverse temperature $1/T$. For small values of $1/T$ an almost linear decrease in δ can be observed.

In silicon carbide, measurements of the elastic constants and the absorption were taken up to a temperature of 1370 K. The temperature dependence of the elastic constants, and the logarithmic decrement δ measured at 0.5 MHz and 3 MHz are shown in Figs. 7 and 8 respectively. The logarithmic decrements first decrease with increasing temperature and exhibit a minimum of about 580 K. δ then increases again, apparently passing through a maximum. Note that the relative change in elastic constants over the temperature range covered is only about 5%.

In addition, we measured the ultrasonic absorption at 5 MHz in white cast iron containing 57.6 wt.%

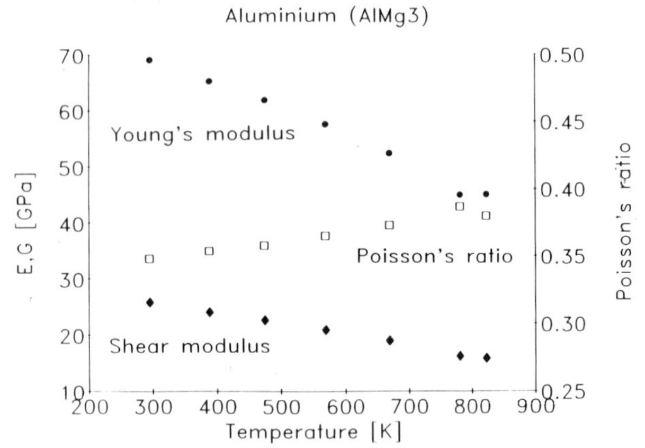

Fig. 4. Elastic moduli and Poisson's ratio μ of an aluminum sample as a function of tempeature up to 850 K. The relative change in G is larger than that in E, leading to an increase in μ.

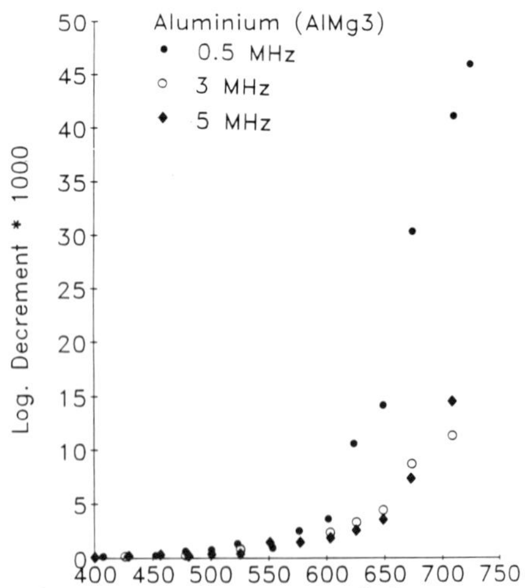

Fig. 5. Logarithmic decrement δ as a function of temperature in aluminum for three frequencies. A strong increase in δ can be observed for temperatures beyond 600 K.

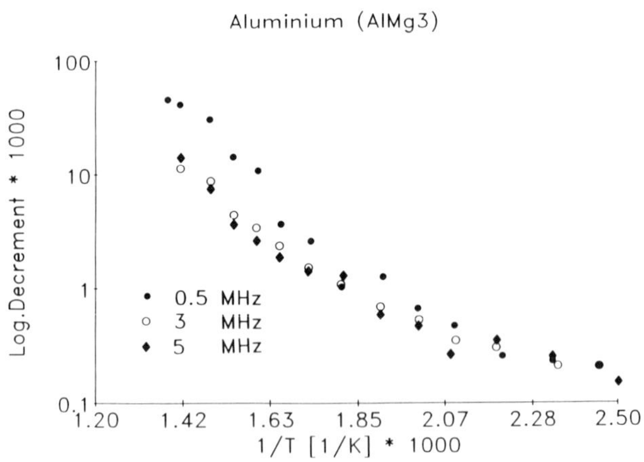

Fig. 6. Logarithmic decrement δ as in Fig. 5 but as a function of inverse temperature $1/T$. The graph shows a linear dependence indicating a thermally activated process with an activation energy of 0.6 eV.

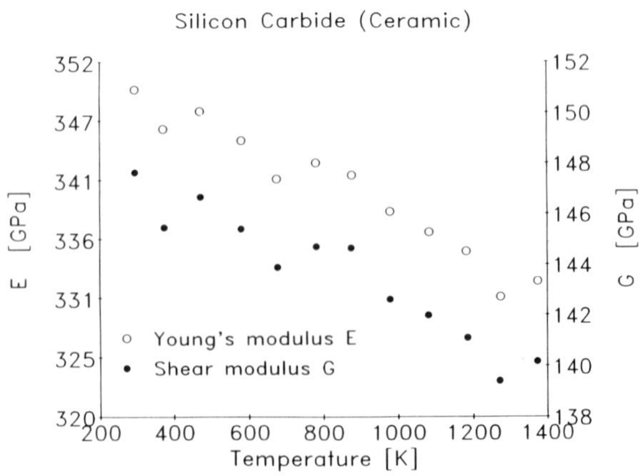

Fig. 7. Elastic moduli E and G in silicon carbide as a function of temperature up to 1400 K.

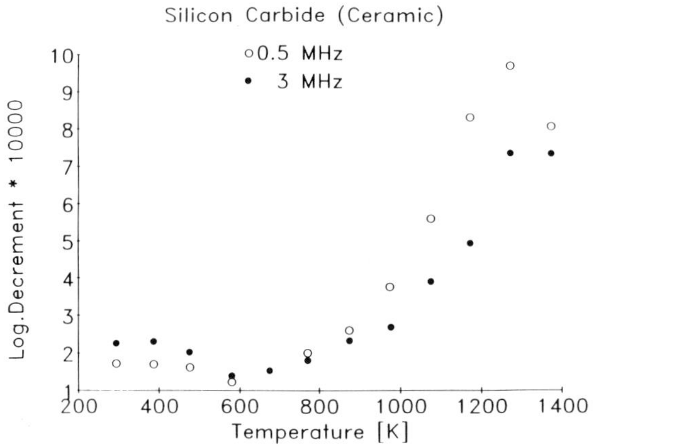

Fig. 8. Logarithmic decrement δ in silicon carbide measured at 0.5 and 3 MHz from room temperature up to 1400 K. There is a clear minimum in δ near 600 K.

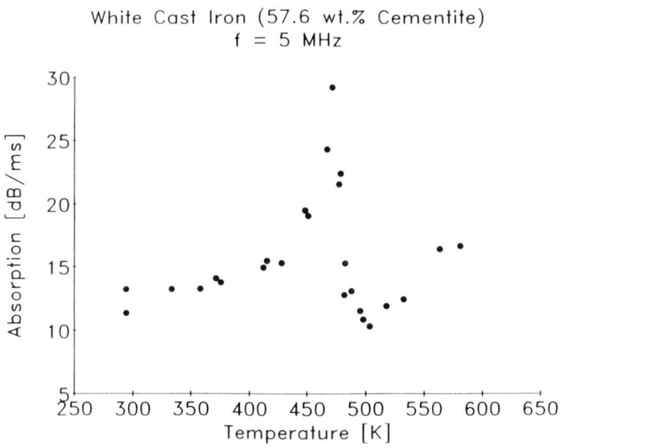

Fig. 9. Ultrasonic absorption at 5 MHz as a function of temperature in white cast iron. The absorption shows a narrow maximum at 480 K typical of a phase transition.

cementite (Fig. 9). A sharp maximum is obtained at 480 K, indicating the occurrence of a phase transition which is also reflected in the data for the elastic moduli as a function of temperature (Fig. 10). A sharp decrease in E occurs again at 480 K. G shows similar behavior, however, the magnitude of the effect is much reduced. Figure 11 displays the frequency dependence of the logarithmic decrement at room temperature.

5. Discussion

If relaxation mechanisms are at the origin of absorption processes, the following relationship holds for the logarithmic decrement δ for a viscoelastic solid:

$$\delta = \pi \frac{\Delta M}{M} \frac{f/f_0}{1+(f/f_0)^2} \tag{3}$$

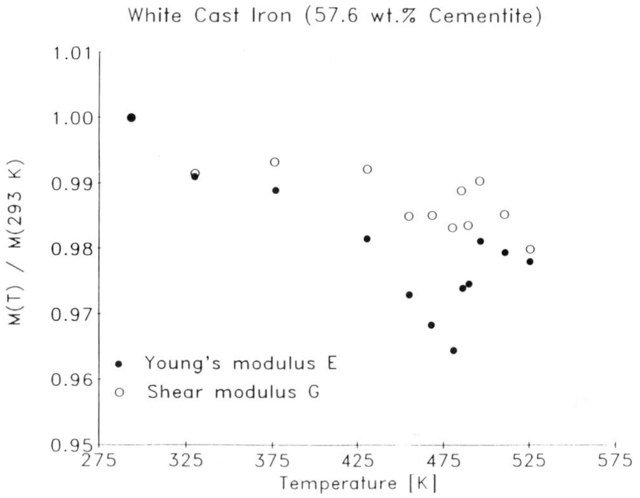

Fig. 10. Corresponding change (see caption of Fig. 9) in the elastic moduli. As can be seen the change in Young's modulus at the phase transition is much stronger than the change in shear modulus G.

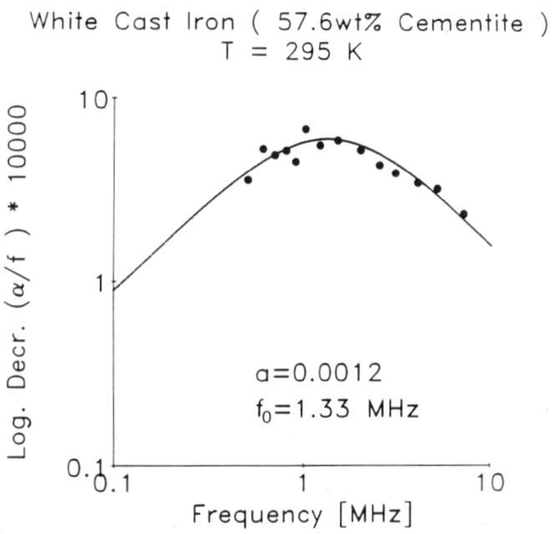

Fig. 11. Logarithmic decrement of white cast iron as a function of frequency from 0.5 to 9 MHz at room temperature. A clear maximum is discernible. In terms of relaxation behavior of a viscoelastic solid, a best fit to eqn. (3) is obtained with a relaxation strength of $\pi\Delta M/M = 1.2\times10^{-3}$ and a relaxation frequency $f_0 = 1.33$ MHz (solid line).

ΔM is the difference between the relaxed and unrelaxed ("anelastic" and "elastic") modulus, M is the unrelaxed modulus, and $f_0 = (2\pi\tau)^{-1}$ is the relaxation frequency. For a thermally activated process, τ follows the relationship $\tau = \tau_A \exp(H/k_B T)$. Here, H is the activation enthalpy, and k_B is the Boltzmann constant. We discuss the measured absorption values in terms of relaxation processes.

In the aluminum sample, the steep increase in α_A around 600 K is accompanied by a decrease in the elastic moduli as a function of temperature which is stronger than linear. Assuming a thermally activated process, one finds an activation energy of about 0.60 ± 0.15 eV by fitting eqn. (3) to the temperature dependence of δ (see Fig. 6). Internal friction measurements at a frequency of 67 mHz [21] revealed similar behavior. The relaxing species were interpreted as vacancy diffusion promoted by dislocations [22].

In silicon carbide, the activation energy is 0.4 eV for the absorption processes exhibiting a maximum near 1300 K (see Fig. 8) with a relaxation strength of $\pi\Delta M/M \approx 2 \times 10^{-3}$.

We presume that the absorption and elastic moduli data for the white cast iron specimen can be interpreted in terms of the phase transition which the cementite phase undergoes from the ferromagnetic to the paramagnetic state at 480 K. The data for both absorption and moduli show behavior typical of second-order phase transitions investigated by ultrasound. Since the change in longitudinal sound velocity is more pronounced, compared with the shear velocity, the coupling mechanism is volume magnetostriction [5]. The data of Young's modulus are in good agreement with the results obtained by a flexural vibration method in a sample of comparable cementite content [23]. Furthermore, the absorption data as a function of frequency in the ferromagnetic state (Fig. 11) can be interpreted as a magnetomechanical relaxation due to the movement of the domain walls in the ultrasonic strain field.

6. Conclusion

We have shown that laser generated ultrasound combined with its interferometric detection offers the possibility of material characerization even at high temperatures. The measurement of reverberation allows new access to accurate data on ultrasonic absorption. By combining in the interpretation the results for ultrasonic velocity and absorption, information which depends on the polarization of the ultrasonic waves in the reverberation field can be discerned, as was demonstrated in the measurement of the phase transition.

Acknowledgment

We should like to thank H. Willems for helpful discussions and we are grateful to the European Commission on Coal and Steel for financial support of part of this work through various research grants.

References

1 J. Krautkrämer and H. Krautkrämer, *Ultrasonic Testing of Materials*, Springer, Berlin, 1977.
2 H. J. McSkimin, in W. P. Mason (ed.), *Physical Acoustics*, Vol. 1A, Academic Press, New York, 1964.
3 R. Truell, C. Elbaum and D. Chick, *Ultrasonic Methods in Solid State Physics*, Academic Press, New York, 1969.
4 H. D. Tietz, *Ultraschallmeßtechnik*, VEB Verlag Technik, Berlin, 1969.
5 H. S. Nowick and B. S. Berry, *Anelastic Relaxation in Crystalline Solids*, Academic Press, New York, 1972.
6 R. DeBatist, *Internal Friction of Structural Defects in Crystalline Solids*, North-Holland, Amsterdam, 1972.
7 C. Zener, *Elasticity and Anelasticity in Metals*, University of Chicago Press, Chicago, IL, 1948.
8 K. Goebbels, in R. S. Sharpe (ed.), *Research Techniques in NDT*, Vol. IV, Academic Press, London, 1980, p. 87, and references cited therein.
9 C. B. Guo, P. Höller and K. Goebbels, *Acoustica, 59* (1985) 112.
10 J. P. Monchalin and J. F. Bussiere, in C. O. Ruud and R. E. Green (eds.), *Non-Destructive Methods for Material Property Determination*, Plenum, New York, 1984, p. 289.
11 M. Deka and N. Eberhardt, in C. O. Ruud and R. E. Green (eds.), *Non-Destructive Methods for Material Property Determination*, Plenum, New York, 1984, p. 135.
12 H. Willems, in D. O. Thompson and D. E. Chimenti (eds.), *Review of Progress in QNDE*, Plenum, 1987, p. 473.
13 R. L. Weaver, in J. D. Achenbach and Y. Rajapakse (ed.), *Solid Mechanics Research for QNDE*, Martinus Nijhoff, Dordrecht, 1987, p. 425.
14 R. L. Weaver, *J. Acoust. Soc. Am., 71* (1982) 1608.
15 J. D. Aussel and J. P. Monchalin, *Ultrasonics, 27* (1989) 165.
16 M. Paul, H. Peukert and W. Arnold, *Proc. 12th World Conf. on NDT, Amsterdam, 1989*, Elsevier, Amsterdam, 1989, p. 261.
17 C. B. Scruby and L. E. Drain, *Laser Ultrasonics—Techniques and Applications*, Adam Hilger, Bristol, 1990.
18 S. Faßbender, B. Hoffmann and W. Arnold, *Mater. Sci. Eng., A122* (1989) 37.
19 J. P. Monchalin, *IEEE Trans. Ultrason., Ferroelectron. Freq. Contr., 33* (1986) 485.
20 M. Paul, *Diploma Thesis*, Physical Department, University of Saarbrücken and IzfP Saarbrücken, 1987, unpublished.
21 J. Friedel, C. Boulanger and C. Crussard, *Acta Metall., 3* (1955) 380.
22 B. Escaig, *Acta Metall., 10* (1962) 829.
23 A. Kagawa, T. Okamoto and H. Matsumoto, *Acta Metall., 35* (1987) 797.

Elastic modulus and interface stress constraint of particle-reinforced composites

X. Ge* and S. Schmauder

Max-Planck-Institut für Metallforschung, Institut für Werkstoffwissenschaft, Seestrasse 92, D-7000, Stuttgart 1 (Germany)

Abstract

A theoretical investigation was carried out of the role of particle-reinforced composites. A model is proposed which describes the deformation behavior of materials based on shear stress constraint and the strain compatibility at the interface of a spherical particle in a cuboid. The model is derived from stress analysis and calculation methods combining series and parallel models with an integral method. Using this method, the elastic properties of composites reinforced with spherical particles can be expressed analytically. Good agreement with experimental data was obtained for Al/SiC, SiC/Al, WC/Co and Co/WC particle-reinforced composites. The constraining effect of shear stress at the interface is also discussed in this model.

1. Introduction

Composites composed of a metal or intermetallic matrix reinforced with particles have special properties which make them potentially attractive for a range of new applications. A great deal of effort [1–5] has been spent in describing the mechanical properties of these materials, particularly in terms of elastic behaviour. In the simple cases, either a constant stress or constant strain is assumed and the resultant properties of the composites can be calculated from the stress–strain properties of the individual phases and volume fractions. For example, the series model assumes that both materials contribute to the composite stiffness in proportion to their respective stiffness and volume fraction:

$$E = E_\alpha f_\alpha + E_\beta f_\beta \tag{1}$$

It has been shown that eqn. (1) provides an upper bound for the composite elastic modulus E in those cases where both constituent materials have the same value of Poisson's ratio [6, 7]. Another model is the parallel model, which describes the compliance in terms of the constituent materials subjected to uniform stress conditions:

$$E = (f_\alpha/E_\alpha + f_\beta/E_\beta)^{-1} \tag{2}$$

Equation (2) provides a lower bound for the equivalent elastic modulus E [6, 7]. Between the upper and lower bounds of eqns. (1) and (2), other work has been done to describe more accurately the mechanical properties of composites: for example, the Hashin–Shtrikman model [6], Paul model [7], Poech–Fischmeister model [8], and others [9]. Furthermore, additional work has been done based on two-dimensional finite element models, in order to calculate the stress–strain distributions in the microstructure of the composites and to obtain the modulus and deformtion of composites [10, 11]. Although these analyses are more exact compared with eqns. (1) and (2), there are still two differences between the actual particle-reinforced composites and the models as follows [12–16]: (1) in the model the shape of a particle is considered to be spherical, and (2) the constraining effect of the stress in the interface between the particle and matrix is not expressed.

In this paper, a new model based on a unit cube with a spherical particle inclusion is proposed. The stress around the particle is assumed to be a uniform shear stress. The model satisfies the crucial requirement of strain compatibility at the interface, and the influence of shear stress at the interface on the elastic modulus is also addressed.

2. Model of particle-reinforced composites

The microstructural model of the materials is based on the symmetric cell model of a uniform composite including four particles. Symmetric cell models have been widely used for this purpose, but most of them approximate a unit cell with a short cylindrical particle

*Permanent address: East China University of Chemical Technology, Shanghai, China.

0921-5093/93/$6.00

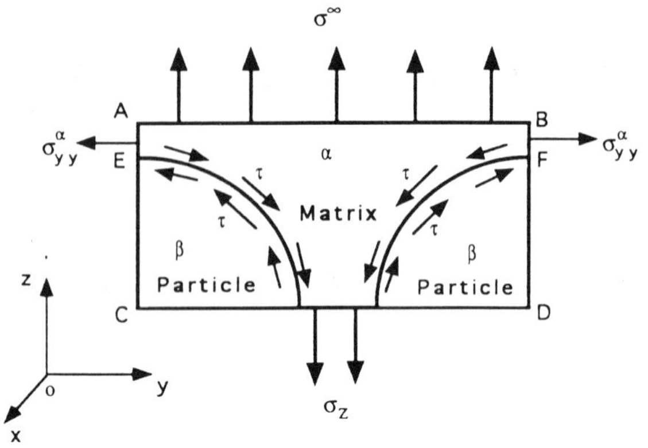

Fig. 1. Scheme of stresses for a cubic model with particles in it.

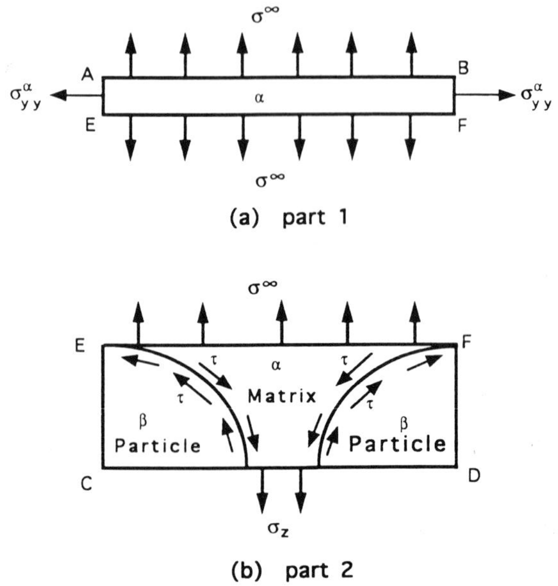

Fig. 2. Scheme of stresses in part 1 and part 2 of the cubic model.

or sandwich structure. The unit cube shown in Fig. 1 is subjected to an applied force σ^{∞} parallel to the cubic axis. A uniform shear stress is assumed at the interfaces which is distributed along the particle–matrix interface. The lateral faces remain planar with zero average normal tractions. The volume fraction of particles f_β is taken as the ratio of the particle volume to the unit cell volume. From Fig. 1, at points E and F as well as on the faces AC and BD, there are respectively

$$\Delta u^{\alpha}=\Delta u^{\beta} \quad \text{at points E and F} \tag{3}$$

$$\int \sigma_{yy} \mathrm{d}A=0 \quad \text{on the faces AC and BD} \tag{4}$$

In eqns. (3) and (4), the total Poisson contraction (elastic and plastic) in the composites has been considered. It follows from eqns. (3) and (4) that both materials will experience lateral tension or compressive stress in the X and Y directions. The model in Fig. 1 can be divided into two parts along the line EF as shown by parts 1 and 2 in Fig. 2. In part 1, according to eqns. (3) and (4), the expression for $\sigma_{yy}{}^{\alpha}$ can be derived:

$$\sigma_{yy}{}^{\alpha}=\frac{f_\beta \nu_\alpha \sigma^{\infty}}{f_\beta(1-\nu_\alpha)+(2/\pi) f_\alpha k} \tag{5}$$

$$k=\frac{E_\alpha}{G_\beta}(1-2\nu_\beta)$$

where k is the ratio of the two material stiffnesses.

And then from Hooke's law in Fig. 2(a), $E_{1c}=\sigma_{\infty}/\varepsilon_{1c}$, the composite Young's modulus of part 1 can be defined:

$$E_{1c}=\left[\frac{1}{E_\alpha}(1-e)\right]^{-1} \tag{6}$$

$$e=\frac{2 f_\beta \nu_\alpha{}^2}{f_\beta(1-\nu_\alpha)+(2/\pi) f_\alpha k} \tag{7}$$

where e is the parameter of shear stress constraint.

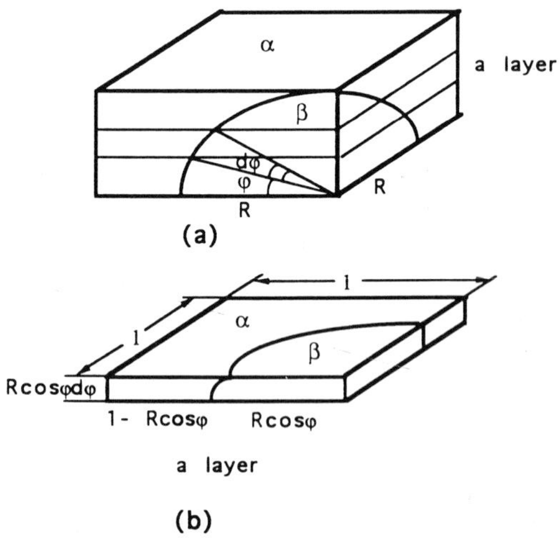

Fig. 3. Dimensions of layer i of part 2.

Obviously, E_{1c} is larger than E_α, because of the influence of the particle and shear stress constraint. For part 2 in Fig. 2(b), the stress distribution with tensile and shear stresses makes the theoretical analysis more complex, so another analysis method is used on only half of part 2 as shown in Fig. 3.

The rectangular cubic model in Fig. 3(a) can be divided into microlayers with constant thicknesses equal to $R\cos\varphi\,\mathrm{d}\varphi$. Every layer is considered as a small parallel model of two materials. For the layer shown in Fig. 3(b), the volume fractions of the component materials are as follows respectively:

$$f_i^{\beta}=\frac{\pi}{4}R^2\cos^2\varphi \qquad f_i^{\alpha}=1-\frac{\pi}{4}R^2\cos^2\varphi$$

The elastic modulus of a layer can therefore be expressed according to eqn. (1):

$$E_{2c}^{\ i}=E_{\alpha}\left(1-\frac{\pi}{4}R^2\cos^2\varphi\right)+E_{\beta}\frac{\pi}{4}R^2\cos^2\varphi \tag{8}$$

All of the layers combine to form the rectangular cube of part 2 with eqn. (2). The volume fraction of every layer is equal to:

$$f_i=\cos\varphi\ \mathrm{d}\varphi \tag{9}$$

From eqn. (2)

$$E_{2c}^{\ -1}=\sum\left(\frac{f_i}{E_{2c}^{\ i}}\right)$$

$$=\int_0^{\pi/2}\frac{\cos\varphi\ \mathrm{d}\varphi}{E_{\alpha}[1-(\pi/4)R^2\cos^2\varphi]+E_{\beta}(\pi/4)R^2\cos^2\varphi} \tag{10}$$

Equation (10) is a mathematical integral equation. If the particle is harder than the matrix, $E_{\beta}\gg E_{\alpha}$ (for example an SiC-reinforced aluminum matrix), and the solution of eqn. (10) can be written as

$$E_{2c}^{\ -1}=\frac{1}{2a_1a_2}\ln\frac{a_2+a_1}{a_2-a_1} \tag{11}$$

$$a_1=\mathrm{R}\left(\frac{\pi}{4}\Delta E\right)^{1/2}\qquad a_2=(E_{\alpha}+a_1^{\ 2})^{1/2}$$

$$\Delta E=E_{\beta}-E_{\alpha}$$

If the particle is softer than the matrix, $E_{\alpha}\gg E_{\beta}$, such as a particle-toughened ceramic or brittle matrix composite, eqn. (10) has another solution:

$$E_{2c}^{\ -1}=\frac{1}{a_1a_2}\arctan\left(\frac{a_1}{a_2}\right) \tag{12}$$

$$a_1=R\left(\frac{\pi}{4}\Delta E\right)^{1/2}\qquad a_2=(E_{\alpha}-a_1^{\ 2})^{1/2}$$

$$\Delta E=E_{\alpha}-E_{\beta}$$

Combining part 1 and part 2 gives a composite elastic modulus as in the series model. So the elastic modulus of particle-reinforced composites can be calculated as follows:

$$E_c=\left(\frac{1-R}{E_{1c}}+\frac{R}{E_{2c}}\right)^{-1} \tag{13}$$

In eqn. (13), R can be replaced by the overall fraction of particles f_{β}. For example, R in the model of Fig. 1 is expressed as

$$R=1.244f_{\beta}^{\ 1/3}$$

For other particle arrangements, the relation of R and f_{β} can be simply derived through an analysis of solid geometry.

3. Comparison with other models and experiments

Figure 4 shows eqn. (13) plotted as a function of volume fraction for the cases of phase α as the matrix and phase β as the matrix with $E_{\alpha}/E_{\beta}=10$. The plots of other models are also presented for comparison. It can be seen that present model gives results very close to the model of Zahl *et al.* [9].

Figure 5 shows a comparison of eqn. (13) with the experimental data for both Al/SiC and WC/Co systems. From Fig. 5, the value of elastic modulus of the WC/Co composites with higher volume fraction of WC is more exact than that for the lower volume fraction composites. That means that the current model is more suitable for evaluating the case of softer particles in a stiffer matrix, or a softer matrix with stiffer particles in higher volume fractions.

4. Effects of parameter shear stress constraint

From eqns. (6) and (7), the value of the elastic modulus of part 1 is greater than its original value because of the constraint of shear stress and the spherical shape of the interface. The *e* parameter can be used to show this change in elastic modulus. Figure 6 shows

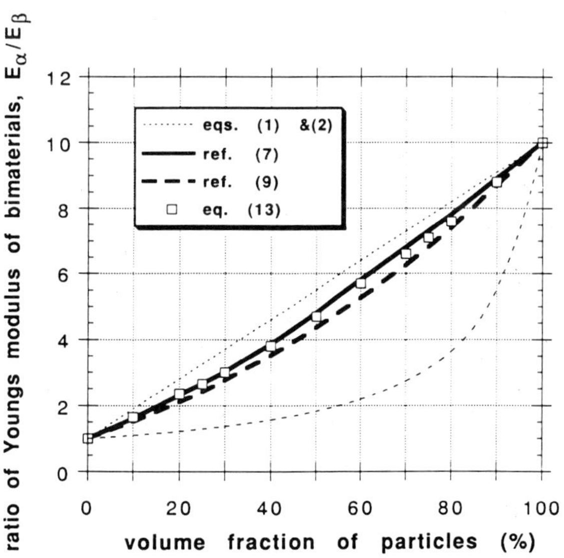

Fig. 4. Comparison of current model with the rule of mixture and other models for $E_{\alpha}/E_{\beta}=10$.

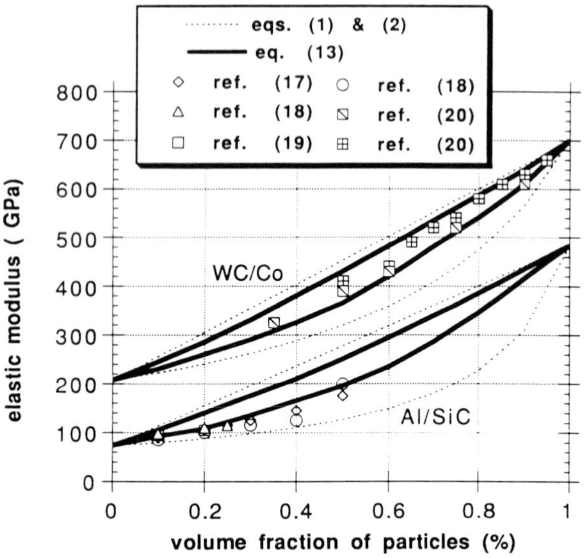

Fig. 5. Comparison of prediction of current model with experimental data for Al/SiC and WC/Co.

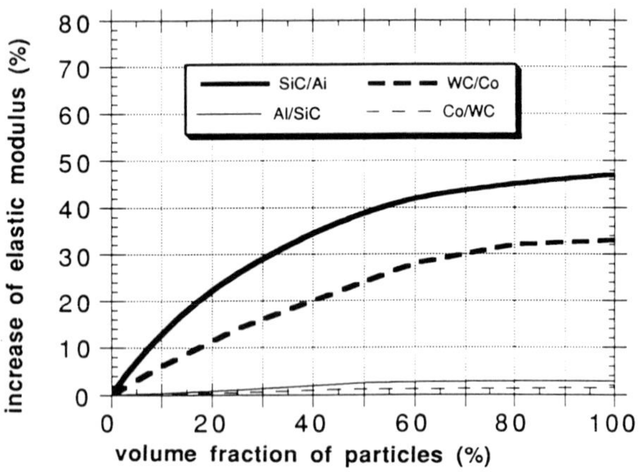

Fig. 6. Increment of elastic modulus of composites due to shear stress constraint at the interface.

the increase in elastic modulus as a function of volume fraction of particles. A large difference between elastic modulus increases for the systems of a softer matrix with harder particles and a harder matrix with softer particles is shown in Fig. 6. The constraining effect on the elastic modulus in the latter system is very small. The ratio of particle and matrix stiffness k, and Poisson's ratio of the matrix can be used to express the constraining effect of the interface. When the value of k is large and ν_α is small, the value of e is small, and the constraint effect is small; this is shown in Fig. 7.

5. Conclusions

A new three-dimensional model which describes the elastic modulus of particle-reinforced composites is

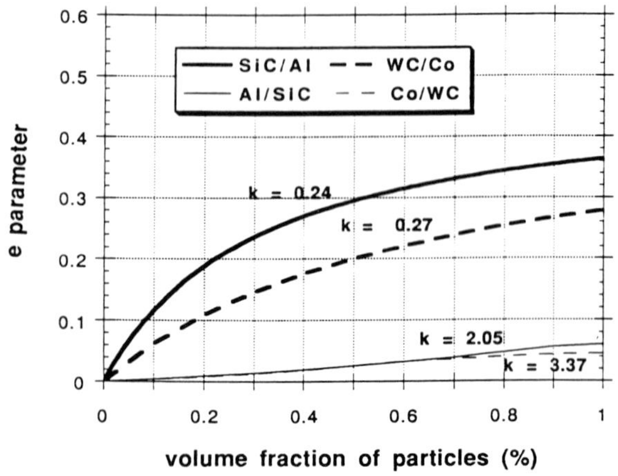

Fig. 7. Shear stress constraint parameter e (in eqn. (7)) *vs.* ratio of stiffnesses of particle and matrix in Al/SiC and WC/Co composites.

proposed. The model is based on the effect of shear stress constraint and strain compatibility at the interface. Using the model expressed in eqn. (13), the elastic properties of composites can be investigated more precisely, and the geometry effect for inclusions of spherical shape can also be discussed with eqn. (7). The experimental data show good agreement with this model. Discussion of the effect of shear stress constraint at the interface shows that the shear stress constraint parameter e is dependent on k (ratio of stiffnesses of the particle and matrix) and Poisson's ratio ν_α. If $k \ll 1$, and ν_α is large, the elastic modulus of the composites will increase.

Acknowledgment

One of the authors, GE, wishes to thank DAAD/K.C. Wong fellowship for supporting him working at MPI. A helpful discussion with Dr M. Poech and Dr P. Lipetzky is gratefully acknowledged.

References

1 H. F. Fischmeister and B. Karlsson, *Z. Metallkd., 68* (1977) 311.
2 D. S. McLachlan, M. Blaszkiewicz and R. E. Newnham, *J. Am. Ceram. Soc., 73* (1990) 2187.
3 K. Cho and J. Gurland, *Metall. Trans. A, 19* (1988) 2027.
4 S. Ankem and H. Margolin, *Metall. Trans. A, 13* (1982) 595.
5 G. Bao, J. W. Hutchinson and R. M. McMeeking, *Acta Metall. Mater., 39* (1991) 1871.
6 Z. Hashin and S. Shtrikman, *J. Mech. Phys. Solids, 11* (1963) 127.
7 B. Paul, *Trans. AIME, 218* (1960) 36.
8 M. H. Poech and H. F. Fischmeister, *Acta Metall. Mater., 40* (1992) 487.

9 D. B. Zahl, S. Schmauder and R. M. McMeeking, Elastic behavior of discontinuously reinforced composites, to be published.
10 B. Karlsson and B. O. Sundström, *Mater. Sci. Eng., 16* (1974) 161.
11 A. Ankem and H. Margolin, *J. Met., 68* (1986) 25.
12 T. Christman, A. Needleman and S. Suresh, *Acta Metall., 37* (1989) 3029.
13 L. B. Ekbom, U. Lindegran and J.-E. Andersson, *Refract. Hard Met., 7*(3) (1988) 2.
14 L. Su and J. Gurland, *Mater. Sci. Eng., 95* (1987) 151.
15 V. C. Nardone and K. M. Prewo, *Scripta Metall., 20* (1986) 43.
16 M. F. Ashby, C. Gandhi and D. M. R. Taplin, *Acta Metall., 27* (1979) 699.
17 D. McDanels, *Metall. Trans. A, 16* (1985) 1105.
18 T. G. Nieh and D. G. Chellman, *Scripta Metall., 18* (1984) 925.
19 A. G. Evans, J. W. Hutchinson and R. M. McMeeking, *Scripta Metall., 25* (1991) 3.
20 H. Doi, Y. Fujiwara and K. Miyake, *Metall. Trans., 245* (1970) 1417.

Inhomogeneity of internal stresses and X-ray elastic constants of ceramic materials

V. I. Trefilov, O. N. Grigoriev and S. M. Kushnerenko

I.N. Frantsevich Institute for Problems of Material Science, Academy of Sciences of the Ukraine, Kiev 252180 (Ukraine)

Abstract

Experimental determination and theoretical calculation of the so-called X-ray elastic constants, which describe the relation between the lattice strain and the stress, were carried out to test the validity of traditional methods of calculation of X-ray elastic constants for heterogeneous ceramic materials. For this investigation materials such as α-Al_2O_3, α-Al_2O_3/ZrO_2, and TiC/α-Al_2O_3/TiB_2 were used. It was found that the values of elastic constants, calculated for multiphase systems, may produce significant errors when used to determine the internal stress. The discrepancy between calculated and experimental values may be utilized as a diagnostic test of inhomogeneity of the material being investigated.

1. Introduction

The mechanical properties of ceramics are known to be influenced by residual stresses. If the X-ray diffraction method is to be used to determine these stresses, one has to know the X-ray elastic constants of the material in question. In the present paper a comparison of calculated and experimental values of the X-ray elastic constants is presented for multiphase ceramic materials.

2. Theory

In a biaxial state of stress with the principal stresses σ_1 and σ_2, lattice strain components in the direction ϕ, ψ are described by the equation for the linear theory of elasticity:

$$\left(\frac{\mathrm{d}(D)}{D_0}\right)_{\varphi,\psi} = \varepsilon_{\varphi,\psi} = \frac{1}{2}\,S_2(\sigma_1\cos^2\varphi + \sigma_2\sin^2\varphi) \times \sin^2\psi + S_1(\sigma_1+\sigma_2) \qquad (1)$$

where the angles ϕ and ψ are defined by the direction of the strain components $\varepsilon_{\varphi\psi}$ in relation to the direction of the principal stress σ_1 and to the surface normal of the specimen respectively; D_0 is the lattice spacing in a stress-free direction; S_1 and $\frac{1}{2}S_2$ are the so-called X-ray elastic constants.

Under the two opposite assumptions introduced by Voigt and Reuss (homogeneous strain and stress), the limiting values of the X-ray elastic constants can be calculated. Under the assumption of Kroner a more realistic value is obtained, which is usually about the mean of the two values obtained using the Voigt and Reuss assumptions. The corresponding relations for cubic and hexagonal crystals can be found elsewhere [1, 2].

In the case of multiphase material under the assumption that the stress is homogeneous in all phases, the X-ray elastic constants of each individual phase are the same as for the single-phase material [3], and the Reuss estimate may be used to obtain the X-ray elastic constants of the individual phases. Provided that the distortions in the system and the stresses in the individual phases are homogeneous, the equation for calculation of the X-ray elastic constants of multiphase material obtained by Evenschor and Hauk [4] can be used.

To measure the X-ray elastic constants, a series of loads is applied in the elastic range. At each load, the interplanar spacing D is measured *vs.* $\sin^2\psi$. The slope β, obtained at each applied stress, is [5]

$$\beta_1 = \sigma_1\bar{S}_2\left(\frac{1+\nu}{E},\ K_i(\psi)\right) + C(\varepsilon_{ij}^{\mathrm{r}}) \qquad (2)$$

Here σ_1 is the applied stress, $(1+\nu)/E$ is the X-ray elastic constant of the homogeneous material and the K_i are the stress interaction constants between grains, which depends on their orientation, and hence ψ [5]. The function $\bar{S}_2(1+\nu)/E$, $K_i(\psi)$ is an average, defined by Noyan [5], of the terms $(1+\nu)/E$ and $K_i(\psi)$. The

0921-5093/93/\$6.00

TABLE 1. Compositions, methods of fabrication and surface treatment of investigated materials

Material	Method of fabrication	Surface treatment
α-Al_2O_3 (3 mol% Y_2O_3)	Hot-pressed at 1600 °C (15 min)	Ground
α-Al_2O_3/(t/m) ZrO_2(32 vol.%)	Hot-pressed at 1650 °C (1 h)	Polished
TiC/α-Al_2O_3(18 vol.%)/TiB_2(8 vol.%)	Hot-pressed at 1880 °C	Ground

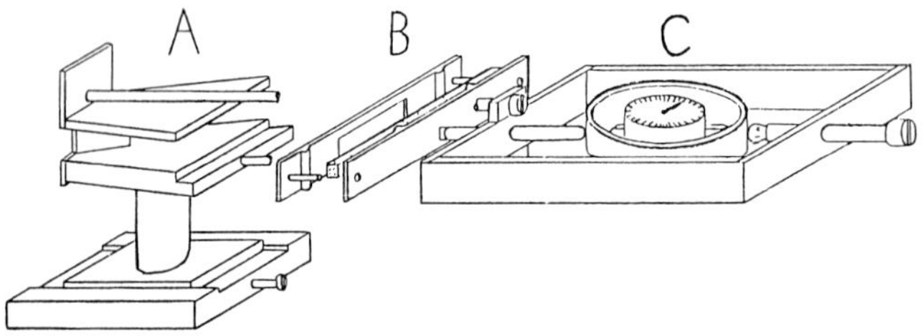

Fig. 1. Four-point bending apparatus mounted at the centre of the diffractometer: A device holder; B four-point bending fixture; C loading cell with strain gauge.

term $C(\varepsilon_{ij}^{r})$ is a complex function of any residual strain (ε_{ij}^{r}) present in the specimen. The slope β is plotted *vs.* the applied stress. The slope of this second plot is related to one of the desired constants.

3. Experimental details

The investigated materials, the methods of fabrication and the surface treatment are given in Table 1.

The experiment was performed with a four-point bending apparatus mounted in an automated diffractometer DRON-3M (Fig. 1). The bending sample is held in place by four supports between two plates of a bending fixture (B). One of the plates is attached to a device holder (A). The construction of the device holder allows accurate positioning of the specimen. The other plate is attached to a load cell (C). A frame of the load cell is attached to the device holder. The force on the sample is transmitted via an elastic ring. The mechanical strain of the ring is indicated by a mechanical strain gauge, which is inside the ring. Before the experiment the relationship between the strain of the ring and the applied load was determined.

To provide greater accuracy of the X-ray measurements, an etalon powder was used: for TiC it was silicon powder, and for Al_2O_3 it was β-SiC.

The deviations of the measured values of the elastic constants owing to counting statistics were analysed.

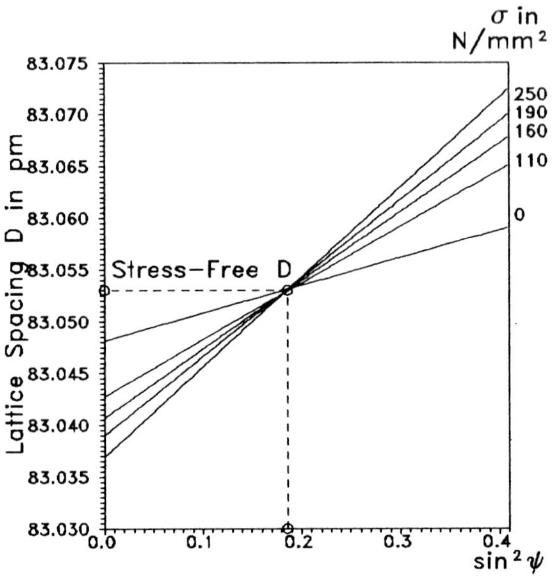

Fig. 2. Experimental determination of X-ray elastic constants of single-phase α-Al_2O_3 (416): lattice spacing distribution *vs.* $\sin^2\psi$ as a function of loading stress.

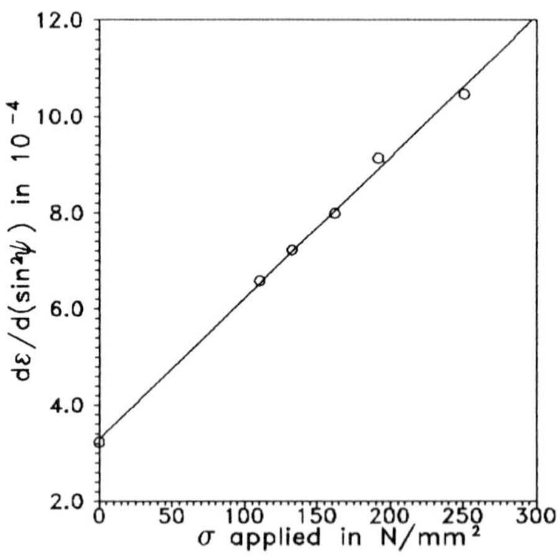

Fig. 3. Experimental determination of X-ray elastic constants of single-phase α-Al_2O_3 (416): slope of lattice strain distribution *vs.* loading stress.

4. Results

The lattice strain distribution measured for single-phase α-Al_2O_3 *vs.* $\sin^2\psi$ is shown in Fig. 2. The *D vs.* $\sin^2\psi$ plot should meet at one intersection. Within the accuracy of the measurement this is fulfilled. Figures 3 and 4 give the corresponding Hooke lines for determination of the X-ray elastic constants.

The results for the measured and calculated X-ray elastic constants are summarized in Tables 2 and 3.

In the case of single-phase Al_2O_3, if the deviation is taken into account, the experimental values of the constants are in reasonable agreement with theoretical estimates and with the results of other authors [8]. For Al_2O_3/ZrO_2 the measured values of the X-ray elastic constants fall between the Reuss and Evenschor limits (homogeneous stress and strain in all phases) within the accuracy of the measurement. However, for $TiC/Al_2O_3/TiB_2$ this is not fulfilled. If the Evenschor estimate is used to determine the residual stresses of $TiC/Al_2O_3/TiB_2$, the calculated values of the residual stresses are about 30% less than the real values. Thus a more accurate model is required to calculate the X-ray elastic constants of multiphase materials.

In accordance with Noyan [5], the difference between the experimental values of the X-ray elastic constants and the values obtained for the homogeneous state of the material are measures of the reaction stresses (and strains) caused by the inhomogeneity of the material in question. For single-phase Al_2O_3 this difference is reasonably negligible, while for Al_2O_3/ZrO_2 and especially for $TiC/Al_2O_3/TiB_2$, where the reaction strains are caused by the elastic interaction of phases with different elastic properties, it has the same order of magnitude as the measured constants. It is known that the extent of inhomogeneity of the internal stresses can influence the micromechanism of fracture of ceramics (microcracking, phase transformation in a crack tip, etc.) and, in consequence, determine the mechanical behaviour of the ceramic materials [9]. Thus the measured X-ray elastic constants may be used not only to identify residual stress but also to investigate the mechanical properties of the material in question.

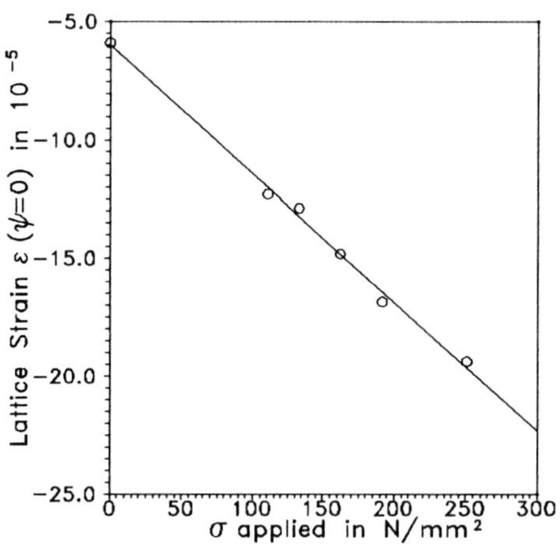

Fig. 4. Experimental determination of X-ray elastic constants of single-phase α-Al_2O_3 (416): intersection of lattice strain distribution $\varepsilon_{\psi=0}$ *vs.* loading stress.

5. Conclusions

In the case of multiphase ceramic materials, use of the existing estimates for calculation of the X-ray elastic constants may not provide the necessary preci-

TABLE 2. Experimental[a] and calculated values of $\frac{1}{2}S_2$ (TPa^{-1})

Material	Plane	Voigt	Reuss	Kroner	Evenschor	Experimental values
Al_2O_3[b]	Al_2O_3 (416)	3.02	3.16	3.09	—	2.93 ± 0.12
Al_2O_3[b]/ZrO_2	Al_2O_3 (416)	3.02	3.16	3.09	3.72	3.81 ± 0.19
TiC[c]/Al_2O_3/TiB_2	TiC (511)	2.74	2.64	2.69	2.58	1.77 ± 0.21

[a]Cu Kα radiation.
[b]Single-crystal elastic constants from ref. 6.
[c]Single-crystal elastic constants from ref. 7.

TABLE 3. Experimental[a] and calculated values of S_1 (TPa^{-1})

Material	Plane	Voigt	Reuss	Kroner	Evenschor	Experimental values
Al_2O_3[b]	Al_2O_3 (416)	−0.569	−0.620	−0.594	—	−0.545 ± 0.031
Al_2O_3[b]/ZrO_2	Al_2O_3 (416)	−0.569	−0.620	−0.594	−0.789	−0.701 ± 0.043
TiC[c]/Al_2O_3/TiB_2	TiC (511)	−0.455	−0.421	−0.438	−0.340	−0.30 ± 0.037

[a]Cu Kα radiation.
[b]Single-crystal elastic constants from ref. 6.
[c]Single-crystal elastic constants from ref. 7.

sion of determination of residual stresses. There is a need for more accurate models for calculation of the X-ray elastic constants of multiphase systems.

If X-ray diffraction analysis of internal stresses is carried out for different structural conditions of ceramic materials, the X-ray elastic constants must be measured for the same structural conditions.

Experimentally determined values of the X-ray elastic constants may be used to investigate the mechanical properties of the material in question.

Acknowledgment

The authors thank Dr. Orlovskaya for supplying the Al_2O_3 and Al_2O_3/ZrO_2 specimens.

References

1 H. Dolle, *J. Appl. Crystallogr., 12* (1979) 489.
2 P. D. Evenschor, W. Frohlich and V. Hauk, *Z. Metallkd., 62* (1971) 1.
3 V. Hauk and H. Kockelman, *Z. Metallkd., 68* (1977) 719–724.
4 P. D. Evenschor and V. Hauk, *Z. Metallkd., 66* (1975) 210–213.
5 I. C. Noyan, *Mater. Sci. Eng., 75* (1985) 95–103.
6 J. B. Wachtman, W. E. Tefft, D. G. Lam and R. P. Stienchfield, *J. Am. Ceram. Soc., 43* (1960) 334.
7 I. D. Klerk, *Rev. Sci. Instrum., 36* (1965) 1540–1545.
8 K. Tanaka, E. Matsui, T. Kurimura and Y. Akiniwa, *J. Soc. Mater. Sci. Jpn., 36* (1988) 810.
9 W. Pompe, in E. Kaldis (ed.), *Current Topics in Materials Science*, Vol. 12, North-Holland, Amsterdam, pp. 205–472.

Effect of pressure on 2-magnon Raman scattering in K_2NiF_4

V. V. Struzhkin*, U. Schwarz, H. Wilhelm and K. Syassen
Max-Planck Institut für Festkörperforschung, Heisenbergstrasse 1, 7000 Stuttgart 80 (Germany)

Abstract

We measured the effect of pressure ($P \leqslant 10$ GPa) on 2-magnon excitations ($T = 10$ K) and on lattice parameters ($T = 300$ K) of K_2NiF_4 by Raman spectroscopy and X-ray diffraction respectively. The results allow us to determine the dependence of the superexchange interaction J on the transition metal to anion distance d. We find that J scales as $J \propto d^{-n}$ with $n = 9.5 \pm 0.5$. Previous measurements of 2-magnon excitations under pressure in NiO ($n \approx 7$–8), La_2CuO_4 and Eu_2CuO_4 ($n \approx 5$–6) are analyzed and an explanation for the trends in the pressure dependence of the superexchange interaction J is proposed.

The pressure dependence of the superexchange integral J in the antiferromagnets La_2CuO_4 and Eu_2CuO_4 [1, 2] reveals an anomalous scaling of J with cation–anion (Cu–O) distance d: $J \propto d^{-5 \pm 2}$. This behavior contradicts the empirical expression $J \propto d^{-11 \pm 1}$ [3, 4], which was derived from an analysis of previous data on the superexchange integral in various antiferromagnetic compounds. K_2NiF_4 is a prototype antiferromagnet with crystal structure analogous to La_2CuO_4. Its Neel temperature is $T_N = 97.1$ K [5]. In K_2NiF_4 the perovskite-like layers of $KNiF_3$ are separated by planes of KF. This makes the antiferromagnetic interaction essentially two dimensional. Two-magnon Raman scattering in K_2NiF_4 is well described by the spin-wave theory [6] in the Green-function formalism for an $S = 1$ antiferromagnet. Thus, the frequency of the 2-magnon Raman line is proportional to the superexchange integral J. We measured 2-magnon Raman scattering under pressure in K_2NiF_4 in order to determine the pressure dependence of J. We also determined the pressure dependence of the lattice parameters of K_2NiF_4 using X-ray diffraction. These measurements allow us to determine the $J(d)$ dependence, which we compare with that of La_2CuO_4 [1], Eu_2CuO_4 [2] and NiO [7, 8]. We propose an explanation for the observed trends in the $J(d)$ dependences in these antiferromagnets, which is based on the theoretical model of ref. 9.

The compound K_2NiF_4 cannot be grown directly from the melt. Samples were prepared in a manner analogous to that described in ref. 10. For Raman measurements good optical quality wafers ($100 \times 100 \times 30$ μm^3) were placed in a gasketed diamond-window high-pressure cell [11]. Condensed helium was used as a pressure transmitting medium in order to ensure the best possible hydrostatic conditions at low temperatures. The pressure in the cell was only changed at room temperature, and then the cell was cooled down to 10 K. Low-temperature Raman spectra were measured with a multichannel spectrometer system.

For X-ray diffraction investigations the crystals were powdered. Samples were placed in a 4:1 methanol–ethanol pressure medium. The powder pattern was observed down to a d-spacing of 1.4 Å using an angle-dispersive diffraction geometry (Debye–Scherrer method, filtered Mo Kα radiation) and a position-sensitive proportional counter system. Experimental errors in absolute lattice parameters are estimated to be about 0.2%. The ruby luminescence method with calibration according to ref. 12 was used to measure the pressure in Raman as well as in X-ray diffraction studies.

The lattice parameters of K_2NiF_4 under normal conditions are $a_0 = 4.008(2)$ Å and $c_0 = 13.064(2)$ Å. The pressure–volume data are well fitted by a Murnaghan equation of state [13] with bulk modulus $B_0 = 56.3(25)$ GPa and its derivative $B_0' = 5.7(10)$. The pressure dependence of d(Ni–F) is given by the pressure dependence of the lattice parameter a which is also well fitted by a Murnaghan-like equation $a(P) = a_0$-$[1 + (K_0'/K_0)P]^{-1/K_0'}$, P is the pressure in gigapascals, using $K_0 = 183(4)$ GPa, $K_0' = 18$.

Figure 1 shows 2-magnon Raman spectra of K_2NiF_4 at several pressures. The pressure dependence of the 2-magnon Raman line is shown in Fig. 2. This depen-

*On leave from: Institute for High Pressure Physics, Russian Academy of Sciences, 142092 Troitsk, Moscow Region, Russian Federation.

0921-5093/93/$6.00

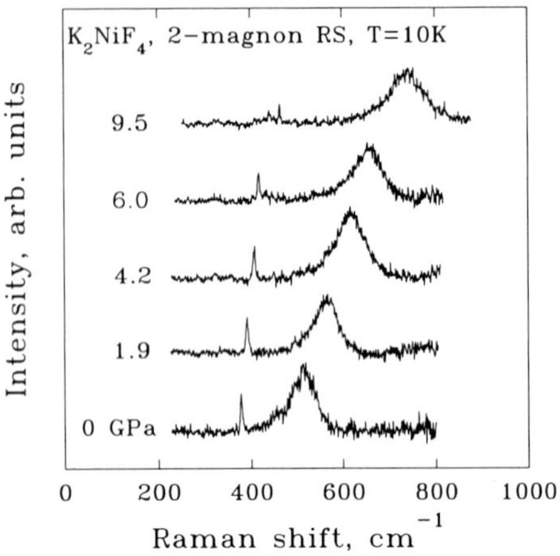

Fig. 1. Raman spectra of K_2NiF_4 ($T = 10$ K) at different pressures, showing the shift of the 2-magnon Raman line.

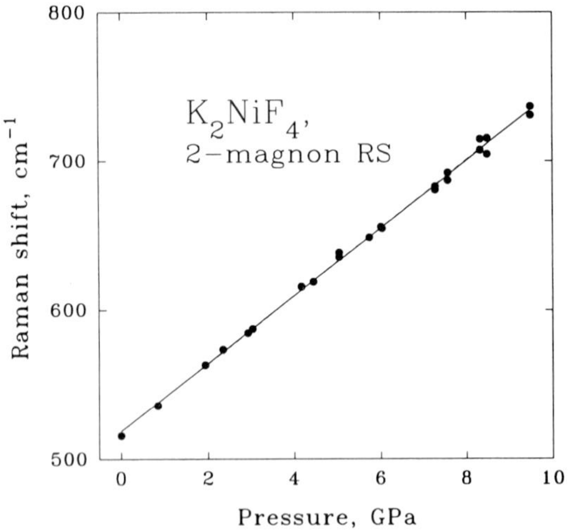

Fig. 2. Frequency of the 2-magnon Raman line in K_2NiF_4 as a function of pressure ($T = 10$ K).

dence is found to be essentially linear (pressure coefficient 22.7 (1.4) cm^{-1} GPa^{-1}).

In order to compare previous results with ours we use the coordinates $x = \ln(d/d_0)$, $y = -\ln(J/J_0)/\ln(d/d_0)$ and show the corresponding data for K_2NiF_4, NiO [7, 8], La_2CuO_4 [1] and Eu_2CuO_4 [2] in Fig. 3. In their interpretation of the pressure dependence of J for NiO, Massey *et al.* [8] used early compressibility data [14]. We have corrected these P–V data using Decker's [15] equation of state for NaCl, and have used corrected values of the bulk modulus of NiO ($B_0 = 170$ GPa), and its pressure derivative ($B_0' = 4$). In the case of La_2CuO_4 we have used the results of Aronson *et al.* [1] without any corrections. For Eu_2CuO_4 the situation with respect to compressibility data is most uncertain. The in-plane linear compressibility of Eu_2CuO_4 is

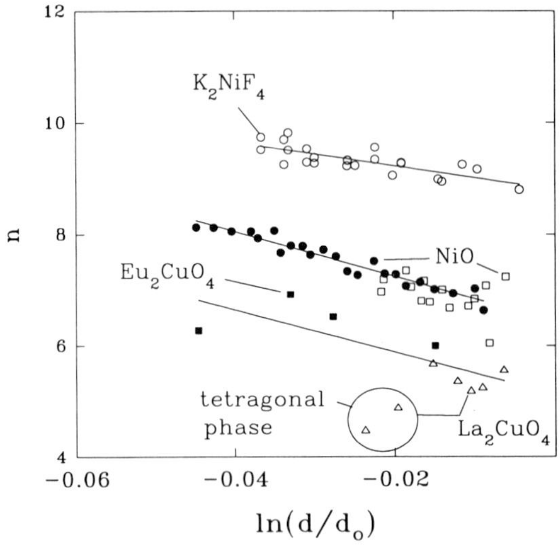

Fig. 3. Logarithmic plot of $n = -\ln(J/J_0)/\ln(d/d_0)$ *vs.* $\ln(d/d_0)$ for several antiferromagnets. For NiO full symbols are from ref. 8, open symbols are from ref. 7.

estimated to be $k = -1.3 \times 10^{-3}$ GPa^{-1} based on the results for Nd_2CuO_4 [16]. Another estimate is to take the linear compressibility of the in-plane lattice parameter a for La_2CuO_4 in the orthorhombic phase [17] $k = -1.5 \times 10^{-3}$ GPa^{-1}, which is in fact quite close to the previous value. Because of this uncertainty the data for Eu_2CuO_4 are to be considered with some caution. As can be seen from Fig. 3, we cannot describe the experimental results with a single scaling law $J \propto d^{-n}$, where $n = 10$–12, as was proposed earlier for the superexchange interaction [3, 4]. This even holds if we take into account the uncertainties in the compressibilities.

The pressure dependence of J can be discussed in terms of the superexchange interaction as derived from an effective Hubbard Hamiltonian [9]:

$$J = \frac{4t^4}{\Delta^2}\left(\frac{1}{U} + \frac{1}{\Delta}\right) \tag{1}$$

Δ is the charge transfer gap, U is the Mott–Hubbard gap, $t \ll \Delta$, U is the p-d matrix element. It is assumed that the Hubbard U is independent of pressure, and that t scales as $t \propto d^{-4}$.

If one places the three insulators K_2NiF_4, NiO and La_2CuO_4 on the well known Zaanen–Sawatzky–Allen diagram [18], see Fig. 4, then La_2CuO_4 and other cuprates will be in the region of the charge-transfer insulators, for which $\Delta < U$, NiO will be close to the intermediate region $\Delta \approx U$, and K_2NiF_4 will be situated in the $\Delta > U$ region of the diagram, being a Mott–Hubbard insulator. Here we have used for La_2CuO_4 and NiO the parameter values from refs. 19 and 20. The Δ and U values for K_2NiF_4 were deter-

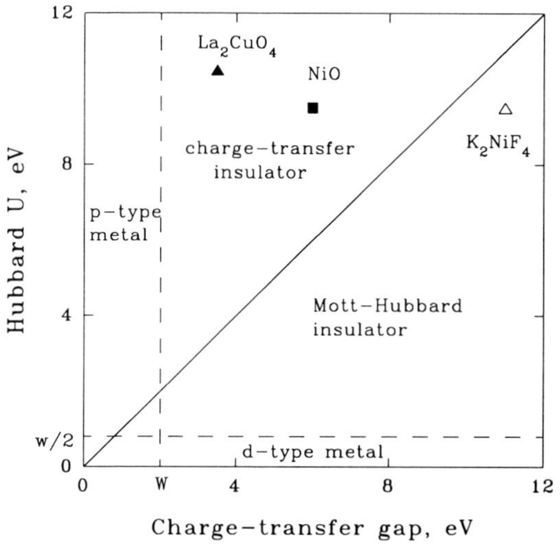

Fig. 4. Zaanen–Sawatzky–Allen diagram. The line $\Delta = U$ divides the charge-transfer insulators from Mott–Hubbard insulators. W is the p-bandwidth, w is the d-bandwidth.

mined using cation and anion systematics as given by Zaanen and Sawatzky [20, 21]. It is clear from Fig. 3 that the deviation from the empirical Bloch law $J \propto d^{-10}$ [4] correlates with the value of the charge-transfer gap, see Fig. 4. More specifically, the power n is lower for lower values of the charge-transfer gap. This correlation becomes plausible in the framework of the ionic model [22], taking into account the Madelung energies of the initial and final states of the charge-transfer transition. Quantitative treatment of Δ is not possible at present, but for model purposes we will use an approach proposed by Ohta *et al.* [23]:

$$\Delta = \Delta E_{\mathrm{mad}}/\varepsilon_\infty - E_{\mathrm{I}}^{\mathrm{i}} - E_{\mathrm{A}}^{\mathrm{i}} - e^2/d \tag{2}$$

ΔE_{mad} is the difference in Madelung energies at cation and anion sites, ε_∞ is the dielectric constant at frequencies higher than the frequency of the charge-transfer transition, $E_{\mathrm{I}}^{\mathrm{i}}$ is the ionization potential of the cation, and $E_{\mathrm{A}}^{\mathrm{i}}$ is the electron affinity of the anion. The e^2/d term accounts for the Coulomb interaction of an electron and a hole, introduced by the charge-transfer transition, d is the cation–anion distance. $E_{\mathrm{I}}^{\mathrm{i}}$ and $E_{\mathrm{A}}^{\mathrm{i}}$ are pressure independent and we expect that the major contribution to the pressure dependence comes from the Madelung potential and the term e^2/d. Coulomb terms scale with distance as d^{-1}, but owing to the term $E_{\mathrm{I}}^{\mathrm{i}} + E_{\mathrm{A}}^{\mathrm{i}}$ in eqn. (2), which partly compensates for the Coulomb terms, the effective scaling of the charge-transfer gap is $\Delta \propto d^{-m}$, where $m > 1$. The value of m becomes larger when absolute values of the Coulomb terms and the term $E_{\mathrm{I}}^{\mathrm{i}} + E_{\mathrm{A}}^{\mathrm{i}}$ become nearly equal in magnitude, thus reducing the charge-transfer gap. This explains qualitatively the systematic variation of the J scaling coefficient with the charge-transfer gap (see eqn. (1)). For a smaller charge-transfer gap the scaling coefficient m is larger and the resulting scaling coefficient n for J is reduced. Moreover, the scaling coefficient m of Δ also varies with d and this makes it impossible to describe the $J(d)$ dependence with the single exponent n.

Summing up, the Bloch empirical law $J \propto d^{-10}$ is approximately valid for K_2NiF_4, but it is not valid for cuprates or for NiO. Using a perturbative expression for the superexchange interaction [19] and an ionic model [23], we find that the pressure dependence of the charge-transfer gap Δ explains the observed trends in the $J(d)$ dependence in the compounds considered here. It is worth noting that recent results on CoO [24], which has $\Delta = 7.2$ eV [20], give a $J \propto d^{-8\pm0.5}$ dependence. This is consistent with the interpretation given in this work, because CoO has a value of charge-transfer gap intermediate between those of NiO and K_2NiF_4, and its scaling coefficient $n = 8 \pm 0.5$ also has an intermediate value between $n = 7.5$ for NiO and $n = 9.5$ for K_2NiF_4.

Acknowledgments

We would like to thank E. Schönherr and S. Lacher for the help in preparing K_2NiF_4 samples. V.V.S. acknowledges financial support from DAAD and the warm hospitality of the Max-Planck Institute staff.

References

1 M. C. Aronson, S. B. Dierker, B. S. Dennis, S.-W. Cheong and Z. Fisk, *Phys. Rev. B, 44* (1991) 4657.
2 M. I. Eremets, A. V. Lomsadze, V. V. Struzhkin, A. A. Maksimov, A. V. Puchkov and I. I. Tartakovskii, *Pis'ma Zh. Eksp. Teor. Fiz., 54* (1991) 376 (*JETP Lett., 54* (1991) 372).
3 L. J. de Jongh and R. Block, *Physica B, 79* (1975) 568.
4 D. Bloch, *J. Phys. Chem. Solids, 27* (1966) 881.
5 R. J. Birgenau, H. J. Guggenheim and G. Shirane, *Phys. Rev. Lett., 22* (1969) 720.
6 P. A. Fleury and H. J. Guggenheim, *Phys. Rev. Lett., 24* (1970) 1346.
7 K. Reimann, K. Syassen, Z. Wang and G. Gegenheimer, *Verh. Dtsch. Phys. Ges., (VI), 25* (1990) 830.
8 M. J. Massey, N. H. Chen, J. W. Allen and R. Merlin, *Phys. Rev. B, 42* (1990) 8776.
9 V. J. Emery and G. Reiter, *Phys. Rev. B, 38* (1988) 4547.
10 S. R. Chinn, H. J. Zeiger and J. R. O'Connor, *Phys. Rev. B, 3* (1971) 1709.
11 G. Huber, K. Syassen and W. B. Holzapfel, *Phys. Rev. B, 15* (1977) 5123.
12 G. I. Piermarini, S. Block, J. D. Barnett and R. A. Forman, *J. Appl. Phys., 46* (1975) 2774.
H. K. Mao, P. M. Bell, J. W. Shaner and D. J. Steinberg, *J. Appl. Phys., 49* 1978) 3276.
13 F. D. Murnaghan, *Proc. Natl. Acad. Sci. USA, 30* (1944) 244.

14. I. Wakabayashi, H. Kobayashi, H. Nagasaki and S. Minomura, *J. Phys. Soc. Jpn., 25* (1968) 227.
15 D. L. Decker, *J. Appl. Phys., 42* (1971) 3239.
16 Y. Tokura, K. Kikuchi, T. Arima and S. Uchida, *Phys. Rev. B, 45* (1992) 7580.
17 M. J. Akhtar, C. R. A. Catlow, S. M. Clark and W. M. Temmerman, *J. Phys. C, 21* (1988) L917.
18 J. Zaanen, G. A. Sawatzky and J. W. Allen, *Phys. Rev. Lett., 55* (1985) 419.
19 M. S. Hybertsen, E. B. Stechel, M. Schlüter and D. R. Jennison, *Phys. Rev. B, 41* (1990) 11068.
20 J. Zaanen and G. A. Sawatzky, *J. Solid State Chem., 88* (1990) 8.
21 J. Zaanen and G. A. Sawatzky, *Can. J. Phys., 65* (1987) 1262.
22 H. Ledbetter and M. Lei, *Physica C, 166* (1990) 483.
J. Kondo, Y. Asai and S. Nagai, *J. Phys. Soc. Jpn., 57* (1988) 4334.
J. Kondo, *J. Phys. Soc. Jpn., 58* (1989) 2884.
23 Y. Ohta, T. Tohyama and S. Maekawa, *Phys. Rev. B, 43* (1991) 2968.
24 V. V. Struzhkin, A. F. Goncharov and K. Syassen, to be published.

Effect of pressure on magnetic excitations in CoO

V. V. Struzhkin*, A. F. Goncharov** and K. Syassen
Max-Planck Institut für Festkörperforschung, Heisenbergstrasse 1, D-7000 Stuttgart 80 (Germany)

Abstract

We investigated the effect of high pressure ($P \leq 17$ GPa) on magnetic excitations in CoO by Raman spectroscopy in the temperature range 10–300 K. In addition to the strong lines at 143, 221 and 296 cm^{-1} observed previously, we also measured weak Raman features at 251, 430 and 486 cm^{-1}, and an asymmetrical broad band centered around 1040 cm^{-1} arising from magnetic excitations. We propose a single-ion effective Hamiltonian for Co^{2+} and fit the energy levels of magnetic excitations using approximate second quantization techniques. The antiferromagnetic exchange interaction constant J_2 for next-nearest neighbors is found to scale with the bond length d according to $d \ln J_2/d \ln d = 8.0 \pm 0.5$.

1. Introduction

The phenomenon of high-temperature superconductivity has led to interest in the physical properties of materials with strong electron–electron correlations (Mott–Hubbard insulators). Transition-metal oxides belong to this class of materials, and they have been extensively studied in recent decades. Many of them are insulators and show antiferromagnetic ordering. In the case of CoO, the Neel temperature T_N is 293 K. The high-temperature NaCl-type structure becomes weakly distorted in the antiferromagnetic phase [1]. The ground state of the Co^{2+} ions is triply degenerate in the paramagnetic phase, if spin-orbit coupling is neglected. This degeneracy is lifted in the antiferromagnetic phase by the lattice distortion and by spin-orbit interaction. The coupling of low-energy states of the Co^{2+} ions by exchange interaction gives rise to magnetic excitations which can be observed by neutron [2, 3], Raman [4, 5], and IR spectroscopy [6, 7].

In this work we present the results of the first experimental study of the pressure dependence of magnetic excitations in CoO, measured by Raman spectroscopy. We calculate the energies of the low-lying magnetic excitations in mean-field approximation and, using the approximate second quantization method, in the form given by Buyers *et al.* [8]. The experimentally measured spectral positions of magnetic excitons are used to fit the parameters of the effective Hamiltonian under pressure. In particular, the pressure dependence of the superexchange integral J_2 between next-nearest neighbors is determined from the data.

*On leave from the Institute for High Pressure Physics, Russian Academy of Sciences, 142092 Troitsk, Moscow Region, Russian Federation.

**On leave from the Institute of Crystallography, Russian Academy of Sciences, Moscow, Russian Federation.

2. Experimental details

We investigated single-crystal CoO samples with a Neel temperature of 290 ± 2 K, as determined by magnetic susceptibility measurements. For the low-temperature Raman measurements at high pressures we used a diamond-anvil cell in combination with a continuous-flow helium cryostat. Small chips of CoO single crystals of $100 \times 60 \times 15$ μm^3 were placed in the cell. Condensed helium served as a pressure-transmitting medium providing the best possible hydrostatic conditions at low temperatures. The Raman spectra were excited by an argon-ion laser ($\lambda_{ex} = 488$ nm and 514.5 nm) and were measured in near backscattering geometry employing a triple spectrograph and a multichannel detector. The pressure was measured by the ruby luminescence technique using the pressure scale of ref. [9].

3. Results and discussion

Figure 1 shows several Raman spectra of CoO taken at various temperatures and almost constant pressure near 2.8 GPa. As the temperature decreases several new bands appear and become sharper for lower temperatures. Three narrow lines near 140, 220 and 296 cm^{-1} were observed previously [5] at normal

0921-5093/93/$6.00

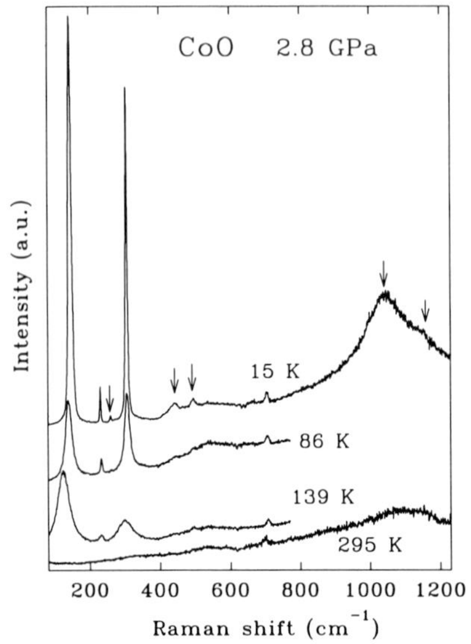

Fig. 1. Raman spectra of CoO at 2.8 GPa measured at different temperatures.

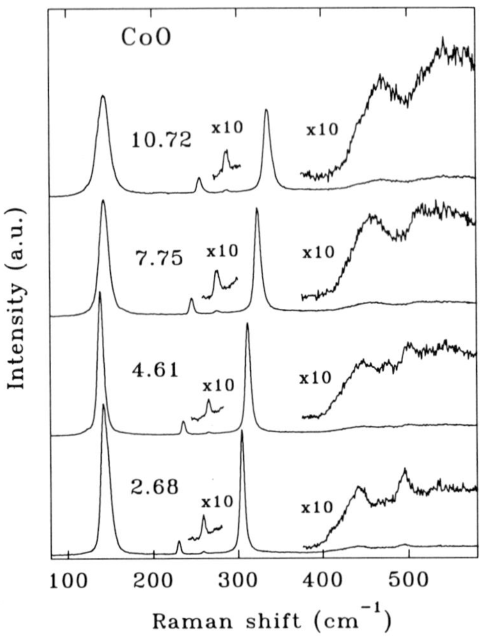

Fig. 2. Low temperature (T=20 K) Raman spectra of CoO at different pressures; numbers refer to the pressure in gigapascals.

pressure and were attributed to magnetic excitations. We also observe weak lines at 251, 430 and 490 cm^{-1}, which are assigned to magnetic excitations because of the temperature dependence of their intensities and correlation with IR absorption [7] and inelastic neutron scattering [3] results. Note that phonon first-order Raman scattering is forbidden according to the selection rules for the centrosymmetric high- and low-temperature structures of CoO. Thus, the weak broad features near 320, 530 and 1030 cm^{-1} as well as a sharper peak at 697 cm^{-1} observed under normal conditions are attributed to two-phonon Raman scattering. This assignment is consistent with the calculated phonon dispersion curves and phonon density of states [10] and neutron scattering data [2]. However, the strong band near 1040 cm^{-1} appears to be of magnetic origin (at least partly), because it becomes much more intense at low temperatures. We shall return to this point below.

Figure 2 shows the low-temperature Raman spectra of CoO at various pressures. The frequencies of magnetic excitations as a function of pressure are shown in Fig. 3. The pressure leads to an increase in frequency of all the observed Raman features except the low-frequency line at 143 cm^{-1}, which shows a small negative pressure coefficient. For this line we also observed a non-regular variation in frequency with pressure which is larger than the experimental uncertainty. Note also the substantial broadening of this band under pressure, which is apparently not due to non-hydrostatic conditions. This behavior could be related to the complex domain structure of CoO and may result from varying stress conditions in different

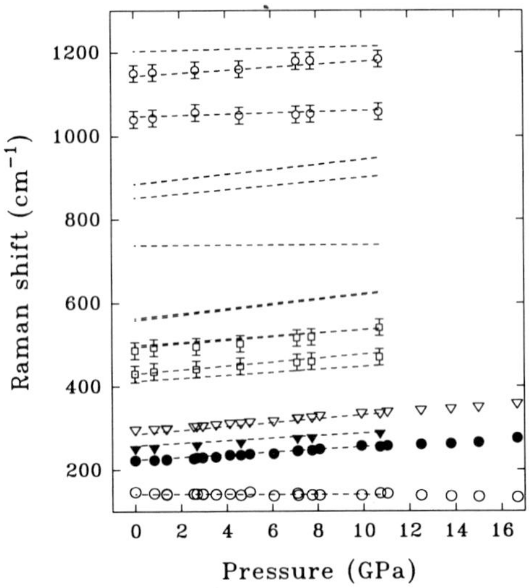

Fig. 3. Frequencies of magnetic excitons in CoO (20 K) as a function of pressure. The dashed lines correspond to the results of fitting a model Hamiltonian (see text) to the experimental data.

domains. It is known that the magnetostrictive component in the spin-anisotropy of Co^{2+} is appreciable [11]. The pressure coefficients for the high-frequency line at 1040 cm^{-1} and its shoulder at 1150 cm^{-1} are found to be much smaller compared with that for the broad high-frequency band seen at room temperature near 1030 cm^{-1}. Thus, we assign these two lines to magnetic excitations.

A treatment of the ground state of the Co^{2+} ion in the weak crystal field limit was presented by Abragam and Pryce [12] for tetragonal and trigonal symmetry and by Thornley *et al.* [13] in the strong-field limit for octahedral symmetry. For CoO in its low-temperature phase, not only the tetragonal distortion is important, but in various calculations [3, 5] the spin-anisotropy term was also found to be quite appreciable. In the case of C_{2h} symmetry (low-temperature phase of CoO) and appreciable single-ion anisotropy [11], we should take into account the terms $D_{xy}(S_x^2-S_y^2)$ and $D_zS_z^2$ [14], so our effective single-ion Hamiltonian reads

$$H_i = C_z l_z^2 + C_{xy}(l_x^2 - l_y^2) - \alpha\lambda(\mathbf{lS}) + D_{xy}(S_x^2 - S_y^2) + D_z S_z^2 + H_{ex} S_z \quad (1)$$

Here $\mathbf{l}$ is the effective orbital momentum, $\mathbf{S}$ is the spin, $\lambda = 180\ \mathrm{cm}^{-1}$ is the free ion spin-orbit coupling constant [15], $C_z l_z^2$ and $C_{xy}(l_x^2 - l_y^2)$ describe the tetragonal and lower symmetry distortions respectively, α is the effective Landé factor, $1 \leqslant \alpha \leqslant 1.5$ according to ref. 12, H_{ex} is the effective mean field term resulting from the average exchange interaction with neighboring spins. We neglect all higher order terms in spin and orbital momentum [11]. After calculation of the mean-field energy levels for the Hamiltonian in eqn. (1), the next step is to take into account the exchange interaction between neighboring spins. This was done as described by Buyers *et al.* [8]. Mean-field energy levels and matrix elements of the spin component S_z and spin raising and lowering operators S^+ and S^- between the ground and excited states of the Hamiltonian in eqn. (1) are used as inputs to build the spin-wave Hamiltonian [8] which is diagonalized to obtain the energy levels of magnetic excitons.

We used the magnetic excitations at 140, 220, 296, 251, 430, 490, 1040 and 1150 cm^{-1} at every pressure to fit the parameters of the effective Hamiltonian. The nearest-neighbor ferromagnetic exchange constant J_1 was set to be zero. The pressure dependence of the fitted parameters C_z, C_{xy}, $\alpha\lambda$, D_{xy}, D_z, J_2 (J_2 is the superexchange constant between next-nearest neighbors) is shown in Fig. 4. We find $J_2 \propto d^{-n}$ with $n = 8.0 \pm 0.5$. We used here the value $B = 170$ GPa for the bulk modulus of CoO, derived from ref. 17 using Decker's equation of state for NaCl [18]. The pressure dependence of J_2 is weaker compared with the experimentally determined pressure dependence of the Neel temperature [16]. Corresponding coefficients are $\mathrm{d}\ln J_2/\mathrm{d}P = 1.60(8) \times 10^{-2}\ \mathrm{GPa}^{-1}$ and $\mathrm{d}\ln T_N/\mathrm{d}P = 2.0(2) \times 10^{-2}\ \mathrm{GPa}^{-1}$.

4. Summary

We have investigated the effect of pressure on magnetic excitations in CoO in the temperature range 10–300 K up to 17 GPa. In particular, we have measured weak Raman features at 251, 430, 486 cm^{-1} and an asymmetrical broad band centered around 1030 cm^{-1} at low temperatures (20 K), which have not been observed previously. A single-ion effective Hamiltonian for Co^{2+} is proposed to fit the energy levels of magnetic excitations using an approximate second quantization technique. The exchange interaction constant J_2 for next-nearest neighbors is found to scale as d^{-n}, where d is the bond length and $n = 8.0 \pm 0.5$. This deviates from the empirical Bloch law ($n = 10$) [19].

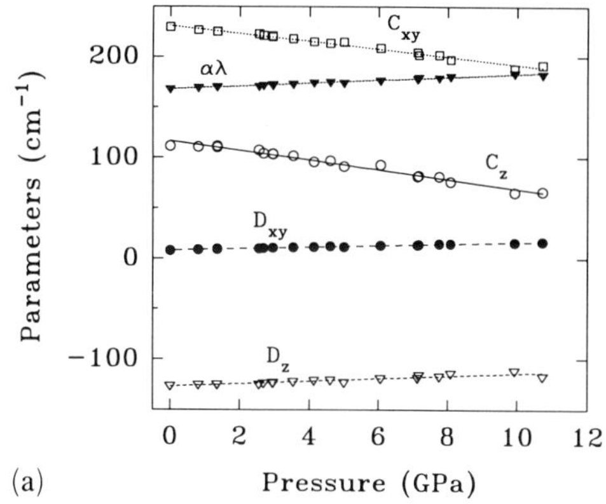

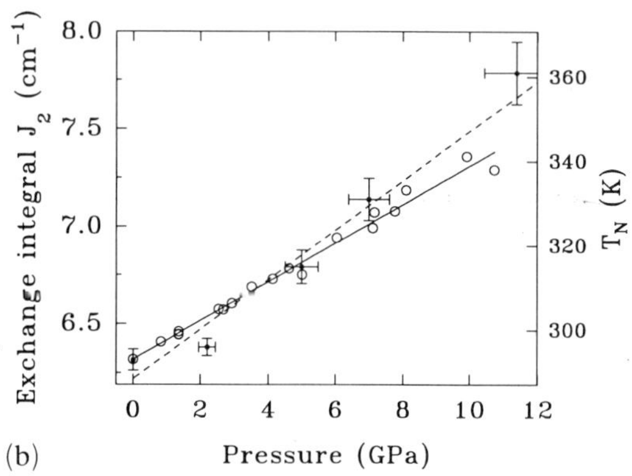

Fig. 4. (a) Parameters of the effective Hamiltonian (see text) as a function of pressure. (b) Superexchange integral J_2 between next-nearest neighbors in CoO as a function of pressure (solid line and points) compared with the pressure dependence of the Neel temperature (dashed line and full circles) according to ref. 16.

Acknowledgments

V.V.S. acknowledges financial support from DAAD. A.F.G. acknowledges support by the A.v. Humboldt Foundation. The authors would also like to thank W. Dieterich and U. Weber for excellent technical assistance.

References

1 S. Saito, K. Nakahigashi and Y. Shimomura, *J. Phys. Soc. Jpn., 21* (1966) 850.
2 J. Sakurai, W. J. L. Buyers, R. A. Cowly and G. Dolling, *Phys. Rev., 167* (1968) 510.
3 V. Wagner, *J. Phys. (Paris), Coll. C6, 39* (1978) C6-822.
4 R. R. Hayes and C. H. Perry, *Solid State Commun., 14* (1974) 173.
5 H.-h. Chou and H. Y. Fan, *Phys. Rev. B, 13* (1976) 3924.
6 M. R. Daniel and A. P. Cracknel, *Phys. Rev., 177* (1969) 932.
7 I. G. Austin and E. S. Garbet, *J. Phys. C, 3* (1970) 1605.
8 W. J. L. Buyers, T. M. Holden, E. C. Swensson, R. A. Cowley and M. T. Hutchings, *J. Phys. C, 4* (1971) 2139.
9 I. F. Silvera and R. J. Wijngaarden, *Rev. Sci. Instrum., 56* (1985) 121.
10 M. S. Kushwaha, *Physica B, 112* (1982) 232.
11 J. Kanamori, *Prog. Theor. Phys. (Kyoto), 17* (1956) 177, 197.
12 A. Abragam and M. H. L. Pryce, *Proc. R. Soc. London, Ser. A, 206* (1951) 173.
13 J. H. M. Thornley, C. G. Windsor and J. Owen, *Proc. R. Soc. London, Ser. A, 284* (1965) 252.
14 Y. Tanabe and H. Kamimura, *J. Phys. Soc. Jpn., 13* (1958) 394.
15 O. Laporte, *Z. Phys., 47* (1928) 761.
16 W. B. Holzapfel and H. G. Drickamer, *Phys. Rev., 184* (1969) 323.
17 R. L. Clarendon and H. G. Drickamer, *J. Chem. Phys., 44* (1965) 4223.
18 D. L. Decker, *J. Appl. Phys., 42* (1971) 3239.
19 D. Bloch, *J. Phys. Chem. Solids, 27* (1966) 881.

Optical studies of polymers in high pressure gas environments

B. J. Briscoe and C. T. Kelly

Particle Technology and Interfacial Phenomena Group, Department of Chemical Engineering and Chemical Technology, Imperial College of Science, Technology and Medicine, Prince Consort Road, Kensington, London SW7 2BY (UK)

Abstract

This paper primarily describes the use of a novel optical high pressure cell to examine the microscopic and macroscopic interactions between supercritical carbon dioxide and a polyurethane elastomer. Fourier transform IR spectroscopy was employed to examine the effect of increasing ambient gas pressure on the molecular conformation of the polymer, while a simple visual experimental configuration was used to quantify the corresponding gas-induced volume changes. A simple first-order analysis of the resulting data has shown a correlation between the induced volume changes in the polymer and changes in the polymer's IR spectrum (associated with changes in the hydrogen bonding which occurs in the polymer) at carbon dioxide pressures of up to 16 MPa.

1. Introduction

Organic polymers are frequently employed in engineering applications where they are exposed to extremely hostile ambient gaseous conditions. The high elasticity of elastomers, for example, makes them suitable for use as seals and hoses in high pressure gaseous environments. It is established that these types of polymers are capable of absorbing large quantities of gas; this capability is particularly enhanced at high pneumatic pressures [1]. This gas sorption process naturally induces swelling in the polymer. A sudden reduction in the ambient gas pressure can cause serious and irreversible mechanical damage to the polymer, a phenomenon often referred to as explosive decompression failure [2].

The study of this problem has led the present authors to consider means of obtaining information regarding the microscopic equilibrium interactions between gases at elevated pressures and elastomers. Conventionally, the effects of high pressure gases on polymers have been investigated by sample inspection and testing before and after exposure to pneumatic stress, but not generally *in situ*. More recently, studies have been carried out *in situ* during gas pressurization and depressurization. However, even these investigations have largely concentrated on yielding information regarding changes in the macroscopic or bulk properties of a polymer (for example, mass and volume changes) as it is exposed to fluctuating gas pressures. This paper describes the use of a purpose built high pressure optical cell which allows novel studies into gas–polymer interactions. Using an optical fibre link between a Fourier transform IR spectrometer and the pressure vessel, spectroscopic data, which reflect the effect of gases on the polymer morphology, have been obtained. A visual experimental configuration of the optical cell also enabled measurement of the volumetric changes produced in polymers during gas mass sorption.

The primary aim of this paper is to describe the experimental components and application of the high pressure optical cell. At a speculative level, the significance of the spectral data obtained of a polyurethane–CO_2 gas system is interpreted within the context of precedence based on the known morphological features of polyurethanes and the way in which hydrogen-bond disruption may occur.

2. Experimental details

2.1. *Materials*

The material of interest in this study was a polyester/MDI (diphenyl methyl-4-4-diisocyanate) based polyurethane elastomer, chain extended with a blend of polyester polyols (Kemira Polymers, Diprane 54 series). The material was supplied in the form of a ready blended isocyanate prepolymer, which was subsequently mixed with the supplied curing agent in the ratio 1:4.7 (parts by weight). Following a thorough mixing of the components, the reaction mixture was degassed under vacuum (5 Torr minimum) until bubbling ceased. The mixture was then cast into the required

0921-5093/93/$6.00

moulds and post cured for 12 h at 80 °C. 15 μm thick films were prepared for the spectroscopic studies, and 8 mm cubes cast for the visual investigations. The cross-link density of the resulting polymer was relatively low. The gas used was carbon dioxide (CO_2) (BOC) and had a nominal purity of 99.5%.

2.2. *High pressure equipment*

The high pressure equipment consisted essentially of the gas feed (cylinder), an intermediate gas receiver, a reciprocating gas booster pump (Charles Madan, Altrincham) and the high pressure optical cell. A detailed description of the pumping system may be found elsewhere [3].

Figure 1 shows the main design features of the high pressure cell. The cell was heated using a heating tape (RS Components, UK) in conjunction with a temperature controller (British Electric Resistance Co., UK). The temperature of the vessel containing the prepared specimens was maintained above the critical temperature of CO_2 (31 °C), at 42 °C.

2.3. *Visual studies*

Sapphire windows (Goodfellow, UK) 16 mm thick and 35 mm in diameter were employed in the optical cell for the visualization studies with a camera (Pulnix TM-90, USA) placed at one of the vessel windows and a white light source at the other. The camera was linked to a TV monitor/video recorder system (Pace, UK) for data collection. The cell was arranged with a total path length of 30 mm and an 8 mm cubic polyurethane sample was placed inside the vessel. The ambient CO_2 pressure was raised in steps of *ca.* 2–3 MPa to a maximum of 16 MPa. The gas–polymer system was allowed to equilibrate at each pressure step for 1 h and the changes in specimen dimensions were subsequently recorded using the TV/video system. The video recordings were then analysed using a calibrated Optomex image analyser (Analytical Measuring Systems, UK) to calculate the exact dimensional changes in the *X* or *Y* directions of the polyurethane specimen, at each gas pressure increment.

2.4. *Spectroscopic studies*

Zinc selenide (ZnSe) windows (Harrick Scientific Corp., USA) 16 mm thick and 35 mm in diameter were employed for the spectroscopic studies because of their transmission range (0.58–21 μm). The high pressure cell was linked to the IR spectrometer (Bomen Ramspec 152, USA) via a transmitter/receiver fibre probe system (Graseby Specac Ltd., UK) and microbeam interface (26400 Bomen Michelson 100, USA). The cell had a total path length of 5 mm with a 15 μm thick polyurethane film supported in the cavity. The experimental procedure assumed for the dimensional changes was adopted. The gas–polymer system was allowed to equilibrate for 1 h at each pressure step before an IR spectrum was recorded. All spectra were obtained at a resolution of 4 cm^{-1} averaged over 750 scans, using a liquid nitrogen cooled MCT (mercury cadmium telluride) detector. The spectra obtained at each gas pressure increment were subsequently examined and the observed changes recorded.

3. Results and discussion

3.1. *Visual studies*

The change in dimension of a polyurethane elastomer, as a function of the increasing pressure, is given in Fig. 2. The data are presented as the percentage linear expansion of the elastomeric sample, pressurized up to *ca.* 16 MPa. It is observed that as the

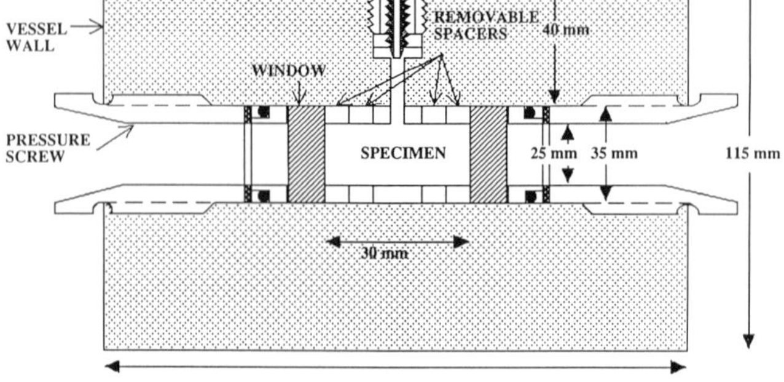

Fig. 1. Schematic representation of the high pressure optical cell.

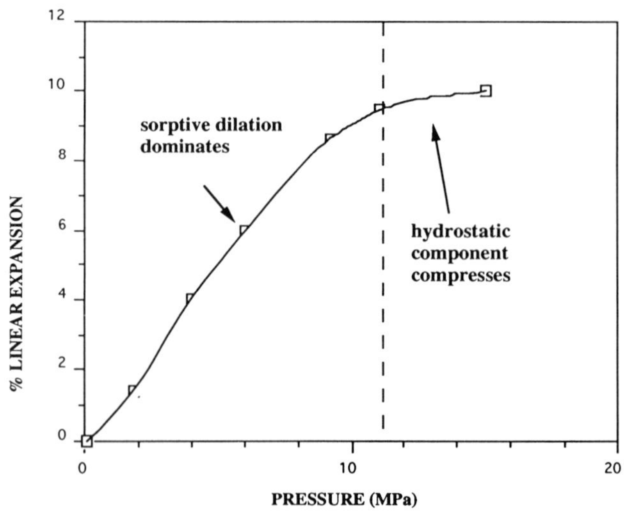

Fig. 2. Linear expansion of the polyurethane as function of increasing gas pressure.

pressure increases up to 10–11 MPa, the polymer dilates by *ca.* 9%. However, at pressures above this level, the hydrostatic effect of the gas begins to dominate, inhibiting gas sorption and as a result the rate of swelling is seen to decrease quite significantly. The dimensional change between 11 MPa and 16 MPa is approximately 0.5%. The sample volume appears to attain a limiting value at higher pressures. Solvent (acetone) swelling studies of polyurethane have indicated that dilation is not inhibited by chemically cross-linked molecular constraints [4]. This trend is similar to that reported by Briscoe and Zakaria [5] for CO_2 induced swelling of a silicone elastomer and Wissinger and Paulaitis [6] for CO_2 induced dilation of PMMA.

3.2. Spectroscopic studies

Fourier transform IR spectroscopy is commonly used as a tool to follow the effect of environmental changes (such as temperature fluctuations and exposure to organic solvents) on the molecular structure and morphology of polymers [7–10]. In polymers, where hydrogen bonding is known to be present, these investigations frequently focus on the regions of the spectra corresponding to the potential hydrogen bonding groups. It is established that polyurethanes are extensively hydrogen bonded and, in the polyurethane investigated here, the N–H of the urethane linkage acts as the proton donor, while the proton acceptor may be either the carbonyl of the urethane or the ester carbonyl of the polyester soft segment. These N–H and C=O groups show absorbance bands composed of two contributions, attributed to either "bonded" or "free" groups, which are known to be affected by changes to the polymer's environment [11, 12]. For a first-order analysis of the spectra produced during this experiment, the N–H region was focused on because of the complexity associated with the peak in the carbonyl stretching region. (The carbonyl stretching band is known to be composed of at least four different contributions: "bonded" urethane C=O, "bonded" ester C=O, "free" urethane C=O and "free" ester C=O.)

Figure 3 shows the N–H stretching region in polyurethane at atmospheric presure, the peak at 3349 cm^{-1} being indicative of "bonded" N–H while the shoulder at 3449 cm^{-1} is associated with "free" N–H groups. Figure 3 also shows the N–H stretching region at 16 MPa; the peak maximum of the "bonded" N–H was seen to shift gradually to a higher frequency, towards the "free" N–H shoulder, as the gas pressure was increased in steps to 16 MPa. Figure 4 shows a plot of this shift in frequency against the pressure, where it was noted that at a pressure of *ca.* 11 MPa the frequency shift appeared to approach a plateau value. Shifts in the bonded N–H peak towards the free N–H peak are generally interpreted as a disruption to the hydrogen bonding [11]. A deconvolution of the N–H stretching bands at atmospheric pressure and at 16 MPa also revealed that the ratio of "bonded" N–H

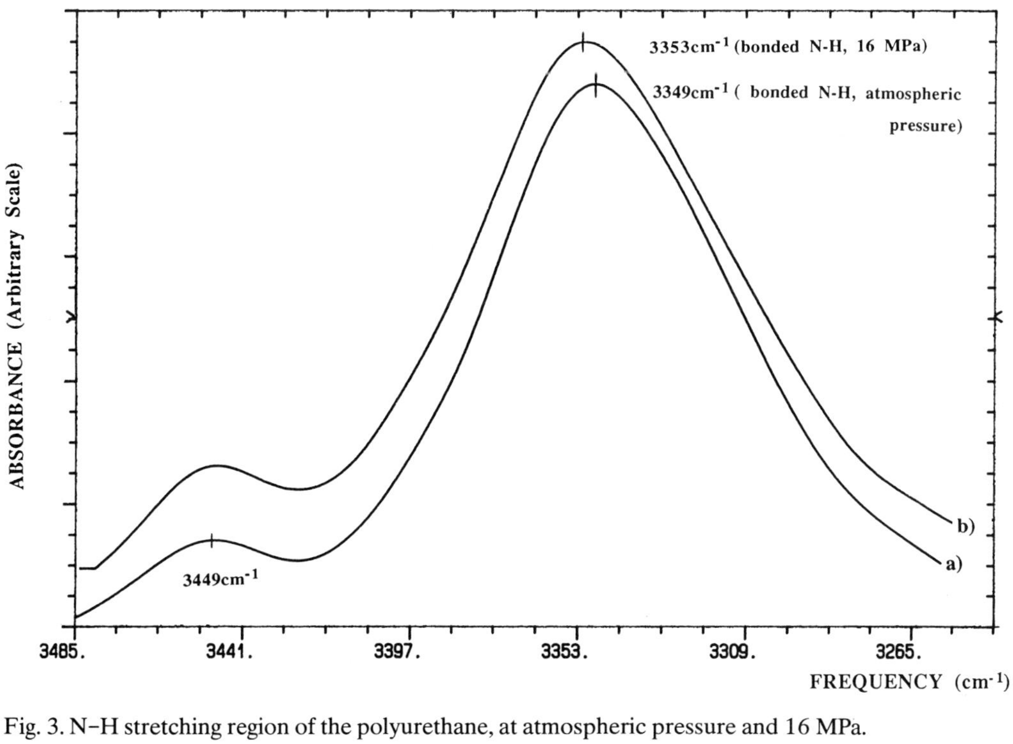

Fig. 3. N–H stretching region of the polyurethane, at atmospheric pressure and 16 MPa.

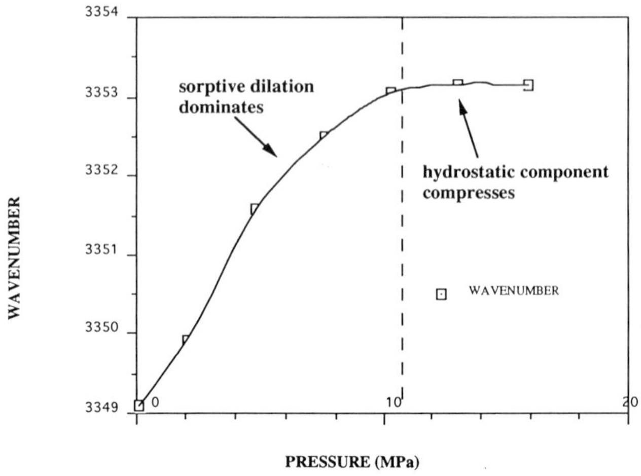

Fig. 4. Shift in frequency for the hydrogen bonded N–H stretching band maximum as a function of increasing carbon dioxide pressure.

groups to "free" N–H groups changed from 9.5:0.5 at atmospheric pressure to 9:1 at 16 MPa. These data are thus consistent with the observed spectral shifts. As the polymer was exposed to increasing CO_2 pressures, a gradual disruption was seen to occur to the extent of the hydrogen bonding. In the following paragraph we speculate on the origins of the hydrogen-bond disruption.

Polyester-based polyurethane elastomers may be regarded as block copolymers containing flexible (soft) segments alternating along the polymer chain with strongly interacting rigid segments. The soft segments consist of the polyester while the rigid segments contain the polyurethane linkage. It is generally believed that these copolymers have a "two-phase" microstructure [13], where the rigid segments separate, in the main, to form discrete highly ordered hard domains within the soft segment matrix. The rigid segments are held together in these domains by hydrogen bond interactions between the N–H of the urethane linkage and the C=O of the urethane. However, hydrogen bonding interactions are also known to occur between the rigid segment N–H groups and the soft segment oxygen (the C=O of the polyester), possibly at either the rigid segment domain interface, or between rigid and soft segments which are not located in domains [11, 13, 14]. The resulting structure of the polymer may possibly contain regions of high order or paracrystallinity (rigid segment domains) and low order, consisting of mainly a soft polyester segment matrix, in which rigid segments may be interdispersed. Hydrogen bonding may exist in both regions.

It is hypothesized tht the imbibed carbon dioxide preferentially infiltrates the low order soft polyester phase of the polyurethane, where the polymer chains have the flexibility to move to accommodate the gas. As the gas pressure increases, further movement of the polymer chain occurs, causing a gradual disruption to the hydrogen bonds which exist in this region. The rigid segment domains can act as filler particles [15], and hence may be largely impermeable to the penetrant gas molecules. On this basis it is assumed here that the significant majority of the hydrogen bonds affected by the gas are those hydrogen bonds located within the low order, soft segment regions. This may account for the size of the shift observed—not all of the N–H groups (*i.e.* those contained in the rigid segment domains) are affected by the imbibed carbon dioxide. It must be stressed, however, that this interpretation is tentative, based on the known morphology of polyurethanes and the fact that a relatively small change to the hydrogen bonding was observed as the gas pressure was increased.

At pressures above 11 MPa, the hydrostatic component of the gas pressure begins to dominate, so that the rate of gas uptake decreases significantly and movement of the polymer chains is inhibited any further. Thus, no further disruption to the hydrogen bonding occurs, and a plateau in the frequency shift against pressure curve is observed.

4. Conclusions

The optical high pressure cell has proven to be a potentially useful tool to characterize interactions between high pressure carbon dioxide and a polyurethane elastomer. Using the cell in a visual configuration, the gas-induced swelling of the polymer has been deduced. Spectroscopic studies have revealed interaction between the gas and the polymer, reflected in the changing nature and extent of the hydrogen bonding which exists in this class of polymers. A correlation has been observed between the extent of swelling and spectroscopic shift data. Dilation of the polymer and the frequency shift associated with disruption of the hydrogen bonding both appear to attain limiting values at a similar pressure of *ca.* 11 MPa. This has been attributed to the hydrostatic component of the gas pressure becoming dominant at pressures above 11–12 MPa, suppressing gas uptake into the polymeric matrix.

We have chosen to interpret the spectral data, in a rather speculative way, as a disruption of the hydrogen bonding which occurs in the softer phase, based on the available information regarding the morphological structure of this class of polymers. However, a degradation of the rigid segment domains *cannot* be positively discounted on the basis of our experimental data. These speculations apart, the paper demonstrates the potential for high pressure spectroscopy and visualization as a means of studying gas–polymer interactions. This is the main purpose of this paper.

Acknowledgments

The authors gratefully acknowledge the financial support provided by Shell Expro. The assistance provided by K. Gurney and A. Pushman in the manufacture and assembly of some of the experimental equipment is also gratefully acknowledged.

References

1 D. H. Ender, *Chemtech, 16* (1) (1986) 52.
2 B. J. Briscoe and S. Zakaria, *Chemtech* (August 1990), 492.
3 H. Mahagerefteh, The behaviour of PTFE at high pressures, *PhD Thesis*, Imperial College, University of London, 1984.
4 B. J. Briscoe and C. T. Kelly, unpublished results, 1991.
5 B. J. Briscoe and S. Zakaria, Gas–polymer interactions at high pressures, *PhD Thesis*, Imperial College, University of London, 1990.
6 R. G. Wissinger and M. E. Paulaitis, *J. Polym. Sci. Polym. Phys. Ed., 25* (1987) 2497.
7 F. M. Fowkes, D. O. Tischler, J. A. Wolfe, L. A. Lannigan, C. M. Ademu-John and M. J. Halliwell, *J. Polym. Sci. Polym. Chem. Ed., 22* (1984) 547.
8 M. M. Coleman, K. H. Lee, D. J. Skrovanek and P. C. Painter, *Macromolecules, 19* (1986) 2149.
9 M. F. Roberts and S. A. Jenekhe, *Macromolecules, 24* (1991) 3142.
10 B. J. Briscoe, P. S. Thomas and D. R. Williams, *Proc. 19th Leeds–Lyon Symp. on Tribology of Thin Films*, Elsevier, 1993.
11 D. I. Bower and W. F. Madams, *The Vibrational Spectroscopy of Polymers*, Cambridge University Press, Cambridge, 1989, p. 256.
12 J. T. Koberstein and I. Gancarz, *J. Polym. Sci. B: Polym. Phys., 24* (1986) 2487.
13 C. Hepburn, *Polyurethane Elastomers*, Applied Science Publishers, London, 1982, p. 65.
14 D. C. Allport and W. H. Janes (eds.), *Block Copolymers*, Applied Science Publishers, London, 1973, p. 443.
15 J. S. McBride, T. A. Massaro and S. L. Cooper, *J. Appl. Polym. Sci., 23* (1979) 201.

Narrowband Landau emission from high purity GaAs layers grown by molecular beam epitaxy

C. Wirner, G. Strasser, C. Steeb and E. Gornik
Walter-Schottky-Institut, Am Coulombwall, D-8046 Garching (Germany)

H. Riechert
Siemens AG, Research Laboratories, D-8000 Munich (Germany)

Abstract

We analysed narrowband Landau emission from a high purity GaAs layer grown by molecular beam epitaxy using a magnetic field tuned InSb detector. The cyclotron resonance and impurity level transitions of the InSb detector are clearly resolved. Comparison of the detector linewidth with results from a fast Fourier transform spectrometer yields a narrow GaAs emitter linewidth of 1.3 ± 0.3 cm^{-1} which is in good agreement with cyclotron resonance transmission data. This indicates very high quality of the GaAs material with very low impurity compensation. Only two types of impurity, silicon and sulphur, are detected in the extrinsic photoresponse spectrum.

1. Introduction

High purity epitaxial layers of GaAs have been extensively studied for use as photodetectors as well as Landau emitters in the far IR [1–3]. The photoresponse of GaAs at low temperatures is dominated by the shallow impurities in the material due to carrier freeze out. Consequently, the peaks in the photosignal yield important information about the type and density of impurities in the high purity material [4]. The linewidths become very narrow (approximately 0.25 cm^{-1}) in layers with very low impurity density [1]. Radiative emission of electrons between Laundau levels ("Landau emission") was first investigated by Waldmann *et al.* in impact ionized high purity GaAs [5]. Lindemann *et al.* [3] found a linear dependence of the emission linewidth on the square root of the total ionized impurity concentration. Therefore, the application of GaAs layers as narrowband radiation sources and high resolution photodetectors in the far IR depends critically on a very low impurity content in the semiconductor material.

In this paper we report narrowband Landau emission from a high purity GaAs layer grown by molecular beam epitaxy which was investigated using an improved magnetic field tunable InSb detector [6]. The linewidth of the emitter is derived by comparing the InSb detector response with results obtained from a fast Fourier transform spectrometer and checking with cyclotron resonance transmission data. In addition the impurities are identified by studying the extrinsic photoresponse of the high purity layer.

2. Emission experiments

The 12 μm thick high purity GaAs layer was grown by molecular beam epitaxy on a semi-insulating GaAs substrate. The layer was silicon doped with a total donor concentration of 2×10^{14} cm^{-3}. Ohmic contacts were made by annealing indium pads at 350 °C in an inert gas atmosphere. At 77 K a carrier concentration of 2.85×10^{14} cm^{-3} and an electron mobility of 1.96×10^5 cm^2 V^{-1} s^{-1} were determined by temperature-dependent Hall measurements.

All emission experiments were performed in a liquid helium cryostat with two magnet coils for the GaAs emitter and the InSb detector system respectively [7]. At constant emitter magnetic field, pulsed electric fields of several volts per centimetre were applied to the GaAs layer to generate Landau emission.

Figure 1 shows the 4.2 K InSb detector signal for five different GaAs emitter magnetic fields. The Landau level and impurity level transitions of the InSb detector are clearly resolved at detector fields higher than 0.3 T which gives first evidence of a narrow emitter linewidth well below the distance of both detector lines (6 cm^{-1}).

0921-5093/93/$6.00

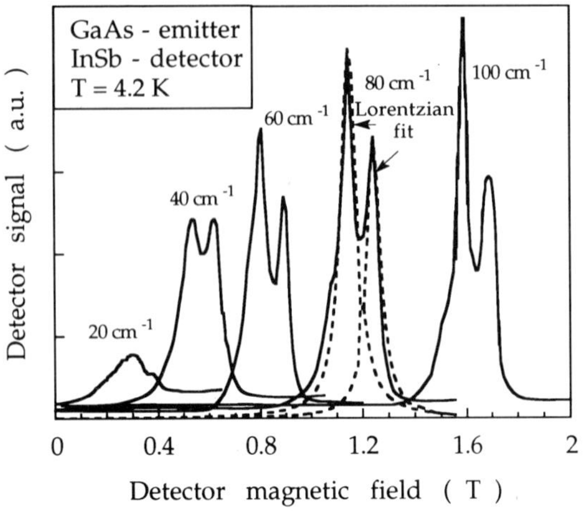

Fig. 1. 4.2 K InSb detector signal (a.u. arbitrary units) at five different wavenumbers of the high purity GaAs emitter: ———, the impurity- and cyclotron resonance lines are clearly resolved; - - -, Lorentzian fit.

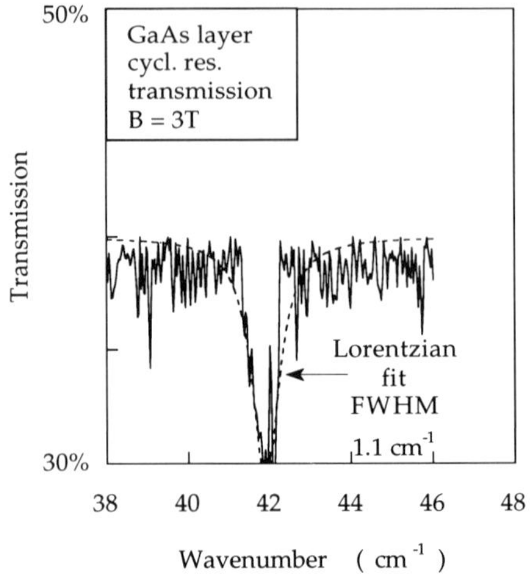

Fig. 3. Cyclotron resonance transmission data of the GaAs emitter at a magnetic field of 3 T, measured with the fast Fourier transform spectrometer (———), Lorentzian fit (- - -).

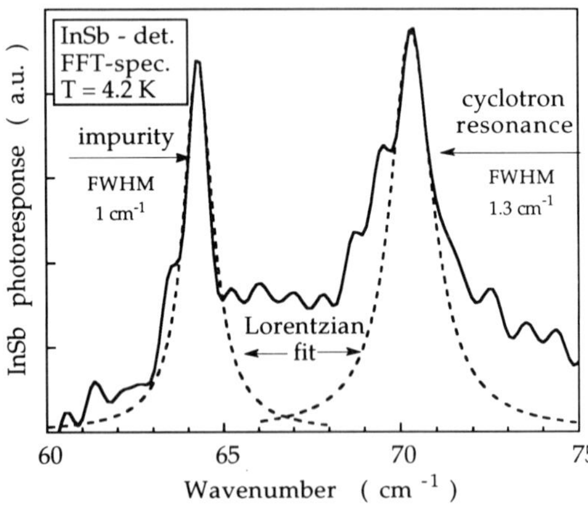

Fig. 2. Photoconductive response (a.u. arbitrary units) of the InSb detector at 4.2 K and a detector magnetic field of 1 T, measured with the fast Fourier transform spectrometer (———), Lorentzian fit (- - -).

At 80 cm^{-1} GaAs emitter wavenumber, a Lorentzian fit gives a full width at half-maximum (FWHM) of 2.6 cm^{-1} for the InSb impurity line and a FWHM of 2.4 cm^{-1} for the InSb cyclotron resonance line.

In order to determine the GaAs emitter linewidth the InSb FWHM was compared with the unbroadened FWHM obtained with a high resolution (0.05 cm^{-1}) fast Fourier transform spectrometer. In this experiment the detector magnetic field was kept constant at 1 T, corresponding to a cyclotron resonance wavenumber of 71 cm^{-1}. The results are displayed in Fig. 2. The impurity and cyclotron resonance lines are now clearly separated, having FWHM values of 1 cm^{-1} and 1.3 cm^{-1} respectively. Consequently, the linewidth of the GaAs emitter has to be of the order of 1.3 ± 0.3 cm^{-1} by subtracting the equivalent FWHM values.

The linewidth $\Delta\omega$ is strongly correlated to the total impurity content N_I due to a $\Delta\omega \propto \sqrt{N_I}$ law first observed by Lindemann *et al.* [3]. From this one obtains a total impurity content of 2.1×10^{14} cm^{-3} in the GaAs layer, which is very close to the silicon donor density. This indicates the high quality of the GaAs material with very low impurity compensation resulting in narrow emission linewidths.

3. Transmission experiments

Studying the cyclotron resonance transmission linewidth is another method for deriving the impurity content in the GaAs layer, since in both cyclotron emission and transmission experiments the same dependence of linewidth on impurity has been observed [4].

Figure 3 displays cyclotron resonance transmission data of the GaAs sample measured with the fast Fourier transform spectrometer. The spectrum was taken at a magnetic field of 3 T, corresponding to a cyclotron resonance wavenumber of 42 cm^{-1}. To observe cyclotron resonance due to free carriers, the sample was biased with voltages above impurity breakdown. From the transmission data one obtains a FWHM of 1.1 cm^{-1} which is of the same order of magnitude as the value measured from Landau emission. This confirms the results of linewidth and impurity content obtained from the emission experiments.

4. Extrinsic photoresponse

From a technological point of view it is very important to obtain more information about the different

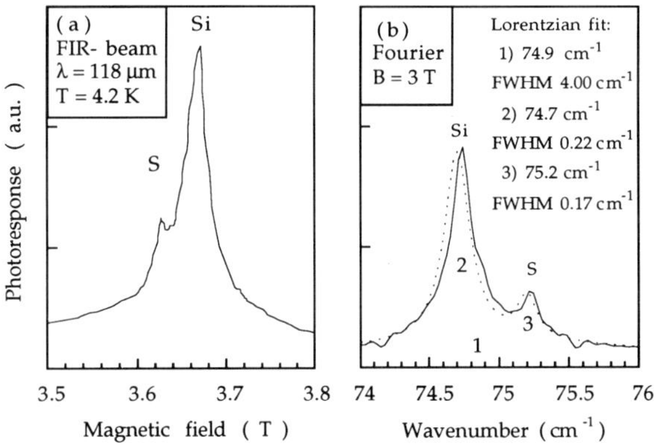

Fig. 4. Photoconductivity spectra (a.u. arbitrary units) of the GaAs layer: (a) taken with a far IR laser beam, (b) taken with a fast Fourier transform spectrometer.

types of impurities present in the GaAs layer. Study of the extrinsic photoresponse at 4.2 K allows the identification of impurities since every type of impurity has a different ionization energy.

Figure 4 displays photoconductivity data taken from a far IR laser beam with a fixed wavelength of 118 μm (a) and a Fourier transform spectrometer with fixed detector magnetic field of 3 T (b). In both spectra a broad peak with one satellite is clearly resolved. The main peak is attributed to the 1s–2p+ transition of silicon by comparing the energetic peak position with theoretical data from Low and coworkers [8–10] and Bose *et al.* [11] which is obvious since the GaAs layer is silicon doped. The satellite peak is attributed to sulphur. No additional impurities were found within the given resolution of the system. Assuming a similar relation between the photoconductivity linewidth and the individual impurity density as in the cyclotron resonance emission experiments, $\Delta\omega_S/\Delta\omega_{Si} \propto (N_S/N_{Si})^{1/2}$, a relative impurity content of sulphur to silicon $N_S/N_{Si} = 0.6$ is derived from the individual linewidths (Fig. 4(b), 2,3). The total linewidth in the photoresponse signal (Fig. 4(b), 1) gives evidence of an impurity density of 1×10^{16} cm^{-3} derived from the $\Delta\omega \propto \sqrt{N_I}$ law. Taking the intensity of the photoresponse signal into account we find that this high doped background impurity is restricted within a thin layer of approximately 0.1 μm between the semi-insulating substrate and the high purity GaAs layer. It might be caused either by imperfections of the surface of the GaAs substrate or by impurities at the beginning of the growth process. The background impurity of the remaining GaAs layer has to be of the order of about 10^{13} cm^{-3} since the total impurity content derived from the individual linewidths is close to the silicon donor density.

5. Conclusion

Landau emission from high purity GaAs layers grown by molecular beam epitaxy was analysed by a magnetic field tunable InSb detector. Comparison of the InSb detector linewidth with that from a high resolution Fourier transform spectrometer yields a narrow linewidth of 1.3 ± 0.3 cm^{-1} for the GaAs emitter. This indicates high quality of the GaAs material with very low impurity compensation and a total impurity content close to the silicon donor density of 2×10^{14} cm^{-3}. The result is confirmed by additional cyclotron transmission data giving linewidths of the same order of magnitude. From a study of the extrinsic photoconductivity two types of impurity, silicon and sulphur, are clearly resolved.

Acknowledgment

This work was sponsored by the Deutsche Forschungsgemeinschaft.

References

1 E. Gornik, *Physica B, 127* (1984) 95.
2 G. E. Stillman, C. M. Wolfe and J. O. Dimmock, *Solid State Commun., 7* (1969) 921.
3 G. Lindemann, E. Gornik, R. Schawarz and D. C. Tsui, *Inst. Phys. Conf. Ser., 56* (1981) 631.
4 G. Strasser, S. Dubois, M. Besson, E. Gornik, G. Weimann, E. Bauser and H. Riechert, *Appl. Surf. Sci., 50* (1991) 261.
5 J. Waldmann, T. S. Chang, H. R. Fetterman, G. E. Stillman and C. M. Wolfe, *Solid State Commun., 15* (1974) 1309.
6 G. Strasser, K. Bochter, M. Witzany and E. Gornik, *Infrared Phys., 32* (1991) 439.
7 E. Gornik, W. Müller and F. Kohl, *IEEE Trans. Microwave, 22* (12) (1974) 991.
8 T. S. Low, G. E. Stillman, A. Y. Cho, H. Morkoc and A. R. Calawa, *Appl. Phys. Lett., 40* (1982) 611.
9 T. S. Low, G. E. Stillman, D. M. Collins, C. M. Wolfe, S. Tiwari and L. F. Eastman, *Appl. Phys. Lett., 40* (1982) 1034.
10 T. S. Low, G. E. Stillman, T. Nakanisi, T. Udagawa and C. M. Wolfe, *Appl. Phys. Lett., 41* (1982) 183.
11 S. S. Bose, B. Lee, M. H. Kim and G. E. Stillman, *Appl. Phys. Lett., 51* (1987) 937.

Author Index of Volume 168 Issue 1

Elsevier Sequoia

Subject Index of Volume 168 Issue 1

PART II

Symposium C
on
Nanophase Materials

Materials Science and Engineering, A168 (1993) 125–129

The synthesis of metal hydrous oxide particles by multiphase electrodispersion

Michael T. Harris, Timothy C. Scott and Charles H. Byers*

Chemical Technology Division, Oak Ridge National Laboratory, Oak Ridge, TN 37831-6224 (USA)

Abstract

A multiphase electrodispersion precipitation reactor called the electric dispersion reactor (EDR) was developed to synthesize ultrafine particles for the production of precursor powders of advanced ceramic materials. A pulsed electric field is employed to create dispersions of micrometer- and submicrometer-scale conducting (aqueous-based) drops in non-conducting (organic) liquids. Species initially dissolved in the organic phase diffuse into the aqueous droplets containing the other reagents. Reaction ensues, leading immediately to precipitation and gelation in the aqueous droplets. The EDR has produced 0.1–5 μm particles of silica, alumina, zirconia, yttria, and composite hydrous oxide particles from metal alkoxide and metal salt precursors. Also considered is the application of the EDR to the control and modification of the morphological characteristics of these particles, with illustrative examples.

1. Introduction

The synthesis of powders comprising ultrafine metal oxide powders is of primary importance in developing advanced ceramics. The particle shape, composition, and surface properties play a major role in determining the ultimate microstructure of the ceramic material [1]. Homogeneous precipitation techniques, such as the hydrolysis of aqueous salt solutions and metal alkoxides [2], have been shown to produce excellent single-component materials; however, achieving stoichiometric homogeneity in mixed metal oxide particles is more difficult. Recently, several techniques have been developed for producing such powders by the formation of microreactor environments that confine precipitation to very small regimes. These techniques include aerosol reactions [3], spray pyrolysis [4], coprecipitation in emulsions [5], and the electrostatic atomization of liquid droplets into air [6]. All these methods not only have certain advantages, but significant deficiencies as well. For example, powder synthesis by electrostatic atomization involves the disintegration of liquid drops, for example of aqueous zirconium acetate in ref. 6, into air from a nozzle at high voltage (10–15 kV). The high potential causes the meniscus at the nozzle to form a cone [7]. The accumulation of electric charge makes the cone surface unstable, causing ejection of fine jets that subsequently break into droplets. These are collected at a grounded substrate by which time the water has evaporated, leaving solid particles of zirconium acetate which is then pyrolyzed to zirconia. The main disadvantage here is that large amounts of energy are required to rupture the droplets. This is the result of high surface tension at liquid–air interfaces.

To overcome the disadvantages of these methods of producing ultrafine ceramic powders, a research program was initiated to investigate the use of high-intensity-pulsed electric fields for controlling the size of droplets dispersed into liquid [8]. The liquid–liquid reaction system significantly reduces the amount of energy required to produce ultrafine powders relative to the liquid–vapor system. Furthermore, particle agglomerations are usually soft (*i.e.* they can easily be broken up) in comparison with aerosol reactor methods and electrostatic atomization of liquid drops in air. Small amounts of surfactants may also be used to minimize agglomeration; this increases the green body density of the precipitate and consequently improves the microstructure of the ceramic.

2. The electric dispersion reactor (EDR)

The EDR is the fusion of two research concepts that focus on ultrafine powder synthesis by homogeneous precipitation of metal alkoxide solutions and electrodispersion in multiphase systems. Our homogeneous precipitation research is directed at evaluating the formation of ultrafine silica, titania, titania–zirconia, and titania–alumina particles by metal alkoxide

*Author to whom correspondence should be addressed.

0921-5093/93/$6.00

hydrolysis. Chemically homogeneous particles of titania and zirconia were synthesized by the hydrolysis of mixed titanium–zirconium alkoxide solutions. However, mixed alumina–titania particles were difficult to form. The formation of other multicomponent (ternary and higher) ultrafine hydrous metal oxide particles with well defined shapes proved to be impractical. Thus, new techniques are required to synthesize multicomponent precursor powders for advanced ceramic applications.

One such method is electrodispersion to produce microemulsions. Of particular interest is the development of a device known as the emulsion phase contactor (EPC) [9]. This device uses a pulsing d.c. electric field to enhance mass transport by dispersing aqueous drops into a relatively non-conducting continuous organic phase. The dispersion consists of micrometer-size droplets [10]; thus, the surface area is very large for interfacial mass transport. In addition to dispersing the drop, the device also promotes coalescence of the aqueous droplets [9], an excellent separation characteristic, but a disadvantage for reactions.

Pulsed or static d.c. fields can be used to cause a liquid droplet to break up by a number of modes, depending on the character of the applied field [11]. Pulsed d.c. fields can accomplish the same extent of droplet dispersion utilizing a lower r.m.s. voltage than steady d.c. fields [11]. Yet another advantage of a pulsed field is that liquid dispersion rates from nozzles can be controlled by properly adjusting the frequency of the pulsed electric field used to break up the liquid stream. Both the drop size and its distribution can be controlled by the nature of the applied field. Micrometer-size droplets formed in this way make electrodispersion an ideal technique for forming hydrous oxide particles in liquid–liquid systems. These droplets are microreactors in which precipitation and gelation occur. Particle formation by electrodispersion requires the minimization of coalescence to control the particle size distribution. The resulting conceptual device is called the electric dispersion reactor (EDR).

Figure 1 depicts a complete flowsheet of the EDR including its fluid-handling and solid-removal system. Early versions of the reactor section were constructed from a 6 inch long and 1 inch internal diameter glass pipe with conventional flanges and a grounded nozzle consisting of an 1/8 inch stainless steel tube with a 5/8 inch PVC tube positioned coaxially to the nozzle, thus forming an annulus through which the continuous organic phase entered the reactor. This arrangement allowed the organic phase to sweep the dispersion past the electrode region, thus reducing unwanted coalescence. The positive electrodes consisted of two $5/8 \times 7/8 \times 1/16$ inch^3 stainless steel plates located in the continuous phase downstream of the nozzle tip.

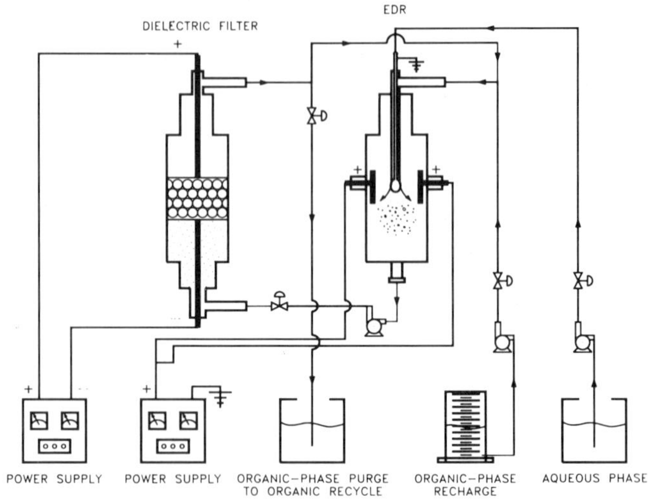

Fig. 1. Continuous EDR process showing the dielectric filtration unit for removal of suspended solids.

To minimize energy usage, coalescence experiments and theoretical computations were conducted to test the effect of the electrode configuration on the dispersion of aqueous droplets by electric fields [12]. While the overall results of this work are in preparation, a design result was that recent configurations involved reducing the cross-sectional area of the EDR. The electrohydrostatic computations pointed to more efficient use of the electric field with a smaller reactor, which also reduced field-induced coalescence by rapidly sweeping the dispersion past the electrode, the region of highest electric field strength. Pulsed fields were generated with a d.c. power supply driving a BK precision model 3030 sweep/function generator and a 12 V ignition coil (Mallory 29625). The high voltage is supplied to the electrode near the nozzle tip. The inlet aqueous stream is grounded, thus completing the circuit. Normally, pulsing frequencies of 1000–5000 Hz with amplitudes of 6–40 kV are used.

In the region just above the nozzle tip, where the continuous phase is introduced into the reactor, rapid flow of the continuous phase minimizes backflow. The electric field strength is highest between the nozzle tip and the electrode, where the dispersion occurs. It is crucial that the material be swept rapidly through this zone to minimize electric-field-induced coalescence and chaining of particles. Good practice demands that precipitation–gelation occur most rapidly in this zone, thus minimizing the effects of coalescence on the final particulates. Beyond this region is the aging zone where precipitation and gelation are completed and loose flocs are formed, which are easily separated from the droplet liquid. The effluent from the reactor is passed through a solid-removal system such as a continuous centrifuge or the dielectric filter which is shown in Fig.

1 [13], where ultrafine particles and water emulsions not removed by gravitational settling can be collected.

3. Chemical methods for particle formation in the EDR

Two methods of chemical synthesis have been explored in the EDR: external and internal gelation. Traditionally, the external gelation method involves the precipitation of aqueous salt solutions and dispersing the precipitate to form a stable sol. Gelation is induced by the removal of water or by reacting the sol with ammonia. The difficult step is dispersing the precipitate to form a stable sol. To circumvent the need to disperse a precipitate into a stable sol, gel-spheres are formed in the EDR by dispersing micrometer-size aqueous droplets containing soluble metal salts into a continuum that contains a precipitation agent such as ammonia, as shown schematically in Fig. 2(a). Metal salts in the droplet react with ammonia transferring from the continuous phase, followed by precipitation of a metal oxide gel-sphere.

Alternately, the metal-containing species (metal alkoxide) begins in the continuum and the dispersed phase contains an aqueous solution of the precipitation agent, and/or ammonia (Fig. 2(b)). This technique allows the formation of hydrous metal oxide shells that upon washing can be collapsed to create high-surface-area wafers [8]. The external gelation technique allows the formation of oxide powders over a wider range of conditions than internal gelation [14] because of the added flexibility of locating reactants in both liquid phases.

4. Experimental details

For the external gelation experiments, concentrated aqueous solutions of $Y(NO_3)_2$, $Al(NO_3)_3$, zirconyl nitrate, and a mixture of $CuCl_2$–$Ba(NO_3)_2$–$Y(NO_3)_3$, are pumped into the EDR as the conductive disperse phase. The relatively non-conducting continuous phase is 2-ethyl-1-hexanol which contains 0.1–0.2 M NH_3, 1%–2% H_2O and 0%–10% ethanol. The volumetric ratio of the disperse phase to continuous phase flow rate ranges from 0.004 to 0.04. Higher volumetric ratios resulted in extensive coalescence in the electrode region. The peak voltage was 20–40 kV and the pulsing frequency ranged from 2000 to 5000 Hz.

Scanning electron micrographs of the particles were taken by drying a drop of the sample on a carbon grid. Samples were both air dried and heated with a spot light to remove residual organic solvent.

5. Results and discussions

Hydrous oxide particles of yttrium, aluminum, zirconium, and a mixture of (1:2:3) yttrium, barium, and copper were synthesized in the EDR by the external gelation method where the metal salt solutions were dispersed in a 2-ethyl-1-hexanol continuous phase containing either 0.12 M NH_3 or 0.12 M NH_3 plus 10 vol.% ethanol. The molar concentrations of corresponding metal salt solutions were 1.1 M $Y(NO_3)_3$, 2.5 M $Al(NO_3)_3$, 1.2 M zirconyl nitrate, and 1.5 M (1:2:3) $Y(NO_3)_3$–$Ba(NO_3)_2$–$CuCl_2$. Figure 3 shows scanning electron micrographs of the particles. The particle sizes range from a few tenths of a micrometer to approximately 5 μm. Hydrous yttrium oxide and the hydrous (1:2:3 Y–Ba–Cu) oxide particles are dense and spherical.

The hydrous oxide zirconium and aluminum particles have holes and/or dimples in them. It was initially thought that these particles collapsed upon rapid evaporation of the excess solvent; however, particles that were air dried over several days had the same shapes. Therefore, the particle shapes were attributed to the viscous oscillations that are damped at a rate

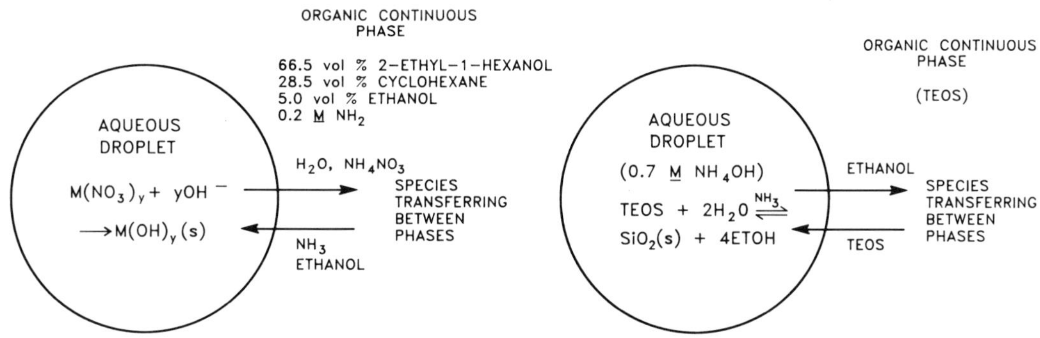

(a) (b)

Fig. 2. Microreactor droplet: (a) formation of hydrous metal oxide particles by the external gelation of metal salt solutions with ammonia in the continuous phase; (b) formation of hydrous oxide particles or shells using metal alkoxides in the continuous phase.

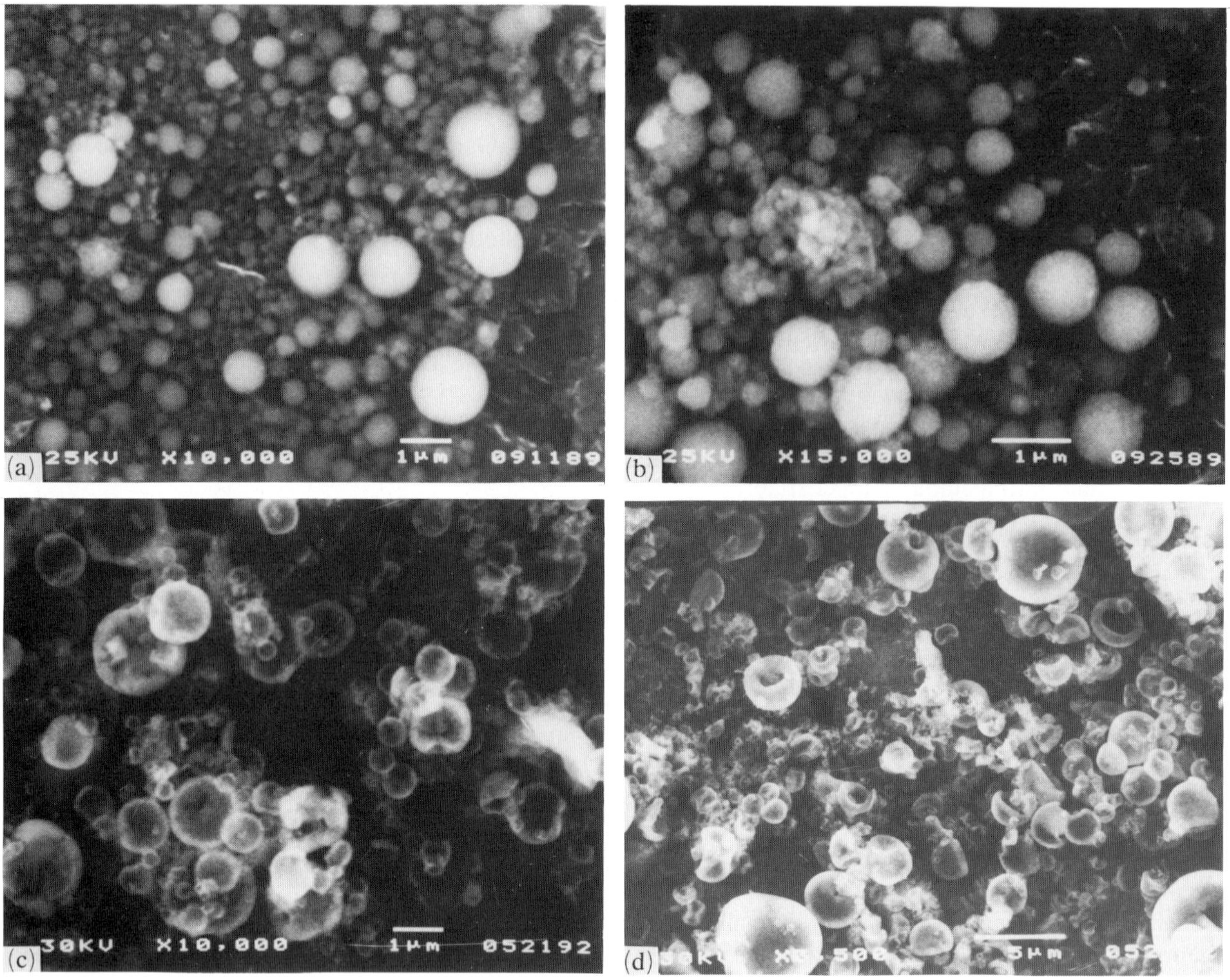

Fig. 3. Hydrous metal oxide particles produced in the EDR by the external gelation of metal salt solutions: (a) 1-2-3 yttrium–barium–copper, (b) yttrium, (c) aluminum, (d) zirconium.

governed by the physical properties of the two phases and the local electric field. Basaran has shown that multi-lobed shapes are observed during drop oscillation [15]. If precipitation–gelation occurs very rapidly, a dimple-like, multi-lobed morphology is obtained as a result of mode coupling during non-linear drop oscillations. Such is the case for the zirconia and alumina particles. Particles of similar shapes have also been observed for hydrous aluminum oxide particles that are formed by dispersing 1 M $Al(NO_3)_3$ solution in the EDR. Thus, the particle shape is not dependent on the metal salt concentration.

Using the internal gelation method, solid spherical hydrous aluminum oxide particles were formed. Thus, the EDR can be used to change the morphology of the particles by properly choosing the method to induce precipitation–gelation. A typical scanning electron micrograph of these particles is shown in Fig. 4. These particles were synthesized by dispersing 0.25–1 M

Fig. 4. Hydrous aluminum oxide particles produced by the internal gelation method: aqueous droplets containing 0.5 M $Al(NO_3)_3$ and 1.1 M HMTA dispersed in 2-ethyl-1-hexanol.

$Al(NO_3)_3$ and 0.8–1.25 M HMTA in 2-ethyl-1-hexanol. The particle size ranges from a few tenths of a micrometer to approximately 10 μm.

6. Conclusions

Experiments have shown that the characteristics of particles produced in the EDR can be changed by carefully selecting the method of precipitation. In the case of external gelation, different particle morphologies are obtained for various hydrous and metal oxide particles. Coupling of non-linear drop oscillations and rapid precipitation kinetics is important.

The production of several hundred grams of hydrous aluminum oxide and zirconium oxide particles in the EDR will allow thorough characterization for porosity, chemical composition, and sinterability.

Acknowledgments

The support by the Division of Advanced Energy Projects, Office of Basic Energy Sciences, US Department of Energy is gratefully acknowledged. We thank W. G. Sisson for his special assistance.

The submitted manuscript has been authored by a contractor of the US Government under contract No. DE-AC05-84OR21400. Accordingly, the US Government retains a non-exclusive, royalty-free license to publish or reproduce the published form of this contribution, or allow others to do so, for US Government purposes.

References

1 W. H. Rhodes, *J. Am. Ceram. Soc., 64* (1981) 19.
2 E. Matijevic, *MRS Bull., 14* (1989) 18.
3 B. J. Ingebrethsen, E. Matijevic and R. E. Partch, *J. Colloid Interface Sci., 95* (1981) 228.
4 C. B. Martin, R. P. Kurosky, G. D. Maupin, C. Han, J. Javadpour and I. A. Aksay, *Ceram. Trans., 12* (1990) 99.
5 W. T. Minehan and G. L. Messing, *J. Non-Cryst. Solids, 121* (1990) 375.
6 E. B. Slamovich and F. F. Lange, *Mater. Res. Soc. Symp. Proc., 121* (1988) 257.
7 C. T. R. Wilson and G. I. Taylor, *Proc. Camb. Philos. Soc., 22* (1925) 98.
8 M. T. Harris, T. C. Scott, O. A. Basaran and C. H. Byers, *Mater. Res. Soc. Symp. Proc., 180* (1990) 853.
9 T. C. Scott and R. M. Wham, *I &EC Res., 28* (1989)
10 T. C. Scott, *AIChE J., 33* (1987) 1557.
11 T. C. Scott, *Sep. Sci. Technol., 25* (1990) 1709.
12 M. T. Harris, W. G. Sisson and O. A. Basaran, *Better Ceramics Through Chemistry, Mater. Res. Soc. Symp. Proc., 271* (1993) 945.
13 R. E. Barker, *PhD Thesis*, University of Tennessee, 1988.
14 J. L. Collins, M. H. Lloyd and R. L. Fellow, *Radiochim. Acta, 42* (1987) 121.
15 O. A. Basaran, *J. Fluid Mech., 242* (1992) 169.

Formation and observation of ultrafine particles

A. R. Thölén
Laboratory of Applied Physics, Building 307, Technical University of Denmark, 2800 Lyngby (Denmark)

Abstract

The nucleation and growth of gas-evaporated particles is discussed and obvious shortcomings in today's knowledge are pointed out. A better knowledge of gas evaporation would improve on the production and give new possibilities in the choice of starting materials for nanostructures. Furthermore, two examples are shown where a wide range of transmission electron microscopy methods are used together to gain maximum information, namely a superstructure of iron oxide particles and contrast from small particles, including adhesion strain fields. By using a combination of phase contrast, amplitude contrast, microdiffraction, microanalysis and image simulation, more information can be revealed from nanostructured materials.

1. Introduction

Nanophase materials are by definition built up from small particles or grains, *e.g.* formed by gas evaporation, by chemical methods, galvanic coating or ball-milling. Many methods exist to investigate these structures but transmission electron microscopy (TEM) of small particles is a very powerful method, owing to its high resolution and microanalysis abilities.

This paper will concentrate on three different aspects. The first part is a discussion of how and where small particles form during gas evaporation—a much used method of making particles in large numbers. Also, some deficiencies in the current knowledge of the process will be pointed out. Secondly, TEM of a highly regular nanostructure, *i.e.* a superlattice of small iron oxide particles, is described. Different electron microscopy methods, with Fresnel diffraction, image simulation, microdiffraction and energy-dispersive X-ray analysis, are all used to reveal the true structure. Thirdly, individual particles of Fe–Ni and pairs of such particles with associated strain fields are observed. Improved ways of analysing the results by a direct comparison between micrograph and image simulation results indicate a new way of finding material parameters.

2. Experimental details

Metal particles (Au,Ag,Ni,Cu,Co,Mg,Fe) were produced by evaporating in a dilute argon gas (5–20 Torr) and subsequently collected on carbon-covered copper grids [1–3]. The iron oxide superstructure was produced by taking a drop of a ferrofluid containing $Fe_{1-x}C_x$ particles, adding it to benzine and allowing it to dry on a carbon-covered electron microscope grid.

When the ferrofluid was exposed to air, the particles oxidized to iron oxide [4, 5]. During evaporation of the liquid, the particles, which had a very narrow size distribution (diameter, 6.9 nm; standard deviation, 0.4 nm), built up a three-dimensional lattice [6, 7]. The particles were covered with a thin layer of surfactant and did not directly touch one another.

Ultrafine FeNi particles, formed by simultaneous evaporation of iron and nickel, were reduced in hydrogen and sintered to give larger particles. These particles were subsequently used for the study of adhesion. The particles were observed in a Philips 430 transmission electron microscope operating at 300 kV.

3. Formation of particles by gas evaporation

In forming small particles in large numbers for nanophase applications, a precise knowledge of particle nucleation and growth is of utmost importance, both in the case of single-phase and multiphase materials. Groups in Japan [8–16] looked at the gas evaporation process and recognized three different zones in the particle fume. The central zone contains relatively few particles, and these are large and often faceted. The shell of the tubular smoke zone holds the majority of the particles, while the outer zone is relatively free from

0921-5093/93/$6.00

particles. Granquist and Buhrman [17] used the log-normal distribution of particle diameters as a starting point and derived a coalescence model for particle growth, *i.e.* when two particles meet, they coalesce or sinter to give a larger particle. However, little is said about the geometry (where the particles are formed), temperature and the time aspects of sintering.

The effect of the geometry and temperature in gold evaporation was considered in ref. 2. The tubular particle fume is essentially closed with a hemispherical cloud below the source (Fig. 1). Depending on the evaporation temperature and gas pressure, its radius is typically 10–15 mm. The shape of the cloud is determined by a balance between the convection current, the evaporation rate and the temperature. The temperature at particle formation is very critical and was measured to be $(0.35–0.40)T_{\mathrm{m}}$, where T_{m} is the absolute melting temperature.

Although the basic mechanisms in particle formation are known, the list of areas where information is lacking about the process is long and delineates important fields for future research on nanostructured materials: nucleation and growth as a function of temperature; metal and gas atom density; the possibility of re-evaporation of particles once formed; the exact mechanisms of sticking and coalescence or sintering; the problem of transferring macroscopic sintering laws to small particles (the need for a united picture of sintering from macrophysics to microphysics); faceting of particles. These areas should be treated with a combination of experimental and theoretical methods, which ought to include results from cluster physics and laser measuring techniques for probing velocity and temperature profiles in a gas. Computer simulation offers another way of research. Some critical temperature effects could also be used, such as the magnetization temperature [2].

Fig. 1. A cloud of gold particles during gas evaporation. It is clearly seen how the particles essentially are formed in a tubular geometry with a hemisphere ending below the source.

4. Contrast from a superlattice

A superlattice of iron oxide particles (particle diameter, 6.9 nm) with a very regular structure [6, 7] is shown in Fig. 2. The particles are covered with a surfactant, so there is no direct adhesion between the particles and, thus, they can more easily find an equilibrium structure. The particles form a close-packed structure with an unknown number of layers and an unknown stacking, and are visible owing to the Fresnel contrast. With a small amount of defocusing, the Fresnel contrast is visible in each individual particle as a fringe close to the rim. With greater defocusing, Fresnel fringes from neighbouring particles start to interfere and the structure of the whole “crystal” is now mirrored instead. The micrographs are thus very sensitive to the amount (and direction) of defocusing. The true structure of the layer stacking is revealed from a comparison between the micrographs and image simulations, microdiffraction and energy-dispersive X-ray analysis.

The defocusing in this study was varied from -3 μm to $+3$ μm, which is much larger than the Scherzer focus used in high resolution electron microscopy (HREM). It was also found essential to align the (close-packed) layers perpendicular to the incoming beam, and this is an obvious parallel to HREM. The images of the superstructure and those found in HREM are also very similar, although the length scales are quite different. In both cases, phase contrast plays an important role.

We have only calculated the contrast for pure hexagonal close-packed (h.c.p.) and face-centred cubic (f.c.c.) structures (up to 10 layers), as the number of possible stackings increases rapidly with the number of layers. Further information is found from microdiffraction, using a very long camera length (6300 mm), and energy-dispersive X-ray analysis. Here, we assume that the signal from an unknown structure is proportional to the number of layers, using a single close-packed layer as a reference and neglecting absorption. This turns out to yield very good results. In Fig. 2, different “grains” with four and six layers of f.c.c. packing can be recognized.

5. Contrast from single particles and from particles with contacting strain fields owing to adhesion

Small particles observed with amplitude contrast in the transmission electron microscope often exhibit extinction rings (Fig. 3). This is the case for particles larger than about 10 nm in bright field close to a Bragg condition. However, the effect is quite pronounced in much smaller particles, using weak beam dark-field conditions [18–20]. The position of these rings depends on the particle size (given by extinction distance), particle shape, material, operating diffraction vector and the deviation w from the Bragg condition. A distortion of the regular ring pattern with extra contact fringes is sometimes observed for two particles in contact (Fig. 4).

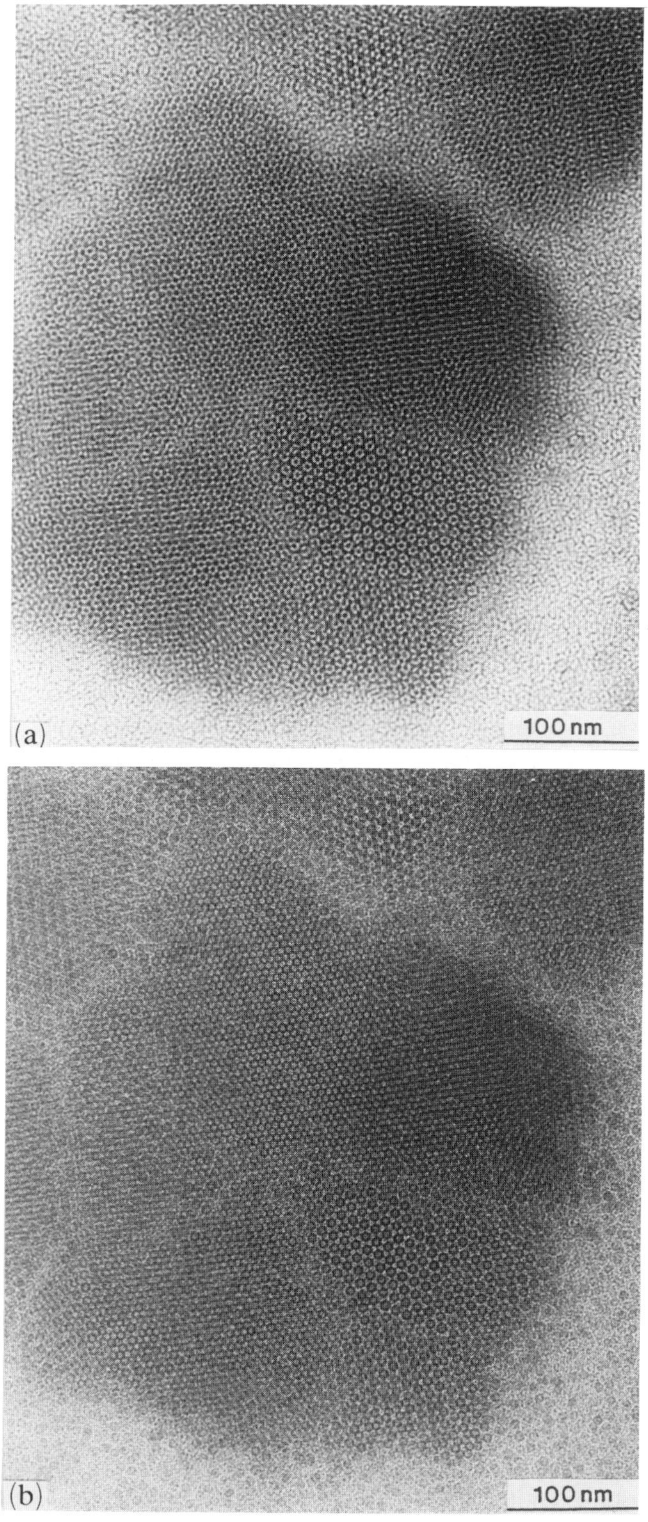

Fig. 2. Superlattice of iron oxide particles: (a) defocusing, + 2300 nm; (b) defocusing, − 1150 nm.

The physical background to this is as follows [21]. When two particles touch, the total energy is lowered by the particles making contact across an area rather than meeting at one point. In this way, the outer surface energy is diminished but a new grain boundary with an associated energy is formed. When the particles move closer to each other, a stress field is also created at the contact zone, with a compressive stress in the centre and a tensile stress further out. The net result is that equilibrium is reached for a finite radius a, given by

$$a = \left(\frac{9\pi(1-\nu^2)R^2\gamma_{\mathrm{eff}}}{4E} \right)^{1/3}$$

where a is the contact radius, R the particle radius, E Young's modulus, ν Poisson's ratio and $\gamma_{\mathrm{eff}}(=2\gamma_s - \gamma_{gb})$ is the effective surface energy, with γ_s the surface energy and γ_{gb} the grain boundary energy. The bending of lattice planes in the stressed contact zone makes the electrons see a non-perfect lattice and this will subsequently lead to amplitude contrast.

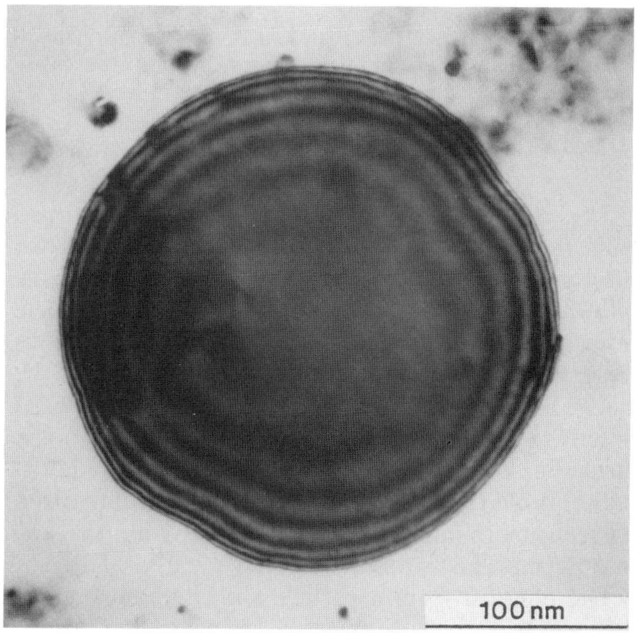

Fig. 3. Fe–Ni particle with extinction contours.

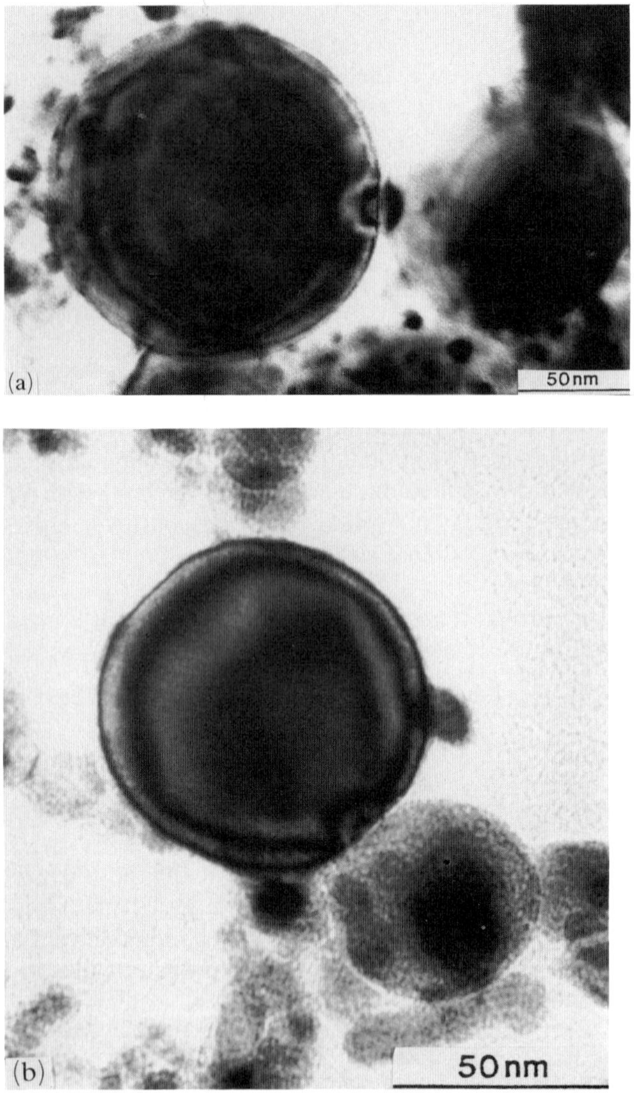

Fig. 4. Adhesion strain fields between Fe–Ni particles. Due to diffraction effects, the strain field is normally only observed in one particle at a time but can alternatively be observed in the other particle by proper tilting.

The contrast calculations, which were performed using the matrix multiplication method [22], had two different goals. The first goal was to obtain a mapping of the adhesion strain fields as a function of the particle radius and surface energy (Fig. 5). The calculations presented in this way give a very clear picture of the change in the parameters and, in principle, can be used directly to determine the surface energy [23]. However, during this work it became clear that much more information could be drawn from a quantitative comparison between micrographs and theory. Therefore, the line we now pursue is as follows.

The contrast from a particle with a certain radius (given in extinction distances ξ_g) is calculated. Moreover, to accept a deviation from the spherical shape,

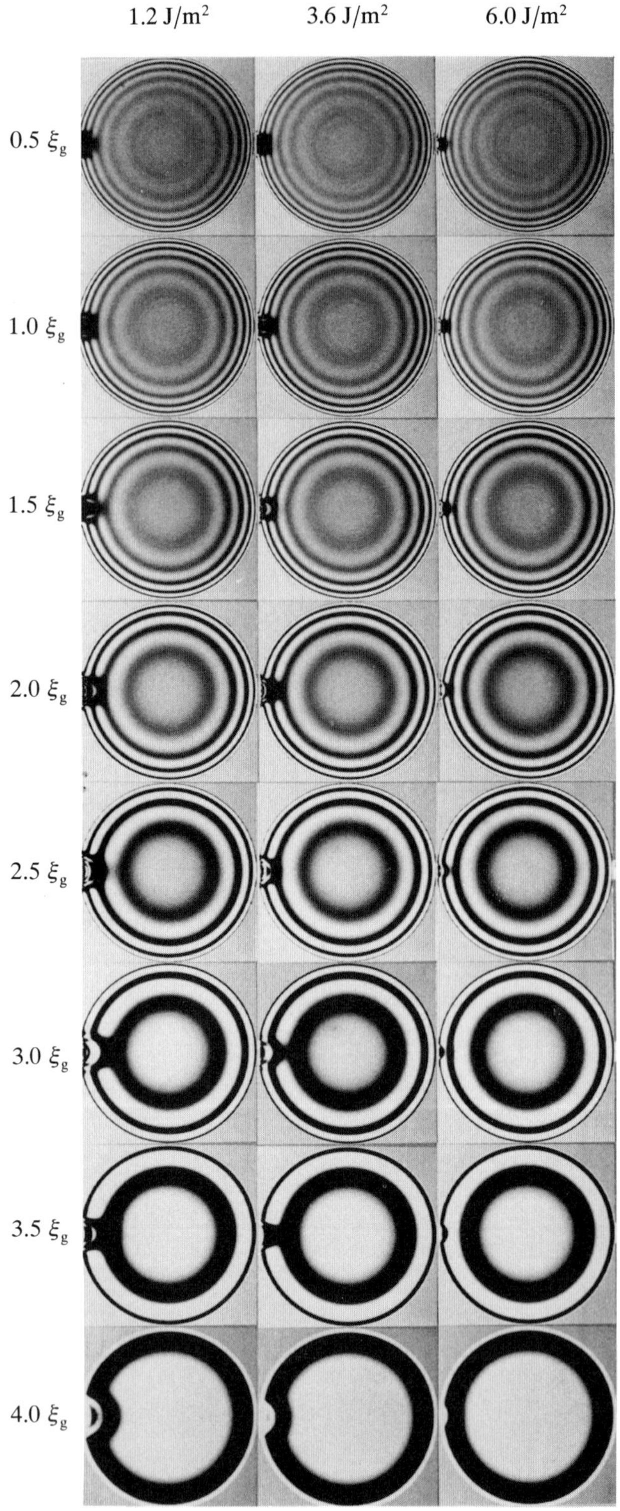

Fig. 5. Simulation of adhesion strain fields. The five rows represent different particle radii ($0.5–4.0\xi_g$) and the three columns represent different surface energy values (1.2, 3.6 and 6.0 J m^{-2}).

ellipsoidal particles are also considered and this shape change could further be built out to handle faceted particles as well. The deviation from the Bragg condition is given by $w = \xi_g s$. The normal and the anomalous absorption are also input parameters, *i.e.* we have in total five parameters describing the particle and the diffraction conditions used. The micrograph of a particle (*e.g.* Fig. 3) is scanned with a TV camera and the digital output is transformed from optical density to "exposure values" with the help of a reference table for the photographic process. A computer program, which works towards a 1:1 correlation between experiment and theory, is written and, on convergence, data on geometry and diffraction parameters are obtained. A very interesting result is expected from these calculations. The radius will be obtained in extinction distances and also can be measured from the micrographs. Thus, it is possible in principle to "measure" the extinction distance and this in fact gives a chemical fingerprint of the particle without any X-ray spectrum needing to be taken.

6. Conclusions

Particle formation and observation in the gas evaporation method is discussed and the conclusion is that a fundamental understanding of these processes is still lacking and that further knowledge would be needed to make better nanostructured materials.

Furthermore, two examples, namely a superlattice of iron oxide particles and small particle contrast including adhesion strain fields, show how much information can be obtained with a combination of different TEM methods, using image simulation, phase and amplitude contrast, microanalysis and microdiffraction. For example, microdiffraction has been used to reveal the orientation between small particles in a network [24]. It is often the combination of different methods which makes TEM such a powerful tool for nanocrystalline research.

References

1 I. Hansson and A. Thölén, *Philos. Mag. A, 37*(4) (1978) 535.
2 A. R. Thölén, *Acta Metall., 27*(1979) 1765.
3 A. R. Thölén, *Phase Transitions, 24–26* (1990) 375.
4 J. van Wontherghem, S. Mørup, S. W. Charles, S. Wells and J. Villadsen, *Phys. Rev. Lett., 55* (1985) 410.
5 J. van Wontherghem, S. Mørup, S. W. Charles, S. Wells and J. Villadsen, *J. Hyperfine Interact., 27* (1986) 333.
6 M. D. Bentzon, J. van Wontherghem, S. Mørup, A. R. Thölén and C. J. W. Koch, *Philos. Mag. B, 60* (2) (1989) 169.
7 M. D. Bentzon and A. R. Thölén, *Ultramicroscopy, 38* (1991) 105.
8 K. Kimoto, Y. Kamiya, M. Nonoyama and R. Uyeda, *Jpn. J. Appl. Phys., 2* (1963) 702.
9 K. Kimoto and I. Nishida, *Jpn. J. Appl. Phys., 6* (1967) 1047.
10 S. Yatsuya, R. Uyeda and Y. Fukano, *Jpn. J. Appl. Phys., 11* (1972) 408.
11 S. Yatsuya, S. Kasukatse and R. Uyeda, *Jpn. J. Appl. Phys., 12* (1973) 1675.
12 S. Kasukabe, S. Yatsuya and R. Uyeda, *Jpn. J. Appl. Phys., 13* (1974) 1714.
13 I. Nishida and K. Kimoto, *Jpn. J. Appl. Phys., 14* (1975) 1425.
14 T. Ohno, S. Yatsuya and R. Uyeda, *Jpn. J. Appl. Phys., 15* (1976) 1213.
15 T. Hayashi, T. Ohno, S. Yatsuya and R. Uyeda, *Jpn. J. Appl. Phys., 16* (1977) 705.
16 Y. Saito, S. Yatsuya, K. Mikama and R. Uyeda, *Jpn. J. Appl. Phys., 17* (1978) 291.
17 C. G. Granquist and R. A. Buhrman, *J. Appl. Phys., 47* (1976) 2200.
18 M. J. Yacamàn and T. Z. Ocana, *Phys. Status Solidi A, 42* (1977) 571.
19 M. J. Yacamàn, K. Heinemann, C. Y. Yang and H. Poppa, *J. Cryst. Growth, 47* (1979) 187.
20 M. J. Yacamàn, L. D. Romeau, A. Gómez and Z. A. Munir, *Philos. Mag. A, 40* (5) (1979) 645.
21 K. E. Easterling and A. R. Thölén, *Acta Metall., 20* (1972) 1001.
22 A. Thölén, *Philos. Mag., 22* (175) (1970) 175.
23 S. A. Scthi and A. R. Thölén, in A. López-Salindo and M. I. Rodriquez-Garcia (eds.), *Proc. EUREM 92 Conf. on Electron Microscopy, Granada, Spain*, Materials Sciences, 1992, Vol. 2, p. 667.
24 M. D. Bentzon, A. Karlen and A. R. Thölén, *Coll. Phys., Coll. C1, Suppl., 51* (1) (1990) C1–89.

Materials Science and Engineering, A168 (1993) 137–140

Electrochemical deposition: a method for the production of artificially structured materials

A. Haseeb*, B. Blanpain, G. Wouters, J. P. Celis and J. R. Roos

Departement Metaalkunde en Toegepaste Materiaalkunde (MTM), Katholieke Universiteit Leuven, de Croylaan 2, B-3001 Leuven (Belgium)

Abstract

The potential of electrochemical deposition (ECD) as a production method for artificially structured materials is discussed. Compared with physical vapour deposition (PVD), ECD technology has some definite advantages. Its low cost, industrial applicability and its ability to produce structural features with sizes ranging from nanometres to micrometres have recently created increased interest in this technology for the production of artificially structured materials. The different ECD methods are briefly reviewed and some specific cases of artificially structured coatings produced by ECD, such as epitaxial multilayered material (Cu/Ni), amorphous material in combination with crystalline material (NiP_x/NiP_y) and alternating hard and ductile materials (NiP_x/Sn), will be highlighted. It will be shown that, by controlling the ECD conditions adequately, a wide range of artificially structured materials with respect to their constituent elements and their microstructural morphology may be produced.

1. Introduction

Ever-increasing demands for materials with novel properties or improved conventional properties have led materials scientists and engineers to pursue materials with ultrafine microstructures. Such materials of major current interest can be grouped into two categories: (i) nanocrystalline materials with structural features extending a few nanometres in three dimensions and (ii) layered nanostructures or multilayers. The materials in group (i) are mainly prepared by gas condensation and rapid quenching methods, while different vacuum techniques and sputtering have been utilized to synthesize the group (ii) materials.

Electrochemical deposition (ECD), in spite of its century-old tradition as one of the most successful industrial processes, has not been considered seriously for the production of materials with ultrafine microstructures until recently [1–4]. ECD offers certain definite advantages over the other techniques for the synthesis of nanostructures. It is an inexpensive method and has a deep-rooted industrial base, thus allowing easy technology transfer. It can offer higher deposition rates and allows the production of free-standing parts (electroforming) of complex shape.

Recent interest in ECD has been mainly for the production of compositionally modulated multilayers (CMMs). There are two major ways to carry out the ECD of multilayers. In the single-bath technique, deposition is carried out in a plating solution containing ions of the different constituents of the multilayer by changing alternately the plating current and potential, sometimes in combination with the modulation of the mass transfer towards the cathode. The other technique, called the dual-bath technique, is conceptually simple and merely involves the deposition of the constituents from separate baths in an alternate manner. The single-bath technique has the advantage that the transfer of the substrate from one bath to another is not involved. However, in the single-bath technique, a suitable plating solution has to be found from which the different components of the multilayer can be deposited. This limits the applicability of the method. In contrast, a dual-bath technique does not suffer from this drawback and offers a wider choice in the selection of components.

In this paper, the results of the preparation and characterization of three different multilayer systems are presented: Cu/Ni multilayers prepared by a dual-bath technique; NiP_x/NiP_y with a stacking of crystalline and amorphous layers, prepared by a single-bath technique; and, finally, some results on the production of

*Present address: Metallurgical Engineering Department, Bangladesh University of Engineering and Technology, Dhaka-1000, Bangladesh.

0921-5093/93/$6.00

NiP_x/Sn multilayers produced with a dual-bath technique.

2. Experimental results and discussion

2.1. Cu/Ni compositionally modulated multilayers

The ECD of Cu/Ni multilayers has been carried out using the dual-bath technique [3]. Multilayers with a sublayer thickness in a wide range from 100 down to 5 nm were deposited and characterized. A cross-sectional scanning electron microscopy (SEM) micrograph of a Cu/Ni multilayer consisting of sublayers 100 nm thick is shown in Fig. 1. Cu sublayers which have been etched appear dark, while bright sublayers represent Ni. Continuous sublayers with a uniform thickness are clearly visible.

The plating parameters, particularly the current density and overpotential, have been found to have a marked effect on the morphology and structure of multilayers. This is illustrated vividly in Fig. 2, which shows a Cu/Ni multilayer that has been deposited at a lower Cu current density of 0.5 A dm^{-2}. A breakdown of the layered structure has occurred. A careful inspection of the micrograph shows that, although the Ni sublayers are continuous, the Cu sublayers are discontinuous: Cu deposits as islands in many areas. The inhibited nucleation of Cu on Ni at low current densities is believed to be the reason for the non-uniform, island structure of the Cu deposits. However, by using appropriate plating conditions, dual-bath ECD can produce multilayers with an epitaxial stacking of sublayers as thin as or less than 5 nm.

The above study illustrates the necessity of selecting property and controlling the plating parameters to achieve a layered nanostructure, and the possibility of achieving nanocomposites with other structural morphologies by ECD, as is depicted in Fig. 3. By using the appropriate plating conditions, one can engineer the microstructure to achieve fine layering on the one hand and a totally granular structure on the other hand.

2.2. NiP_x/NiP_y: a crystalline/amorphous multilayered system

The electrochemical deposition of Ni–P alloys from aqueous solutions was first described by Brenner [6].

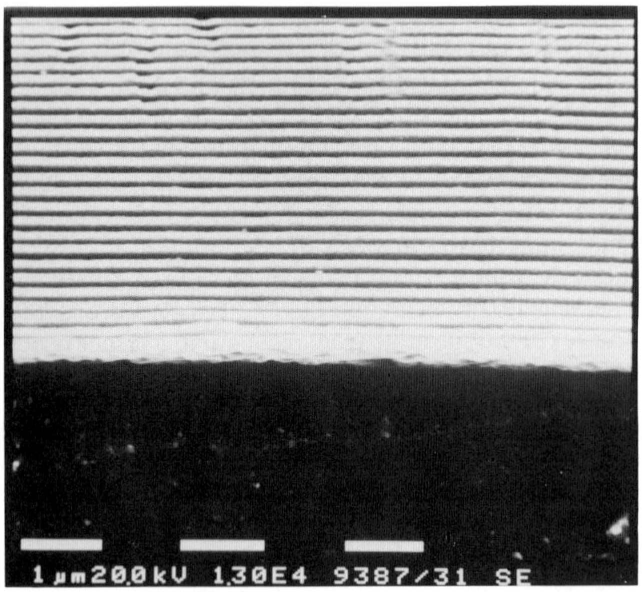

Fig. 1. Cross-sectional scanning electron micrograph of electrochemically deposited Cu/Ni multilayers with sublayers 100 nm thick by the dual-bath technique. Deposition conditions: 2 A dm^{-2} for both Cu and Ni depositions; 30 °C; Cu bath—90 g l^{-1} $CuSO_4 \cdot 5H_2O$, 200 g l^{-1} H_2SO_4 and 10 ml l^{-1} EX-10 (Harshaw); Ni bath—400 g l^{-1} $Ni(NH_2SO_3)_2$ and 30 g l^{-1} H_3BO_3.

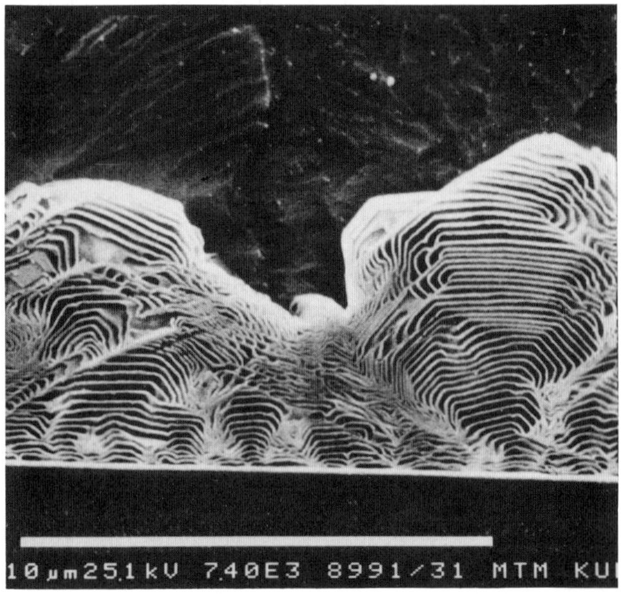

Fig. 2. Cu/Ni structure deposited at a Cu deposition current density of 0.5 A dm^{-2} and an Ni deposition current density of 2 A dm^{-2}.

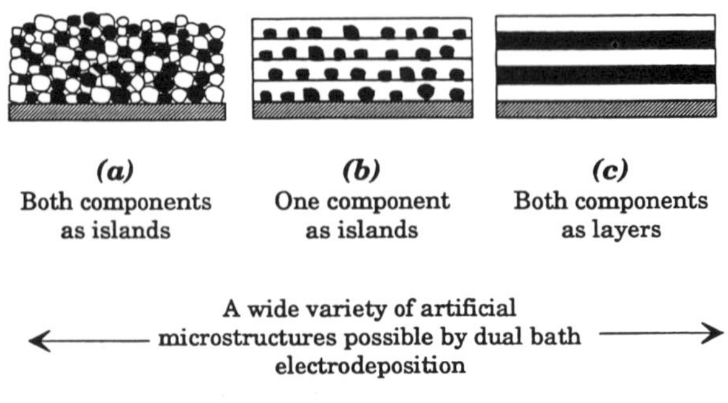

Fig. 3. Schematic diagram showing the range of microstructures obtainable by dual-bath ECD.

With the proper selection of bath composition and/or plating conditions, the P content in the Ni–P alloy can be varied from a few atomic per cent up to 40 at.%. It has been well documented that electrodeposited Ni–P has an amorphous structure above a certain P content [7]. The value at which this transition from a polycrystalline to an amorphous alloy occurs varies somewhat according to different reports, but generally it is accepted that, at 18 at.% P, the Ni–P alloy is fully amorphous. Therefore, the production of crystalline/amorphous NiP_x/NiP_y multilayers by ECD should be feasible. A dual-bath technique developed by Goldman *et al.* [8] using a Ni and Ni–P bath was shown to produce CMMs with a modulated wavelength down to 2.1 nm, as was verified with small-angle X-ray diffraction (XRD) measurements. The possibility of using a single-bath process is discussed below.

In a single-bath process, only the plating conditions can be manipulated to generate a multilayered structure. In practice, this is realized by imposing potentiostatic or galvanostatic pulses in the form of a periodic function. In this work, a simple periodic block function between two current densities levels was selected. At low current density, the P content of the ECD Ni–P alloy coatings is high and, if the current density is increased, an Ni–P alloy with low P content is deposited [9].

A plating bath was selected with a large compositional range under fixed current density conditions, as shown in Fig. 4. The duration of the galvanostatic pulses was controlled to obtain multilayered structures with modulation wavelengths ranging from 2 to 50 nm, after correction for the current efficiency. The compositional modulation was observed with Auger electron spectroscopy (AES) depth profiling down to a modulation wavelength of 20 nm (Fig. 5). This finding was confirmed with XRD experiments, where a sharp peak corresponding to the microcrystalline Ni–P could be observed for the sample with a modulation wavelength of 20 nm. At lower wavelengths, a broad diffraction peak more typical of the average composition of the deposit was measured.

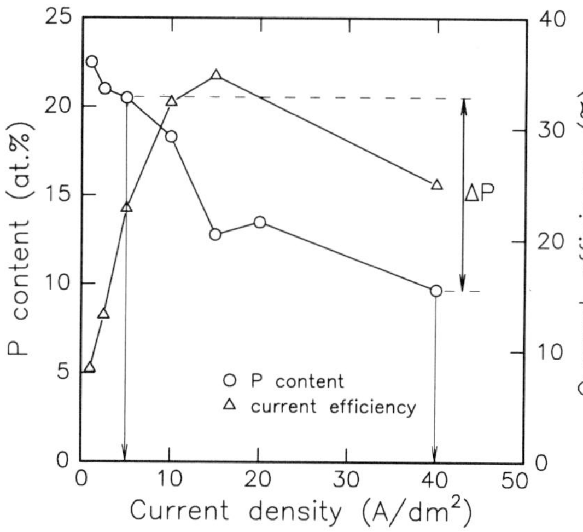

Fig. 4. P content and current efficiency as functions of the current density of the plating bath used for the deposition of NiP_x/NiP_y multilayers from the alternating deposition at 5 and 40 A dm^{-2} (65 °C, 150 g l^{-1} $NiSO_4 \cdot 6H_2O$, 50 g l^{-1} $NiCl \cdot 6H_2O$, 290 g l^{-1} H_3PO_4, 30 g l^{-1} H_3PO_3).

Single-bath plating of crystalline/amorphous NiP_x/NiP_y multilayered coatings has been achieved down to a modulation wavelength of 20 nm. Specific engineering of the galvanostatic pulse signal in a single-bath electrodeposition could be attempted to produce NiP_x/NiP_y multilayers with even smaller wavelengths.

2.3. *NiP/Sn: the production of hard/soft multilayered systems*

The alternation of soft and hard sublayers could be an attractive route to improve the wear properties, owing to the lubricating effect of the soft sublayer between the hard sublayers [10, 11]. For that reason, the electrolytic synthesis of amorphous hard Ni–P alloy layers interlaced with soft Sn layers was attempted using the dual-bath technique.

Similarly to the Cu/Ni multilayers, a smooth, layered structure could be obtained for sublayer thicknesses as small as 200 nm for Sn deposition at 2 A dm^{-2}, whereas, for a lower current density (0.5 A dm^{-2}), a wavy pattern was observed. However, for sublayer thicknesses lower than 200 nm, no planar growth process (a primary requirement for the synthesis of good quality CMMs) was established. Therefore, the electrocrystallization process was investigated. Three different growth modes can occur, depending on the process parameters and on the substrate–deposit system [12]: island growth (Volmer–Weber), layer growth

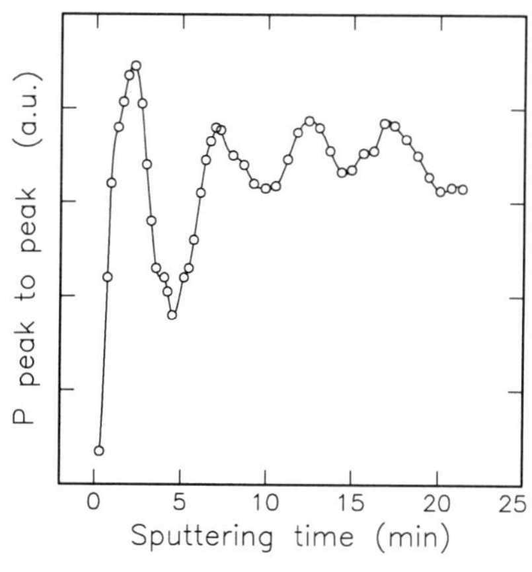

Fig. 5. AES depth profile of the P signal of an NiP_x/NiP_y multilayer with a modulation wavelength of 20 nm.

(Frank and van der Merwe) and layer-plus-island growth (Stranski–Krastanov).

From electrochemical measurements for the deposition of Sn on the amorphous Ni–P alloy, it appeared that the polarization curve exhibits a region that is similar to an "underpotential deposition", *i.e.* a given cathodic current is measured at a potential lower than that for Sn deposition on Sn (Fig. 6). These measurements were coupled with a metallographic investigation. When the potentiodynamic polarization is ceased after the "underpotential deposition" region and also after deposition of a thin Sn layer (50 nm or smaller) on an NiP substrate, nodules are observed on top of the surface (Fig. 6). It appears that the concentration of the nodules increases with the extension of the deposition time of Sn and/or with increasing current density of the Sn deposition. By chemical activation of the NiP surface before Sn deposition in a sulphuric acid + hydrochloric acid solution, the amount of nodules is diminished.

At present, the nature of the nodules is still a point of discussion. The several analytical techniques used for analysis of those nodules gave no decisive answer. Microprobe (wavelength-dispersive spectroscopy) measurements showed some oxygen signal for some of the nodules. However, a Rutherford backscattering (RBS) measurement showed Ni on the surface after an excepted 50 nm Sn deposit. Along with the metallographic investigation (Fig. 6), the height of the Ni signal at the surface position in the RBS measurement suggested an outgrowth of the underlying amorphous Ni–P into nodules caused by the deposition of Sn.

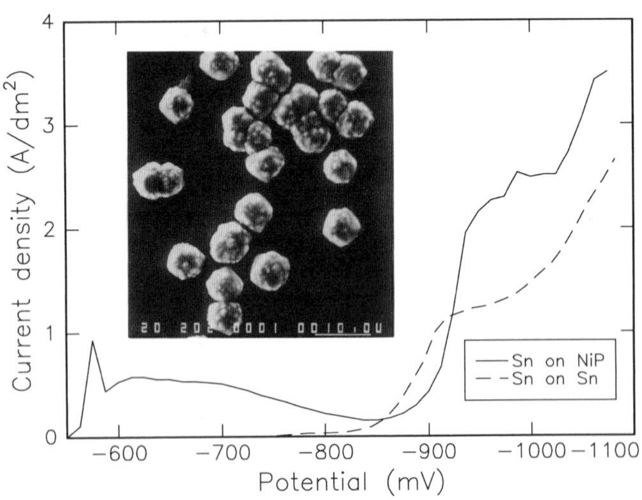

Fig. 6. Polarization curves for deposition of Sn on Sn, and Sn on NiP. The SEM micrograph inset shows the surface after Sn deposition on Ni–P (0.5 A dm^{-2}, 2.37 s, expected thickness of 10 nm; Ni–P bath—150 g l^{-1} $NiSO_4 \cdot 6H_2O$, 90 g l^{-1} $NiCL \cdot 6H_2O$, 45 g l^{-1} H_3PO_4, 40 g l^{-1} H_3PO_3; Sn bath—30 g l^{-1} $SnSO_4$, 170 g l^{-1} H_2SO_4, 27 ml l^{-1} CULMO additive).

3. Conclusions

ECD has certain technological and economic advantages compared with vacuum techniques for the production of artificially structured materials. For three different materials systems, ECD can be a useful technique for the production of nanostructured materials. Through the control of the deposition conditions, materials with structural features exhibiting large ranges in size and morphology may be obtained. A successful development of nanostructured materials with ECD demands an understanding of the nucleation processes (dual- and single-bath technique) and electrokinetics (single-bath technique), in addition to good control of the plating conditions.

ECD has been shown to be a good candidate for the production of controlled nanostructured materials. The possibility of combining different elements and of achieving structural sizes ranging from a few nanometers to several micrometers opens up a new avenue for the development of artificially structured materials.

Acknowledgments

This work has been partially supported by a KULeuven research grant to AH in the framework of IUAP (no. 4) and GOA contracts of the Belgian government. One of the authors (BB) was supported as a senior research assistant with the National Fund for Scientific Research (Belgium), and GW acknowledges the financial support of the IWONL.

References

1 U. Cohen, F. B. Koch and R. Sard, *J. Electrochem. Soc., 130* (1983) 1987.
2 D. S. Lashmore and M. P. Dariel, *J. Electrochem. Soc., 135* (1988) 1218.
3 J. P. Celis, A. Haseeb and J. R. Roos, *Trans. Inst. Met. Finish., 70* (3) (1992) 123.
4 A. Haseeb, J. P. Celis and J. R. Roos, *Trans. Met. Finish. Assoc. Ind., 1* (3) (1992) 15.
5 A. Haseeb, *PhD Thesis*, Faculty of Engineering, Katholieke Universiteit Leuven, Belgium, 1992.
6 A. Brenner, *Electrodeposition of Alloys: Principles and Practices*, Academic Press, New York, 1963.
7 G. S. Cargill, *J. Appl. Phys., 41* (1970) 12.
8 L. M. Goldman, B. Blanpain and F. Spaepen, *J. Appl. Phys., 60* (1986) 1374.
9 J. R. Roos, J. P. Celis and E. Bredąel, *Surf. Coat. Technol., 58* (1993) 63–71.
10 U. Cohen, *Proc. Electrochem. Soc. Meet., Denver, CO, 1981.*
11 R. Weil, C. C. Nee and J. W. Chang, *Metall. Trans. A, 19* (1988) 1569.
12 E. Bauer, *Z. Kristall., 110* (1958) 372.

Materials Science and Engineering, A168 (1993) 141–148

Structure and properties of ultrafine-grained materials produced by severe plastic deformation

R. Z. Valiev, A. V. Korznikov and R. R. Mulyukov

Institute for Metals Superplasticity Problems, Russian Academy of Sciences, Khalturina 39, Ufa 450001 (Russian Federation)

Abstract

Strain–heat methods of obtaining ultrafine-grained (UFG) metallic materials with grain sizes as small as 20 nm and peculiarities of their structure are considered. It is shown that intercrystalline boundaries are the main element of the structure of UFG materials and that they are typically in a non-equilibrium state. The formation of a special grain boundary phase, *i.e.* a thin near-boundary layer with high dynamic activity of atoms, has been found. This unusual structure leads to the manifestation of promising new elastic, strength, superplastic, damping and magnetic properties of UFG materials.

1. Introduction

Investigations conducted in recent years testify that severe plastic deformation (SPD) can be an effective method of producing ultrafine-grained (UFG) (submicrometer and nanocrystalline) materials [1–5]. In contrast to traditional methods, namely gas condensation with subsequent consolidation [6, 7] and ball-milling [8], in the process of SPD, the treatment of ingots and massive semi-products are conducted.

The method under consideration has a number of advantages. First, one can produce massive samples (sheets, rods), which is very useful for mechanical testing. Secondly, no residual porosity is found in the samples produced. Thirdly, it is possible to use electron diffraction microscopy for more complete investigation of the structure of UFG materials. Recent studies [5, 9–11] were able to reveal the non-equilibrium state of grain boundaries (GBs) and the formation of specific near-boundary layers (the so-called GB phase) in UFG materials. They also found changes in some of their fundamental parameters, such as the elastic moduli, diffusion coefficient, *etc.* All these have provided an opportunity to consider the structure and properties of these materials, this being the purpose of the present work.

This paper reports the major results of the structure and properties of UFG metals and alloys: pure Cu, Ni and Fe, alloys on Al, Fe and Mg base, and an Ni_3Al intermetallic compound. The investigations of the elastic properties, microhardness, strength, damping and superplasticity reported in the present work have been performed simultaneously with careful structural studies using transmission electron microscopy (TEM), high resolution electron microscopy (HREM), Mössbauer spectroscopy, magnetometry and dilatometry.

2. Experimental procedure

The Al alloys Al–1.5%Mg and Al–4%Cu–0.5%Zr (model alloys) and a commercial alloy 1420 (Al–5%Mg–2%Li–0.1%Zr) were used for this study. Prior to the deformation, the cast ingots were subjected to homogenization annealing. In addition, another material was used, namely an Mg alloy (Mg–1.5%Mn–0.3%Ce) which is an α-Mg based solid solution with a small content of Mg_9Ce and β-Mn particles. The purities of Cu, Fe and Ni were not less than 99.97%. The Ni–Al had an Al content of 22.0–23.0 at.%. To produce the UFG structure, the alloys were subjected to SPD at room temperature by torsion straining under a pressure of the order of gegapascals or equal-channel angular pressing, the true logarithmic strain being in both cases 4–7 [1, 5, 9, 12].

The TEM studies were conducted on a JEM-2000EX electron microscope. The room temperature measurements of the microhardness (HV) were performed on a Shimadzu microhardness tester, using a Vickers diamond pyramidal indenter with a load of 0.05 kg and the duration of testing was 15 s. The mechanical properties also were studied by tensile testing in a universal testing machine of the Instron type, with a constant strain rate. The Mössbauer spectroscopy, magnetometry and dilatometry measurements are described in detail in refs. 10, 11 and 13.

0921-5093/93/$6.00

3. Results

3.1. Formation of UFG structure

The study of the highly strained materials behavior [14] indicates that, under these conditions, the rotation modes of deformation play an active role and result in the formation of a severely misoriented fragmented structure. Different methods of treatment, *e.g.* shear strain under high pressure [1, 11, 12], special types of pressing [9], rolling, forging [2, 4, 15] and others, can be used to obtain high strains. Brittle materials are treated at elevated temperature which is subsequently lowered, while ductile materials are treated at room temperature.

Let us consider, for example, the TEM observed structure of the Mg alloy (Mg–1.5%Mn–0.3%Ge) strained to $e=6$ at room temperature (Fig. 1(a)) [12]. The complex diffraction contrast of the sample structure can be seen and the use of the dark-field method allows us to determine the size of individual fragments, which is approximately 100 nm. The diffraction pattern provides evidence for the existence of large-angle misorientations in the fragments. The formation of a UFG structure occurred during further annealing of a highly strained state (Fig. 1(b)).

The problem of the minimal size of grains which can be obtained by the strain–heat method is not clear yet. Usually, SPD leads to the formation of a UFG structure with a grain size of 100–200 nm [5]. However, this size can be reduced by varying the parameters of deformation (*i.e.* increasing pressure or decreasing temperature). The degree of alloying, especially by interstitial atoms, considerably contributes to this as well. For example, it was shown that, under similar conditions of SPD, structures with grain sizes of 20 and 150 nm are formed in steel with 1.2% C and pure Fe respectively. A very small grain size (less than 50 nm) also can be obtained in ordered alloys and intermetals [5, 15, 16].

3.2. Structure of UFG materials

The main specific feature of the structure of UFG materials is the existence of non-equilibrium (unrelaxed) GBs, depending on the treatment conditions. A typical example of this is given in Fig. 2, where there are presented TEM photographs of the alloy Al–4%Cu–0.5%Zr [9] with the grain size of about 200 nm after severe deformation ($e=7$) (Fig. 2(a)) and after additional annealing at 160 °C for 1 h (Fig. 2(b)). In both cases, a granular type of structure and large-angle GBs with random misorientations exist. This is confirmed by the diffraction patterns of the foil area of 0.5 μm^2, with a large number of spots arranged in circles, and data on the direct estimation of the parameters of grain misorientation.

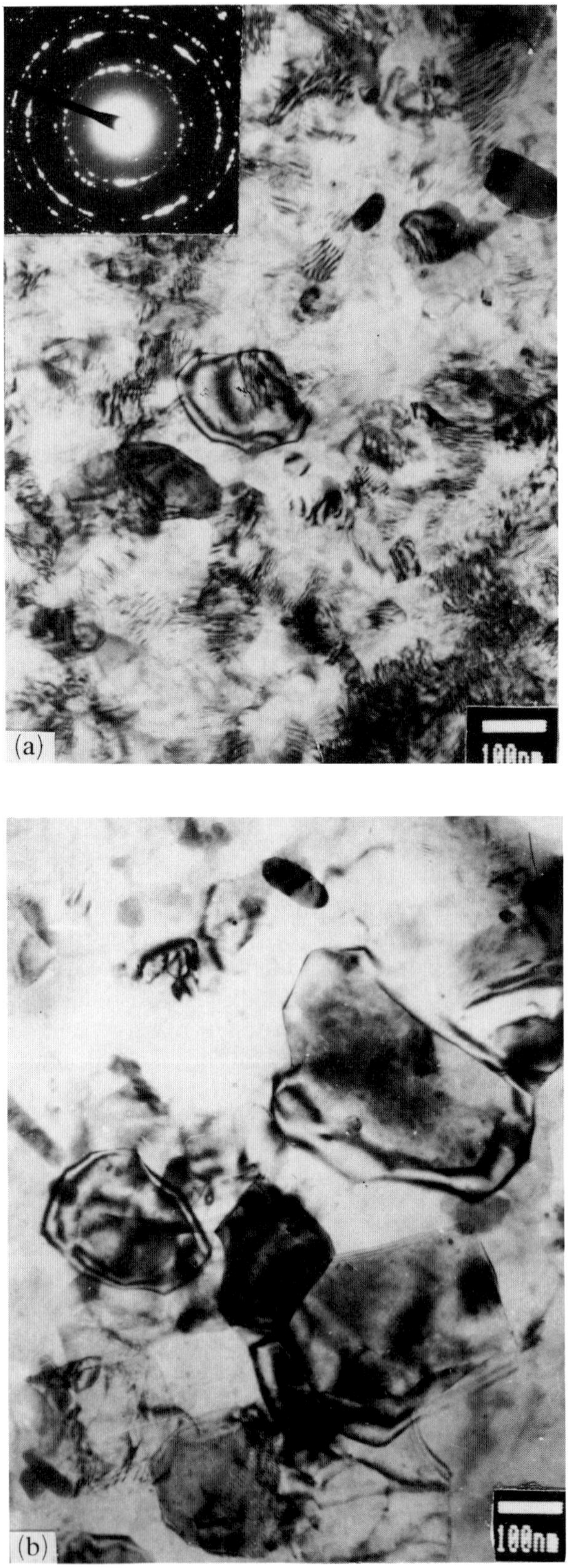

Fig. 1. Structure of the Mg alloy Mg–1.5%Mn–3.2%Ce observed by TEM: (a) after deformation at $e=6$ and $T=20$ °C; (b) after additional annealing at 140 °C for 2 h [12].

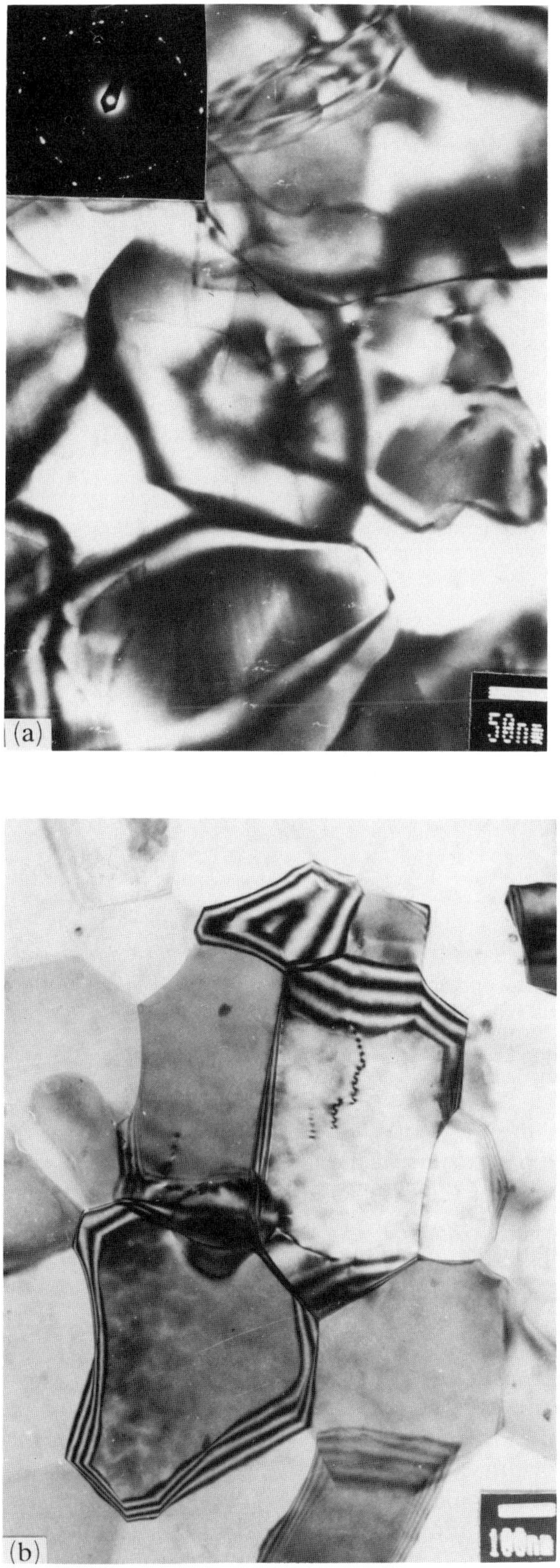

Fig. 2. Transmission electron micrographs and diffraction patterns of the structure of the Al–4%Cu–0.5%Zr alloy: (a) after deformation at $e=7$ and $T=20$ °C; (b) after additional annealing at 160 °C for 1 h [9].

At the same time, the structures of both these states differ considerably. One can observe a specific "diffusive" diffraction contrast of inclined GBs in the alloy after SPD (compare Figs. 2(a) and 2(b)). This contrast is a typical feature of non-equilibrium GBs with a higher energy and long-range stresses [17], and has been observed in different UFG materials [5, 11, 22]. Other evidence for the non-equilibrium state of GBs is given by the occurrence of extinction contours inside grains (Fig. 2(a)), which are linked with elastic stresses originated from GBs. On increasing the annealing temperature, one can observe a gradual decrease in the extinction contours in grain interiors and the simultaneous appearance of usual GBs with a typical banded contrast (Fig. 2(b)), which is very often also associated with some grain growth.

HREM investigation of UFG materials [18] enabled us to observe two main peculiarities of the GB structure. First, the crystallographic width of the GBs, as in usual materials, is narrow and does not exceed 1 nm. It should be noted, however, that the analogous conclusion has been recently reached for typical nanocrystals produced by the gas condensation method [19, 20]. Secondly, near a GB there is an elastically distorted layer, about 10 nm in width, where the value of the elastic strains is rather high and is 3%–5%. The existence of this layer is responsible for the appearance of "diffusive" contrast observed by TEM.

The results of Mössbauer investigations are of great interest as well. They provide information not only about GBs but also about the near-boundary region. In refs. 10 and 21, the Mössbauer spectroscopy was performed on UFG Fe (99.97% purity). Fe has a hyperfine magnetic structure which is easily resolved, making it suitable for Mössbauer experiments. It has been shown that the UFG Fe spectrum is a superposition of two magnetically split subspectra with considerably different electric and magnetic hyperfine structure (HFS) parameters [10]. This testifies to the existence of two different states of Fe atoms in the UFG material studied. The parameters of the first, more intense subspectrum practically coincide with those for coarse-grained α-Fe. The second subspectrum is probably related to a specific state of some Fe atoms near GBs. Following from the data obtained, we found the thickness of the near-boundary atomic layer (GB physical width) to be 8.4 ± 1.5 nm. The appearance of strictly fixed HFS parameters for the atoms under consideration proves the existence of a 'grain boundary phase' (GBP) coexisting with a grain phase (GP). The results obtained showed [21] that the Debye temperature of the GBP is 200 °C lower than that of the GP. The observed difference between these phases' parameters is associated first with a higher dynamic activity of GBP atoms as compared with that of GP. This fact can be attributed to the change in an atom's behavior in the elastically distorted near-boundary volume.

A schematic illustration of the structure of the UFG material is presented in Fig. 3 [5, 22]. Its main peculiarity is the presence of an elastically distorted layer with a high dynamic activity of atoms (GB phase) near the GBs. The thickness of the GBP (the physical width of a GB) is determined by the degree of the non-equilibrium state of GB structure. In intermetallic compounds, the GBP can be disordered [22]. The dilatometrical measurements of UFG samples directly testify that a decreased density is a specific feature of the GB phase [11, 13]. The GB phase evidently also contributes to the changes in such fundamental parameters of UFG materials as the saturation magnetization, Curie temperature, diffusion coefficient *etc.* [5]. In nanocrystalline materials with a grain size less than 10 nm, the width of the GBP is close to the crystal size, which is why high dynamic properties can be manifested here by all atoms of the crystalline lattice.

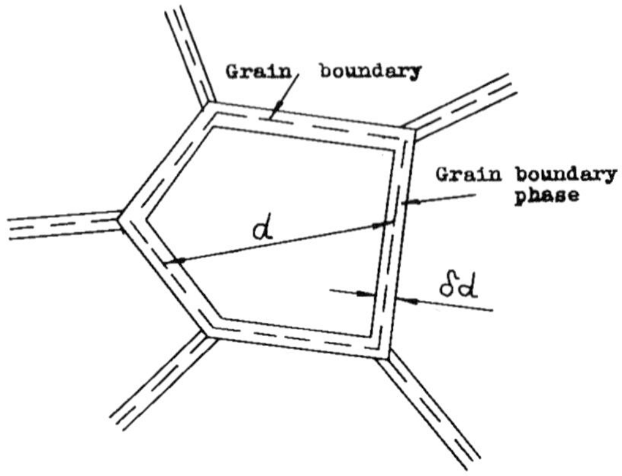

Fig. 3. Schematic illustration of the structure of the UFG material. δd, physical width of grain boundaries [5, 22].

3.3. Properties of UFG materials

UFG materials possess a number of unsual physico-mechanical properties. The data of structural analysis are the basis for the understanding of their origin.

3.3.1. Elasticity

Recently, in ref. 23, the elastic properties of Cu (99.98% purity) have been studied using an ultrasound method. The samples were subjected to severe deformation up to $e=4$ and subsequently annealed for 1 h at various temperatures within the interval 100–350 °C. The grain growth took place from 0.2–4 μm after annealing at higher than 175 °C. The measurements of the rate of longitudinal and transverse waves allowed us to estimate the values of Young's modulus E and of the shear modulus G, and their dependence on the annealing temperature (Fig. 4). For both curves within the region 150–175 °C, an abrupt change in the E and G values occurs: for large grain sizes, the elastic properties depend weakly on these sizes, whereas for small grain sizes, as in typical nanocrystals [24], the elastic properties are considerably lower than in coarse-grained samples.

The structural investigations showed that the observed variations of the elastic parameters were caused by the change in the GB structure state in the region of very small grain sizes. Actually, in the Cu samples subjected to severe deformation and subsequently annealed at temperatures up to 150 °C, the mean grain size was about 0.2 μm and non-equilibrium GBs prevailed. At the same time, in the samples annealed at 175 °C and higher, the GBs were mainly at equilibrium.

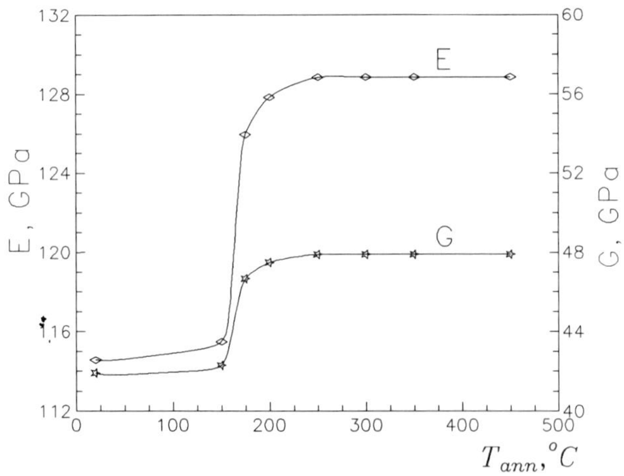

Fig. 4. Dependences of Young's modulus E and the shear modulus G on annealing temperature in UFG Cu [23].

On the basis of the data obtained in ref. 23, the elastic parameters of the GBs were estimated. Taking into account the crystallographic width of the boundary (equal to 1 nm for the equilibrium GBs) we obtained the value of the elastic modulus $E_{GB}{}^{e}=0.16E$, where E is Young's modulus of a single crystal and the shear modulus $G_{GB}{}^{e}=0.12G$. These values of $E_{GB}{}^{e}$ and $G_{GB}{}^{e}$ are close to the results in ref. 25 and look quite plausible.

In the case of non-equilibrium GBs, values of the moduli close to these can be obtained if we assume that their width is 4 nm, *i.e.* it considerably exceeds the crystallographic width of GBs. Therefore, in the case of non-equilibrium GBs, one should speak about some thickness of GB phase, as it has been mentioned above.

3.3.2. Damping properties

The investigations of the amplitude dependence of internal friction, using the method of the direct torsion pendulum, have been carried out on the Cu samples produced by the method described above, the mean grain size being from 0.2 to 50 μm [5]. The value of the defect of the shear modulus also has been measured. From the results given in Fig. 5 and the TEM investigations, it is seen that the smaller the grain size and the more non-equilibrium the GBs are, the higher are the background of the internal friction and, consequently, the damping properties at low strains. In Cu with a mean grain size of 0.2 μm and non-equilibrium GBs, these values are higher by a factor of 4–5 than those in a coarse-grained sample. At the same time, changes of δ_f closely correlate to the changes in the elastic properties. However, an abrupt decrease in $\Delta G/G$ is observed at the transformation of non-equilibrium GBs into an equilibrium state. Consequently, one can conclude that a large internal friction in SMG materials at low strains is caused by a large volume fraction of non-equilibrium GBs.

It is possible that the cause of this effect is connected with a number of reasons. First, there is the difference between the elastic moduli of the grain and GB phases, and, as a result, the scattering of waves during the propagation of elastic vibrations. The second reason may be a quasi-elasticity, which is conditioned by the processes of relaxation at non-equilibrium GBs. The substantial effect of relaxation in GBs on internal friction is well known. Taking into account the specific character of non-equilibrium GBs and the elevated diffusion ability, in particular, one can expect the enhancement of relaxation processes in these GBs.

The traditional methods of obtaining high damping materials (the selection of the alloy composition, the use of the alloys with magnetomechanical dissipation of energy, *etc.*) are accompanied by a decrease in the strength properties. The formation of the UFG structure along with the rise in damping properties increases the strength characteristics, as shown below.

3.3.3. Strength

The decrease of the mean grain size is a traditional method of increasing the material strength. However, with UFG materials, this problem remains unsolved as yet. Let us consider, for example, the deformation behavior of the alloy Mg–1.5%Mn–0.3Ge with submicrometer (d = 0.3 μm) and fine (d = 10 μm) grains [9]. Figure 6 shows three typical stress–strain curves for the alloy at 20 °C. Curves 1 and 2 refer to UFG structures distinguished by the states of GBs: non-equilibrium (curve 1) and relatively equilibrium (curve 2) states. In these cases, the flow stresses are quite similar but there is a sharp difference in the elongations to failure. However, the curves show no signs of pronounced strain hardening and this is what principally distinguishes the UFG alloy behavior from that of the alloy with the grain size of 10 μm (curve 3).

Much recent work [5, 9, 26] has studied the problem of the Hall–Petch dependence in UFG materials. As with the nanocrystals produced by the method of gas condensation [27, 28], the results are ambiguous. At times, the deviation from the Hall–Petch dependence is observed, while at other times this is not seen. In this case, the recent work [26] will be useful.

In this work, the Hall–Petch dependence in the UFG Al–1.5%Mg alloy with different degrees of a non-

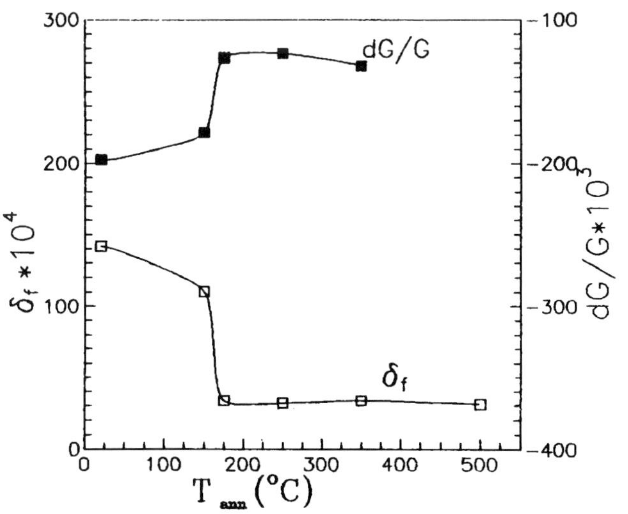

Fig. 5. Dependences of background of the internal friction δ_f and defect of the shear modulus dG/G on annealing temperature in UFG Cu.

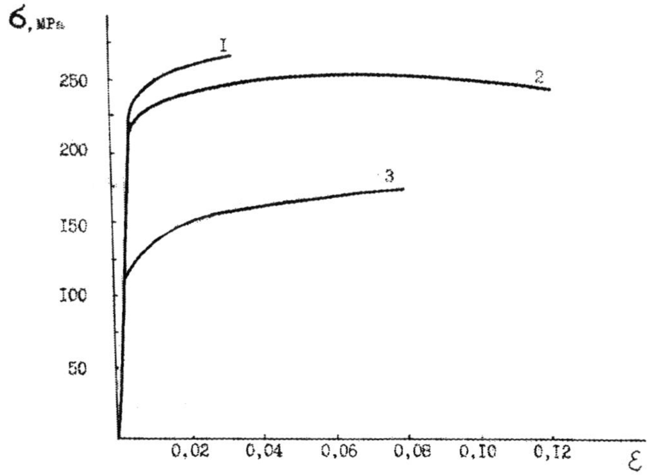

Fig. 6. True stress–strain curves of the Mg alloy Mg–1.5%Mn–0.3%Ce with T = 20 °C and $\dot{\varepsilon} = 5 \times 10^{-4}$ s^{-1} [9]: 1, UFG structure, d = 0.3 μm, non-equilibrium GBs; 2, UFG structure, d = 0.3 μm, equilibrium GBs; 3, alloy with the mean grain size of 10 μm.

equilibrium state of GBs is investigated (Fig. 7). As can be seen, the change in the GB state can lead to a noticeable change in the microhardness, while the grain size remains the same. This effect is even more visible in the Ni_3Al intermetallic compounds [22]. Furthermore, the degree of deviation from the linear dependence $\sigma - f(d^{-1/2})$ evidently depends on the homological temperature of the alloys [5]. At the same time, it is interesting that the maximum strength properties are often observed in samples with grain sizes of 0.1–0.2 μm. Thus, for example, in Table 1 there are presented the highest characteristics of some UFG alloys. It is seen that, in the Al–Mg–Li alloy, on decreasing the grain size from 10 μm to 0.2 μm, the yield stress rises from 330 to 550 MPa, *i.e.* by more than 60%. However, in the case of UFG materials, one should take into account the effect of the GB state as well. Thus, for example, owing to GB transition from an equilibrium state to a non-equilibrium state, the increase in the yield stress from 600 to 945 MPa and from 755 to 925 MPa was observed in the steels Fe–0.03%C–1.3%Mn–0.15%Si–0.2%V–0.2%Cr and Fe–0.2%C–1%Mn–0.5%Si–0.9%V respectively.

The examples presented indicate that the decrease of the grain size to submicrometer dimensions results in rather high strength states in metals and alloys. However, at the same time, any further decrease of the grain size does not often produce a considerable effect, this being connected with the increase of the role of GB processes.

3.3.4. Toughness

The refinement of the granular structure is known to lead to an increase of the toughness in metals. A similar tendency is observed on the transformation to a UFG structure. The dependence of the impact strength KCU on the temperature of testing for two states of the steel Fe–0.03%C–1.3%Mn–0.2%V (d=0.7 μm and 30 μm) is given in Fig. 8. It can be seen that the specific work of failure in UFG samples is noticeably larger than that in coarse-grained samples, especially in the region of low temperature. Evidently, in the case of smaller grain sizes, GBs become effective barriers to crack propagation.

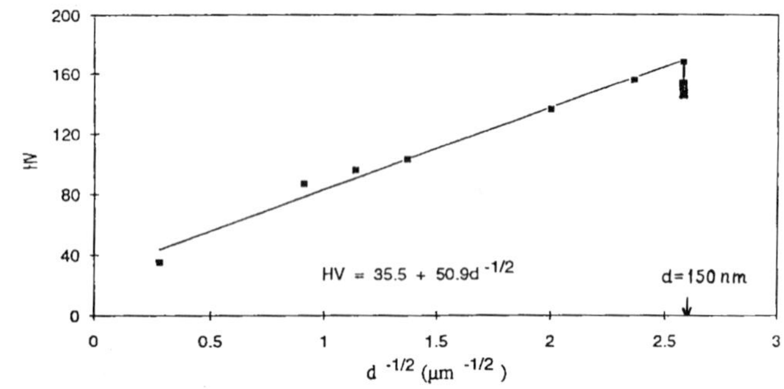

Fig. 7. Microhardness *vs.* on the mean grain size in the Al–1.5%Mg alloy (deformation temperature, 20 °C). Scattering of HV values for d=150 nm corresponds to different degrees of GB non-equilibrium state [26].

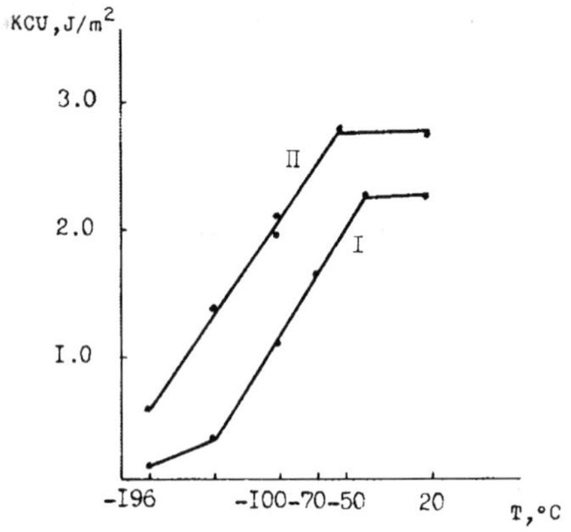

Fig. 8. Work of failure of the steel Fe–0.03%C–1.3%Mn–0.15%Si–0.2%V–0.2%Cr, depending on temperature of testing: I, d=12 μm; II, d=0.7 μm.

TABLE 1. Values of yield stress ($\sigma_{0.2}$), tensile strength stress (σ_b) and maximum elongation to failure (δ) in the UFG alloys

Alloy	Grain size d (μm)	$\sigma_{0.2}$ (MPa)	σ_b (MPa)	δ (%)	Ref.
Al–5%Mg–2%Li–0.2%Sc–0.1%Zr	10	330	450	5	9
	0.2	550	600	6	
Al–2%Mg–1%Cr–0.9%Zr	2	280	300	14	30
	0.2	350	390	18	
Steel	12	300	465	26	5
(Fe–0.03%C–1.3%Mn–0.15%Si–0.2%V–0.2%Cr)	0.7	925	945	13	
Ti–6.5%Al–3.2%Mo–0.3%Si	5	—	1050	20	29
	0.4		1310	15	

TABLE 2. Parameters of superplasticity and mechanical properties of some UFG alloys

Alloy	Mean grain size (μm)	Temperature of deformation (°C)	Strain rate, $10^4\dot{\varepsilon}$ (s^{-1})	Parameters $m=\frac{d\ln\sigma}{d\ln\dot{\varepsilon}}$	Stress for $\varepsilon=40\%$ (MPa)	Relative elongation to failure, δ (%)	Ref.
Ti–6.5%Al–3.2Mo–	5.0	650	7	0.12	460	50	29
–0.3%Si	0.1	650	7	0.46	155	500	
Al-4%Cu–0.5%Zr	8.0	500	3	0.5	130	800	3
	0.3	220	3	0.48	270	>250	
Mg–1.5%Mn–	10	400	6.4	0.44	10	300	9
–0.3%Ge	0.3	160	6.4	0.42	30	>150	

3.3.5. Superplasticity

Let us consider now the behavior of UFG materials at elevated temperatures. In ref. 9, the temperature dependence of the Mg alloy yield stress when the grain sizes were 0.3 and 10 μm has been studied. In the case when $d=10$ μm, a drop in the $\sigma_{0.2}$ value occurs at 330–350 °C, and at 400 °C the alloy displays typical superplastic features. At the same time, when $d=0.3$ μm, an abrupt softening takes place in the temperature range 100–130 °C, and the superplastic effect is realized at 180 °C (Table 2). The given data indicate that the UFG alloy displays typical superplastic properties at 180 °C, which are similar to those of the alloy with the grain size of 10 μm observed at 400 °C. Similar regularities are observed in other UFG alloys (Table 2).

The electron microscopy studies indicated that the structural changes occurring during deformation of UFG alloys are also typical of superplastic materials. After deformation, the grains remain equiaxial, there are no dislocations in the structure and some grain growth is observed. For example, in the Al–Cu–Zr alloy, after 100% tensile straining, the grain size was 0.5 μm compared with 0.3 μm in the initial state [3]. Hence, during the deformation of UFG alloys, a typical superplastic behavior can be observed. Since, in this case, the superplastic temperature decreases by 200–250 °C, we can speak of the so-called low temperature superplasticity of UFG materials. The origin of this effect is associated with a considerable acceleration of GB diffusion during the deformation of UFG alloys, when the diffusion coefficient is 1.5 orders of magnitude larger than that under 'conventional' deformation [3].

4. Conclusions

The development of methods of SPD has lately made it possible to use them to obtain a superfine-grained structure with a grain size of about 0.1 μm (100 nm) or less (even 20 nm). The investigations conducted indicate that metals and alloys in the UFG state possess substantially changed fundamental parameters, namely the Curie and Debye temperatures, elastic moduli and diffusion coefficients. All these variations are caused not only by small grain sizes but also by the extremely large area of GBs and the high degree of GBs in a non-equilibrium state.

In fact, UFG alloys represent a new class of materials with unusual physical and mechanical properties, namely abnormally high strength and damping properties, lower temperature superplasticity, elevated toughness and high magnetic properties [5]. The combination of SPD with powder metallurgy can be considered as one of the most promising methods of SPD utilization. One can expect that this combination of methods will permit us to obtain new alloys of unusual compositions [31] and to produce new nanocrystalline metal–ceramic composites. Although all these questions need more generalized investigations, taking into account the good technology of developed methods of UFG materials preparation, one may expect their great practical use in future.

References

1 N. A. Smirnova, V. I. Levit, V. P. Pilyugin, R. I. Kuznetzov, L. A. Davydova and R. A. Sazonova, *Met. Phys. Metall., 61* (1986) 1170.

2 A. Lasalmoni and J. L. Stryel, *J. Mater. Sci., 21* (1986) 1837.

3 R. Z. Valiev, O. A. Kaibyshev, P. N. Kuznetzov, R. Sh. Mussalimov and N. K. Tsenev, *Dokl. Akad. Nauk SSSR, 301* (1988) 864.

4 O. Kaibyshev, R. Kaibyshev and G. Salishchev, *Proc. Int. Conf. on Recrystallization—92, San-Sebastian, Spain, Mater. Sci. Forum, 113–115* (1993) 423.

5 R. Z. Valiev, A. K. Korznikov and R. R. Mulyukov, *Met. Phys. Metall., 4* (1992) 70.

6 R. Birringer and H. Gleiter, in R. W. Cahn (ed.), *Encyclopedia of Materials Science and Engineering, Supplement 1,* Pergamon, Oxford, 1988, p. 339.

7 F. H. Froes and C. Suryanarayana, *J. Met., 6* (1989) 12.
8 C. C. Koch and Y. S. Cho, *Nanostruct. Mater., 1* (1992) 207.
9 R. Z. Valiev, N. A. Krasilnikov and N. K. Tzenev, *Mater. Sci. Eng., A137* (1991) 35.
10 R. Z. Valiev, R. R. Mulyukov and V. V. Ovchinnikov, *Philos. Mag. Lett., 62* (1990) 253.
11 R. Sh. Musalimov and R. Z. Valiev, *Met. Phys. Metall., 9* (1992) 72; *Scr. Metall. Mater., 12* (1992) 1685.
12 R. Z. Abdulov, R. Z. Valiev and N. A. Krasilnikov, *Mater. Sci. Lett., 9* (1990) 1445.
13 Kh. Ya. Mulyukov, S. V. Khafizov and R. Z. Valiev, *Phys. Status Solidi A, 133* (1992) 447.
14 V. V. Rybin, *High Plastic Deformations and Metals Fracture*, Metallurgia, Moscow, 1986, p. 270 (in Russian).
15 G. A. Salishchev, R. M. Imayev, V. M. Imayev and N. K. Gabdullin, *Mater. Sci. Forum, 113–115* (1993) 613.
16 E. N. Teitel, M. A. Uimin, A. E. Ermakov, A. V. Shangurov, V. A. Barinov, G. M. Makarova, R. I. Kuznetzov, V. P. Pilyugin and V. N. Gundyrev, *Met. Phys. Metall., 7* (1990) 95.
17 R. Z. Valiev, V. Yu. Gertsman and O. A. Kaibyshev, *Phys. Status Solidi A, 97* (1986) 11.
18 G. Kapelski and R. Z. Valiev, to be published.
19 G. J. Thomas, R. W. Siegel and J. A. Eastman, *Scr. Metall. Mater., 24* (1990) 202.
20 W. Wunderlich, Y. Ishida, R. Maurer, *Scr. Metall. Mater., 24* (1990) 403.
21 R. Z. Valiev, R. R. Mulyukov, V. V. Ovchinnikov and V. A. Shabashov, *Scr. Metall. Mater., 25* (1991) 12.
22 R. Z. Valiev, *Proc. NATO ASI, Mechanical Properties of Ultrafine-grained Materials*, to be published.
23 N. A. Akhmadeev, R. Z. Valiev, R. R. Mulyukov, N. P. Kobelev and Ya. M. Soifer, *Sov. Solid State Phys., 34* (1992) 3155.
24 D. Korn, A. Morsch, R. Birringer, W. Arnold and H. Gleiter, *J. Phys., 49* (1988) 769.
25 D. Wolf and M. D. Kluge, *Scr. Metall. Mater., 24* (1990) 907.
26 R. Z. Valiev, R. Chmelik, F. Bordeaux, G. Kapelski and B. Baudelet, *Scr. Metall. Mater., 27* (1992) 855.
27 A. H. Chokshi, A. Rosen, J. Karch and H. Gleiter, *Sci. Met., 23* (1989) 1679.
28 G. E. Fougere, J. R. Weertman, R. W. Seigel and S. Kim, *Scr. Metall. Mater., 26* (1992) 1879.
29 O. R. Valiakhmetov, R. M. Galeev and G. A. Salishchev, *Met. Phys. Metall., 10* (1990) 204.
30 R. Z. Valiev and N. K. Tsenev, in T. G. Langdon, H. D. Merchant, J. G. Morris and M. A. Zaidi (eds.), *Hot Deformation of Aluminium Alloys*, The Minerals and Materials Society, Warrendale, PA, 1991, p. 319.
31 A. V. Korznikov, I. M. Safarov, D. V. Laptionok and R. Z. Valiev, *Acta Metall. Mater., 39* (1991) 3193.

Materials Science and Engineering, A168 (1993) 149–152

Designing a high energy ball-mill for synthesis of nanophase materials in large quantities

D. Basset, P. Matteazzi* and F. Miani

Dipartimento di Scienze e Tecnologie Chimiche, Università degli Studi di Udine, Via Cotonificio 108, I-33100 Udine (Italy)

Abstract

Nanophase materials can be synthesized by high energy milling. However, processing of large quantities finds a limitation in the devices available commercially, which are usually designed mainly for comminution and for laboratory-scale operations. We propose the design of a high energy, high capacity ball-mill which can be easily scaled up. An experimental device capable of processing up to about 250 g of powders with about 5 kg of balls has been realized. Impact velocities of about 3.5 m s^{-1} has been measured with operating frequencies of 17 Hz. The synthesis of nanophase iron carbides (assumed as a test system) is shown to be feasible with the new mill, with kinetics comparable with the best obtained in a Spex mill.

1. Introduction

Besides alloys [1], various materials, such as intermetallic compounds [2, 3], transition metal borides [4], silicides and carbides [5, 6], have been obtained by mechanical activation (mechanochemical synthesis or mechanosynthesis) through ball-milling. It has been shown recently that reduction reactions also are feasible by mechanical activation in ball-mills, such as the reduction of various oxides with aluminum to obtain nanocrystalline metal–alumina composites [7].

The potential of nanophase materials for atomic-scale engineering is somehow limited by the availability of large quantities for compaction. Among the synthesis methods of such materials, high energy milling is promising at a production scale. Vibratory and planetary ball-mills (widely used as high energy mills) available commercially are generally not designed for the specific task of synthesizing nanophase materials in large quantities.

In the present work, we present the criteria adopted in the design of a high energy, high capacity mill and comparative results obtained for the mechanosynthesis of iron carbides.

2. Design of the high energy, high capacity mill

It appears that several criteria must be met to design such a mill: (a) high impact velocities [8, 9] (or at least comparable with those in the actual high energy mills, *i.e.* about 3 m s^{-1}) of the grinding means and, therefore, a high energy available for each impact; (b) high impact frequency for each grinding mean to enhance reaction rates and productivity; (c) easy scale-up of the mill capacity.

In the commercially available mills, at least one criterion is not satisfied, such as the following: planetary and vibratory (such as the Spex, with an oscillation amplitude of about 20 mm or more) mills find a capacity limit in the control of inertial forces; tumbling and attritor mills are limited in the impact frequency for each grinding mean; attritor and vibratory (with oscillation amplitude up to about 5 mm) mills have low velocities of grinding means.

To meet the requirements of the above criteria, we designed the mill shown in Fig. 1, which is basically comprised of a vial, a system to compensate the inertial forces and a driving mechanism.

The driving system comprises a crank and connecting rod, powered by an asynchronous electric engine (3 kW). To choose the frequency and amplitude of the vial movement, we performed measurements of the ball speeds in the Spex mill: the speeds, as described in ref. 10, were estimated from the width of the impact craters of the balls on copper targets, and were found to be less than 3 m s^{-1}.

Therefore, the vial in the new mill has been designed to oscillate along its axis "H" in Fig. 1, with a frequency 17 Hz and amplitude 30 mm, guided by low friction sliding blocks. This gave rise to a speed (measured as above [10]) of the balls of the order of 3.5 m s^{-1} (criterion (a)), with a maximum vial acceleration of 180

*Author to whom correspondence should be addressed.

0921-5093/93/$6.00

m s^{-2} and maximum speed of 1.7 m s^{-1}. Also, the requirement for the impact frequency of each ball has been satisfied (close to 17 Hz) and, in any case, can be easily adjusted. The vial is composed of hardened carbon steel. A shaped top-plate (with an O-ring for internal atmosphere control) is provided (Fig. 1) to avoid preferential impact zones of the grinding means (carbon steel balls).

To fulfil the requirements of criterion (c), it is necessary to provide dynamic compensation of inertial forces (becoming critical during the scale-up), which is proposed in Fig. 1 by 4+4, counteracting, preloaded springs. The dynamic control of inertial forces, which is critical especially when large weights (vial plus charge) are involved, is presented in Fig. 2, where an elastic constant of 2.5 kg mm^{-1} is assumed for each spring. It is shown that the actual load on the driving system is reduced to a value close to zero during the entire cycle. The data in Fig. 2 are given under the hypothesis of a maximum total load of 16 kg, of which 5 kg is the vial charge consisting (for example) of 250 g of powders to be processed with a powder-to-ball weight ratio 1:20.

The dynamic compensation of inertial forces shown in Fig. 2, for a particular case, can be easily applied for the scale-up of the mill (1) by increasing the size of the vial and (2) designing the spring system to withstand the increased loads.

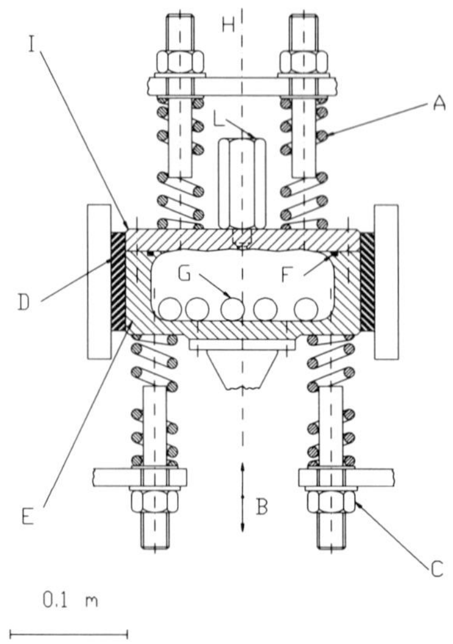

Fig. 1. Sectional view of the high energy, high capacity mill: A, spring; B, driving force (sinusoidal-like); C, hut for spring preloading; D, Teflon sliding block; E, vial; F, O-ring; G, grinding balls; I, shaped top-plate; L, fluids in-out (vacuum, inert or reactive gas); H, oscillation axis.

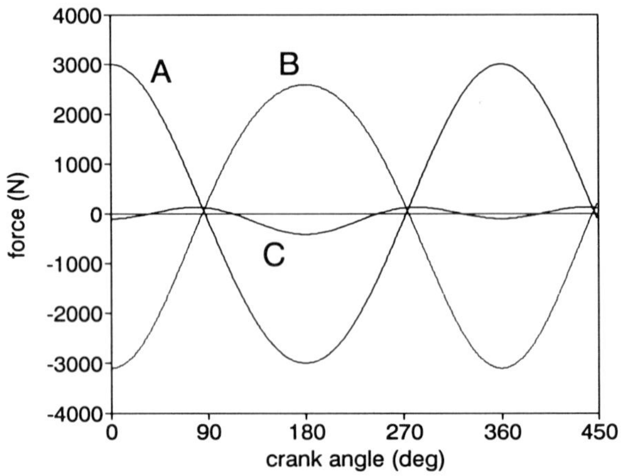

Fig. 2. Behaviour of the net dynamic force in the direction of the oscillation axis (positive sign, up) for each cycle as a function of crank angle: A, elastic forces; B, inertial forces; C, A+B, *i.e.* actual load at the big end of the connecting rod.

3. Experimental details

As a test experiment to verify the efficiency of the new mill, we choose to study the kinetic of synthesis of nanophase iron carbides in comparison with the results previously obtained [11] with a commercial mill (Spex 8000 mixer/mill; ball-to-powder ratio (BPR), 20:1; inert atmosphere; powder charge, 3 g; discontinuous milling with milling steps of 2.5 h and a wait of 0.5 h). An elemental powder mixture with composition $Fe_{75}C_{25}$ was milled at room temperature in the new mill described above, using the following conditions: BPR, 20:1; total amount of powder, 70 g; balls diameter, 25.4 mm. We utilized a carbon steel vial and balls. The starting materials were iron (99%, 60 μm) and graphite (45 μm). The powders of iron and graphite, in the atomic proportions 3:1, were introduced into the vial and sealed under air. The powders were sampled for structural examination with time intervals of 2.5 h (with a wait [11] time of at least 0.5 h). The milling times in the text are referred to as effective milling times (*i.e.* not including the waiting times).

The milled powders were characterized by Mössbauer spectroscopy and X-ray diffraction (XRD). The Mössbauer spectra were obtained in transmission geometry at room temperature and the spectra were analysed with a hyperfine field distribution (HFD) [11]. Isomer shift δ values are calculated with respect to α-Fe. The XRD patterns were collected using Co Kα radiation ($\lambda = 0.1789$ nm) and the X-ray lines broadening was analysed [12] to separate broadening owing to crystalline size from broadening owing to strain.

4. Results and discussion

The XRD patterns obtained after milling the elemental iron and graphite powders in the new mill showed, in the first stages of milling, the broadening of iron lines (crystal size of about 11 nm after 2.5 h of milling). With further milling, the (121, 221) lines of orthorhombic Fe_3C can be resolved in the pattern starting from 5 h. After 7.5 h (and moreover after 10 h) of milling, the XRD pattern revealed the appearance and growth of other components: *i.e.* hexagonal ε-Fe_2C; χ-$Fe_{2.5}C$ monoclinic Hägg carbide. The crystal size of Fe_3C after 10 h of milling was 9 ± 1 nm.

The Mössbauer spectra presented in Fig. 3 for 2.5 and 10 h were analysed (assuming [11] $\delta = 0.3$, 0.2 and 0 mm s^{-1} for the respective field intervals 0–95, 95–270 and 270–400 kG) by an HFD (reported in the right-hand side of Fig. 3) attributing [11] (a) field interval 185–215 kG to cementite Fe_3C, (b) intervals 0–95 kG, 130–170 kG and 230–270 kG to ε-carbide, a disordered hexagonal carbide (either magnetic or superparamagnetic); (c) intervals 95–130 kG, 170–185 kG and 215–230 kG to Hägg carbide χ-$Fe_{2.5}C$, (d) field components between 270 and 300 kG to isolated carbon multiplets and (e) the field interval 300–340 kG was finally attributed to α-Fe.

Previous experiments performed [11] with the Spex mill showed quite similar structural features at the different synthesis stages. In particular, also in this case, the mechanosynthesis of iron carbides can be attributed [11] to a mechanism in which an Fe–C (martensite-like) alloy is formed and aged during milling to form carbon clusters and hexagonal (or close to hexagonal) carbides, which may be transformed finally into cementite (or cementite and Hägg carbide).

The HFD was analysed to extract the relative area (%RA) of the various components as above and with α-Fe. In Fig. 4, the results are presented, along with the results obtained [11] with the Spex mill, in a similar milling condition (same BPR and discontinuous milling, but different charge weight (3 g), atmosphere and size of balls). The kinetics of synthesis of the various carbides are quite comparable between the two mills (Figs. 4(b) and 4(c)), as are the kinetics of disappearance of the iron (Fig. 4(a)). The carburization of the iron is also almost completed in about 10 h with the new mill.

5. Conclusions

A high capacity, high energy mill has been developed using the concept of the dynamic control of inertial forces by counteracting elastic systems. In this

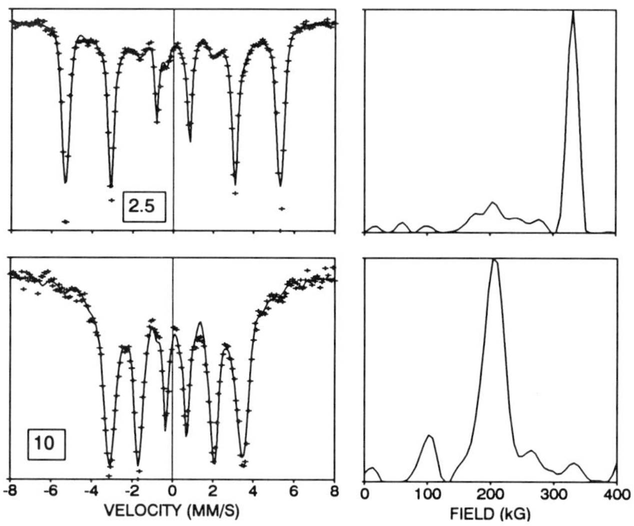

Fig. 3. Mössbauer spectra (left) and HFDs (right) of milled (new mill) powders with PBR of 20:1 charge of 70 s. Numbers are milling times.

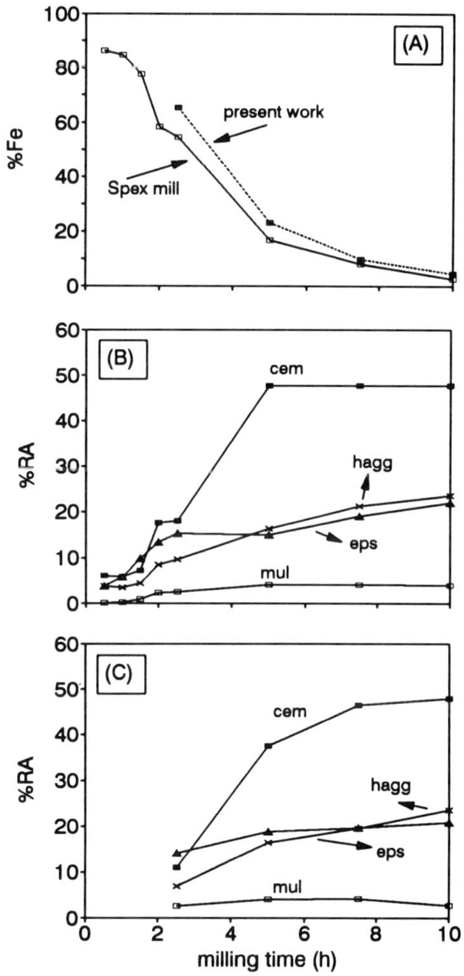

Fig. 4. (A) %Fe (%RA) *vs.* milling time for milled $Fe_{75}C_{25}$ in the Spex and new mill (present work). Behaviours of hFD components as %RA *vs.* milling time for the (B) Spex and (C) new mill: cem, cementite; Hagg, Hägg carbide; eps, ε-carbide; mul, isolated carbon multiplets.

way, a mill has been constructed and successfully tested for the synthesis of nanophase iron carbides in batches of 70 g in 10 h. The reaction rates and mechanisms, together with the final crystal sizes, are comparable with those obtained with other high energy mills. The design of the mill allows an easy scale-up of its capacity.

Acknowledgments

This work was financially supported by EC Contract SC1-CT91-0668 (Science Program).

References

1 J. S. Benjamin, *Met. Powder Rep., 45* (2) (1990) 122.
2 L. Schultz, J. Wecker and E. Hellstern, *J. Appl. Phys., 61* (1987) 3583.
3 P. Matteazzi and G. Le Caër, *Proc. Adv. Mater., 1* (1991) 135.
4 A. Calka and A. P. Radlinski, *J. Less Common Met., 161* (1990) L23.
5 G. Le Caër, P. Matteazzi, E. Bauer Grosse, A. Pianelli and E. Bouzy, *J. Mater. Sci., 25* (1990) 4726.
6 P. Matteazzi and G. Le Caër, *J. Am. Ceram. Soc., 74* (6) (1991) 1382.
7 P. Matteazzi and G. Le Caër, *J. Am. Ceram. Soc., 75* (10) (1992) 2749.
8 Y. Chen, R. Le Hazif and G. Martin, *Solid State Phenom., 23–24* (1992) 271.
9 D. R. Maurice and T. H. Courtney, *Metall. Trans. A, 21* (1990) 289.
10 P. Matteazzi, D. Basset and F. Miani, *Mater. Sci. Eng.*., in press.
11 P. Matteazzi, F. Miani and G. Le Caër, *J. Hard Mater., 2* (3–4) (1991) 171.
12 P. Matteazzi and G. Le Caër, *Mater. Sci. Eng., A149* (1991) 135.

Materials Science and Engineering, *A168* (1993) 153–155

Catalytic properties of mechanosynthesized nanophase iron carbides

F. Miani, P. Matteazzi*, G. Dolcetti, A. Lutman and A. Trovarelli
Dipartimento di Scienze e Tecnologie Chimiche, Via Cotonificio 108, I-33100 Udine (Italy)

Abstract

Nanophase iron carbides can be obtained by high energy milling (mechanosynthesis). The hydrogenation of CO_2 catalyzed by mechanosynthesized iron carbides (at different milling stages) is studied from the point of view of activity, selectivity and stability during the reaction. The catalytic properties are attractive and competitive with those of other catalytic materials. The catalytic behaviour is shown to be stable during testing for 24 h. Structural characterization of the catalysts after the reaction is also reported.

1. Introduction

Among the synthesis methods of nanophase materials, mechanical activation (mechanochemical synthesis or mechanosynthesis) is a promising method for the processing of large quantities of materials. The mechanosynthesis of nanophase intermetallic compounds [1, 2], carbides and silicides [3, 4] and metal–oxide composites [5] has been shown to be feasible.

Nanocrystalline materials are a new, promising class of materials for their enhanced properties, which also arise from the high density of grain boundaries and defects [6]. These characteristics, which exceed by far traditionally chemically synthesized materials [7], naturally lead to their use as catalysts being considered.

The hydrogenation of CO_2 is now a current research topic in catalysis [8, 9], as environmental issues point out the necessity to widen the field of research from the original investigations on the hydrogenation of CO [10].

The hydrogenation of CO_2 catalysed by mechanosynthesized nanophase iron carbides [11] is discussed in this work from the point of view of the stability of the catalysts during the reaction itself.

2. Experimental details

Mechanosynthesized iron carbides were obtained in a high energy ball-mill (Spex 8000 mixer/mill) using a WC vial and balls. The starting materials were 5.6 g Fe in powders (99%, 60 μm) and 0.4 g graphite (45 μm), giving the composition $Fe_{75}C_{25}$. The powders were introduced into the vial which was sealed in a pure N_2 atmosphere. The WC balls weighted 60 g and a ball-to-powder ratio (BPR) of 10 to 1 was obtained. The milling times were 2.5, 7.5, 15 and 22.5 h. From previous studies [12], in these conditions, a significant presence of carbides is detected even after 5 h.

The catalytic experiments were performed in a conventional stainless steel flow microreactor, operating at atmospheric pressure, at 523 K, for periods of analysis lasting up to 24 h. The amount of catalysts used were 0.4–0.6 g and the ratio of H_2 to CO_2 gas mixture was 3 to 1.

We observed the structure of both as-prepared and used catalysts by Mössbauer spectroscopy and X-ray diffraction (XRD). The Mössbauer spectra were obtained in transmission geometry at room temperature. The spectra were analysed with a hyperfine field distribution (HFD) [12]. Isomer shift δ values were calculated with respect to α-Fe. The XRD patterns were collected using Co Kα radiation ($\lambda = 0.1789$ nm). The X-ray line broadening was analysed to separate broadening owing to crystallite size from broadening owing to strain, according to the method described in ref. 5.

3. Results and discussion

3.1. Mechanosynthesis of nanophase iron carbides

The XRD patterns revealed the appearance (for the longer milling times) of the following carbides: orthorhombic Fe_3C, hexagonal ε-Fe_2C, χ-Fe_5C_2 monoclinic Hägg carbide. The crystallite sizes of mechanosynthe-

*Author to whom correspondence should be addressed.

0921-5093/93/$6.00

sized Fe_3C (about 13 nm after 7.5 h of milling) are reported in Table 1.

A more detailed analysis can be made by consideration of the Mössbauer spectra reported in Fig. 1 for the case of a milling time of 15 h before and after the reaction. These spectra have been analysed by a HFD, attributing [12, 13] the different field intervals (with δ in the range 0.3–0 mm s^{-1}) to the following components, as represented in Fig. 2: cementite Fe_3C; ε-carbide (assumed to be of composition Fe_2C [12]); Hägg carbide χ-Fe_5C_2; isolated carbon multiplets (Fe–C alloy [12]); α-Fe. The HFD was analysed to extract the relative area (%RA) of the various components and the results are presented in Table 1.

TABLE 1. Phase composition from Mössbuaer spectra (given as %Fe on total Fe) and crystallite size (from XRD line broadening analysis)

Milling time (h)	Fe (%)	Fe_3C (%)	ε-Fe_2C (%)	χ-Fe_5C_2 (%)	Carbon multiplets (%)	Crystallite size (nm)
As prepared						
2.5	73.8	8.0	11.0	5.0	2.2	9[a]
7.5	31.6	24.4	21.2	16.0	7.0	13[b]
15	9.0	31.0	30.4	23.1	6.6	13[b]
22.5	2.0	38.9	26.0	29.1	4.0	10[b]
Used catalysts						
2.5	56.3	18.8	9.4	15.3	0.2	9[a]
7.5	16.5	53.2	10.7	18.4	1.2	13[b]
15	3.8	56.1	16.5	22.2	1.5	13[b]
22.5	2.2	44.5	19.5	32.8	1.0	16[b]

[a] α-Fe.
[b] Fe_3C.

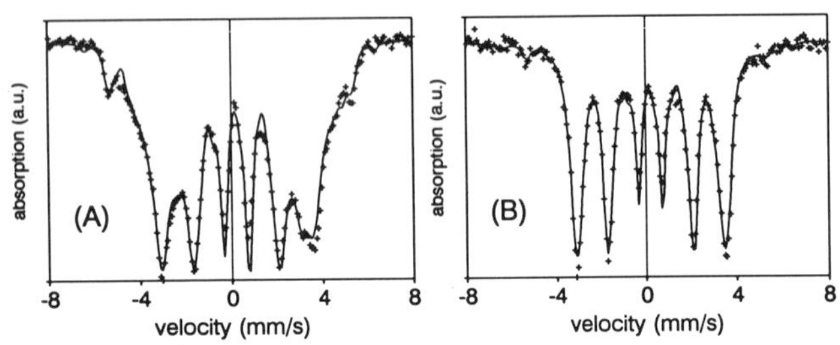

Fig. 1. Mössbauer spectra of mechanosynthesized iron carbides (15 h milling): (a) as-prepared; (b) used catalyst.

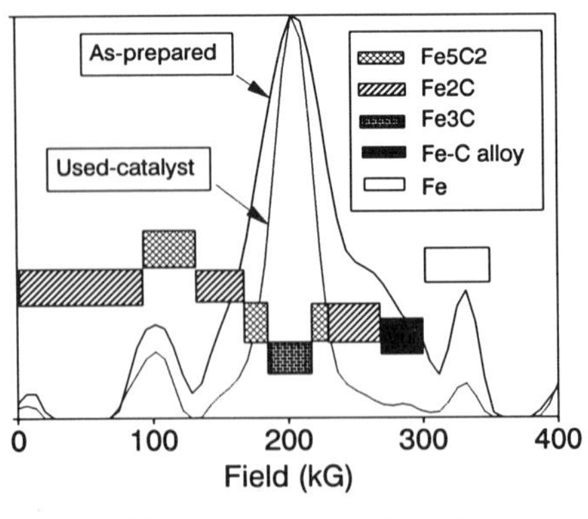

Fig. 2. HFD extracted from Mössbauer spectra of Fig. 1: ——, as prepared; ....., used catalyst.

We remark here that, while the stoichiometry of carbides in which the interstitial atoms are tetrahedrally coordinated (Fe_3C, Hägg carbide) is well established, for carbides in which the interstitial atoms are octahedrally coordinated (O-carbides), the situation is less clear [12, 14]. For the ε-carbides, for the sake of simplicity, we refer to Fe_2C, while a more accurate analysis might bring the presence of other carbides with a close stoichiometry, which is often indicated as $Fe_{(2+x)}C$, where x may assume values up to 0.5 [14].

3.2. *Catalytic activity*

In a previous investigation of CO_2 hydrogenation catalysed by mechanosynthesized nanophase iron carbides [11], it was shown that these materials are active catalysts in the range of temperatures and pressures examined. The results reported are similar to those found in the literature for iron catalysts at atmospheric pressure.

We wish now to focus our attention on the structural evolution of the catalysts during the reaction time, compared with the catalyst performance. In Table 2 are reported the catalytic properties of mechanosynthesized carbides. CO and CH_4 are the main products (CO being the prevalent), while appreciable amounts of higher hydrocarbons are produced on milled powders. Extending the milling period from 15 to 22.5 h, *i.e.* after the total carburization of iron, does not improve the catalytic activity, so the maximum value is for the 15 h sample.

The conversion of these samples has been followed as a function of time, until a constant value of catalytic activity was reached. It turns out that longer milling times affect the initial behaviour of the catalyst, changing the time in which steady state conditions are reached. Samples D and E are not much affected by the reaction time, while samples B and C show a marked increase in production of both CO and CH_4, thus leading to an overall increase in the rate of hydrocarbon formation. The much greater degree of carburization during the reaction of samples B and C could account for this behaviour. Samples D and E indeed have a higher percentage of carbides which are less affected by the reaction.

The structural observations reported in Tables 1 and 2 lead to several remarks in the comparison of used and as-prepared catalysts. Despite many major changes in the relative proportion of the phases, the catalyst

TABLE 2. CO_2 hydrogenation[a] over mechanosynthesized iron carbides: rates and product distribution

Sample	Milling time (h)	Surface area ($m^2 g^{-1}$)		Selectivity based on hydrocarbons							
				CO[b]	CH_4[b]	C_1	C_2	C_3	C_4	C_5	C_6[c]
A	0	2.4	i[d]	29	1	94.1	4.7	1.2	—	—	—
			s[d]	27	<1	97.0	2.6	—	—	—	—
B	2.5	6	i	99	7	81.8	11.5	5.3	1.4	0.4	—
			s	136	22	75.6	14.3	7.2	1.9	0.9	0.1
C	7.5	7.5	i	96	10	78.1	13.5	6.1	1.6	0.6	0.1
			s	138	18	77.5	14.1	6.0	1.6	0.7	—
D	15	3	i	142	12	78.8	14.7	5.0	0.8	0.6	0.1
			s	169	14	81.5	13.0	4.1	0.9	0.5	—
E	22.5	4.2	i	72	5	75.9	15.3	5.9	2.0	0.7	0.2
			s	68	6	76.6	15.5	5.7	1.3	0.6	—

[a]Reaction conditions: GHSV = 15000–20000 h^{-1}; before CO_2 introduction, the catalyst was conditioned by purging with H_2 for 1 h to remove any oxygen-containing gases and to ensure complete removal of any surface iron oxide present, which would interfere with hydrocarbon formation.
[b](Mol. CO or CH_4 formed)/(g catalyst × h) × 10^5.
[c]C_n indicates paraffins or olefines with n carbon atoms.
[d]i, initial conditions; s, steady state conditions.

nanostructure is not profoundly altered (Table 1) by the effluent stream and the processing temperature (523 K), as is proved also by the stability of the catalytic behaviour. For instance, the crystal sizes of all the processed samples is comparable with those of the as-prepared materials. We observe that the most stable phases are Fe_3C and Fe_5C_2, whose contribution is not decreased in any treated sample, while ε-Fe_2C is less stable. This ε-Fe_2C carbide may in fact transform to cementite [12], as can be seen from the HFD in Fig. 2, in which we observe also the narrowing of the distribution around the mean field of cementite (205 kG), mainly owing to the transformation of ε-carbides. The Fe–C (martensite-like [12]) alloy is drastically decreased in all the samples after the catalytic treatment, probably owing to thermal aging. The overall carbon content in the carbides generally does not change (*i.e.* close to $Fe_{75}C_{25}$), except for the initial, less-carburized powders (2.5 and 7.5 h).

4. Conclusions

Mechanosynthesized iron carbides have been shown to be quite active and stable catalysts for CO_2 hydrogenation. Enhanced selectivity for hydrocarbons was also reported. The catalytic properties, for this reaction, appear to be in competition with those of much studied and often much rarer catalytic materials, such as noble metals dispersed on high area supports [8].

The catalytic properties of nanophase materials, as also suggested by the results of the present work, appear attractive and should be considered for further investigation (other materials and reactions).

References

1 P. Matteazzi and G. Le Caër, *Adv. Mater., 1* (1992) 135.
2 L. Schultz, J. Wecker and E. Hellstern, *J. Appl. Phys., 61* (1987) 3583.
3 G. Le Caër, E. Bauer-Grosse, A. Pianelli, E. Bouzy and P. Matteazzi, *J. Mater. Sci., 25* (1990) 4726.
4 P. Matteazzi and G. Le Caër, *J. Am. Ceram. Soc., 74* (1991) 1382.
5 P. Matteazzi and G. Le Caër, *J. Am. Ceram. Soc., 75* (10) (1992) 2749.
6 H. Gleiter, *Nanostruct. Mater., 1* (1992) 1.
7 R. W. Chorley and P. W. Lednor, *Adv. Mater., 3* (1991) 474.
8 A. Trovarelli, C. Mustazza, G. Dolcetti, J. Kaspar and M. Graziani, *Appl. Catal., 65* (1990) 471.
9 F. Solymosi, *J. Molec. Catal., 65* (1991) 337.
10 G. Henrici-Olivè and S. Olivè, *The Chemistry of the Catalyzed Hydrogenation of Carbon Monoxide*, Springer, Berlin, 1984.
11 A. Trovarelli, P. Matteazzi, G. Dolcetti, A. Lutman and F. Miani, *Appl. Catal., 95* (1993) L9–L13.
12 P. Matteazzi, F. Miani and G. Le Caër, *J. Hard Mater., 2* (1991) 171.
13 D. Basset, P. Matteazzi and F. Miani, *Mater. Sci. Eng., 168* (1993) 149–152.
14 M. Pijolat, V. Perrichon and V. Bussiere, *J. Catal., 107* (1987) 82.

Materials Science and Engineering, A168 (1993) 157–160

Enhanced consolidation of ball-milled Nb–Al intermetallic powders

K. Mino, H. Yoshizawa and Y. Saitoh

Research Institute, Ishikawajim-Harima Heavy Industries Co. Ltd, Toyosu, Kot-ku, Tokyo 135-91 (Japan)

Abstract

Nb_3Al and Al_3Nb compounds were milled in a high energy ball–mill to refine the grain structures to 10 nm in size. Die-press tests showed enhanced consolidation by ball-milling at temperatures as low as $0.4T_m$ and $0.5T_m$ for Al_3Nb and Nb_3Al respectively. Using the hot isostatic pressing mechanism of Ashby and coworkers, the densification rate was calculated to specify the process. Diffusion from interparticle boundary was negligibly small. This implies that the formation of fine-grained structures is responsible for the enhanced consolidation, rather than the refinement of the powder size. Good agreement was found between the measured and calculated density values found using the power-law creep rate equation.

1. Introduction

A novel processing route via a powder metallurgy needs to be developed for intermetallic compounds to overcome the inherent difficulty in their fabrication. The formation of nanometer-sized structures by ball-milling is one such attractive technique. Although consolidation of ball-milled powders maintaining non-equilibrium structures has been studied to some extent [1–3], few papers have been dealt with the effects of nanostructures on consolidation. It has been briefly reported [4] that conventionally brittle ceramics became ductile at low temperatures by generating a crystal size of a few nanometers. It is also described that the ductility seems to originate from the diffusional flow of atoms along the intercrystalline interfaces.

In the present study, crushed pieces of Nb_3Al and Al_3Nb castings were ball-milled to produce fine-grained structures about 10 nm in size. The milled powders were characterized using ordinary means, such as thin-foil transmission electron microscopy (TEM) and differential scanning calorimetry (DSC). The powders were die-pressed at temperatures below $0.55T_m$ (where T_m is the melting temperature) to investigate the effects of ball-milling on consolidation.

2. Experimental procedures

Nb–26.4at.%Al (designated as Nb_3Al for simplicity, although Nb_2Al exists) and Nb–74.9at.%Al (Al_3Nb) ingots were prepared by plasma arc melting and crushed into small pieces of 1–3 g. About 30 g of the crushed pieces and 10 stainless steel balls 11 mm in diameter were loaded into a vibratory ball-mill. Ball-milling for 50 h in an argon atmosphere increased the oxygen content in the Nb_3Al and Al_3Nb powders from the original content in the ingots of less than 0.05 at.% to 1.1 at.% and 2.5 at.% respectively.

The thin foils for TEM were prepared by ion-milling the powders mounted in a nickel-plating film. The thermal stability was examined in a Netzch 404 type differential scanning calorimeter. The consolidation tests were performed using a 50 kN vacuum press, loading the powders into a cylindrical cavity 8 mm in diameter in superalloy dies.

3. Results and discussion

3.1. Structural refinement of powders by ball-milling

Figures 1(a) and 1(b) show cross-sectional micrographs of the Nb_3Al powders ball-milled for 11 ks and 260 ks respectively. Small particles adhering to the original ingot piece are seen in Fig. 1(a). Some of the as-polished sections of the powder milled for 260 ks looked like an unprocessed powder. However, chemical etching revealed well-processed fine structures, as shown in Fig. 1(b). Almost the same observations were made for Al_3Nb.

The thin-foil TEM observations shown in Fig. 2 demonstrated that the ball-milled powders had a crystalline structure with a grain size of about 10 nm. The selected area diffraction (SAD) pattern changed from dotted circles to broken rings as the milling continued for longer durations. An amorphous phase of

0921-5093/93/$6.00

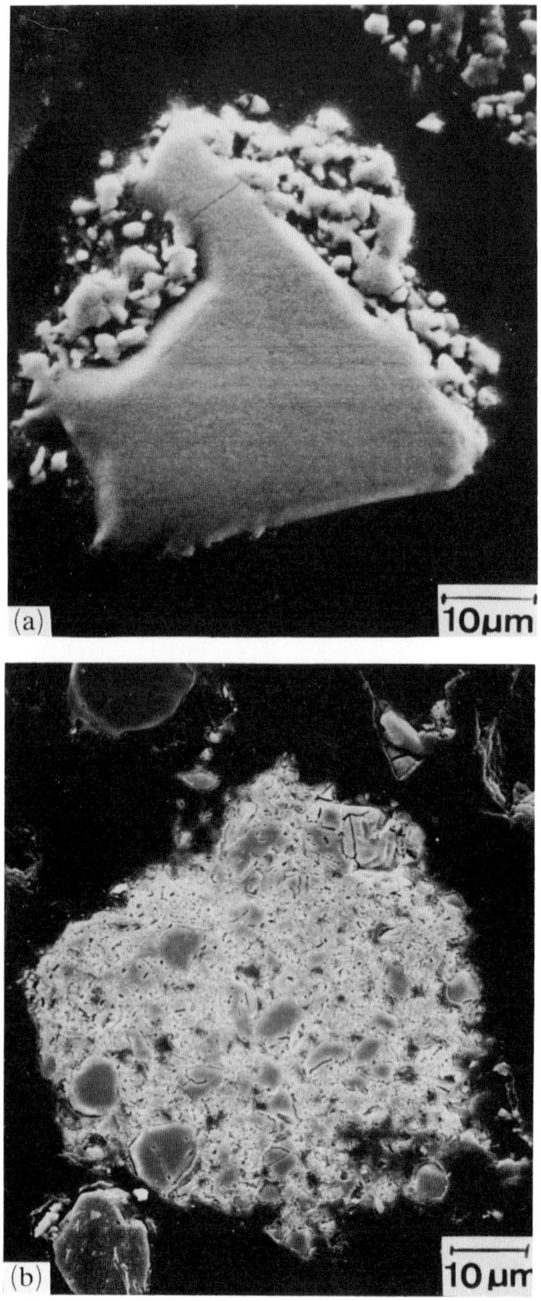

Fig. 1. Scanning electron micrographs of the cross-sections of the Nb_3Al powders ball-milled for (a) 11 ks and (b) 260 ks.

Al_3Nb has been reported to be formed by mechanical alloying of elemental powders of aluminum and niobium powders [5]. However, it was not found in the present milling of the compounds.

An exothermic peak was found at temperatures around $0.4T_m$ in the DSC scans of the ball-milled Nb_3Al powders. The heat release was about 2 kJ mol^{-1} for the range of heating rate utilized here (0.033–1 K s^{-1}). Assuming that the phase change obeys first-order kinetics, the Kissinger plot gave an activation energy of 340 kJ mol^{-1}. This value is somewhat less than the activation energy (400 kJ mol^{-1}) for the self-diffusion of niobium [6].

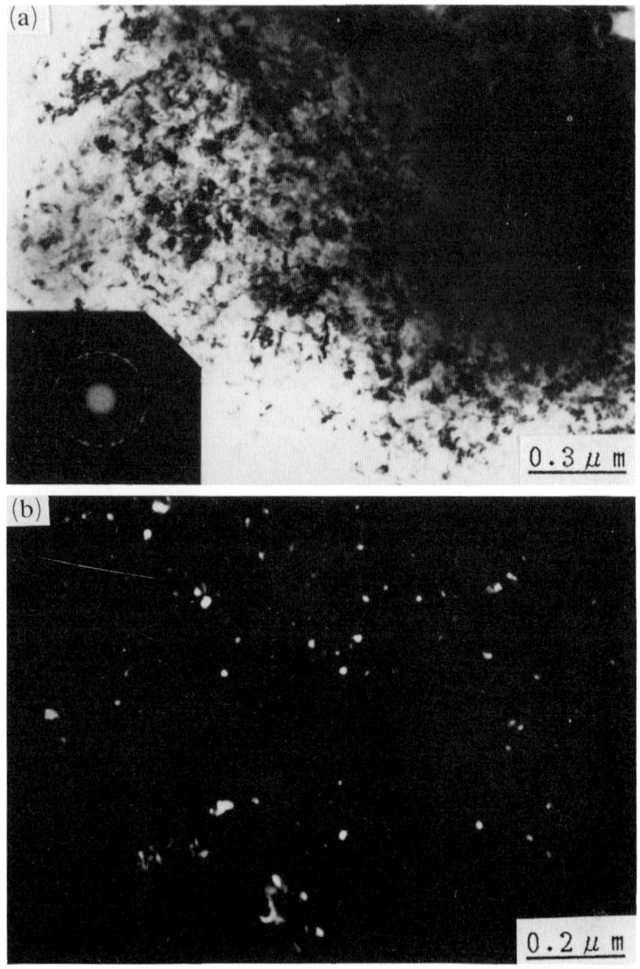

Fig. 2. (a) Bright-field and (b) dark-field transmission electron micrographs of the ball-milled powders.

An X-ray diffraction analysis, carried out on the ball-milled Nb_3Al powders before and after heat release, revealed the formation of a new phase during ball-milling. The lattice constant was calculated to be 0.327 nm, which is very close to that for pure niobium of 0.33 nm. Since the diffraction peak disappeared on heating above the DSC peak temperature, the DSC peak is considered to be due to an annihilation process of the non-equilibrium phase. The formation of the non-equilibrium phase has been reported in mechanically alloyed Nb_2Al [7]. Studies using an X-ray diffraction technique showed the formation of a niobium solid solution with a b.c.c. structure, while electron diffraction patterns displayed ordered B2-type superlattice spots [8].

3.2. Consolidation tests of the ball-milled powders

The powders milled for 1.8 and 180 ks were used for consolidation tests. Most of the products from 180

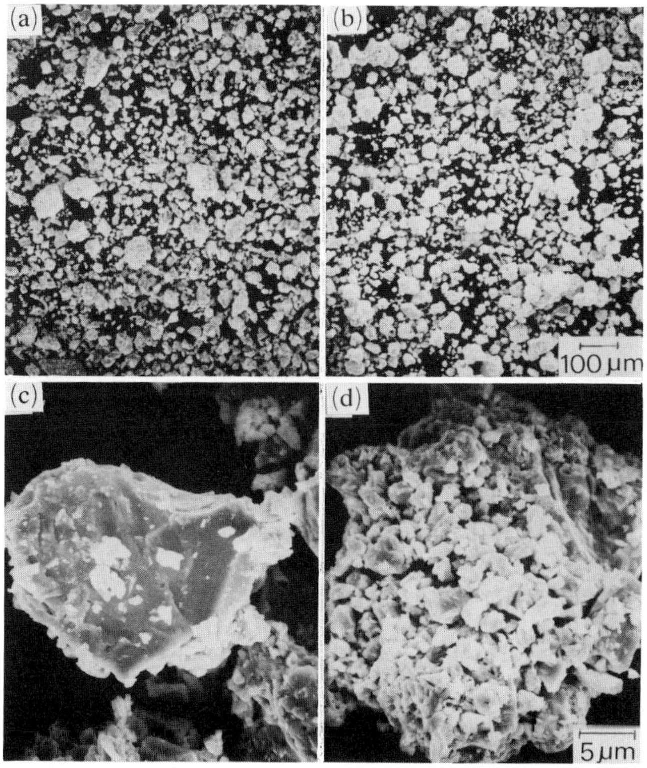

Fig. 3. Nb_3Al powders used for die-press tests: (a), (c) milled for 1.8 ks; (b), (d) milled for 180 ks.

ks of milling were accumulated in the bottom corner of the mill jar. The powders less than 100 μm in size were sieved for samples. Two different sizes of powders are seen in Fig. 3, *i.e.* fine powders less than 10 μm in size and larger powders 30–100 μm in size. The larger particles obtained for 1.8 ks of milling were mostly unprocessed pieces, while those by 180 ks of milling were composed of fine particles ranging from 0.2 to 5 μm, as shown in Fig. 3(d). These powders were consolidated under 600 MPa. The results are shown in Fig. 4. The pressure was held for 1.8 ks for Nb_3Al and 0.3 ks for Al_3Nb. The enhancement by ball-milling appeared at $0.4T_m$ (750 K) and $0.5T_m$ (1120 K) for Al_3Nb and Nb_3Al respectively.

3.3. Mechanism of enhancement in consolidation

B_3Al has the crystallographic structure A15. Since the yield stress of conventionally processed Nb_3Al is more than 2000 MPa at $T=0.6T_m$ [9], it will be interesting to examine the enhancing mechanism for consolidation at lower temperatures and lower stresses. The hot isostatic pressing (HIP) mechanism map constructed by Ashby and coworkers [10] was considered to be useful for this purpose.

The densification processes are classified into (1) power-law creep, (2) Nabbaro–Herring and Coble creep, (3) diffusion from interparticle boundaries and (4) plastic yielding. The densification rate equations for $D<0.9$ with $D_0=0.65$ are described as follows:

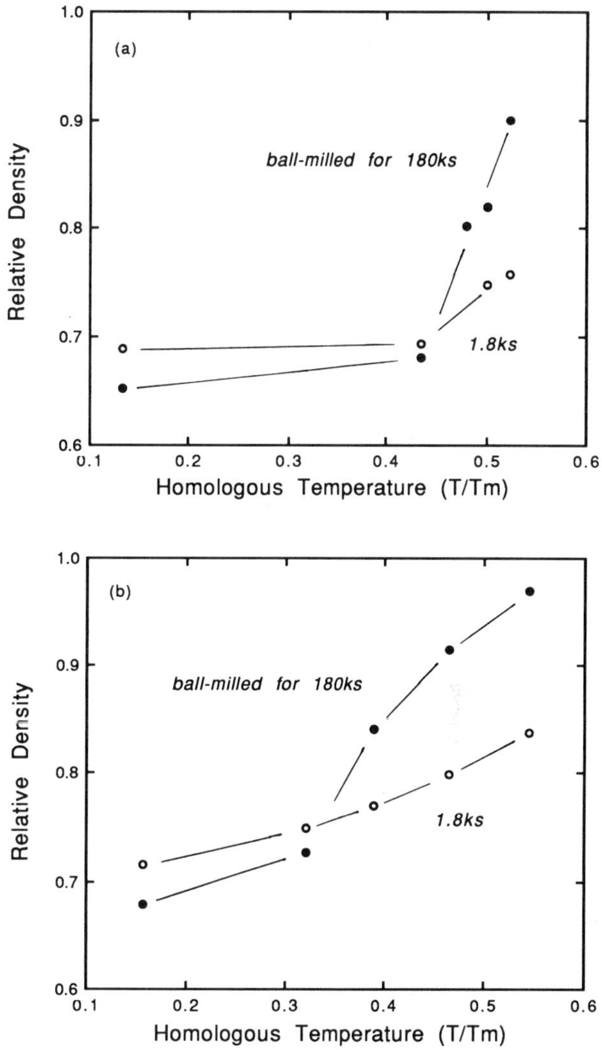

Fig. 4. Effect of ball-milling on consolidation under 600 MPa: (a) Nb_3Al, $t=1.8$ ks; (b) Al_3Nb, $t=0.3$ ks.

process (1)

$$D^{16/3}(D-D_0)^{5/2}\mathrm{d}D/\mathrm{d}t=0.007AP^3 \text{ for } n=3 \tag{1}$$

process (2)

$$D^{4/3}(D-D_0)^{1/2}\mathrm{d}D/\mathrm{d}t=20(\delta D_b/G^3)(\Omega P/kT) \tag{2}$$

process (3)

$$(D-D_0)^2\mathrm{d}D/dt=5(\delta D_b/\bar{R}^3)(\Omega P/kT) \tag{3}$$

Here, D is the relative density and D_0 is the D value after finishing the instantaneous process (4). The grain boundary diffusion term δD_b was supposed to dominate at low temperatures, such as in the present tests, and is an atomic volume, A is the creep rate constant, G is the grain size and $\bar{R}$ is the powder size. It will be realized that the densification rate is strongly dependent on the grain size and powder size. P is an applied pressure for HIP and the above equations are considered to be applied to the present pseudo-HIP conditions.

The time taken to reach $D=0.85$ wsa calculated and compared for processes (2) and (3), substituting $G=10^{-8}$ m and $\bar{R}=10^{-6}$ m into the above equations. The time for process (2) was $4\times10^{-27}/\delta D_b$, while that for process (3) was $1\times10^{-27}/\delta D_b$. This indicates that the interparticle boundary diffusion was negligible. Using the activation energy determined from DSC tests, the time to reach $D=0.86$ at 1123 K by process (2) was calculated to be 0.2–20 ks. Here, δD_0 was assumed to be 10^{-15}–10^{-13} m^3 s^{-1}. The experimental value of 5.4 ks lies within the calculated values.This implies that Coble creep may be the main densification process.

Figure 5(a) compares the calculated density *vs.* time curve with the measurements. Here, the diffusion coefficient in eqn. (2) was chosen, so that the measured density for the press time of 5.4 ks lies on the calculated curve. The measured values for 1.8 ks and 0.6 ks are much larger than those expected from the line. However, better coincidence of the experimental with calculated values was found, as shown in Fig. 5(b), by using the power-law rate equation (eqn. (1)) with $n=3$. Also, the pressure dependence was quite accurately described.

These results are considered to indicate that significant creep occurs in the nanostructure powders at relatively low temperatures. Since both processes (1) and (2) occur simultaneously, both their equations should be summed up for discussion. However, this effect is considered to be not essential but only leaving some modifications of n in eqn. (1).

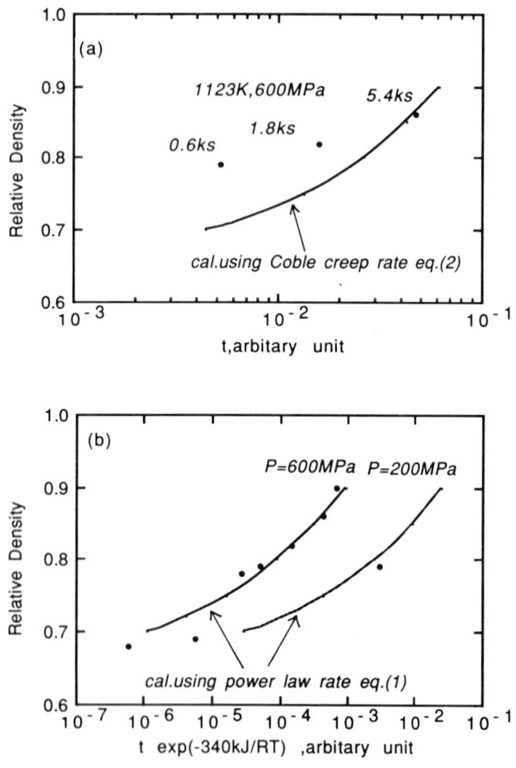

Fig. 5. Comparison of experimental data on the densification of ball-milled Nb_3Al with densification rate equations: (a) Coble creep mechanism; (b) power-law rate equation.

4. Conclusions

Ball-milling was found to enhance the consolidation of Nb_3Al and Al_3Nb powders at temperatures as low as $0.4T_m$ to $0.5T_m$. The fine-grained structures about 10 nm in size rather than the small powder size were shown to have significantly increased low temperature creep and this was responsible for the enhanced consolidation.

Acknowledgment

Part of this work was performed under the management of IHI as part of the R&D Institute of Metals and Composites for Future Industries, supported by the New Energy and Industrial Technology Development Organization.

References

1 T. J. Tianen and R. D. Schwarz, *J. Less-Common Met., 140* (1988) 99.
2 H. Esaki, K. Ameyama and M. Tokizane, *Mater. Sci. Technol., 5* (1989) 369.
3 D. K. Kim and K. Okazaki, *Mater. Sci. Forum, 88–90* (1992) 553.
4 J. Karch, R. Birringer and H. Gleiter, *Nature, 330* (1987) 556.
5 Z. Peng, C. Suryanarayana and F. H. Froes, *Scr. Metall. Mater., 27* (1992) 475.
6 C. English, Niobium, Proc. Sympo., ed. by H. Stuart, *Trans. Metall. Soc.-AIME* (1984) 239.
7 E. Hellstern, L. Shultz, R. Bormann and D. Lee, *Appl. Phys. Lett., 53* (1988) 1399.
8 T. N. Marieb, A. D. Kaiser, S. R. Nutt, D. L. Anton and D. M. Shah, *Mater. Res. Soc. Symp. Proc., 213* (1991) 329.
9 T. Kumagai, S. Hanada and S. Saito, in O. Izumi (ed.), *Proc. Int. Symp. on Intermetallic Compounds—Structure and Mechanical Properties*, Japan Institute of Metals, Tokyo, 1991, p. 1039.
10 A. S. Helle, K. E. Easterling and M. F. Ashby, *Acta Metall., 33* (1985) 2163.

Thermodynamic properties of nanocrystalline Ni_3Al-based alloys prepared by mechanical attrition

S. Suriñach, J. Malagelada and M. D. Baró

Grup de Física de Materials, Departament de Física, Universitat Autònoma de Barcelona, 08193 Bellaterra (Spain)

Abstract

In this paper, the thermodynamic properties of nanocrystalline Ni_3Al-based alloys prepared by mechanical attrition are reported. A broad exothermic transformation occurs on heating the nanocrystalline samples. The stored energy of cold work has been evaluated by measuring the enthalpy release during heating. The values obtained far exceed the energies achieved by conventional deformation processes. The specific heat increases by more than 5%, indicating changes in the vibrational and configurational part of the entropy.

1. Introduction

Recently, it has been shown that nanocrystalline materials and compounds can be synthesized by mechanical attrition [1–3]. It has been proved that, by ball-milling, the deformation is localized at the early stage in shear bands, where the nanometre-sized grains nucleate, and that further ball-milling results in an extremely fine-grained microstructure with randomly oriented grains separated by large-angle grain boundaries [4].

Because of the small crystallite size, these materials consist of two components with comparable volume fractions: a crystalline component comprising all atoms located in the grains and an interfacial component formed by the grain boundaries [5]. If the atomic structure of nanocrystalline materials differs from the structures of crystals, the structure-sensitive properties of nanocrystalline materials are expected to be different from those of the chemically identical substances in the crystalline state [6, 7].

The aim of this paper is to report on the thermal behaviour and on the specific heat measurements of a metastable nanocrystalline Ni_3Al-based alloy.

2. Experimental methods

The metastable, nanocrystalline, completely disordered Ni_3Al-based alloy was prepared by mechanical attrition, following the procedure described in ref. 8. The integral breadths of the diffraction peaks were used to evaluate the average size of the corresponding diffraction domains and the microstrain; for more details, see ref. 9. The thermal stability of the metastable nanophase formed during ball milling was tested by studying the behaviour on heating in a computerized differential scanning calorimeter (Perkin–Elmer DSC7) under a pure Ar atmosphere. The heating was carried out in an unsealed Au pan, using both continuous-heating and isothermal regimes. The continuous-heating experiments were performed at scan rates in the range 5–160 K min^{-1}. The calibration procedure of the apparatus [10] ensures that the temperature measurements have less than 0.05% error and that the enthalpy output measurement has 2% error. The isothermal annealings were carried out in the scanning calorimeter by taking the samples to the annealing temperature at a heating rate of 40 K min^{-1} and cooling, at the maximum rate available (about 200 K min^{-1}), to room temperature after the annealing.

The specific heat c_p as a function of the temperature was measured and the error in the heat capacity measurement was 2%, using a synthetic sapphire standard for calibration.

3. Results and discussion

3.1. *Reordering of the metastable nanocrystalline Ni_3Al-based powder*

On heating the milled samples, reordering, accompanied by grain growth of the crystalline phase from an average initial size below 12 nm, is observed. In Fig. 1, the differential scanning calorimetry (DSC) curve of the as-milled sample relative to the same sample heat treated at high temperature, and probably recrystal-

0921-5093/93/$5.00

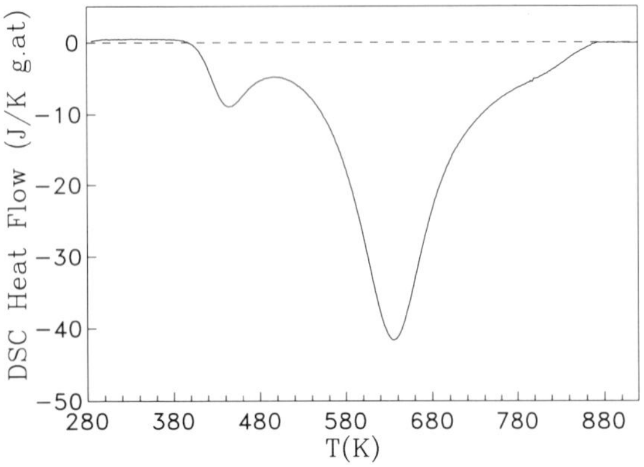

Fig. 1. DSC curve at 40 K min^{-1} from a sample milled for 42 h relative to the same sample annealed at high temperature. The increase in specific heat of the as-milled sample is seen as a positive deviation from the zero of the signal.

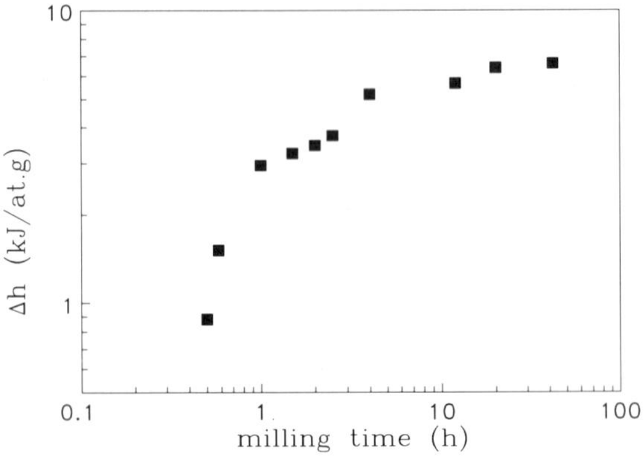

Fig. 2. Enthalpy released in the exothermic process for different samples.

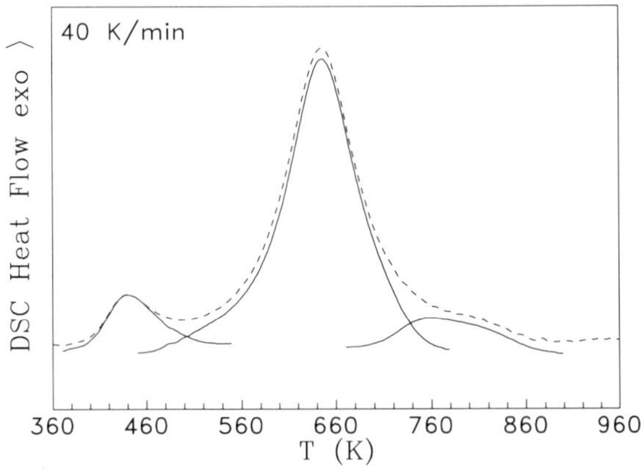

Fig. 3. Possible deconvolution of the exothermic process for the sample milled for 20 h.

lized, is shown. An exothermic transformation occurs over a wide (100–600 °C) range of temperatures.

This transformation is composed of different processes, as is evident from the complex shape of the calorimetric signal. Similar DSC curves were obtained in the first work, in which a disordered nanocrystalline structure is reported to be obtained in Ni_3Al by mechanical milling [11] and also for Ni_3Al vapour-deposited thin films [12]. These results suggest that reordering of this compound is not affected by the way in which the metastable, disordered, nanocrystalline structure is achieved. Also, they indirectly confirm for Ni_3Al the analogies between the metastable and nanocrystalline structures produced by mechanical attrition and other techniques, such as certain thin-films deposition methods, already reported for other systems [13].

All these transformations are irreversible: on repeating a heating DSC scan with the same specimen, no transformation is detected any more. These second scans, performed without changing the sample configuration, were then used as the baseline for estimating the total enthalpy output from the integrated area of the peaks. The enthalpy release increases with the milling time and, even long after the long-range order has disappeared, no evident steady value is achieved. In this sense, the contribution of grain growth to this enthalpy release cannot be disregarded. According to different authors [4, 6], the enthalpic contribution from grain growth in nanocrystalline elemental powders shows values double those expected for the enthalpy stored in the fully equilibrated grain boundaries of such a refined microstructure. In this case, approximately 40% of the enthalpy release can correspond to grain growth. The values presented in Fig. 2 belong to samples milled with Cr steel media, but this behaviour seems to be unaffected by a change in the milling conditions.

Identification of the processes which may contribute to the complex calorimetric signal has been attempted. For this purpose, isothermal annealing, coupled with further continuous-heating experiments (which allowed the effect of annealing to be checked), were performed. In this way, three stages have been identified (Fig. 3), with the second, more intense, signal being associated with the simultaneous evolution of ordering and grain growth [14]. This result is again in agreement with those already reported for both milled powders [11] and thin films [12].

Harris *et al.* [12] regarded the low temperature peak as being determined by the restablishment of short-range order. This hypothesis is clearly proved by the results obtained from electron energy-loss, fine structure measurements on Ni_3Al thin films, vapour deposited in the disordered nanocrystalline state and then annealed for several hours at 150 °C [15]. In fact, an excellent agreement was found with the plane-wave calculations used to simulate the environment of each atomic species, when the ordered configuration was adopted. The ordered clusters must spread over only a few atomic spacings, as no superlattice reflections were detected by conventional X-ray and electron diffraction. Further proof of this attribution of the low temperature exothermic peak to short-range reordering is given indirectly by the DSC curves reported for elemental milled powders [16]. In fact, they show a complex exothermic process very similar to those we observed for our milled powders. The only substantial difference is the lack of the low temperature peak; however, in view of its suggested origin, it cannot be present, as no chemical reordering can occur in pure metal powders.

No clear and conclusive interpretation is proposed for the high temperature, weaker component of the overall transformation. It is also the more difficult phenomenon to deal with, since it overlaps markedly with the second process. By considering the temperature range over which it occurs and the low enthalpy associated with it, it can be inferred that the high temperature component might be due to annealing-out of dislocations, usually not mobile at temperatures as low as those of the first two processes. Against this dislocation-related interpretation are the observations mentioned already as reported for Ni_3Al nanocrystalline thin films. As said, the structure of the DSC curves is exactly the same as that of the milled powders. It is difficult to determine why a high density of dislocations should be present in these thin films in such a concentration as to affect their reordering kinetics.

3.2. *Specific heat*

The increase in c_p on going from the original, premilled powder to the nanocrystalline state is about 5%. We report a value of $c_p = 25.5$ J g^{-1} at^{-1} K^{-1} at 300 K for the original powder, which presented an average grain size of 70 nm, and an increase of this c_p by 1.1 J g^{-1} at^{-1} K^{-1} at the same temperature for the as-milled powder (in the nanocrystalline state). The increase in the specific heat at the beginning of the run can be seen clearly from Fig. 1.

At room temperature, the equilibrium, and even the metastable nanocrystalline Ni_3Al, though ground for 42 h in Cr–steel media, are paramagnetic [17]. Therefore, the specific heat of the metastable nanocrystalline Ni_3Al alloys, in the temperature range between 300 and 450 K, is due fundamentally to the thermally induced variation of the vibrational and configurational entropy of the materials, the electronic and magnetic contributions being negligible. Taking into account that the vibrational and configurational entropy of the original microcrystalline state and the nanocrystalline structure should be different, an enhancement of c_p for the nanocrystalline Ni_3Al alloys could be expected, as was found experimentally.

This argument implies that the increase in c_p is principally due to the interfacial disorder. Thus, grain growth should decrease the specific heat of the metastable nanocrystalline materials to values near to that measured for the original material. In fact, this was observed when the metastable nanocrystalline Ni_3Al alloys were annealed in the DSC at 950 K to produce grain growth, where the grain size after annealing was about 40 nm and the c_p value found for such annealed alloy was 0.5 J g^{-1} at^{-1} K^{-1} larger than that obtained for the microcrystalline material. Our results agree well with those recently reported for pure metals and intermetallic phases [6, 18], as well as those for nanocrystalline materials obtained from crystallization of amorphous Ni–P alloys [19].

4. Conclusions

The DSC experiments show that, when heating metastable nanocrystalline Ni_3Al alloys, a broad exothermic transformation occurs. The deconvolution of the complex calorimetric signal reveals three stages, the second being associated with the simultaneous evolution of the ordered phase.

The enthalpy released during the heating of metastable nanocrystalline Ni_3Al alloy has been measured; it could be that 40% of it was associated with grain growth and the rest with chemical reordering.

Specific heat measurements in the temperature range 300–450 K revealed that the c_p values of the metastable nanocrystalline Ni_3Al alloys were about 5% higher than those in the microcrystalline state. The enhancement of the c_p values in the metastable nanocrystalline Ni_3Al alloys might be attributed to changes in the vibrational and configurational parts of the entropy, probably owing to the complicated structure of the grain boundaries.

Acknowledgments

This work was supported by CICYT (Project MAT92-0501). The technical assistance of S. Briata and R. Taverna is greatly indebted.

References

1 H. J. Fecht, E. Hellstern, Z. Fu and W. L. Johnson, *Metall. Trans. A, 21* (1990) 2333.
2 M. Oehring and R. Bormann, *J. Phys., 51* (1990) C4–169.
3 J. S. C. Jang and C. C. Koch, *Scr. Metall. Mater., 24* (1990) 1599.
4 J. Eckert, J. C. Holzer, C. E. Krill and W. L. Johnson, *J. Mater. Res., 7* (1992) 1751.

5 J. Rupp and R. Birringer, *Phys. Rev. B, 36* (1987) 7888.
6 H. Gleiter, *Prog. Mater. Sci., 33* (1989) 223.
7 H. Gleiter, *Adv. Mater., 4* (1992) 474.
8 J. Malagelada, S. Suriñach, M. D. Baró, S. Gialanella and R. W. Cahn, *Mater. Sci. Forum, 88–90* (1992) 497.
9 H. P. Klug and L. E. Alexader, *X-Ray Diffraction Procedures*, Wiley, New York, 1974, pp. 661.
10 S. Suriñach, M. D. Baró, M. T. Clavaguera-Mora and N. Clavaguera, *J. Non-Cryst. Solids, 58* (1983) 209.
11 J. S. C. Jang and C. C. Koch, *J. Mater. Res., 24* (1990) 498.
12 S. R. Harris, D. H. Pearson, C. M. Garland, B. Fultz, *J. Mater. Res., 6* (1991) 2019.
13 K. L. Breitbach, L. S. Chumbley, *Scr. Metall. Mater., 25* (1991) 2553.
14 M. D. Baró, S. Suriñach, J. Malagelada, M. T. Clavaguera-Mora, S. Gialanella and W. R. Cahn, *Acta Metall. Mater., 41* (1993) 1065.
15 J. K. Okamoto, *TMS Student Paper Competition*, 1991.
16 J. Eckert, J. C. Holzer, C. E. Krill and W. L. Johnson, *Mater. Sci. Forum, 88–90* (1992) 505.
17 A. R. Yavari, P. Crespo, E. Pulido, A. Hernando, G. Fillion, P. Lethuillier, M. D. Baró and S. Suriñach, in A. R. Yavari (ed.), *Ordering and Disordering in Alloys*, Elsevier, Amsterdam, 1992, p. 12.
18 E. Hellstern, H. J. Fecht, Z. Zu and W. L. Johnson, *J. Appl. Phys., 65* (1989) 305.
19 K. Lu, J. T. Wang and W. D. Wei, *J. Phys. D, 25* (1992) 808.

Thermal stability of ultrafine-grained metals and alloys

A. Kumpmann, B. Günther and H.-D. Kunze

Fraunhofer Institute für Angewandte Materialforschung, Lesumer Heerstrasse 36, 2820 Bremen 77 (Germany)

Abstract

Grain growth in ultrafine-grained elemental metals (Cu, Ag, Pd) and alloys has been studied by differential scanning calorimetry and transmission electron microscopy. The samples were prepared via the inert gas condensation technique, followed by uniaxial high pressure compaction. Abnormal grain growth is observed in all the pure elemental samples. The onset of this secondary recrystallization is at or slightly below the recrystallization temperature known from the respective conventionally cold-worked metals. The activation energies for grain growth were determined by the Kissinger method and correspond to the typical values for grain boundary diffusion. For samples measured directly after preparation, a two-step process was found. The low temperature reaction can be attributed to relaxation processes, because no microstructural changes can be observed. The reaction at higher temperatures is due to abnormal grain growth. Principally the same behaviour was found in samples with increased residual porosity (up to 15 vol.%) and concentration of substitutional solutes (up to 5 at.% Au). Gaseous impurities (oxygen) increase the onset temperature for abnormal grain growth. Normal grain growth is observed in duplex microstructures, as demonstrated using AgCu as a model system.

1. Introduction

Ultrafine-grained materials are thermodynamically unstable, owing to the high amount of free energy stored in the grain boundaries. Therefore, grain growth at low temperatures might be expected. Until recently, only normal grain growth at elevated temperatures (above the 'normal' recrystallization temperatures) was reported [1–4].

2. Experimental details

Ultrafine-grained elemental and alloy powders were produced by the inert gas condensation method. Cu, Ag, AgCu and AgAu wire evaporated from a joule-heated W crucible. Pd was evaporated using a high voltage electron beam. The metal gas condensed in a He atmosphere of 10^2 Pa and deposited on a liquid-nitrogen-cooled steel container. The powder was continuously scraped off, collected and transferred into an anvil-and-piston device that was evacuated and loaded externally with 2.8 GPa. Disc-shaped samples with a diameter of 15 or 30 mm and a thickness of about 0.5–1 mm were prepared. More details about this preduction process are described elsewhere [5, 6].

The microstructure was investigated by transmission electron microscopy (TEM) using a Phillips 300 kV-TEM. The TEM specimens were prepared by ultrathin sectioning. Thin slices of the material (about 60 nm) were cut off with a diamond knife and are collected on a water surface. These slices were caught on a C-coated TEM grid. By using this preparation technique, large sample areas of up to 0.1 mm^2 were accessible for visual inspection. Differential scanning calorimetry (DSC; Perkin–Elmer DSC 7) was used to determine the grain growth temperature and free enthalpies. The activation energies for the exothermic reactions were determined by the Kissinger method [7].

3. Results and discussion

3.1. Grain growth in elemental metals

The microstructures of the ultrafine-grained powder compacts (with primary particle size over 15 nm) were examined directly after compaction (Fig. 1). Mass densities above 95% of the theoretical density were measured by Archimedes' principle. In the Cu specimens, which were stored at ambient conditions for a few days, abnormal grain growth was found. This observation was confirmed by DSC measurements. Directly after compaction, an exothermic peak occurs at about 100 °C. After storage for five days under ambient conditions, this reaction has almost vanished. For samples which were stored at −20 °C, the reaction is still found after five days (Fig. 2). For this reaction, an activation energy of 83 kJ mol^{-1} was determined (Table 1). Avrami analysis of an isothermal DSC scan shows that the reaction involves two stages. The reason for this two-step process is as follows.

0921-5093/93/$5.00

Ag powder compacts are more thermally stable than is the Cu specimen. The onset of abnormal grain growth occurs at about 100 °C (Fig. 3). A typical DSC scan (heating rate, 10 K min^{-1}) of an ultrafine-grained Ag powder compact (60 mm) after one month is depicted in Fig. 4(a). The exothermal reaction can be attributed to the abnormal grain growth, as provided by TEM observations. The reaction takes place at basically the same temperature as recrystallization in 90% cold-rolled Ag (Fig. 4(b)). In contrast, the DSC scan (heating rate, 60 K min^{-1}) of an Ag specimen measured directly after compaction (Fig. 5) clearly shows two reactions occurring at about 120 and 210 °C. The activation energies are 63 kJ mol^{-1} and 100 kJ mol^{-1} respectively. No microstructural changes were observed after heat treatment at 175 °C for 15 min. This indicates that the low temperature reaction is due to grain boundary relaxation or a

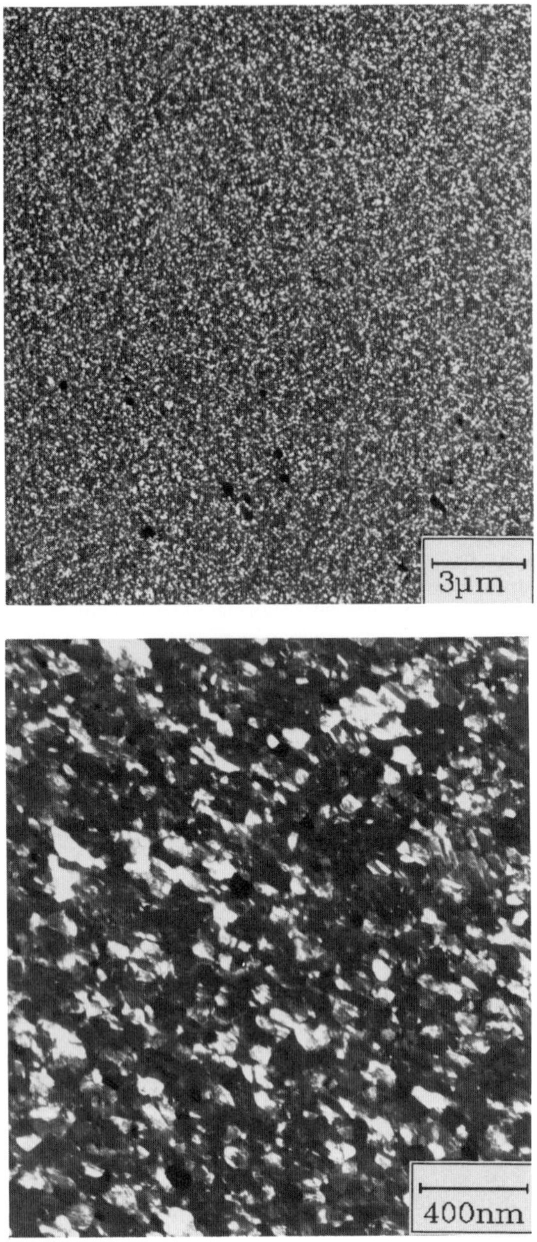

Fig. 1. TEM images of ultrafine-grained Ag powder compacts (dark field).

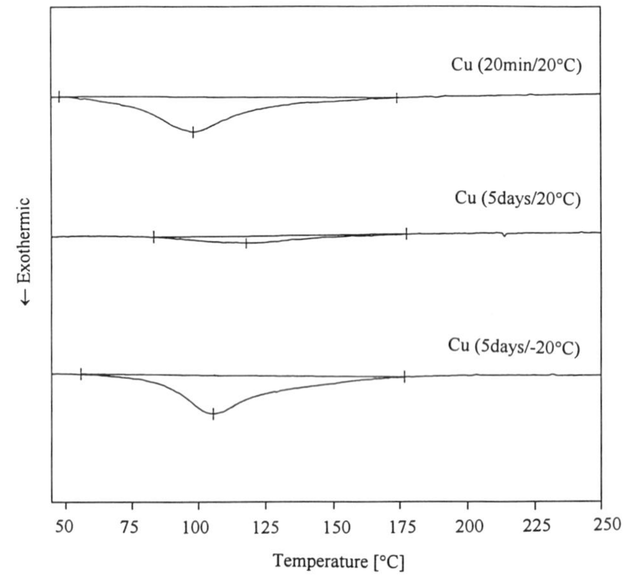

Fig. 2. DSC measurements of a Cu powder compact: (a) directly after preparation; (b) after storage under ambient conditions for five days; (c) after storage at − 20 °C for five days.

TABLE 1. Literature values of the activation energies for self-diffusion and activation energies determined in this work

Material	Activation energy for volume self-diffusion [8] (kJ mol^{-1})	Activation energy for grain boundary self-diffusion [8] (kJ mol^{-1})	Activation energies determined in this work (kJ mol^{-1})
Cu	≈ 200	102–107	83
Ag	≈ 190	85–90	101 63
Ag–7at.%O	—	—	202
Ag–5at.%Au	—	—	85
Ag–63at.%Cu	—	—	97

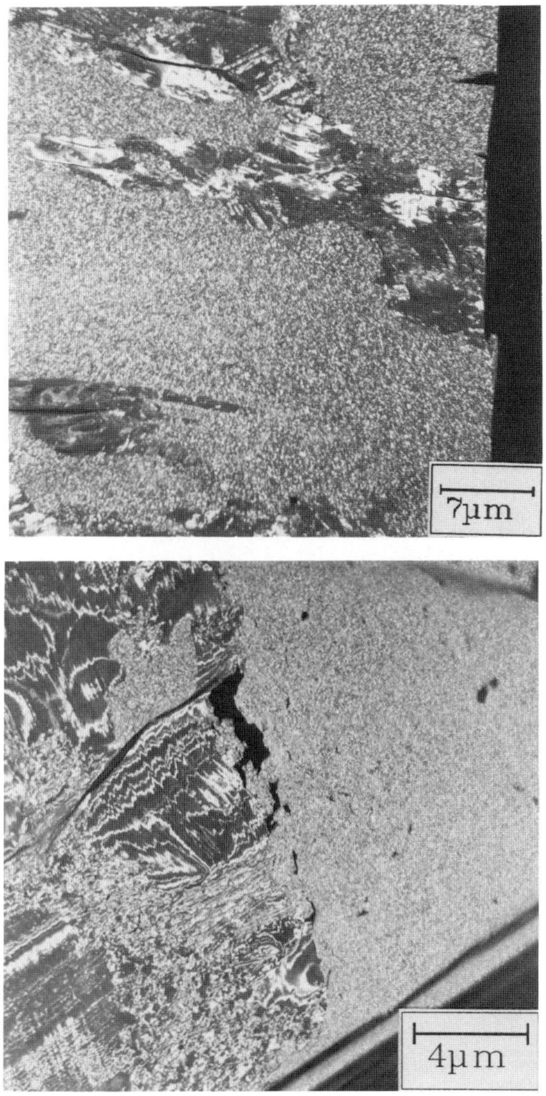

Fig. 3. TEM images of Ag (200 °C, 15 min) and Pd (untreated).

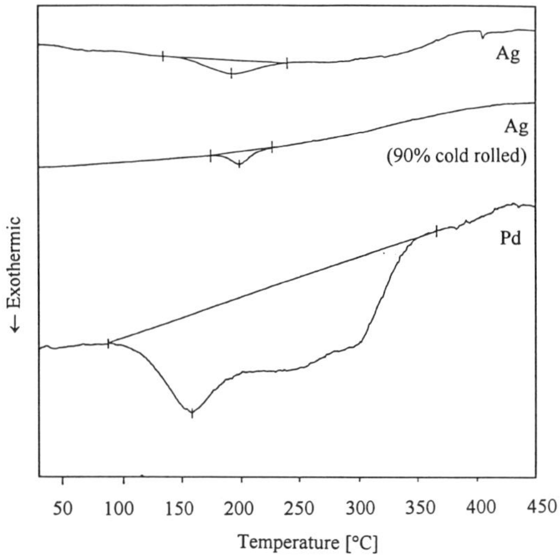

Fig. 4. DSC measurements of pure elemental metals: (a) Ag, 60 nm; (b) Ag, coarse grained, 90% cold rolled; (c) Pd, 16 nm.

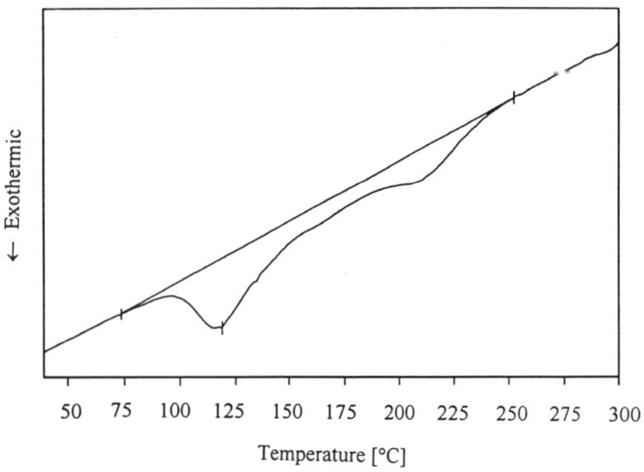

Fig. 5. DSC scan of an Ag powder compact measured directly after preparation.

TABLE 2. Properties of the unconsolidated and compacted powders

Material	Specific surface area A_s ($m^2\,g^{-1}$)	Powder grain size (nm)	Density (%)	Free enthalpy ΔH ($J\,g^{-1}$)	$\Delta H/A$ ($J\,m^{-2}$)
Ag	6	60	98	3	0.5
Cu	11	40	98	5	0.45
Pd	26	16	95	16	0.6
Ag–7at.%O	6	60	96	2.4	0.4
Ag–5at.%Au	6.5	60	97	2.5	0.4
AgCu (37/63)	16	$\leqslant 30$	95	35	—

recovery process. The reaction at higher temperatures is clearly caused by abnormal grain growth. The different grain growth temperatures seen in Figs. 4(a) and 5 are due to the different heating rates. Pd nanopowder compacts show the same behaviour (Figs. 3 and 4(c)).

The integral enthalpies ΔH of the different powder compacts and the specific surface areas A_s of the uncompacted powders are listed in Table 2. $\Delta H/A_s$ is of the order of the typical grain boundary energies. This indicates that the exothermic reactions are indeed basically due to the release of grain boundary energies owing to grain growth.

To check the effect of porosity on grain growth. Ag samples were produced by means of a reduced compaction pressure (less than 1 GPa). Although porosities of up to 15 vol.% measured, grain growth was still abnormal and the temperature for grain growth unchanged compared with the gas-tight Ag samples (density more than 95%).

3.2. *Grain growth in Ag alloys*

It is well known that grain growth is inhibited by second-phase particles and substitutional elements [9]. Therefore, different Ag alloys were prepared as model systems to evaluate the effect on the grain growth behaviour. To check the grain growth in solid solutions, Ag samples containing 5 at.% Au were prepared. TEM observations and DSC scans showed abnormal grain growth at basically the same temperatures and with a similar activation energy (85 kJ mol^{-1}) as occurred in pure Ag powder compacts (Fig. 6).

To investigate the effect of grain boundary contamination, Ag powders were exposed to O_2 for 24 h leading to about 7 at.% O being present in the compacted samples. Nevertheless, abnormal grain growth was found at temperatures of about 250 °C, which correlates with the decomposition temperature of Ag_2O. Thus, grain boundary pinning by an oxide phase seems to be present.

As a model system for a duplex microstructure, an ultrafine-grained Ag–63at.%Cu powder compact was prepared (Fig. 7). Ag and Cu are almost immiscible at room temperature. Owing to the rapid cooling condition, a certain amount of solute atoms are to be expected to the nanoparticles. A DSC scan of the Ag Cu sample is shown in Fig. 6. An exothermic reaction occurs between 70 and 400 °C, with a peak at about 320 °C. The TEM observations of this specimen show that normal grain growth occurs above 250 °C; abnormal grain growth was not found. The broad exothermic reaction process below the grain growth temperature is attributed to dissolution processes in the rapidly quenched binary alloy. An activation

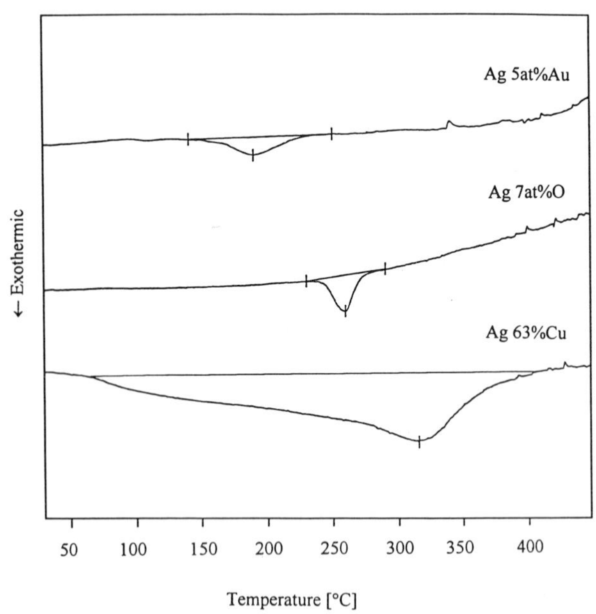

Fig. 6. DSC scans of different ultrafine-grained Ag alloys: (a) Ag–5at.%Au; (b) Ag–7at.%O; (c) AgCu (37/63 at.%).

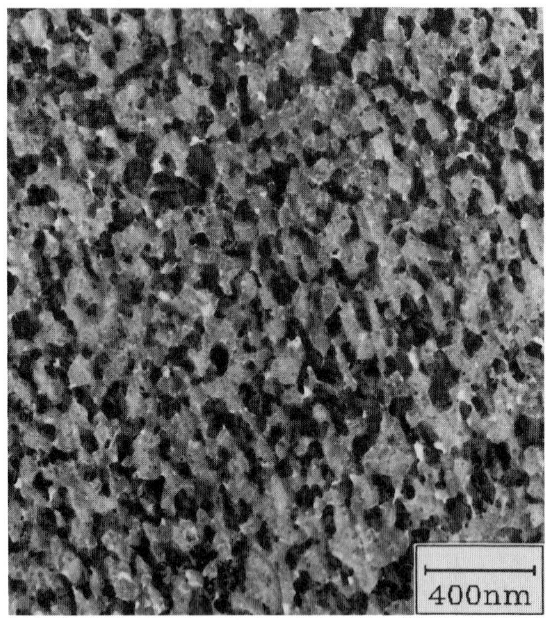

Fig. 7. Transmission electron micrograph of an AgCu powder compact (dark phase, Ag; bright phase, Cu).

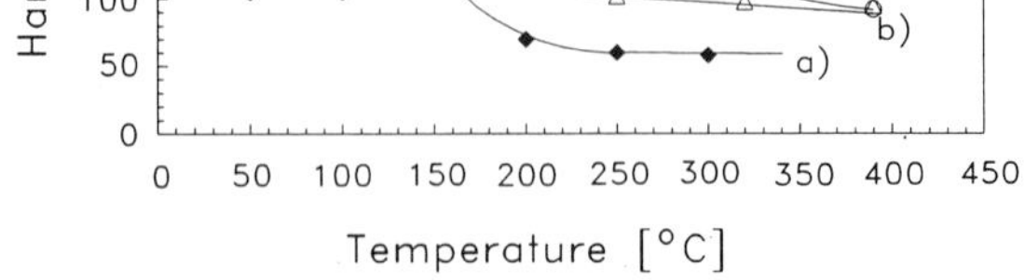

Fig. 8. Hardness values of different ultrafine-grained powder compacts and a cold-worked sample: (a) Ag, coarse grained, 90% cold rolled; (b) Ag, 60 nm; (c) 60 nm, 7 at.% O; (d) AgCu (37/63 at.%).

energy of 97 kJ mol^{-1} was determined. The Vickers hardness of the duplex structure exceeds that of ultra-fine-grained Ag by a factor of 2. Heat treatments up to 250 °C increase the hardness, owing to precipitation (Fig. 8). These results show that the stability of the duplex structure is superior to the stabilities of all the other samples investigated.

4. Summarizing remarks

In ultrafine-grained elemental noble metals, abnormal grain growth occurs at room temperature or slightly above, independent of the porosity (up to 15%) and concentration of impurity solutes (up to 5 at.% Au). Oxygen contamination of the grain boundary increases the thermal stability but grain growth is still abnormal. Normal grain growth was found in an Ag Cu duplex microstructure only. The temperature for grain growth is increased by more than 150 °C as compared with that with elemental metals.

References

1 L. C. Chen and F. Spaepen, *Nanostruct. Mater., 1* (1992) 59–64.
2 K. Lu, *Scr. Metall. Mater., 25* (1991) 2047–52.
3 S. K. Ganapathi, D. M. Owen and A. H. Chokshi, *Scr. Metall. Mater., 25* (1991) 2699–2704.
4 J. Eckert, J. C. Holzer, C. E. Krill, III, and W. L. Johnson, *J. Mater. Res., 7* (7) (1992) 1751–61.
5 B. Günther and A. Kumpmann, *Nanostruct. Mater., 1* (1992) 27–30.
6 R. W. Siegel, *Mater. Res. Soc. Bull., 15* (10) (October 1990) 60–67.
7 H. E. Kissinger, *Anal. Chem., 29* (1957) 1702–1706.
8 I. Kaur *et al., Handbook of Grain and Interphase Boundary Diffusion Data*, Ziegler Press, Stuttgart, 1989.

Materials Science and Engineering, A168 (1993) 171–176

Materials synthesis on the basis of ultrafine refractory compound powders

T. Millers

Institute of Inorganic Chemistry of the Latvian Academy of Sciences, 34, Miera Str., LV-2169 Salaspils-1 (Latvia)

Abstract

This article deals with results of investigations devoted to plasma–chemical synthesis of refractory compounds and their composites in the form of ultrafine powders, as well as their application to obtain various materials. It is shown that powders synthesized in plasma frequently have specific physical and chemical properties, and that they can be successfully applied to obtain high performance ceramics, refractory ceramics, cutting tools, thermal-conducting ceramics, and various coatings, as well as in the modification of metal structure and properties.

The high refractory compounds, *e.g.* carbides, oxides, nitrides, borides, and composites based on these compounds possess physicochemical properties which allow materials to be produced which have very extensive possibilities for their application in various branches of technology. Therefore, the interest in these compounds grows continuously. The interest in compounds in an ultrafine-grained state is also rising continuously. Refractory compounds and their composites are no exception in this respect.

An ultrafine medium is one wherein a particle dimension becomes comparable with the correlation scale of some physical phenomena or with the characteristic length of some transport processes. Therefore, this is a medium where any size effects become apparent. The particle size of ultrafine powders is in a range of a few nanometres.

Various methods can be used to obtain the substances in an ultradispersed state. One of the ways in which this problem may be solved seems to be the gas-phase synthesis of the above-mentioned powders, using plasma technology in particular. Advantageous characteristics of the plasma itself (high temperature, high heat and mass exchange rates, concentrations of charged and excited particles) and of conditions in the plasma-chemical reactor (high gradients of gas flow rate and temperature, energy concentration and others) make it possible to obtain compounds of predetermined chemical composition, in the form of highly dispersed particles from the different initial materials. This is achieved through the transformation of the particles to the gas or vapour state, followed by the chemical reaction of the synthesized end-product and the physical process of condensed particle formation.

The development of plasma–chemical synthesis of different compounds is dealt with in many countries. This problem is also studied in the Institute of Inorganic Chemistry of the Latvian Academy of Sciences. Up to now, the main possibility of obtaining practically all the refractory compounds by plasma–chemical synthesis has been demonstrated. The great advantage of this method is the subsequent possibility of obtaining not only the individual compounds but also the homogeneous composites of these compounds with predetermined proportion of components. In the Institute of Inorganic Chemistry of the Latvian Academy of Sciences, the plasma–chemical synthesis technology for more than 40 compounds and their composites has been elaborated, and experimental production of more than 20 compounds has been realized. The compositions of the synthesized compounds and composites are shown in Table 1.

Plasma technology makes possible the transformation of any raw material into ultradispersed powders, and ensures high technological output and one-stage synthesis of special-purpose compounds of predetermined composition, with the degree of impurities being determined mainly by the purity of the raw materials. Supposedly, besides the small particle size, the powders obtained in plasma–chemical synthesis possess a set of other specific physicochemical characteristics [1].

Examination of the chemical composition of compounds obtained shows that the composition corre-

TABLE 1. Synthesized compounds and composites

Compound	Raw materials	Composites	Raw materials
TiN	Ti, N_2, NH_3	TiN + Me(Ni,Fe)	Ti, Me, $Me(CO)_x$, N_2
ZrN	Zr, N_2, NH_3	TiN + AlN	Ti, Al, N_2
HfN	Hf, N_2, NH_3	TiN + TiB_2	Ti, B, N_2
Si_3N_3	Si, N_2, NH_3	TiN + Si_3N_4	Ti, Si, N_2, NH_3
AlN	Al, N_2, NH_3	$Si_3N_4 + Me_xO_y$	Si, Me_xO_y, NH_3
NbN	Nb, N_2, NH_3	(Me–Al, Y, Mg)	
ZrC	$ZrCl_4$, C_xH_y, H_2	Si_3N_4 + AlN	Si, Al, N_2, NH_3
TaC	$TaCl_5$, C_xH_y, H_2	$ZrO_2 + Al_2O_3$	Zr, Al, O_2
HfC	$HfCl_4$, C_xH_y, H_2	Si_3N_4 + SiC	Si, C_xH_y, NH_3, N_2
SiC	Si, SiO_2, C_xH_y	TiC_xN_y + Me	Ti, TiO_2, C_xH_y
CaB_6	CaO, B, N_2	Me(Fe, Ni, Mo, W)	N_2, NH_3, Me
BaB_6	BaO, B, N_2	$Al_2O_3 + Me_2O$	Al, Me_2CO_3, O_2
LaB_6	La_2O_3, B, N_2	(Me–Li, Na)	
Al_2O_3	Al, Al_2O_3, O_2	SiO_2–Li_2O	SiO_2, Li_2CO_3
SiO_2	Si, SiO_2, O_2		
TiO_2	Ti, $TiCl_4$, O_2		
ZrO_2	Zr, ZrO_2, O_2		
Y_2O_3	Y_2O_3, O_2		
TiC_xN_y	Ti, TiO_2, C_xH_y, N_2, NH_3		

sponds to or is close to the stoichiometric composition. If the system under investigation has several phases with different chemical compositions, it is possible to obtain separately one or other phase, or to obtain the phase which is predominant. The plasma method is characterized by the fact that, in the course of synthesis, no contamination of the final product takes place when so-called non-electrode plasma generators are used. Taking into consideration that, in plasma–chemical synthesis, the final compound is obtained in accordance with the vaporization–condensation system, some redistribution of impurities with respect to the radius of a separate particle is possible, which can be observed in the experiments.

In plasma–chemical processes, the description of particles becomes complicated, because their growth takes place in complex, gaseous dynamic, high temperature flows simultaneously with chemical reactions, which proceed in spatially non-homogeneous concentration and temperature fields. In accordance with the small amount of summarized literature data [1], highly dispersed powders of some refractory compounds and metals, obtained by the plasma–chemical method, have logarithmically normal distributions. By changing the technological parameters of the process and annealing the reaction products, the degree of dispersion of the metal powders can be varied over a sufficiently wide range. Usually, the specific surface area of the products is 10–100 $m^2\ g^{-1}$ and can be considerably higher.

The particles obtained possess a definite geometric shape (sphere, cube, polyhedral plaque, etc.), which is determined by the composition and properties of the synthesized compound, by the formation mechanism of the condensed particles and by some other factors (Figs. 1 and 2). The particles forms are often monocrystals. If the composite is formed from different compounds, the synthesis products can be the mechanical mixture of the separate phases. They can also form particles in which one phase is covered by a layer of the other phase (such particles are characteristic for the Ti–Al–N system (Fig. 3)) or can be of a more complicated shape (as for the Al–Cr–N system (Fig. 4). Taking into consideration the active interaction between the majority of the synthesized oxygen-free compounds with oxygen and water vapour, when in contact with the atmosphere, and the subsequent decrease in quality of the materials prepared from such powders, methods for preventing such effects have been determined. These methods differ for different compounds.

The study of the physicochemical properties of the ultrafine powders synthesized shows that they are mainly crystalline. However, the crystal lattice has many defects—vacancies in the metallic and non-metallic sublattices, displacement of atoms from their equilibrium positions in the lattice, microdeformations and others. The distortion of the lattice depends on the composition of the compound formed as well as on the conditions of synthesis [1–4].

For example, the structural characteristics of titanium nitride synthesized in an r.f. plasma from the elements are considered. Figure 5 shows the dependence of the crystal lattice parameter on the average

Fig. 1. Electron photomicrograph of titanium nitride particles.

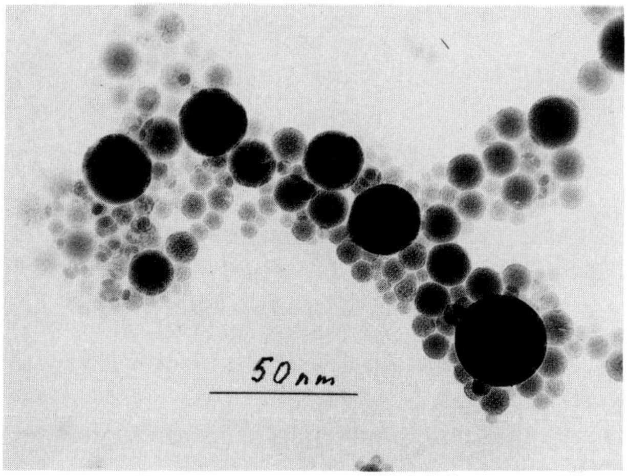

Fig. 2. Electron photomicrograph of alumina particles.

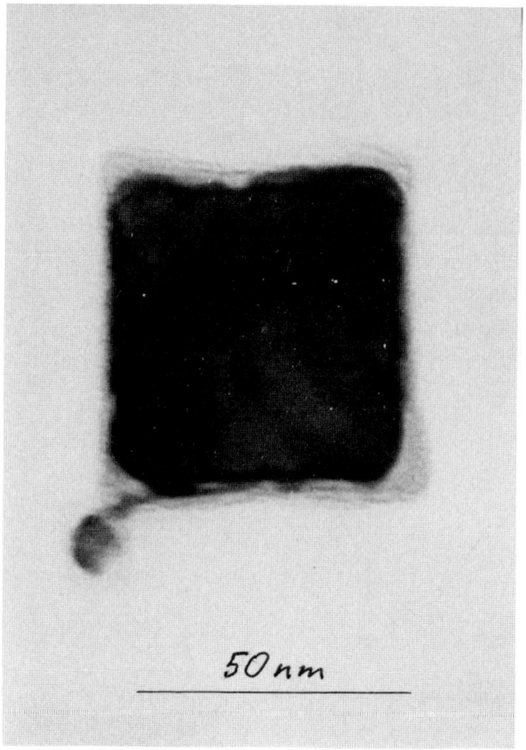

Fig. 3. Electron photomicrograph of particles formed in the Ti–Al–N system.

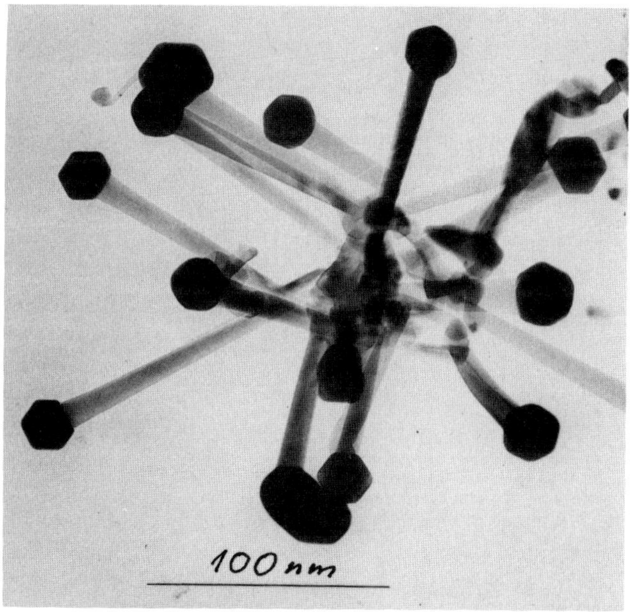

Fig. 4. Electron microphotograph of particles formed in the Cr–Al–N system.

size of the titanium nitride particles. The crystal lattice parameter of the highly dispersed samples is considerably lower than that of the coarse-dispersed and bulky samples. The essential changes dependent on the degree of dispersion are also observed from some other structural characteristics (Fig. 6).

During thermal treatment, initially the elimination of the structure defectiveness takes places, and the mass transfer, growth of the particles and sintering of samples begins. The change in the specific surface area of highly dispersed titanium nitride on heating is demonstrated in Fig. 7, the change in the lattice parameter is shown in Fig. 8 and the change in the crystallite size and microdeformations of the same sample is shown in Fig. 9. The data reported show distinctly that noticeable changes in the sample structure occur even at temperatures of about 1000 °C. Such changes are not characteristic for coarse-dispersed powders obtained by the other methods. Such differences and

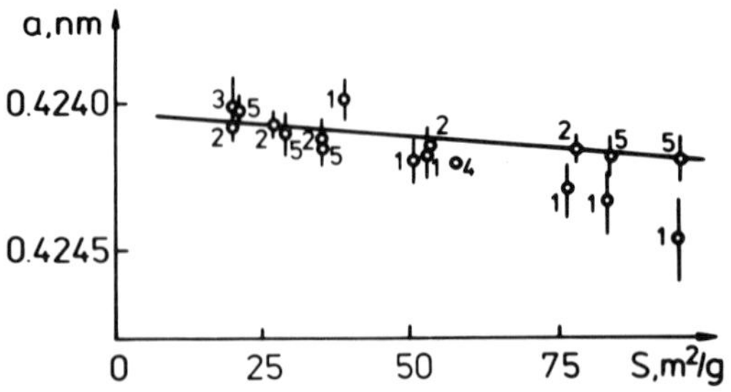

Fig. 5. Dependence of titanium nitride unit cell parameter (a) on average specific surface area (S) for different samples.

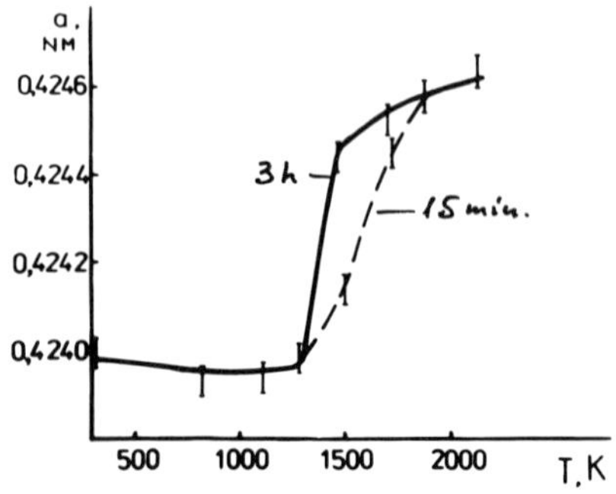

Fig. 8. Change in unit cell parameter (a) of titanium nitride on t heating.

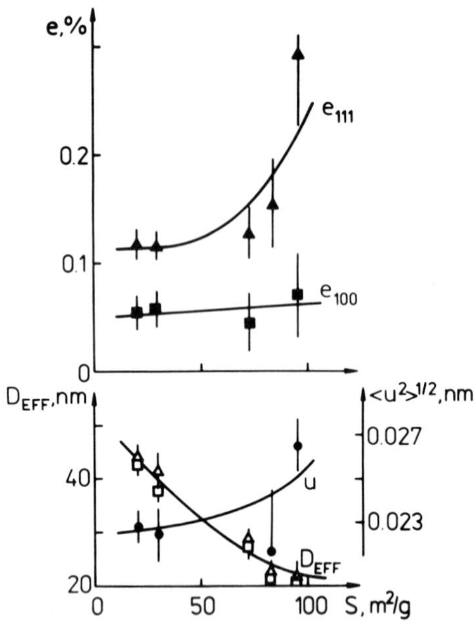

Fig. 6. Dependence of titanium nitride microdeformations (e), average size of crystallites (D_{eff}) and atomic displacement (u) on the specific surface area (S) of sample.

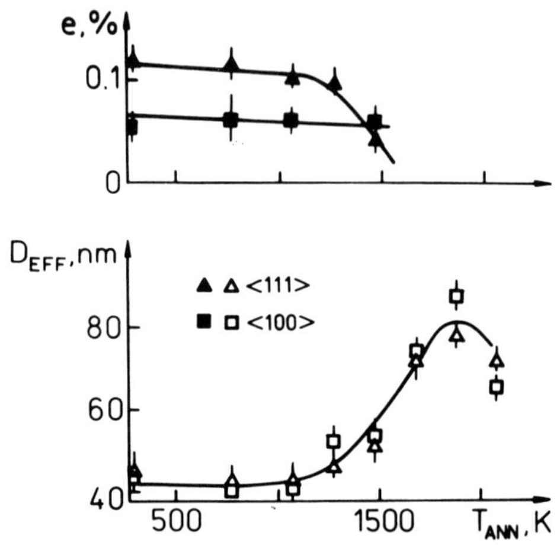

Fig. 9. Change in microdeformations (e) and crystallite size (D_{eff}) titanium nitride on heating.

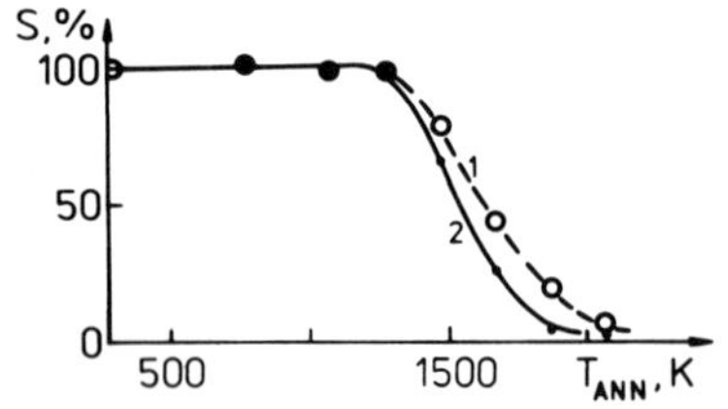

Fig. 7. Change in specific surface area (S) of titanium nitride on heating.

changes of the structure and morphology of ultrafine particles by heat treatment are also observed in the case of other compounds, except that the temperature at which changes begin and their magnitudes are different.

The use of such powders for production materials allows us to realize their principal advantages as follows:

(1) the decrease in sintering temperature, in some cases by 400–600 °C;

(2) the possibility of obtaining material with finely grained structure of if regulating the process of formation of material structure;

(3) the active interaction of the surface (which is formed naturally or artificially) with the matrix, where powder is doped to modify its properties (metal, polymer, glass).

There are a great amount of investigations about the application of ultrafine powders for the preparation of materials and the results confirm the forecasting about the efficiency of such an application [5–9]. Let us consider the example in which ultrafine silicon nitride powder (specific surface area, 32 $m^2\ g^{-1}$; 0%–1.6%) and the composite Si_3N_4–Y_2O_3, synthesized in plasma, are used. For comparison, a mixture of coarse-dispersed powder of silicon nitride (specific surface area, 10.5 $m^2\ g^{-1}$; 0%–1.2%) and yttria, obtained by grinding in a mill is used. Two methods are used to prepare the materials: hot pressing (1850 °C) and two-stage sintering under a small nitrogen pressure. The results obtained are shown in Table 2 and 3.

The experimental results show that the size of powder particles, amount and method of additive introduction, as well as the sintering conditions are more important for the preparation of the high quality ceramics than are the phase composition and the amount of oxygen in the starting powders.

The method of yttria addition determines the β-Si_3N_4 crystallization mechanism, which in turn affects the grain morphology and densification. However, the starting powder properties (amount of free silicon and oxygen, specific surface area etc.) and sintering conditions affect the degree of crystallization and phase composition of the yttrium-containing compounds. Ceramics obtained from the composite Si_3N_4–Y_2O_3 and synthesized by the plasma method possess better mechanical properties.

At present, the highly dispersed refractory compound powders prepared by the plasma-chemical method are being used in certain practical applications (Table 4). Specific characteristics of ultrafine powders, obtained by plasma synthesis, permit the creation of

TABLE 2. Characteristics of hot-pressed silicon nitride-based materials

Starting powder	Additive Y_2O_3 (%)	Density (g cm^{-3})	Porosity (%)	K_{1c} (MPa m$^{1/2}$)	Flexural strength (kg cm^{-2})	
					20 °C	1400 °C
Ground mixture	10	3.1–3.2	0.9–1.3	3–5	47–62	17–19
Synthesized in plasma	8–10	3.3–3.4	0	6–10	65–90	55–78

TABLE 3. Properties of silicon nitride-based materials[a]

Starting powder	Additive Y_2O_3 (%)	1750 °C			1900 °C		
		Density (g cm^{-3})	Flexural strength (MPa m$^{1/2}$)	K_{1c} (MPa m^{-1})	Density (g cm^{-3})	Flexural strength (MPa m$^{1/2}$)	K_{1c} (MPa m^{-1})
Ground mixture	10	3.02	450	—	3.30	800	8.5
Synthesized in plasma	10.2	3.28	710	8.9	3.35	820	12.0

[a]Sintering at normal pressure at 1750 °C and 5 MPa at 1900 °C.

TABLE 4. Application of ultrafine powders

Field of application	Composition	Technological or technical effects
Engineering ceramics	Si_3N_4–Y_2O_3, Al_2O_3, MgO, TiN–AlN, Si_3N_4–TiN, ZrO_2–Al_2O_3, ZrO_2–Y_2O_3	Improvement of mechanical strength, heat resistance, wear resistance
Materials for cutting tools	BN–TiN, Al_2O_3–TiN, Si_3N_4–TiN, Y_2O_3, Al_2O_3 TiC_xN_y–Fe, Ni	Improvement of strength, wear resistance, impact resistance
Heat-conducting ceramics	AlN–Y_2O_3	Improvement of thermophysical characteristics
Modification of metals structure and properties	Al, Al alloys, Ni–Cr, Fe, Ni Fe, Ni	Size reduction of grains and improvement of mechanical properties
Modification of polymers structure and properties	Fluoroplastics, polyethilene and others	Improvement of wear resistance and strength
Galvanic coatings	Ni, Cr, Sn and others	Improvement of wear resistance and other characteristics

materials, based on these powders, with specific applications and with high performance properties. Such powders now are successfully used for productions of (1) materials for cutting tools, (2) refractory ceramics and (3) ceramics with high thermal conductivity.

Materials made from the composites Al_2O_3–TiN and Si_3N_4–Y_2O_3, Si_3N_4–Al_2O_3 or Si_3N_4–MgO are used for turning and milling steels of different varieties and hardness, cast iron, non-ferrous metals and special alloys. A positive technical effect and increased performance properties are results of the specific behaviour of ultrafine powders in the process of sintering and structure development (fine grains, self-reinforcement).

Materials for cutting tools are also created on the basis of finely dispersed carbonitrides (titanium, niobium) and metals (nickel, iron, chromium, molybdenum). In addition, the creation of refractory ceramics for different types of engine has been of great interest in recent years. Indeed, this method of powder composition synthesis is also used in the fabrication of ceramics with high thermal conductivity, *e.g.* ceramics based on aluminium nitride (150 W m^{-1} K^{-1} and more).

The application of refractory compounds as ultrafine powders as modifiers for metals and alloys gives the opportunity to alter the structure of castings in the manner mentioned above. This then allows one to increase both the strength and the plasticity of the cast metal at low expense in comparison with the possibilities of traditional alloying methods.

The change of the alloy structure as a result of modification increases the strength limit by 10%–30% and the plasticity by 1.5–2 times. The addition of less than 10% of finely dispersed refractory compounds, *e.g.* titanium, aluminium, boron nitride and others, in the form of powder or fibre, to the different polymers markedly changes their properties. For example, the doping of fluoroplastics with less than 1% of ultrafine powder essentially increases their cold resistance and wear resistance. Such materials are used in aircraft applications. Doping of polyethylene with 1%–10% of finely dispersed powder increases its wear resistance by a factor of 10 or more, while the addition of a few per cent of fibre increases the failure strength by 3–4 times.

When tin is electrodeposited from potassium pyrophosphate electrolyte containing ultrafine powder of titanium nitride or niobium carbonitride, the compositional coatings were obtained. They contain up to 80% dispersive phase and have a high wear resistance, improved electroconductivity and solderability, and, in certain conditions, also superconductivity.

One effective application of the ultrafine powders is the electrochemical coating of Cr–Sn composite, which provides high mechanical strength and especially chemical stability.

References

1 T. N. Millers and A. A. Kuzjukevičs, *Prog. Cryst. Growth Charact., 16* (1988) 367.
2 V. F. Petrunin, Y. G. Andreew, T. N. Millers, J. P. Grabis, A. G. Jermolajev and F. M. Zelenjuk, *Powder Metall., 8* (1984) 12 (in Russian).
3 S. V. Gurov, V. N. Troitcky and A. L. Dorosinski. *Izv. Akad. Nauk. SSSR. Neorgan. Mater., 17* (3) (1981) 431.
4 V. F. Petrunin, V. A. Pogonin, L. I. Trusov, A. S. Ivanov and V. N. Troitcky. *Izv. Akad. Nauk. SSSR, Neorgan. Mater., 17* (1) (1981) 59.
5 V. I. Torbov, V. N. Troitcky and A. Z. Rahmatulina. *Powder Metall., 12* (1979) (in Russian).
6 R. A. Andrijevski, *J. Mendel. VHO, 36* (2) (1991) 137.
7 V. V. Panickina, *J Mendel. VHO, 36* (2) (1991) 170.
8 V. J. Novikov, L. I. Trusov, V. N. Lapovok and T. P. Geleisvili, *Powder Metall., 5* (1984) 28 (in Russian).
9 I. Zalite, G. Boden, Chr. Schubert, A. Lodzina, J. Plitmanis and T. Millers, *Latv. J. Chem., 2* (1992) 153.

Nanocomposite ceramic powder production by laser-induced gas-phase reactions

E. Borsella, S. Botti, R. Alexandrescu*, I. Morjan*, T. Dikonimos-Makris**, R. Giorgi** and S. Martelli**

ENEA, Area INN, Dip. Sviluppo Tecnologie di Punta, PO Box 65, 00044 Frascati, Rome (Italy)

Abstract

A comparative study of laser-produced nanocomposite ternary Si–C–N powders, obtained from different SiH_4, amines and NH_3 mixtures, is presented. It is shown that the final composition of powders is correlated with and may be controlled by the experimental parameters. Preliminary results on laser irradiation of SiH_4, NH_3 and $(CH_3)_3Al$ mixtures demonstrate the possibility of producing nanosized powders composed of a matrix of Si_3N_4 with a different content of Al.

1. Introduction

Laser gas-phase pyrolysis is a method already known for preparing nanosize clusters in a condensed phase. Ultrafine powders, such as Si_3N_4 and SiC [1, 2], have been produced by CO_2 laser irradiation of mixtures containing SiH_4 and NH_3 or acetylene respectively. Recently, it was demonstrated that, in the laser-driven synthesis, (i) Si_3N_4–SiC particles can be homogeneously mixed [3] and (ii) the chemical composition can be controlled as a function of the reaction parameters [4, 5].

The homogeneous mixing and controlled chemical composition of powders are particularly indicated in the densification process of composite systems, where different sintering behaviours could characterize the matrix and the second-phase particles [6]. Similar requirements appear in the process of the incorporation of the usual sintering aids, such as Al. The nitride and oxynitride class of ceramics requires a homogeneous dispersion of Al to reduce the intergranular phase by which the high temperature properties of the structural ceramics are degraded. At the same time, in mixed ceramics of the Si–Al–N type, the uniformity and purity of the intergranular phase prevail.

In this work, we review our results on the laser synthesis of ternary Si–C–N powders, produced from mixtures of SiH_4 with di-methylamine (DMA) and NH_3. The possibility of adjusting the C/N ratio in the powders by properly altering the experimental parameters is stressed.

We also present novel data about the synthesis of mixed Si_3N_4–Al powders, obtained by CO_2 laser irradiation of SiH_4, NH_3 and trimethylaluminium (TMA) mixtures. Our results show the possibility of varying the Al content to some extent in the powders by adjusting the process parameters (such as the reactant gas flows and laser power).

2. Experimental details

The experimental set-up for laser synthesis of ceramic powders has been described elsewhere [2, 4, 5]. Briefly, a CW CO_2 laser, emitting in resonance with SiH_4 excitation at a frequency of 944.19 cm^{-1}, was focused at the centre of a reaction cell onto the reactant gas stream. Mixtures of SiH_4, DMA and NH_3, or SiH_4, TMA and ammonia were used for the production of ternary Si–C–N and mixed Si_3N_4–Al powders, respectively. To avoid spontaneous reactions, the reactant gas mixtures entered through a system of two coaxial nozzles (1–3 mm in diameter), while a coaxial stream of Ar was used for the confinement of the reaction zone and for entraining the particles to the cell exit. The powders produced in laser-induced reactions were captured in a microfibre filter located between the reaction cell and the vacuum pump.

*Permanent address: Institute for Atomic Physics, POB. MG-6, Bucharest, Romania.

**Permanent address: ENEA, Area INN, Dip. Proc. Chim. Mat. non Met. e relat. Trat., PO Box 2400, AD 00100 Rome, Italy.

0921-5093/93/$6.00

The reaction temperatures were estimated by an optical pyrometer.

Different analytical techniques were used to identify both the gaseous and solid final products: mass spectrometry and IR absorption spectrophotometry were used as on-line diagnostics to monitor the gas composition during the photo-induced reactions; powders have been characterized by IR spectrophotometry, chemical analysis, X-ray photoelectron spectroscopy (XPS), X-ray-induced Auger electron spectroscopy (XAES) and X-ray diffraction (XRD).

XPS and XAES analyses were performed using an Escalab Mk II VG system at a base pressure of 10^{-9} mbar. For a semi-quantitative elemental analysis, the sensitivity factors on the photoelectron peaks were used [5]. Chemical information on Si compounds was obtained by measuring the Auger parameter between the Si 2p and Si KLL peak energies.

The XRD patterns were recorded with a conventional Bragg–Brentano diffractometer, equipped with a Ge monochromator on the incident beam. The selected radiation was Cu Kα ($\lambda = 0.154$ nm) [4].

3. Si–C–N powders

The CO_2 laser irradiation of SiH_4–DMA and SiH_4–DMA–NH_3 mixtures leads to composite Si–C–N powder formation. The composition and morphology of the powders may be adjusted by varying the reaction parameters [4], the total pressure and the reactant relative concentrations being the most important ones.

In the case of binary SiH_4 and DMA mixtures, the composition of the synthesized powders was investigated on the basis of XPS and XAES analysis. It has been found that decreasing the $[SiH_4]/[DMA]$ ratio has the same effect on the final powder composition as that of increasing the total gas pressure. In other words, there is a decrease in the relative O content, an increase in the total $Si_3N_4 + SiC$ content with respect to SiO_2, a decrease in the Si_3N_4 content with respect to the total $Si_3N_4 + SiC$ content, and a decrease in the free Si content.

The best compromise between experimental constraints allowed us to obtain powders with a C/N ratio of about 0.5. However, it has been established by Riedel *et al.* [6] that Si_3N_4–SiC systems can be sintered up to high densities, provided that the SiC content in the powder mixture does not exceed about 24 wt.%. NH_3 was added to the mixture of SiH_4 and DMA with the aim of increasing the N content in the synthesized powders.

A comparison between powders obtained from binary (sample 1) and ternary gas mixtures (sample 2) under similar conditions ($P = 400$ Torr, $[SiH_4]/[DMA] = 3.4$) may be inferred from the XAES Si KLL line spectra shown in Fig. 1. The addition of NH_3 ($[SiH_4]/[NH_3] = 0.75$) leads to an increase in the N/(N + C) quantity (Table 1) and to a decrease in the O and free Si contents in sample 2 with respect to sample 1 (Fig. 1).

At variance with the results obtained with binary gas mixtures, no crystalline phase was detected in samples obtained from ternary gas mixtures. Figure 2 displays the amorphous features of sample 5 (obtained from the ternary gas mixture), as compared with the crystalline structure of sample 1 (obtained from the binary gas mixture) produced at the same total pressure ($P = 400$ Torr).

Finally, we may stress that a C/N ratio close to the value required for ceramic applications (C/N = 0.25) [6] has been reached only in the case of ternary SiH_4/DMA/NH_3 mixtures, by an appropriate choice of the experimental parameters (Table 1, sample 4). One may observe from Table 1 that a large variety of C/N values (where C and N refer to the total C and Ni contents in the samples) is offered by the laser method of preparing Si–C–N composite powders.

4. Si_3N_4–Al powders

We have performed preliminary investigations on the laser synthesis of mixed Si_3N_4–Al powders, starting from SiH_4, NH_3 and TMA gaseous mixtures. At this

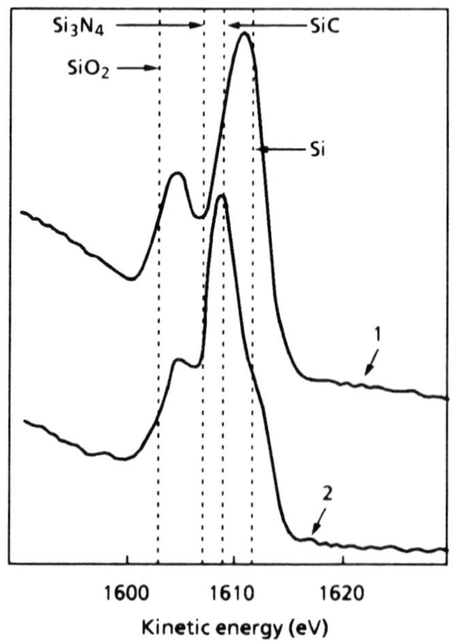

Fig. 1. XAES Si KLL line spectra of powders obtained from SiH_4 and DMA mixture (sample 1), and from ternary SiH_4, DMA and NH_3 mixtures (sample 2), under the same experimental conditions ($P = 400$ Torr; $[SiH_4/DMA] = 3.4$).

TABLE 1. Experimental conditions, reaction temperatures and XPS elemental analysis for samples obtained from SiH_4–DMA and SiH_4–DMA–NH_3 gaseous mixtures

No.	SiH_4 (sccm)	DMA (sccm)	NH_3 (sccm)	P (Torr)	T (°C)	XPS (at.%)				
						Si	C	N	O	C/N
1	62	18		400	1125	38	14	8	40	1.75
2	62	18	82	400	1010	33	8	28	33	0.31
3	62	18		600	1400	37	24	14	25	1.71
4	62	18	82	600	1075	35	6	27	32	0.22
5	42	48	82	400	800	30	16	15	39	1.11

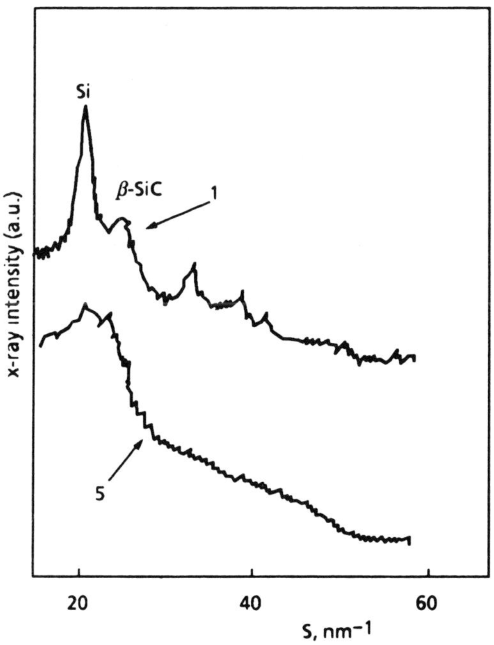

Fig. 2. XRD spectra of powders obtained from binary (sample 1) and ternary (sample 5) gaseous mixtures, at fixed pressure (P = 400 Torr).

stage, our aim was to identify the experimental parameters which favour (i) a variable content of the incorporated Al, (ii) a reduced C level, which is considered a detrimental impurity for operating microstructures, and (iii) a minimized spontaneous reaction of TMA with the other gaseous reactants.

The TMA (saturated vapour pressure of 12 Torr at 25 °C) was vaporized by bubbling with the carrier gas (Ar), with a variable flow rate of 40–120 sccm (standard cubic centimetres per minute). In another set of experiments, the natural established flow of TMA (1.5–2 sccm) was used. The main problem arising in our experimental system was connected with the tendency of the TMA vapours to react spontaneously with NH_3 at the nozzle exit, leading to the complex $(CH_3)_3Al{\cdot}NH_3$ [7], which is a white, crystalline solid. Thus, the process productivity could be affected by incomplete dissociation under the laser radiation and by the shift of the reaction routes.

The main parameters which determine the laser-induced process are the relative concentrations of reactants and the laser intensity. The total pressure was kept constant and equal to 600 Torr.

The results obtained in a series of experiments, in which the TMA carrier flow was varied, are presented in Table 2. The last column lists the elemental composition of powders, as obtained by the semi-quantitative XPS analysis. As pointed out in Table 2, the Al content of the powders is affected by the relative concentration of TMA, the laser intensity and the total flow. By lowering the total gas flow (sample 8), through the decrease of the Ar flow in the mixture, the relative concentration of TMA increased and greater TMA decomposition occurred. Besides crystalline α-Si_3N_4 and β-Si_3N_4, the X-ray analysis of sample 8 (with the highest amount of Al) revealed an unknown phase, whose features could be attributed to a solid solution of Al in Si and which is now under investigation.

It is worth mentioning that, except for sample 10, the Si/N content of this series of samples (referring to the total amount of Si and Ni, as detected by XPS) shows a rather substoichiometric but constant value (see Table 2). Considering that the on-line mass spectrometric analysis of the reactants generally indicated a low depletion of SiH_4, together with a high consumption of TMA and NH_3, it is probable that, besides the main laser reaction, the increased rate of vaporization facilitates the spontaneous complex formation, followed by secondary reaction routes [7].

It was found that by decreasing the vaporization rate of TMA most of these problems could be avoided. Furthermore, powders produced under these conditions are characterized by a rather high quantity of incorporated Al and a very low C content, as revealed by the chemical analysis of the samples. In the second series of samples (Table 3), the flow of TMA is imposed in the system by its saturable vapour pressure. The low C content of samples taken in these conditions

TABLE 2. Experimental conditions, reaction temperatures and results of XPS elemental analysis for samples obtained from SiH_4–NH_3–TMA mixtures

No.	SiH_4 (sccm)	NH_3 (sccm)	Ar (slm)	TMA+Ar (sccm)	P_{las} (Watt)	T (°C)	XPS (at.%) Si	C	N	O	Al	Si/N[a]
6	60	120	2.5	40	300	1200	27.4	14	41.1	16.8	0.7	0.66
7	120	240	4.0	50	500	1500	27.7	10.4	46	15.3	0.6	0.60
8	60	120	2.0	50	250	1050	24.5	16	36.1	20.4	2.9	0.67
9	80	120	4.0	70	500	1450	31.1	5.7	44.7	16.5	1.9	0.69
10	80	120	3.0	120	300	1350	28.9	11.5	36.8	21.1	1.8	0.78

[a]Si/N ratio in Si_3N_4 is 0.754.

TABLE 3. Experimental conditions, reaction temperatures and results of elemental chemical analysis for samples obtained from SiH_4–NH_3–TMA mixtures (at 600 Torr total pressure)

No.	SiH_4 (sccm)	NH_3 (sccm)	Ar (slm)	TMA+Ar (sccm)	P_{las} (Watt)	T (°C)	Chemical analysis (wt.%) Si	C	N	O	Al	Si/N[a]
11	120	160	6	2	100	1450	69.1	0.02	33.5	2.6	2.3	1.82
12	120	160	6	2	300	1750	60.1	0.058	37.5	2.6	2.0	1.61
13	120	160	6	2	500	1830	62.7	0.12	32.2	2.3	2.0	1.94
14	120	160	6	2	700	1850	63.4	0.13	27.6	1.8	2.3	2.29

[a]Si/N ratio in stoichiometric Si_3N_4 is 1.5.

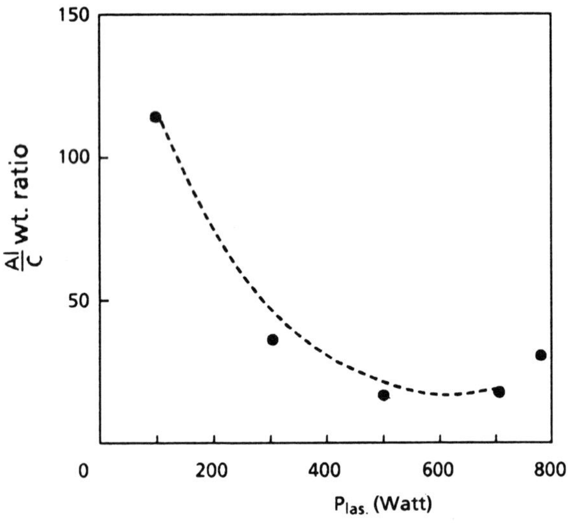

Fig. 3. Al/C ratio in powders *vs.* laser intensity at a total pressure of 600 Torr and SiH_4/NH_3 ratio of 1.5.

is probably due to a shift of the reaction towards C-containing molecular fragments (such as CH_4, C_2H_4, HCN, etc.), which were identified in the residual gas by mass spectrometry and IR spectroscopy. From Table 3, an increase in the C content of powders produced at higher laser intensities may be observed. Consequently, the Al/C ratio experiences a marked decrease when

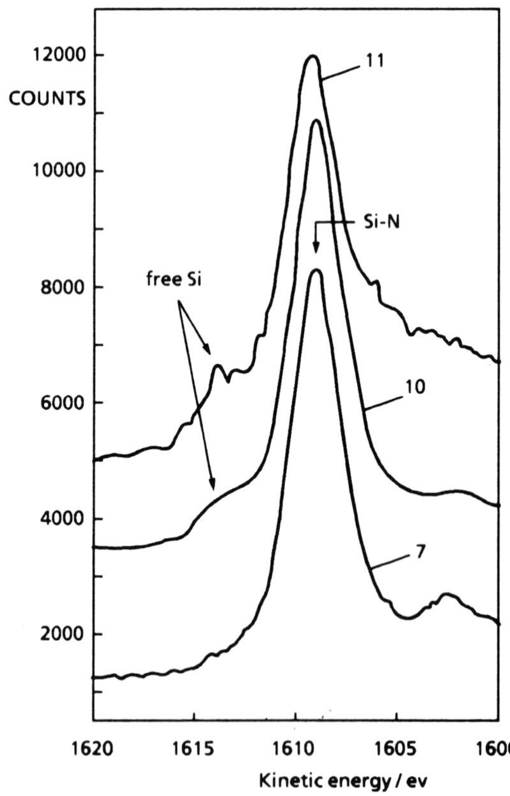

Fig. 4. XAES Si KLL line spectra of Si_3N_4–Al mixed powders, obtained under different experimental conditions. Sample 11: P_{las} = 100 W; $[SiH_4]/[NH_3]$ = 3.4. Sample 10: P_{las} = 300 W; $[SiH_4]/[NH_3]$ = 2/3. Sample 7: P_{las} = 500 W; $[SiH_4]/[NH_3]$ = 1/2.

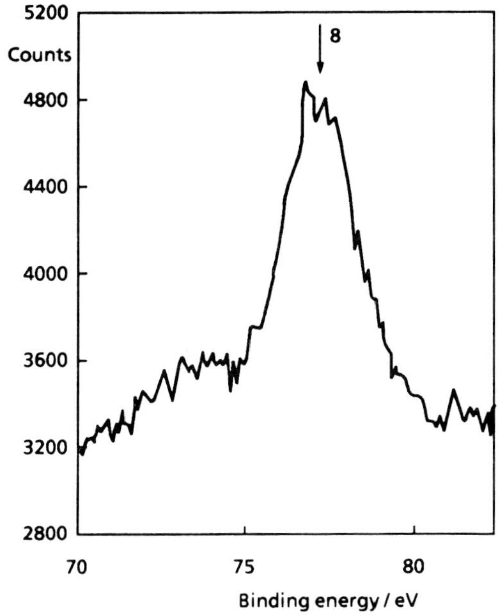

Fig. 5. XPS Al 2p line of the sample with the highest amount of incorporated Al (sample 8).

the laser power increases from 100 to 400–500 W (Fig. 3). The same trend, showing SiC formation in powder produced at increased reaction temperature, was found in ternary Si–C–N powders, obtained from gaseous mixtures of SiH_4–NH_3–amines [4].

Samples 11–14 are characterized by an Si–N ratio exceeding the stoichiometric value of 1.5, for the presence of free Si, as confirmed by XAES and XRD analyses. In Fig. 4, the XAES Si KLL line for powder 7 shows a single peak, whose Auger parameter indicates that it should be attributed to Si_3N_4. Free Si appears in samples 10 (Table 2) and 11 (Table 3).

The XPS Al 2p peak of sample 8 (see Fig. 5) shows a double structure, which is now under investigation. The chemical shift between these two peaks is larger than that observed between metal and oxidized Al; therefore, the main peak could arise from a different bond, involving C or N.

Further studies on the identification of chemical bonds should be connected with the not yet well understood decomposition of TMA at high temperatures [8] and with the subsequent reaction routes.

In conclusion, we have shown that the process of laser synthesis offers the possibility of obtaining Si–C–N and Si–N–Al powders with variable composition, by adjusting the process parameters.

References

1 R. W. Cannon, S. C. Danforth, J. H. Flint, J. S. Haggerty and R. A. Marra, *J. Am. Ceram. Soc., 65* (1982) 324.
2 E. Borsella, R. Fantoni, S. Piccirillo, R. Ceccato and S. Enzo, *J Mater. Res., 5* (1990) 143.
3 M. Cauchetier, O. Croix, M. Lance, C. Robert and M. Luce, in G. de With, R. A. Terpstue and R. Metselaer, *Euro-Ceramics*, Elsevier, London, 1989, Vol. 1, pp. 130–134.
4 R. Alexandrescu, I. Morjan, E. Borsella, S. Botti, R. Fantoni, T. Dikonimos-Makris, R. Giorgi and S. Enzo, *J. Mater. Res., 6* (1991) 2442.
5 E. Borsella, S. Botti, R. Fantoni, R. Alexandrescu, I. Morjan, C. Popescu, T. Diknonimos-Makris, R. Giorgi and S. Enzo, *J. Mater. Res., 7* (1992) 2257.
6 R. Riedel, M. Seher and G. Becker, *J. Euro. Ceram. Soc., 5* (1983) 113.
7 T. Mole and E. A. Jeffrey, *Organoalluminium Compounds*, Elsevier, Amsterdam, 1972.
8 N. Suzuki, C. Anayama, K. Masu, K. Tsubouchi and N. Mikoshiba, *Jpn. J. Appl. Phys., 25* (1986) 1236.

Microstructure of chemically vapour codeposited SiC–TiC–C nanocomposites

R. Hillel and M. Maline
IMP-CNRS, Université de Perpignan, 66860, Perpignan Cedex (France)

F. Gourbilleau and G. Nouet
LERMAT, URA-CNRS 1317, ISMRA, 14050 Caen (France)

R. Carles and A. Mlayah
URA-CNRS 74, Université Toulouse III, 31062 Toulouse Cedex (France)

Abstract

SiC–TiC–C nanocomposites were chemically vapour codeposited onto a graphite substrate under atmospheric pressure, in the temperature range 950–1150 °C, using the $TiCl_4$–C_4H_{10}-SiH_2Cl_2–H_2 gas system. SiC, TiC and excess carbon molar compositions were calculated using determination by electron probe microanalysis–wavelength-dispersive spectroscopy of the atomic percentages of Si, Ti and C after having identified SiC and TiC by Auger electron spectroscopy. The ⟨220⟩ preferential orientation corresponding to β-SiC + cubic TiC, revealed by X-ray diffraction, decreased with deposition temperature and almost no texture remained at 1050 °C. The apparent crystallite size along the ⟨220⟩ direction was about 10–20 nm. Micro-Raman spectra showed that the Raman size of the excess carbon did not exceed a few nanometres and that TiC and SiC were structurally disordered. Using transmission and high resolution electron microscopies, codeposits were found to consist of needles of SiC + TiC embedded in a poorly crystallized SiC matrix. They were stacking faulted and 10–15 nm broad. The needles were predominantly ⟨220⟩ oriented perpendicular to the substrate, whereas secondary needles were growing in {111} planes with ⟨112⟩ average direction.

1. Introduction

Combining two or more different ceramics through dispersed composites offers the promise of tailoring properties. Combinations such as SiC–TiC are examples providing enhancement of mechanical properties or protection against oxidation. However, most of these mixtures, fabricated by high-temperature and high-pressure sintering, are found to consist of grains of several micrometres which limits their applications under extreme conditions (stress and temperature). So recently, increasing interest has focused on structures with nano-sized particles, because these materials are expected to provide improved properties compared with composites presenting coarser structures [1, 2]. Among the available techniques for the preparation of nanocomposites [3] chemical vapour deposition (CVD) is a very attractive method for codepositing ceramics which are very intimately dispersed under near equilibrium conditions [4]. Taking into account previous works concerning CVD codeposition in the ternary system Ti–Si–C [5–8], we have prepared SiC–TiC–C nanocomposites. Owing to the drastic reduction in grain size and the strong imbrication of the phases, characterization of these materials required cross-checking of different data from complementary microanalytical techniques.

2. Experimental procedure

Deposits were prepared by conventional thermal CVD, under atmospheric pressure [9]. The reactor chamber consisted of a vertical silica tube (ϕ = 5.0 or 6.2 cm, L = 20 cm). The codeposition was onto mechanically polished polycrystalline graphite substrate (thickness 1 mm) heated in the range 950–1150 °C by an electric current, the substrate shape being designed to obtain a uniformly heated zone of 25 × 5 mm^2. Vaporized liquid $TiCl_4$, gaseous C_4H_{10} and SiH_2Cl_2 in H_2 gas were used as reactant sources at different concentrations. The H_2 gas flow rate was 30 l h^{-1} or 40 l h^{-1}. Samples of different compositions were prepared by varying both the codeposition temperature and concentrations of the input gas species. The codeposition time was 0.5 h. Coating thicknesses ranged from 100 to 600 μm, depending on the experimental conditions.

0921-5093/93/$6.00

3. Techniques of coating characterization

Si, Ti and C elemental compositions were determined by electron probe microanalysis–wavelength dispersive spectrometry (EPMA-WDS, Camebax equipped with ODPb wavelength-dispersive spectrometer, $I = 70$ to 100 nA, $V = 10$ kV, standards SiC, TiC). Phases were identified using Auger electron spectroscopy (AES, Riber SIA-200, $I = 200$ to 300 nA, $E_p = 3$ to 5 keV) and Raman spectroscopy (Dylor, 457.9 nm line of an Ar^+ laser, spectral resolution 5.7 cm^{-1}, spectra recorded in backscattering geometry on the cross-section of the codeposits, area of the spot beam was about 1 μm^2). X-ray diffraction (XRD, Philips PW 1710, Co $K\alpha$ 1.7902 Å, 40 kV, one-circle camera, the diffractometer being operated through a continuous $[\omega, 2\omega]$ rotation) was used to identify the texture and to evaluate the crystallite size. The microstructure was studied by transmission electron microscopy (TEM, Jeol 200 CX operating at 200 kV) and by high resolution electron microscopy (HREM, Topcon EM002B operating at 200 kV) equipped with an EDS analyser (Kevex).

4. Results

4.1. Auger spectroscopy and EPMA–WDS

Auger spectra recorded for the codeposits, were examined [10] taking into account Auger peak positions of the Si LVV, C KVV and Ti LMM transitions for the reference compounds Ti, Si, SiC, TiC, $TiSi_2$, $Ti_5Si_3(C)$, and Ti_3SiC_2 belonging to the Ti–Si–C ternary system [11]. From these comparisons, we determined the presence of only SiC and TiC. These phases having been identified, molar compositions were calculated from determination of the atomic percentages of Si, Ti and C by EPMA-WDS.

4.2. XRD

XRD patterns of the codeposits obtained above 1000 °C exhibited only a strong (220) reflection attributed to both SiC–3C and TiC. The corresponding calculated value of cubic cell parameter was about 0.434 nm, which fitted well those of β-SiC and $TiC_{0.97}$ coexisting with free carbon. With decreasing temperature, the (220) reflection became less apparent while the intensities of the (111), (200) and (311) reflections increased. The apparent crystallite size along the ⟨220⟩ direction decreased with temperature from about 20 to 10 nm.

4.3. Raman spectroscopy

Raman spectroscopy is able to give information on both the chemical bonding and microstructure in highly disordered materials. Owing to the large difference in atomic masses of C, Si and Ti, three types of vibrational bonds were expected: Ti–C, Si–C and C–C. Moreover, in amorphous-like materials, particularly in covalently bonded materials, the Raman spectrum can be viewed as a smoothed image of the density of vibrational states [12]. Indeed, the Raman selection rules, which only allow the activation of long-wavelength modes ($q = 0$ phonons) in a single crystal, are no longer satisfied in a disordered material. Breakdown of the translational invariance allows activation of $q \neq 0$ modes. The higher the disorder is, the more numerous these processes are, and the Raman spectrum evolves progressively from a discrete spectrum to a continuous spectrum. So, by performing a lineshape analysis [13], one can determine the distance (or "Raman size") over which the vibrational modes are localized by disorder effects.

Figure 1 shows the spectra recorded from three SiC–TiC–C nanocomposites. In the range 450–750 cm^{-1}, the spectrum reflects the calculated density of vibrational states relative to $TiC_{0.95}$ [14], especially in case (b). However, the optical branches (calculated positions 560 and 670 cm^{-1}) are shifted toward higher frequencies by about 40 cm^{-1}, certainly because of structural disorder. The "camel-back" shape of the

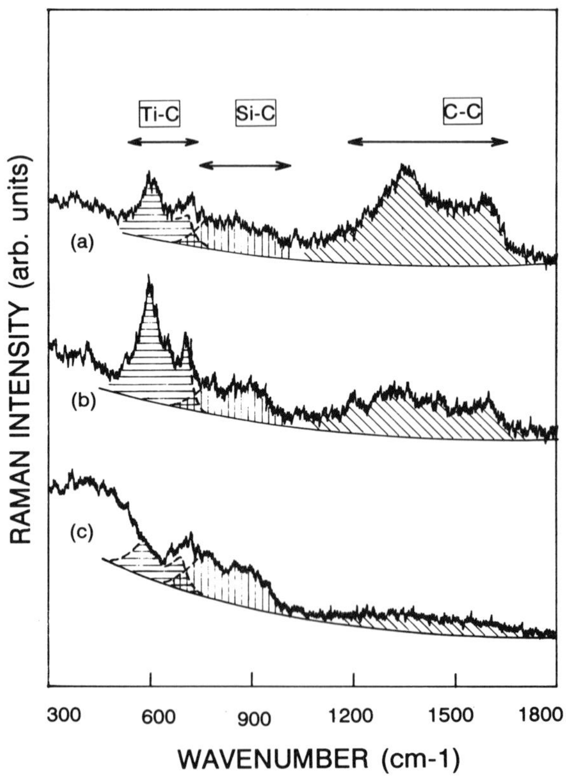

Fig. 1. Raman spectra of SiC–TiC–C nanocomposites: (a) $SiC_{0.81}TiC_{0.11}C_{0.08}$, (b) $SiC_{0.80}TiC_{0.17}C_{0.03}$, (c) $SiC_{0.93}TiC_{0.06}C_{0.01}$.

homonuclear C–C bonds is characteristic of graphitic carbon (*i.e.* sp^2 hybridization). The ratio between the integrated intensity of the lower frequency component (centred at 1350 cm^{-1}) over that of the higher component (centred at 1560 cm^{-1}) has often been used to evaluate the amount of disorder in various carbon phases [15]. From the present data the Raman size does not exceed a few nanometers, *i.e.* the carbon phase is highly dispersed in the materials and/or present as nanosized agglomerates. The optical Si–C band (650–1000 cm^{-1}) is hardly distinguishable because of the large disorder of the SiC phase. In contrast, the spectra of CVD SiC (Fig. 2(a), (b)) prepared at the same deposition temperature, clearly show acoustical and optical disorder-activated structures (DATA, DALA, DATO and DALO) whose shape can be analysed by comparison with the one-phonon density of states of crystalline SiC [16] (Fig. 2(c)). In that sense, the enlargement of these bands with decreasing temperature (spectra (a)–(c) in Fig. 2) is attributed to the reduction in grain size of the SiC phase.

If one assumes that the different bond polarizabilities are not affected by matrix effects or by resonance phenomena, one can evaluate the relative amounts of these bonds. Figure 1 shows that the scattering efficiency of the different bonds is very different. In fact the Si–C bonds which are the most numerous give the smallest scattering contribution, whereas the less numerous C–C bonds appear very efficient for the Raman effect. Consequently, a small amount of homonuclear carbon bonding is easily detected by the Raman probe. From the ratio of the integrated intensities of the different bands in spectrum (a) recorded from $SiC_{0.81}TiC_{0.11}C_{0.08}$, we can deduce the ratio of the corresponding bond polarizabilities: $\alpha(C–C)/\alpha(Si–C) \approx 40$. Therefore, using this value we can evaluate the amount of excess carbon in other samples: 0.03 and 0.01 for (b) and (c) respectively (Fig. 1). A check of the procedure was performed by evaluating the amount of TiC: from spectrum (a) one deduces $\alpha(Ti–C)/\alpha(Si–C) \approx 7$ and thus amounts of TiC of 0.19 and 0.07 from spectra (b) and (c) respectively, in good agreement with the result of chemical analysis (0.17 and 0.06).

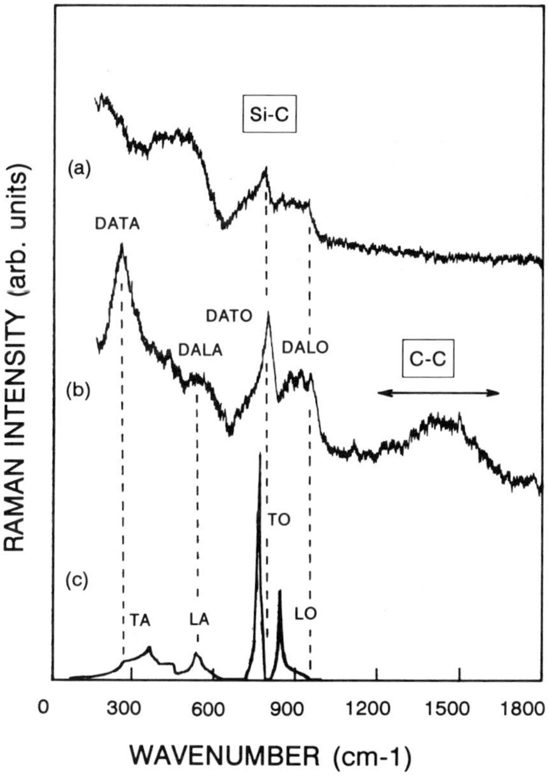

Fig. 2. Raman spectra of chemically vapour deposited SiC and SiC–C: (a) SiC deposited at 1000 °C, (b) $SiC_{0.90}C_{0.10}$ deposited at 1100 °C, (c) One-phonon density of states along the [111] direction in β-SiC.

4.4. TEM and HREM

The cross-sectional technique was used to determine the structural evolution of the material during deposition, by TEM. Figure 3(a) shows the first deposition stage with randomly oriented needles. With an increase in deposit thickness, heat evacuation causes the growth of needles predominantly perpendicular to the substrate (primary needle PN). These needles are oriented along the ⟨220⟩ direction, in agreement with the texture revealed by XRD. At 55° from this ⟨220⟩ direction, some secondary branches are growing in the {111} planes with ⟨112⟩ average direction (secondary needle SN) (Fig. 3(b)). Bright field observation (Fig. 3(b)) in Bragg condition reveals the presence of a small black line along the axis of the needle with small islands disoriented with respect to the others around this axis. This disorientation is confirmed by the electron diffraction pattern showing broadening spots (Fig. 3(c)). We can also observe in bright (Fig. 3(b)) and dark fields (Fig. 3(c)), slight disorientation between the secondary and primary needles.

HREM images were recorded without any objective aperture along the ⟨220⟩ SiC axis parallel to the substrate. We used micrographs taken at the Scherzer focus of 39 nm. Figure 3(d) shows a secondary needle with several stacking faults embedded in a poorly crystallized SiC matrix. The width of needles is in the range 10–15 nm. EDS analysis of the needles and the matrix revealed that titanium is localized in the needles only, whereas silicon is found in both the zones studied.

5. Conclusions

We prepared SiC–TiC–C nanocomposites by chemical vapour codeposition onto a graphite substrate under

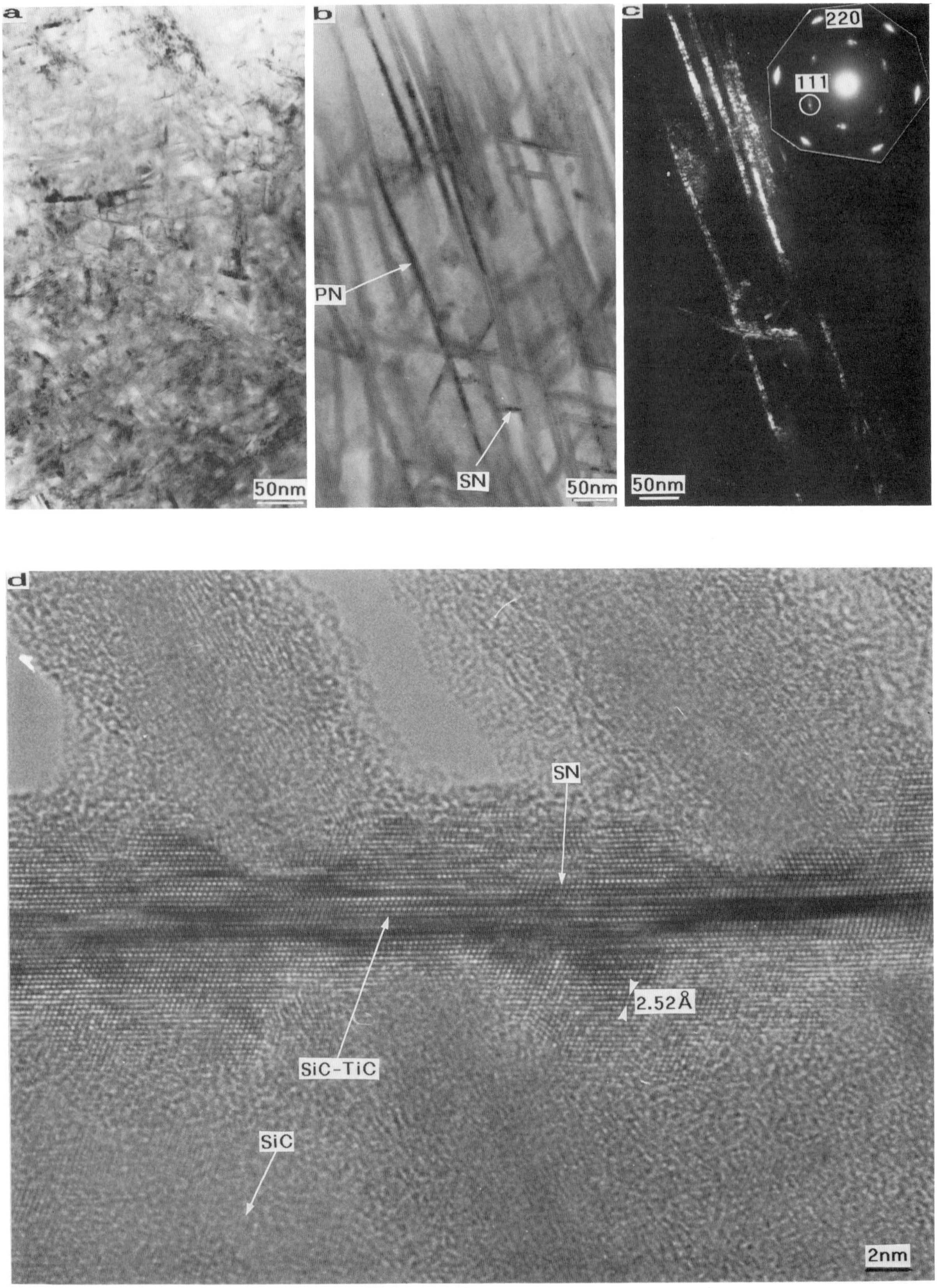

Fig. 3. TEM and HREM images: (a) First deposit stage, (b) microstructure of the deposit with primary needles (PN) and secondary needles (SN), (c) dark field image with the (111) spot diffraction selected, (d) HREM image of a secondary needle, projection along ⟨220⟩ β-SiC.

atmospheric pressure, in the temperature range 950–1150 °C, using vaporized liquid $TiCl_4$, gaseous C_4H_{10} and SiH_2Cl_2 in H_2 gas as reactant sources.

Different characterization techniques show that the materials codeposited at 1150 °C exhibit a strongly textured ⟨220⟩ microstructure which decreases with the temperature of deposition. They consist of TiC-rich SiC needles embedded in a poorly crystallized SiC matrix of small crystalline size. The Raman size of the excess carbon does not exceed a few nanometers. Primary needles, stacking faulted and 10–15 nm broad, are ⟨220⟩ oriented perpendicular to the substrate whereas secondary branches grow in the {111} planes with ⟨112⟩ average direction.

References

1 R. Roy, in J. A. Park and A. G. Evans (eds.), *Ceramic Microstructure '86, Materials Science Research*, Vol. 21, Plenum, New York, 1987, pp. 25–32.
2 R. W. Siegel, in F. E. Fujita (ed.), *Physics of New Materials*, Springer, Heidelberg, 1992, p. 1–19.
3 A. J. Burggraff, *J. Eur. Ceram. Soc., 9* (1992) 245.
4 T. Hirai and M. Sasaki, in P. Vincenzini (ed.), *Advanced Structural Inorganic Composites*, Elsevier, Amsterdam, 1991, pp. 541–544.
5 D. P. Stinton and W. J. Lackey, *Ceram. Eng. Sci. Proc., 6* (1985) 707.
6 T. Goto and T. Hirai, in G. W. Cullen and J. M. Blocher (eds.), *Proc. Xth Int. Conf. on CVD, Honolulu*, Electrochemical Society, 1987, pp. 1070–1079.
7 M. Touanen, F. Teyssandier, M. Ducarroir and J. L. Derep, *Mater. Sci. Eng., A147* (1991) 239.
8 C. Kawai and M. Miyake, *J. Ceram. Soc. Jpn., 99* (1991) 1088.
9 M. Maline, R. Hillel and R. Berjoan, *J. Mater. Sci., 27* (1992) 6187.
10 M. Maline, M. Ducarron, F. Teyssandia, R. Hillel, F. J. J. Vanloo and W. Wakelkamp, *Surf. Sci., 286* (1993) 82–91.
11 W. J. J. Wakelkamp, J. J. Van loo and R. Metselaar, *J. Eur. Ceram. Soc., 32* (1990) 567.
12 R. Schuker and R. W. Gammon, *Phys. Rev. Lett., 25* (1970) 222.
13 H. Richter, Z. P. Wang and L. Ley, *Solid State Commun., 39* (1984) 625.
14 L. Pintschovius, W. Reichardt and B. Scheerer, *J. Phys. C, 11* (1978) 1557.
15 F. Tuinstra and J. L. Koennig, *J. Chem. Phys., 53* (1970) 1126.
16 S. Baroni, 1992, private communication.

Materials Science and Engineering, A168 (1993) 189–197

Synthesis and properties of nanophase materials

Richard W. Siegel

Materials Science Division, Argonne National Laboratory, Argonne, IL 60439 (USA)

(Received February 12, 1993)

Abstract

Nanophase materials, with their grain sizes or phase dimensions in the nanometer size regime, are now being produced by a wide variety of synthesis and processing methods. The interest in these new ultrafine-grained materials results primarily from the special nature of their various physical, chemical, and mechanical properties and the possibilities of controlling these properties during the synthesis and subsequent processing procedures. Since it is now becoming increasingly apparent that their properties can be engineered effectively during synthesis and processing, and that they can also be produced in quantity, nanophase materials should have considerable potential for technological development in a variety of applications. Some of the recent research on nanophase materials related to their synthesis and properties is briefly reviewed and the future potential of these new materials is considered.

1. Introduction

Nanophase materials are three-dimensionally modulated, synthetic materials with average grain, phase, or other structural domain sizes below 100 nm. Interest in them has increased rapidly during the past few years with successive demonstrations that their properties are different from and frequently superior to those of conventional materials that have phase or grain structures on much coarser size scales [1]. These materials can be synthesized by the assembly of physically or chemically derived atom clusters or by means of mechanical deformation processes. Each of these methods has its advantages, but *in situ* consolidation of gas-condensed clusters [2, 3] appears to be fundamentally the most flexible method presently available.

It is important to understand that nanophase materials are only one of the much broader class of nanostructured materials artificially synthesized with microstructures modulated in zero to three dimensions on length scales less than 100 nm that it has become possible to create over the past few years. The various types of nanostructured materials share three common features: atomic domains spatially confined to less than 100 nm, significant atom fractions associated with interfacial environments, and interactions between their constituent domains. It is these three features that largely determine their unique properties. These materials can therefore be loosely classified [4] according to their modulation dimensionality: zero-dimensionality atom clusters (with any aspect ratio from unity to infinity, including equiaxed clusters as well as filaments or tubules); one-dimensionally modulated multilayers; two-dimensionally modulated ultrafine-grained overlayers (coatings) and buried layers; three-dimensionally modulated cluster assemblies, including nanophase materials.

Nanophase materials, through control of the sizes of the constituent structural components of which they are synthesized, can incorporate a number of size-related effects in condensed matter ranging from electronic effects (so-called "quantum size effects") caused by spatial confinement of delocalized valence electrons and altered cooperative ("many body") atom phenomena, such as lattice vibrations or melting, to the suppression of such lattice-defect mechanisms as dislocation generation and migration in confined grain sizes. It appears likely that the emerging possibilities of assembling artificially size-selected entities, such as atom clusters for example, into new materials with unique or improved properties may have a revolutionary impact on our ability to engineer materials with controlled optical, electronic, mechanical, and chemical properties for a wide variety of useful technological applications in the future.

The present paper is intended as a brief review of some of the ways in which nanophase materials can be synthesized and how the controlled synthesis and processing of these new materials can lead to prescribed size dependent properties. More extensive reviews have recently appeared elsewhere [5–7]. Emphasis here will be placed on the special opportunities

0921-5093/93/$6.00

presented by our ability to create nanophase materials assembled from atom clusters of metals and ceramics synthesized by means of the gas-condensation method [3, 6, 8], since this method appears to be the most generally applicable of the presently available avenues for producing nanophase materials, and thus appears to have very broad technological potential.

2. Nanophase material synthesis

2.1. Cluster assembly

There is a wide variety of chemical and physical methods for the production of atom clusters or ultrafine powders that can be assembled to synthesize nanophase materials. One of the traditional methods for synthesizing ultrafine powders or colloidal suspensions has been chemical precipitation. Several of these methods, including sol-gel synthesis [9, 10] and the inverse micelle method [11] for example, have been successfully applied to the synthesis of nanometer sized clusters with narrow size distributions. Additional references to the variety of interesting methods available can be found elsewhere [1, 2, 12–16] and in the pages of these Proceedings. Invariably, in each of these chemical precipitation methods, the synthesized clusters are coated by residual surface layers (often intentionally introduced to cap the clusters at a given size or to keep them from agglomerating and forming large pores) from the solutions in which they are formed. These surface layers create few, if any, problems in the study of isolated clusters. However, if the clusters are to be assembled into bulk solids, then these surface layers can become contaminants in the grain boundaries of the cluster-consolidated material, and processing and properties can be adversely affected. Such potential problems can be avoided in the gas-condensation method for synthesizing nonophase materials.

The study of the gas condensation of ultrafine particles or atom clusters has had a rather long history stretching back to the formation and use of "smokes", such as carbon or bismuth "blacks", for a variety of applications [17]. Scientific research into the controlled production of ultrafine particles by means of the gas-condensation method has been more recent [8, 18–20]. The application of these ideas in recent years [3, 21–24] to the synthesis of a variety of nanophase metals and ceramics has built upon this broad scientific and technological base and the previously assembled knowledge of powder metallurgy and ceramics.

The generation of atom clusters via gas condensation proceeds by evaporating a precursor material, either an element or compound, in a gas maintained at a low pressure, usually well below 1 atm, in an apparatus [25] such as that shown in Fig. 1. The evaporated atoms or molecules lose energy via collisions with the gas atoms or molecules and undergo a homogeneous condensation to form atom clusters in the highly supersaturated vicinity of the precursor source. In order to maintain small cluster sizes, by minimizing further atom or molecule accretion and cluster–cluster coalescence, the clusters once nucleated must be removed rapidly from the region of high supersaturation. Since the clusters are already entrained in the condensing gas, this is readily accomplished by setting up conditions for moving this gas. Such gas motion has generally been driven by natural convection under the combined action of gravity and the temperature difference between the precursor source and a cooled thermophoretic cluster collection surface. However, a forced gas flow can be used instead of natural convective flow, which should be able to yield significant advantages in terms of both cluster size control and process efficiency.

There are only three fundamental rates, which function relative to one another, that essentially control the formation of the atom clusters in the gas-condensation

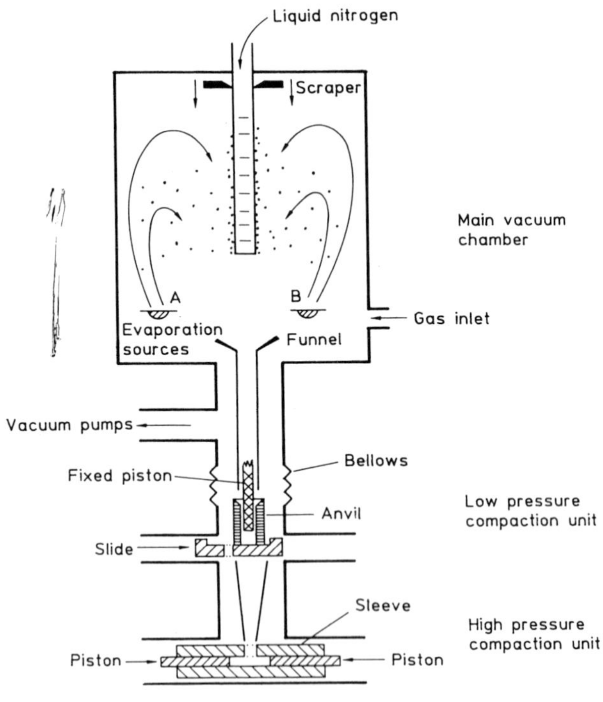

Fig. 1. Schematic drawing of a gas-condensation chamber for the synthesis of nanophase materials. Precursor material evaporated from sources A and/or B condenses in the gas and is transported via convection to the liquid-nitrogen filled cold finger. The clusters are then scraped from the cold finger, collected via the funnel, and consolidated first in the low-pressure compaction unit and then in the high-pressure compaction unit, all in vacuum [25].

process [6]. They are (1) the rate of supply of atoms to the region of supersaturation where condensation occurs, (2) the rate of energy removal from the hot atoms via the condensing medium, the gas, and (3) the rate of removal of the clusters once nucleated from the supersaturation region. Other factors can also affect the clusters finally collected, particularly those that result in significant cluster–cluster coalescence, but these three rates represent the core of the process. Accordingly, the smallest cluster sizes for a given precursor are obtained for a low evaporation rate and condensation in a low pressure of a light inert gas, such as helium (Fig. 2). These conditions lead to a lower supersaturation of precursor atoms in the gas, slower removal of energy from the evaporated atoms (via the lighter gas atoms at lower pressure), and more rapid convective gas flow owing also to the lower gas pressure. The rapid gas flow is significant, since it guarantees a shorter dwell time of the gas-entrained condensed clusters in the supersaturated region in which, if they remained, they would grow further.

The apparatus shown in Fig. 1, which uses such a process for the synthesis of nanophase materials via the *in situ* consolidation of gas-condensed clusters, consists of an ultrahigh-vacuum (UHV) system fitted with two resistively heated evaporation sources, a cluster collection device (liquid-nitrogen filled cold finger) and scraper assembly, and *in situ* compaction devices for consolidating the powders produced and collected in the chamber. Before making the powders, the UHV system is first evacuated by means of a turbomolecular pump to below 10^{-5} Pa and then back-filled with a controlled high-purity gas atmosphere at pressures of about a few hundred pascals. For producing metal powders this is usually an inert gas, such as helium, but it can alternatively be a reactive gas or gas mixture if, for example, clusters of a ceramic compound are desired.

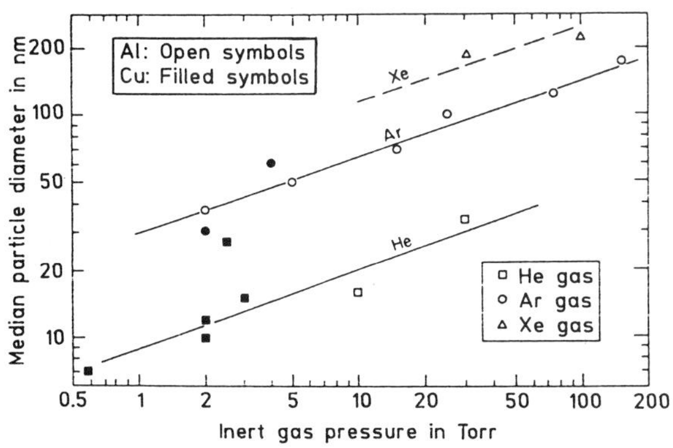

Fig. 2. Median particle diameter *vs.* pressure of helium, argon or xenon gas for clusters of aluminum and copper formed via gas-condensation [8]. The straight lines only serve as a guide.

The clusters that are collected via thermophoresis on the surface of the cold finger form very open, fractal structures. They are held on the collector surface rather weakly, via van der Waals type forces, and are easily removed from this collection surface by means of a Teflon scraper. Upon removal, the clusters fall like "snow" from the surface and are funneled into a set of compaction devices (see Fig. 1) capable of consolidation pressures up to about 1–2 GPa, in which the nanophase samples are formed at room temperature, or at elevated temperatures if needed. The pellets formed in this conventional research apparatus are typically about 9 mm in diameter and 0.2–1.0 mm thick, depending on the amount of material made (usually a few hundred milligrams) and the experiments to be performed. The sizes of these samples have been more a matter of laboratory convenience than any real limitation of the gas-condensation method itself. All of the fundamental rates involved in the cluster synthesis can be significantly increased above those presently in use in the laboratory and scaled up production rates leading to samples with greatly increased sizes are clearly feasible. The scraping and consolidation are performed under UHV conditions after removal of the inert or reactive gases from the chamber, in order to maximize the cleanliness of the particle surfaces and the interfaces that are subsequently formed. Also, any possibility of trapping remnants of these gases in the nanophase compact is minimized by consolidation in vacuum. It should be noted in this regard that the total surface area of the nanophase powders produced in a given run is so great that, for a residual gas pressure of less than 10^{-5} Pa in a volume the size of the UHV chamber used, little gas contamination of the cluster surfaces would be expected.

Since the as-collected gas-condensed clusters are generally aggregated in rather open fractal arrays [18, 19], their consolidation at pressures of 1–2 GPa is easily accomplished, even at room temperature. The difficulties in consolidating the hard equiaxed agglomerates of fine powders resulting from conventional wet chemistry synthesis routes are mostly avoided. The sample densities resulting from cluster consolidation at room temperature have ranged up to about 97% of the theoretical density for nanophase metals and up to about 75%–85% of the theoretical density for nanophase oxide ceramics. The remaining "green-state" porosity represents (at least in part) a manifestation of powder agglomeration leading to void-like flaws. Fortunately, it appears that these can be removed by means of cluster consolidation at elevated temperatures and pressures without significant attendant grain growth.

Most of the atom clusters assembled into nanophase materials to date have been generated from joule-heated evaporation sources. However, such sources have limitations that need not be suffered, since a wide variety of other sources are also available. The primary limitations are source–precursor incompatibility, temperature range, uniformity and control, and dissimilar evaporation rates for different constituents in an alloy or compound precursor. Each of these limitations can be avoided by a host of alternative sources which have been developed over the years of ultrafine particle research [20], but which are only now beginning to enter the field of nanophase materials synthesis. Among other sources for bringing atom supersaturations into a condensing gas medium that have been successfully used to produce clusters or ultrafine particles are joule-heated ovens [26–28], sputtering [29–33], electron-beam heating [34–37], laser ablation [38], and plasma methods [39, 40]. It should be clear that this wide variety of evaporation methods will allow for greatly increased flexibility in the use of refractory or reactive precursors for clusters and will be especially useful as we begin in the future to synthesize technological quantities of more complex multicomponent or composite nanophase materials.

2.2. Mechanical attrition

Nanophase structures can also be generated by means of severe mechanical deformation. Mechanical attrition produces its nanostructures, not by cluster assembly, but by means of the structural decomposition of coarser-grained structures induced by the high stored energy density from the mechanical deformation. It appears that nanometer size grains nucleate within the shear bands of heavily deformed materials converting a coarse-grained structure to nanophase. The heavy deformation is usually effected by means of high-energy ball milling [41–43], but can as well result from surface wear phenomena [44] and other methods for introducing high densities of deformation, such as high-energy shear process and other manifestations of severe deformation as described in these Proceedings. In mechanical attrition, therefore, the individual grains are never isolated clusters and much of the synthesis and processing flexibility is lost relative to the gas-condensation method. Nevertheless, ultrafine grain sizes can be readily accessed by this rather straightforward method, as shown in Fig. 3. High-energy ball milling is capable of producing commercial quantities of material, albeit with some contamination problems from the sources of mechanical work, but for those applications in which careful control of purity or grain morphology is not important, and for which the desired properties are accessible in the mechanically attrited nanostructures, it can be a useful source of these materials. The variation with grain size as a function of ball-milling time in palladium [45] shown in Fig. 3 clearly reveals that grain sizes down into the nanometer regime can be accessed, although this is more easily accomplished in relatively harder materials. Mechanical attrition exercised at low temperatures ("cryomilling") can extend the range of applicability, as shown by Luton *et al.* [46] in their work on dispersion-strengthened aluminum.

The predominant feature of nanophase materials synthesized by any method is their ultrafine grain size below 100 nm and, hence, the large fraction of their atoms that reside in grain boundaries or interfaces. For example, as indicated in Fig. 4, a nanophase material with a 5 nm average grain size will have from about 27% to 49% of its atoms associated with grain boundaries, assuming a simple grain boundary picture and an average grain boundary thickness of about 0.5–1.0 nm (*ca.* 2–4 nearest-neighbor distances). This percentage

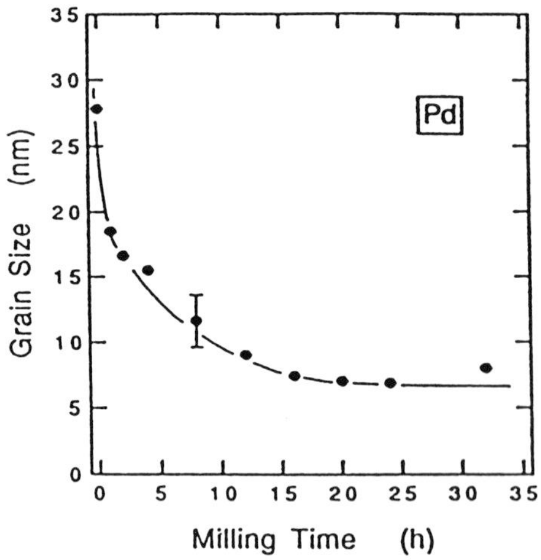

Fig. 3. Average grain size, determined by X-ray line broadening, of palladium as a function of ball-milling time [45].

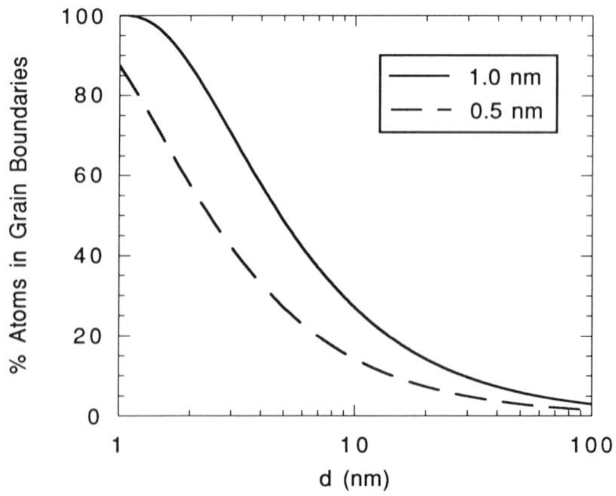

Fig. 4. Percentage of atoms in grain boundaries of a nanophase material as a function of grain diameter, assuming that the average grain boundary thickness ranges from 0.5 to 1.0 nm (*ca.* 2–4 atomic planes wide) [47].

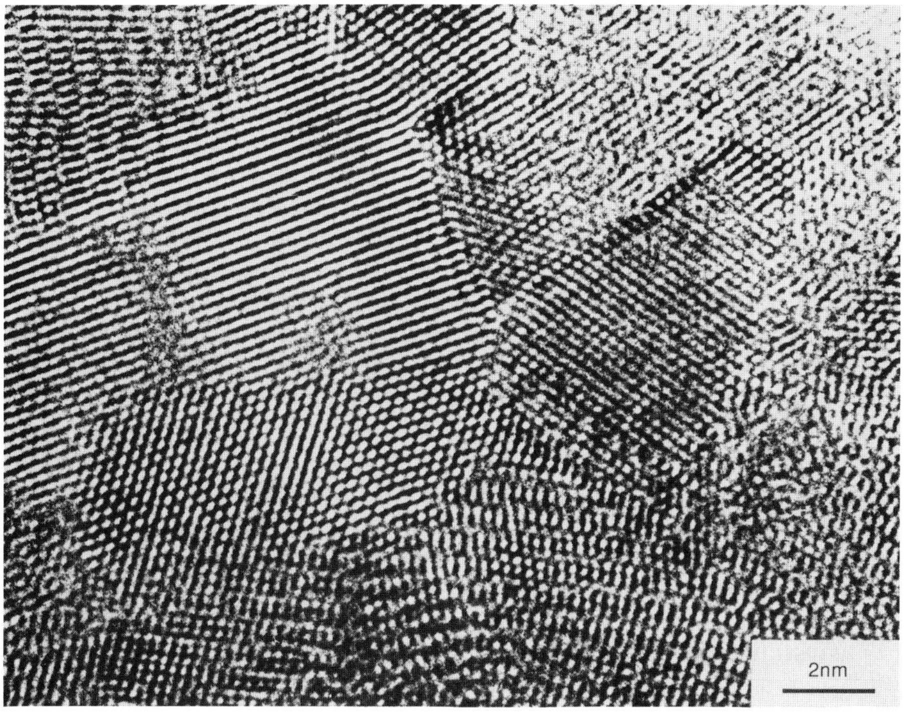

Fig. 5. High resolution transmission electron micrograph of a typical area in nanophase palladium [53].

falls to about 14%–27% for a 10 nm grain size, but is as low as 1%–3% for a 100 nm grain size. The interface volume fraction is, of course, essentially negligible for conventional grain sizes of 1 μm and above. The simple model upon which Fig. 4 is based assumes only that the grain volumes scale as a length cubed and that the grain boundary volume includes the junctions between and among their boundaries [47]. The properties of nanostructured materials are thus expected to be strongly influenced by both their small (and hence confined) grain sizes and the nature (atomic and electronic structure) of their internal boundaries, simply because of the very large number density of these interfaces.

Considerable effort has gone into the elucidation of the structure of nanophase grain boundaries [5, 48, 49]. Investigations of nanophase TiO_2 by Raman spectroscopy [50–52] and of nanophase palladium by high resolution transmission electron microscopy [53, 54] indicate that the grain boundary structures in these materials are essentially low energy configurations that are rather similar to those in coarser grained, conventional materials [55]. The nanophase grain boundaries, shown in Fig. 5 contain short-range ordered structural units representative of the bulk material and distortions that are localized to about ± 0.2 nm on either side of the grain boundary plane. These conclusions are also consistent with the results from complementary small-angle neutron scattering measurements [56–58], and with the expectations for conventional high angle grain boundaries from condensed matter theory [59, 60]. Whether or not there are indeed some local structural deviations in nanophase boundaries compared with conventional scale grain boundaries is still an open question [5, 49], although it is becoming clear that significant lattice strains are a reality in nanophase materials as prepared by both cluster-consolidation and mechanical-attrition methods [61, 62]. However, no firm experimental or theoretical evidence presently exists which indicates that the structures of nanophase grain boundaries are actually different from their coarse-grained counterparts.

3. Material properties

Nanophase materials have a variety of properties that are considerably different and improved in comparison with those of conventional coarser-grained structures. These result from the impact of various combinations of the three imporant aspects (size, composition, and interfaces) of their nature. A few examples will be presented here to try and give a sense of how the interplay among these aspects can significantly alter some useful technological properties of these materials. A number of other examples can be found in recent reviews cited previously, for example refs. 5 and 7, and in these Proceedings.

The chemical reactivity of nanophase materials, with their high surface areas compared with conventional materials, can be rather striking. Since gas-condensed, high-surface-area clusters can be assembled via consolidation in the method described in this paper, there can be an excellent degree of control over the total available surface area in these self-supported ensembles. Also, composition control can be readily achieved [51, 52]. Recently completed measurements [63] of the decomposition of H_2S over lightly consolidated nanophase TiO_2 at 500 °C demonstrate the enhanced chemical reactivity of nanophase materials rather well. Figure 6 shows the catalytic reaction rate for sulfur removal from H_2S via dissociative adsorption for nanophase TiO_2 (rutile) compared with that for a number of other commercially available forms of TiO_2 having either the rutile or anatase structure. It can be seen that the nanophase sample was far more reactive than any of the other samples tested, both initially and also after extended exposure to the H_2S. Indeed, only the nanophase titania demonstrated a non-zero reaction rate for sulphur removal after 7 h. This greatly enhanced and sustainable activity was shown to result from a combination of unique and controllable features of the nanophase material, its high surface area combined with its rutile structure and its oxygen deficient composition.

Quite limited research has been carried out so far on the physical properties of cluster-assembled nanophase materials. However, there appear to be interesting prospects, based on what little has been accomplished and on the expectations for confined systems of atoms in which the sizes of constituent domains fall below the critical length scales pertinent to a given property. Figure 7 shows one example of how the ability to control the porosity in the synthesis of nanophase ceramics will apparently lead to some rather interesting optical applications in the near future. It was recently demonstrated [64] that the oxidation step in the synthesis of cluster-consolidated nanophase Y_2O_3 can be so controlled that the resulting "green-state" porosity of 25%–35% has a size distribution sufficiently small and narrow that the as-consolidated material is effectively transparent. A more recent demonstration of this processing control is shown in Fig. 7. This ability to reduce the size of the porosity in such material to well below

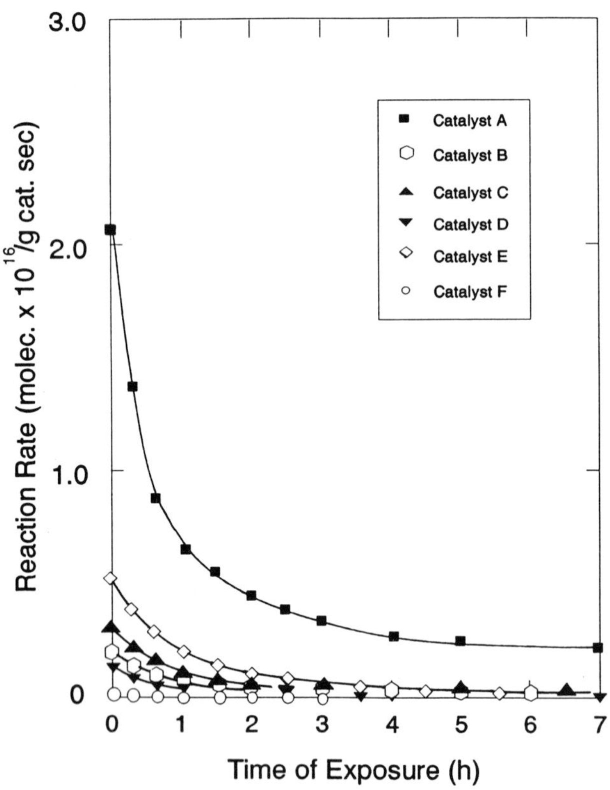

Fig. 6. Reaction rates for H_2S decomposition as a function of exposure time at 500 °C for nanophase TiO_2 compared with that from several commercially available TiO_2 materials and a reference (catalyst A, 76 m^2 g^{-1} nanophase rutile; B, 61 m^2 g^{-1} anatase; C, 2.4 m^2 g^{-1} rutile; D, 30 m^2 g^{-1} anatase; E, 20 m^2 g^{-1} rutile; F, reference alumina) [63].

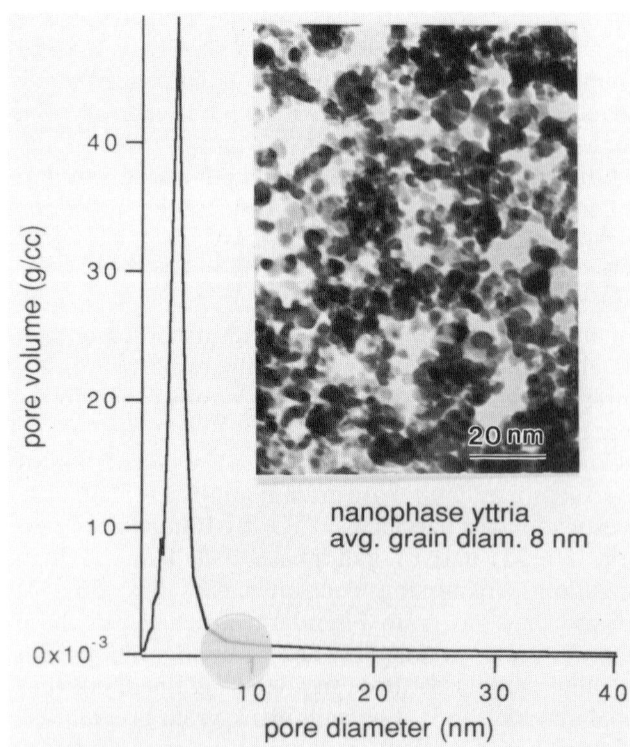

Fig. 7. Pore size distribution in nanophase Y_2O_3 (the powder is shown in the inset transmission electron micrograph) consolidated at room temperature under 136 MPa pressure to a density of about 40% of the theoretical value. The transparent sample is shown on the axis at about its average grain diameter, 8 nm (courtesy of J. C. Parker, Nanophase Technologies Corporation).

the wavelengths of visible or other radiation, coupled with the possibilities for efficient doping of these materials should lead to a variety of interesting optical properties and related applications.

The predominant mechanical property change resulting from reducing the grain sizes of nanophase metals is the significant increase in their strength [61, 65, 66]. Figure 8 shows the microhardness and yield-stress results for several samples of nanophase copper compared with results for coarser-grained samples. The smallest grain size (6 nm) sample is seen to exhibit about a 500% increase in microhardness over an annealed coarser-grained (*ca.* 50 μm) sample. Nanophase palladium samples with 5–10 nm grain sizes have also been observed to exhibit up to about a 500% increase in hardness over coarser-grained (*ca.* 100 μm) samples [65], with concomitant increases in their yield stress as well. The common strengthening behavior found in nanophase copper and palladium indicates that this response is generic to nanophase metals, at least those with an f.c.c. structure. The likelihood that such mechanical behavior is more broadly generic to nanophase metals is indicated by the observations that nanophase metals and alloys produced via mechanical attrition also exhibit significantly enhanced strength. For example, Koch and coworkers [67, 68] have found hardness increases by a factor of about 4–5 in nanophase iron, but only by a factor of about 1.2 in nanophase Nb_3Sn, when the grain size drops from 100 nm to 6 nm.

The increased strength observed in ultrafine-grained nanophase metals, although apparently analogous to conventional Hall–Petch strengthening observed with decreasing grain size in coarser-grained metals, must result from fundamentally different mechanisms. It can be expected that as nanophase grain sizes decrease and fall below the critical length scale for a given mechanism to operate, the associated property will be significantly changed. The grain sizes in the nanophase metals considered here are smaller than the necessary critical bowing lengths for Frank–Read dislocation sources to operate at the stresses involved and smaller also than the normal spacings between dislocations in a pile-up. It is therefore clear that an adequate description of the mechanisms responsible for the increased strength observed in nanophase metals will need to accommodate to the ultrafine grain-size scale in these materials. As this scale is reduced, and conventional dislocation generation and migration become increasingly difficult, it is apparent that the energetic hierarchy of microscopic deformation mechanisms will become successively accessed. Thus, easier mechanisms (such as dislocation generation from Frank–Read sources) will become frozen out at sufficiently small grain sizes and more energetically costly mechanisms will become necessary to effect deformation. Hence, grain confinement appears to be the dominant cause for the increased strength of nanophase metals. An additional contribution to the observed strengthening may be the elastic strain accommodation [61] resulting from cluster consolidation.

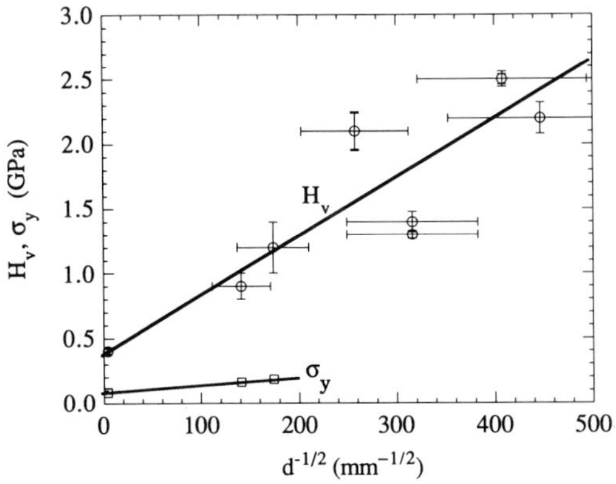

Fig. 8. Vickers microhardness of as-consolidated nanophase copper samples, ranging in average grain size d from 6 to 50 nm, compared with similar measurements from an annealed conventional 50 μm grain size copper sample. The 0.2% offset yield stress of two nanophase copper samples and a coarse-grained sample are also shown as a function of $d^{-1/2}$ [61].

Also in the area of mechanical behavior, it has been observed that nanophase ceramics are easily formed [23, 69–71] and nanoindenter measurements on nanophase TiO_2 [72] and ZnO [73] have recently demonstrated that there is a dramatic increase in strain rate sensitivity with decreasing grain size. Since this strong grain-size dependence is found for sets of samples in which the porosity changes very little, it appears that this increased tendency toward ductility is an intrinsic grain-size dependent property of these ultrafine-grained ceramics. The strain rate sensitivity values at the smallest grain sizes yet investigated (12 nm in nanophase TiO_2 and 7 nm in ZnO) thus indicate ductile behavior of these nanophase ceramics, as well as a significant potential for increased ductility at even smaller grain sizes and elevated temperatures. The enhanced strain rate sensitivity at room temperature found in the nanophase ceramics TiO_2 and ZnO appears to result from increased grain boundary sliding in this material, aided by the presence of porosity, ultrafine grain size, and probably rapid short-range diffusion as well, although this mechanism is yet to be demonstrated. This behavior appears therefore to be dominated by the presence of the numerous interfaces in these materials and the very short diffusion distances involved in effecting the necessary atomic healing of incipient cracks for grain boundary sliding to progress at the strain rates utilized without fracturing the sample. Extrapolating from this apparently generic behavior, one can expect that grain boundary sliding mechanisms, accompanied by short-range diffusion assisted healing events, would be expected increasingly to dominate the deformation of nanophase materials at sufficiently small grain sizes. Enhanced forming and even superplasticity in a wide range of nanophase materials, including intermetallic compounds, ceramics, and semiconductors might become a reality. Consequently, increased opportunities for high deformation or superplastic near-net-shape forming of a very wide range of even conventionally rather brittle and difficult to form materials could result.

4. Conclusions and prospects

Considerable opportunities now exist for synthesizing nanophase materials with a variety of new architectures at nanometer length scales from atomic or molecular precursors via the assembly of atom clusters and by the many other techniques now becoming available. However, the real future of nanophase materials, and other nanostructures as well, will depend on our ability to change significantly for the better the properties of materials by structuring them artificially on these nanometer length scales and on developing economic and environmentally responsible methods for producing these materials in commercially viable quantities. Based on the limited knowledge that has already been accumulated, the future appears to hold great promise for technological applications of nanophase materials, but much work remains to be done.

The examples presented in Section 3 and those that can be found in these Proceedings have demonstrated that, even in our crude initial attempts to change and control the properties of materials via nanostructuring, it is already possible to alter dramatically a variety of technologically important properties of materials in this manner. Since cluster "building blocks" can already be assembled with some modicum of control over the porosity in the nanophase compact and the composition and defect structure of the material, significant positive effects on both the chemical reactivity of a self-supported nanophase "catalyst" and the optical transparency of a nanophase ceramic "window" can result. It is also apparent that the strength of normally very ductile pure metals can be significantly increased simply by reducing their grain sizes into the range where dislocation generation and migration becomes confined and difficult. In addition, normally very brittle oxide ceramics can be rendered more ductile and formable by reducing their grain sizes into the range where grain boundary sliding becomes facilitated by the high number density of internal interfaces and the rapid atomic diffusion therein. Such examples suggest that even more exciting opportunities may be available as we learn to assemble more sophisticated multifunctional composite nanostructured materials in the future for prescribed applications.

Acknowledgments

This work was supported by the US Department of Energy, BES-Materials Sciences, under Contract W-31-109-Eng-38. The submitted manuscript has been authored by a contractor of the US Government under Contract No. W-31-109-ENG-38. Accordingly, the US Government retains a non-exclusive royalty-free license to publish or reproduce the published form of this contribution, or allow others to do so, for US Government purposes.

References

1 B. H. Kear, L. E. Cross, J. E. Keem, R. W. Siegel, F. Spaepen, K. C. Taylor, E. L. Thomas and K.-N. Tu, *Research Opportunities for Materials with Ultrafine Microstructures*, National Academy, Washington, DC, 1989, Vol. NMAB-454.

2 R. P. Andres, R. S. Averback, W. L. Brown, L. E. Brus, W. A. Goddard III, A. Kaldor, S. G. Louie, M. Moskovits, P. S. Peercy, S. J. Riley, R. W. Siegel, F. Spaepen and Y. Wang, *J. Mater. Res., 4* (1989) 704.

3 H. Gleiter, in N. Hansen, A. Horsewell, T. Leffers and H. Lilholt (eds.), *Deformation of Polycrystals: Mechanisms and Microstructures*, Risø National Laboratory, Roskilde, 1981, p. 15.

4 R. W. Siegel, *Nanostruct. Mater., 3* (1993) 1.

5 H. Gleiter, *Proc. Mater. Sci., 33* (1989) 223.

6 R. W. Siegel, in R. W. Cahn (ed.), *Materials Science and Technology*, Vol. 15, VCH, Weinheim, 1991, p. 583.

7 R. W. Siegel, in M. Nastasi, D. M. Parkin and H. Gleiter (eds.), *Mechanical Properties and Deformation Behavior of Materials Having Ultra-Fine Microstructures*, Kluwer, Dordrecht, 1993, p. 509.

8 C. G. Granqvist and R. A. Buhrman, *J. Appl. Phys., 47* (1976) 2200.

9 L. L. Hench and D. R. Ulrich, *Science of Chemical Processing*, Wiley, New York, 1986.

10 L. C. Klein, *Annu. Rev. Mater. Sci., 15* (1985) 227.

11 M. L. Steigerwald and L. E. Brus, *Annu. Rev. Mater. Sci., 19* (1991) 471.

12 I. A. Aksay, G. L. McVay and D. R. Ulrich (eds.), *Processing Science of Advanced Ceramics, Mater. Res. Soc. Symp. Proc., 155* (1989).

13 P. C. Rieke, P. D. Calvert and M. Alper (eds.), *Materials Synthesis Utilizing Biological Processes, Mater. Res. Soc. Symp. Proc., 174* (1990).

14 R. S. Averback, D. L. Nelson and J. Bernholc (eds.), *Clusters and Cluster-Assembled Materials, Mater. Res. Soc. Symp. Proc., 206* (1991).

15 G. M. Whitesides, J. P. Mathias and C. T. Seto, *Science, 254* (1991) 1312.

16 A. H. Heuer, D. J. Fink, V. J. Laraia, J. L. Arias, P. D. Calvert, K. Kendall, G. L. Messing, J. Blackwell, P. C. Rieke, D. H. Thompson, A. P. Wheeler, A. Veis and A. I. Caplan,

Science, 255 (1992) 1098.
17 A. H. Pfund, *Rev. Sci. Instrum., 1* (1930) 397.
18 K. Kimoto, Y. Kamiya, M. Nonoyama and R. Uyeda, *Jpn. J. Appl. Phys., 2* (1963) 702.
19 A. R. Thölén, *Acta Metall., 27* (1979) 1765.
20 R. Uyeda, *Prog. Mater. Sci., 35* (1991) 1.
21 R. Birringer, U. Herr and H. Gleiter, *Suppl. Trans. Jpn. Inst. Met., 27* (1986) 43.
22 R. W. Siegel and H. Hahn, in M. Yussouff (ed.), *Current Trends in the Physics of Materials*, World Scientific, Singapore, 1987, p. 403.
23 R. W. Siegel, S. Ramasamy, H. Hahn, Z. Li, T. Lu and R. Gronsky, *J. Mater. Res., 3* (1988) 1367.
24 J. A. Eastman, Y. X. Liao, A. Narayanasamy and R. W. Siegel, *Mater. Res. Soc. Symp. Proc., 155* (1989) 255.
25 R. W. Siegel and J. A. Eastman, *Mater. Res. Soc. Symp. Proc., 132* (1989) 3.
26 K. Sattler, J. Mühlbach and E. Recknagel, *Phys. Rev. Lett., 45* (1980) 821.
27 K. Sattler, *Ultramicroscopy, 20* (1986) 55.
28 K. Sattler, *Z. Phys. D, 3* (1986) 223.
29 H. Oya, T. Ichihashi and N. Wada, *Jpn. J. Appl. Phys., 21* (1982) 554.
30 S. Yatsuya, K. Yamauchi, T. Kamakura, A. Yanagada, H. Wakaiyama and K. Mihama, *Surf. Sci., 156* (1985) 1011.
31 S. Yatsuya, T. Kamakura, K. Yamauchi and K. Mihama, *Jpn. J. Appl. Phys., 25* (2) (1986) L42.
32 H. Hahn and R. S. Averback, *J. Appl. Phys., 67* (1990) 1113.
33 G. M. Chow, R. L. Holtz, A. Pattnaik, A. S. Edelstein, T. E. Schlesinger and R. C. Cammerata, *Appl. Phys. Lett., 56* (1990) 1853.
34 S. Iwama, E. Shichi and T. Sahashi, *Jpn. J. Appl. Phys., 12* (1973) 1531.
35 S. Iwama, K. Hayakawa and T. Arizumi, *J. Cryst. Growth, 56* (1982) 265; 66 (1984) 189.
36 S. Iwama and K. Hayakawa, *Surf. Sci., 156* (1985) 85.
37 B. Günther and A. Kumpmann, *Nanostruct. Mater., 1* (1992) 27.
38 A. Matsunawa and S. Katayama, in C. Albright (ed.), *Laser Welding, Machining and Materials Processing, Proc. ICALEO '85*, IFS, 1985, p. 205.
39 K. Baba, N. Shohata and M. Yonezawa, *Appl. Phys. Lett., 54* (1989) 2309.
40 M. Uda, *Nanostruct. Mater., 1* (1992) 101.
41 E. Hellstern, H. J. Fecht, Z. Fu and W. L. Johnson, *J. Appl. Phys., 65* (1989) 305.
42 C. C. Koch, J. S. C. Jang and S. S. Gross, *J. Mater. Res., 4* (1989) 557.
43 H. J. Fecht, E. Hellstern, Z. Fu and W. L. Johnson, *Adv. Powder Metall., 1–2* (1989) 111.
44 S. K. Ganapathi and D. A. Rigney, *Scripta Metall. Mater., 24* (1990) 1675.
45 J. Eckert, J. C. Holzer, C. E. Krill III and W. L. Johnson, *J. Mater. Res., 7* (1992) 1751.
J. Eckert, J. C. Holzer, C. E. Krill III and W. L. Johnson, *Mater. Res. Soc. Symp. Proc., 238* (1992) 745.
46 M. J. Luton, C. S. Janath, M. M. Disko, S. Matras and J. Vallone, *Mater. Res. Soc. Symp. Proc., 132* (1989) 79.
47 R. W. Siegel, *Annu. Rev. Mater. Sci., 21* (1991) 559.
48 R. Birringer and H. Gleiter, in R. W. Cahn (ed.), *Encyclopedia of Materials Science and Engineering*, Suppl. Vol. 1, Pergamon, Oxford, 1988, p. 339.
49 R. W. Siegel, in D. Wolf and S. Yip (eds.), *Materials Interfaces: Atomic-Level Structure and Properties*, Chapman and Hall, London, 1992, p. 431.
50 C. A. Melendres, A. Narayanasamy, V. A. Maroni and R. W. Siegel, *J. Mater. Res., 4* (1989) 1246.
51 J. C. Parker and R. W. Siegel, *J. Mater. Res., 5* (1990) 1246.
52 J. C. Parker and R. W. Siegel, *Appl. Phys. Lett., 57* (1990) 943.
53 G. J. Thomas, R. W. Siegel and J. A. Eastman, *Mater. Res. Soc. Symp. Proc., 153* (1989) 13.
54 G. J. Thomas, R. W. Siegel and J. A. Eastman, *Scripta Metall. Mater., 24* (1990) 201.
55 R. W. Siegel and G. J. Thomas, *Ultramicroscopy, 40* (1992) 376.
56 J. E. Epperson, R. W. Siegel, J. W. White, T. E. Klippert, A. Narayanasamy, J. A. Eastman and F. Trouw, *Mater. Res. Soc. Symp. Proc., 132* (1989) 15.
57 E. Jorra, H. Franz, J. Peisl, G. Wallner, W. Petry, R. Birringer, H. Gleiter and T. Haubold, *Philos. Mag. B, 60* (1989) 159.
58 J. E. Epperson, R. W. Siegel, J. W. White, J. A. Eastman, Y. X. Liao and A. Narayanasamy, *Mater. Res. Soc. Symp. Proc., 166* (1990) 87.
59 S. R. Phillpot, D. Wolf and S. Yip, *MRS Bull., XV* (10) (1990) 38.
60 D. Wolf and J. F. Lutsko, *Phys. Rev. Lett., 60* (1988) 1170.
61 G. W. Nieman, J. R. Weertman and R. W. Siegel, *J. Mater. Res., 6* (1991) 1012.
62 R. Valiev, in M. Nastasi, D. M. Parkin and H. Gleiter (eds.), *Mechanical Properties and Deformation Behavior of Materials Having Ultra-Fine Microstructures*, Kluwer, Dordrecht, 1993, p. 303.
63 D. D. Beck and R. W. Siegel, *J. Mater. Res., 7* (1992) 2840.
64 G. Skandan, H. Hahn and J. C. Parker, *Scripta Metall. Mater., 25* (1991) 2389.
65 G. W. Nieman, J. R. Weertman and R. W. Siegel, *Scripta Metall., 23* (1989) 2013.
66 G. W. Nieman, J. R. Weertman and R. W. Siegel, *Scripta Metall. Mater., 24* (1990) 145.
67 J. S. C. Jang and C. C. Koch, *Scripta Metall. Mater., 24* (1990) 1599.
68 C. C. Koch and Y. S. Cho, *Nanostruct. Mater., 1* (1992) 207.
69 J. Karch, R. Birringer and H. Gleiter, *Nature, 330* (1987) 556.
70 J. Karch and R. Birringer, *Ceram. Int., 16* (1990) 291.
71 H. Hahn, J. Logas, H. J. Höfler, P. Kurath and R. S. Averback, *Mater. Res. Soc. Symp. Proc., 196* (1990) 71.
72 M. J. Mayo, R. W. Siegel, A. Narayanasamy and W. D. Nix, *J. Mater. Res., 5* (1990) 1073.
73 M. J. Mayo, R. W. Siegel, Y. X. Liao and W. D. Nix, *J. Mater. Res., 7* (1992) 973.

Materials Science and Engineering, A168 (1993) 199–203

Elaboration and thermal stability of (alumina, aluminosilicate/iron, cobalt, nickel) magnetic nanocomposites prepared through a sol-gel route

V. Vendange and Ph. Colomban*
ONERA, BP 72, 92322 Châtillon (France)

Abstract

Optically clear monolithic (OCM) gels of alumina and aluminosilicates can be prepared by hydrolysis–polycondensation of alkoxides. Subsequent heating induces transformation firstly into OCM mesoporous glasses and then into dense translucent or optically clear ceramics, the densification occurring at low temperature. Immersion of alumina or aluminosilicate gels and glasses in aqueous solutions containing transition metal (cobalt, iron, nickel) ions leads to diffusion of these ions into the water within the pores of the monoliths. After heat treatment under hydrogen, metallic nanoprecipitates are obtained. The size of the precipitates, as studied by transmission electron microscopy, ranges from 5 to 70 nm depending on the host matrix. The thermal stability in air of these materials was studied by differential thermal and analysis X-ray diffraction. The oxidation of porous composites occurs above 600 °C in air. Magnetic parameters such as coercitive field, magnetization, etc. were also measured.

1. Introduction

The preparation of compounds using sol-gel methods not only improves their quality (purity, homogeneity, etc.) but also allows the synthesis of new materials. New phases can thus be obtained by inorganic polymerization through the hydrolysis–polycondensation of alkoxides [1]. This sol-gel route leads to powders, films or optically clear monoliths (OCM) [1, 2]. Ceramic–ceramic composites have already been prepared [2], even on the nanometre scale [3, 4]. Metal–ceramic microcomposites have been also synthesized. Roy and Roy studied the influence of gelation parameters and subsequent thermal treatment on the fine structure of ceramic–metal materials (using Al_2O_3, SiO_2, ZrO_2 with copper, platinum, tin and nickel) [5]. Petrullat *et al.* reported the elaboration of nickel particles in a silica matrix. They studied the possibility of densifying the powder by cold or hot pressing [6]. This problem was also approached by Breval *et al.* with nickel–alumina composites in order to evaluate their mechanical properties [7]. Glass–metal nanocomposites involving iron, nickel and copper in a silica matrix have been prepared as films on glass substrates by sol-gel routes and their electrical and optical properties have been investigated by Chatterjee and Chakravorty [8, 9]. To the best of our knowledge, magnetic ceramic nanocomposites have only been synthesized by Shull *et al.* [10]. In this paper, we report results concerning the elaboration of magnetic ceramic nanocomposites using an infiltration process followed by thermal treatment at medium temperature. Alumina and mullite are used as matrices and cobalt, nickel and iron as metallic components.

*To whom correspondence should be addressed.

2. Experimental techniques

For transmission electron microscopy (TEM) investigations, samples were crushed in an agate mortar with acetone, and particles in suspension were deposited onto a carbon-coated grid. These were observed using a JEOL 200 CX. Differential thermal analysis (DTA) was performed in air between 300 °C and 1400 °C using a Dupont instrument. Precipitate structures were analysed by X-ray diffraction using a Philips diffractometer (Cu Kα radiation). Magnetic parameters were measured with a hysteresismeter purchased from SIIS (Buc, France, CNRS patent). The metal content (in weight) was measured by plasma emission after melting samples with sodium peroxide and dissolving in 10% HCl. The metal content as determined includes metallic phase as well as its oxides.

3. Results

3.1. Elaboration of nanocomposites

The method used to make ceramic–metal composites is a two-stage process. The first stage is the

0921-5093/93/$6.00

preparation of a porous host matrix by a sol-gel route. The second stage involves the introduction of metallic precipitates into the pores.

3.1.1. Porous host framework

Alumina OCM gels (a few cubic centimetres in volume) were prepared by slow hydrolysis of (i) pure s-aluminium butoxide (Alfra-Ventron) by contact with ambient moisture through a small leak at the bottle stopper (duration more than 6 months, batch volume 20 cm^3, bottle volume 1000 cm^3), (ii) s-aluminium butoxide diluted in hexane or acetone (alkoxide-to-solvent volume ratio 1/1, batch volume 40 cm^3, bottle volume 70 cm^3), the atmospheric moisture reaching the surface of the solution through a plastic cover [11]. Alkoxide–solvent mixtures were prepared in a water-free glove box using water-free solvents. The time needed to obtain dense and sonorous monoliths is between 2 and 4 weeks (in acetone) and one year (in hexane). The duration increases with the volume-to-surface ratio of the solution. However, batches prepared simultaneously in the same way gel after various lengths of time: the scattering of values reaches 30%–50%.

Mullite ($3Al_2O_3$–$2SiO_2$) OCM gels were prepared by the same methods using s-aluminium butoxide and silicon methoxide (Fluka). With hexane (1/1) as solvent, 6 months are needed to obtain OCM gels. Other samples were prepared with acetone (alkoxide-to-solvent ratio 1/1) and an additive. The additive was a non-ionic emulsifier (Tween 80) for samples called batch 20;2 and a non-ionic surfactant (Arlacel 80) for samples called batch 20;4. These products were purchased from Imperial Chemical Industries (Kortenberg, Belgium) and were added with an additive-to-solvent volume ratio of 2/50. Gelation is obtained after 3 weeks. The monoliths were then dried at 70 °C for 10 h.

The thermal behaviour and structures of these host matrices have been studied previously [11, 12]. With different conditions of hydrolysis and composition, we obtain mesoporous (2–10 nm) (most samples) or microporous (less than 2 nm) (mullite with additive and acetone for example) materials. The specific surface area and the sample volume decrease regularly on heating up to the nucleation temperature, whereas the mean pore size increases. Xerogels treated at 600 °C are called "glasses".

3.1.2. Metallic ion infiltration and precipitate formation

Introduction of the metallic phase is performed by immersion of the gel or glass monoliths in aqueous solutions of $Co(NO_3)_2 \cdot 6H_2O$, $Ni(NO_3)_2 \cdot 6H_2O$ or $Fe(NO_3)_3 \cdot 9H_2O$. Diffusion of ions occurs into the water within the pores of the monoliths [13]. After heat treatment under hydrogen at 600 °C, metallic nanoprecipitates are obtained.

The metal content increases with the concentration of the solution used to infiltrate the monoliths. The best result is obtained with xerogels soaked in the gel state (without heat treatment) [14]. For example 33% in weight of cobalt can be introduced into a host matrix of alumina prepared without solvent and heat treatment. Alumina prepared in acetone and soaked in the glass state in an iron nitrate solution contains up to 25 wt.% iron.

3.2. Topology and microstructure

Typical electron micrographs obtained of nanocomposites of various matrices and metals are shown in Fig. 1. All the matrices are amorphous after heat treatment at 800 °C, which favours observation of the metallic precipitates. The metallic precipitate structure, studied by X-ray diffraction, is cubic (f.c.c. for nickel and cobalt, c.c. for iron). Crystallization of alumina and mullite is observed at 1250 °C and 1000 °C respectively. Figure 1(g) shows cobalt precipitates inside a crystallized mullite matrix. Nanocomposites remain porous up to their sintering temperature (roughly equal to the crystallization temperature), as observed for pure matrices [12]. The size of the metallic precipitates depends on the initial pore structure of the host matrix.

In Fig. 1 (a), (b), (c) crystallized cobalt and nickel precipitates may be seen to be enclosed in a porous matrix of mullite prepared in acetone and an additive. Spherical precipitates are scattered in an amorphous matrix. The maximum volume ratio is about 10%, indicating that the percolation threshold is not reached. Micrographs 1(a) and 1(b), taken at the same magnification, show that the precipitate size can vary within the same sample (batch 20;2). With heat treatment at 800 °C under hydrogen, the smaller precipitate size lies between 40 and 70 nm and the larger size is 300 nm. The diameters of nickel precipitates observed in a similar matrix (batch 20;4) range between 30 and 60 nm at 600 °C (Fig. 1(c)). The broad distribution of precipitate sizes is also obvious in the dark field micrograph (d), which shows cobalt precipitates enclosed in amorphous mullite prepared in hexane. Indeed, the size of these precipitates varies from 5 to 50 nm. Precipitate scattering, which depends on the porous structure of the matrix before infiltration, is less homogeneous in this sample. The origin of this topology can be found in diffusion theory, which indicates the formation of islands near the diffusion front [15].

In the dark field micrograph (Fig. 1(e)), some cobalt precipitates isolated from the mullite matrix prepared in hexane can be seen which have an octahedral shape and facets. Nickel particles with particular shapes such

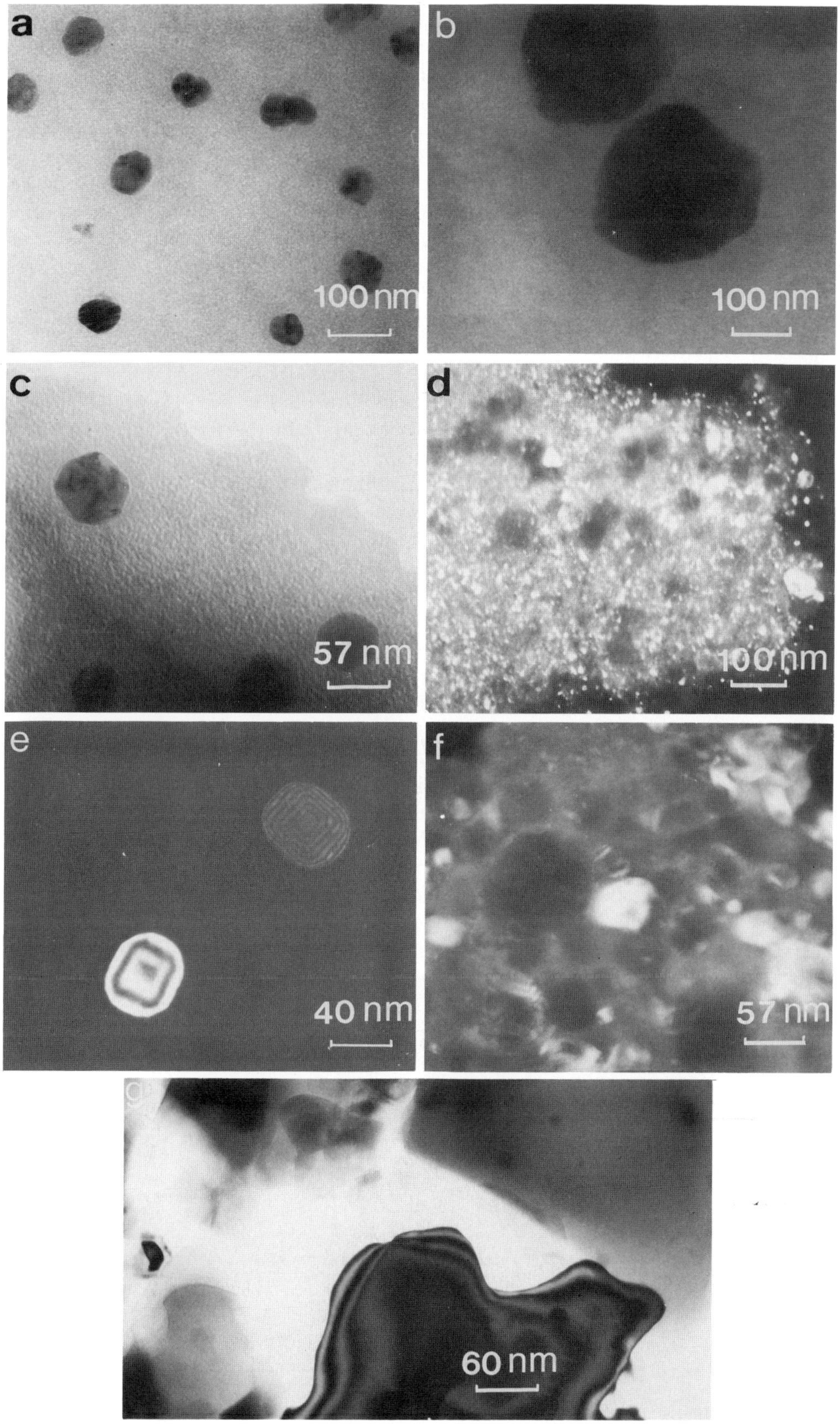

Fig. 1. TEM bright field images for (a), (b) cobalt–mullite (batch 20;2, infiltration at the gel state), and (c) nickel-mullite glass (batch 20;4, infiltration at the mesoporous glass state); composites obtained after H_2 reduction at 800 °C and 600 °C respectively. Dark field images of cobalt precipitates (d) enclosed in or (e) isolated from mullite prepared in hexane and (f) of iron–alumina composite prepared in hexane. The temperature reduction was 800 °C for (d), (e), (f). Comparison with a crystallized matrix is given in (g) for a cobalt–mullite composite prepared in hexane and annealed at 1300 °C under H_2.

as pentagonal [16] and icosahedral [17] have been previously observed by Brieu and Gillet. The mean pore size of these particles prepared by reduction of a nickel salt in organic liquid media is about 30 nm. Spherical ultrafine (10 nm) metallic powders of several transition metals (iron, cobalt, copper, silver) were also prepared by Girardin and Maurer. They used a technique based on evaporation by a high pressure plasma followed by cryogenic condensation [18].

Only the iron precipitates seem to be poorly dispersed and they have no particular geometry. This point can be related to the particular behaviour of the iron salt with respect to the porous host lattice: an insertion reaction inside the Al–O–Si framework is assumed [13]. Figure 1(f) illustrates the precipitate distribution encountered with an alumina matrix prepared in hexane and annealed at 800 °C.

3.3. *Thermal stability*

In order to understand the oxidation process of these nanocomposites, DTA has been performed in air on mesoporous nanocomposites. Figure 2 shows curves for nanocomposites containing nickel and iron in various mullite matrices. Oxides formed as the temperature increased were identified from both their colour and radiocrystallographic analysis.

Figures 2(a) and 3 show the evolution of nickel precipitates in a mullite prepared in hexane and annealed under hydrogen at 1000 °C. Figure 3(a) is the X-ray diffractogram of the starting sample which is dark and magnetic. Strong peaks of nickel (111, 200, 220) and peaks of the quasi-amorphous mullite which has a lacunar spinel structure ("γ-Al_2O_3") can be seen. At 450 °C in air, the oxide NiO is formed and its characteristic diffraction lines (101, 012, 104) appear on the diffractogram (b). At this point, a first broad exothermic peak is observed on the DTA trace (Fig. 2(a)). However, the intensity of lines from nickel is still significant in X-ray diffraction, and the sample retains some magnetic properties at this temperature. These properties are weaker at 600 °C where nickel peaks are no longer observed, and the sample remains darkly coloured (Fig. 3(c)). At 800 °C, the second exothermic peak is over, the sample becomes green–yellow and non-magnetic. NiO peaks are strong (Fig. 3(d)). At higher temperatures, there is reaction between the mullite type matrix and NiO, inducing a deviation of the DTA signal. In Fig. 3(e) strong $NiAl_2O_4$ peaks are visible, the formation of which gives the sample a vivid blue colour.

Oxidation of iron precipitates in a mullite-glass (treated at 600 °C before infiltration) prepared in acetone and an additive (batch 20;4) was studied and its DTA curve is shown in Fig. 2(b). At 650 °C in air, a first oxidation occurs and the dark sample becomes brown. Conservation of magnetic properties confirms the formation of magnetite (Fe_3O_4). At 800 °C, the rust-coloured sample is no longer magnetic. Fe_2O_3 (maghemite, ASTM 16-653) is formed. Iron is incorporated into the mullite phase when exothermic nucleation occurs (about 1000 °C).

Progressive formation of Co_3O_4, as the magnetism is lost was observed above 600 °C for ceramics containing cobalt precipitates. There is a reaction between the matrix and the oxide inducing a brilliant blue colour.

3.4. *Magnetic properties*

Figure 4 illustrates various kinds of curves encountered with alumina containing cobalt or iron precipitates. The hysteresis of the cobalt-inserted compound is wider than that of the corresponding iron-containing composite, in accordance with the usual behaviour. As the samples are partially oxidized, the magnetization at saturation is rather weak (50% and 20%), compared

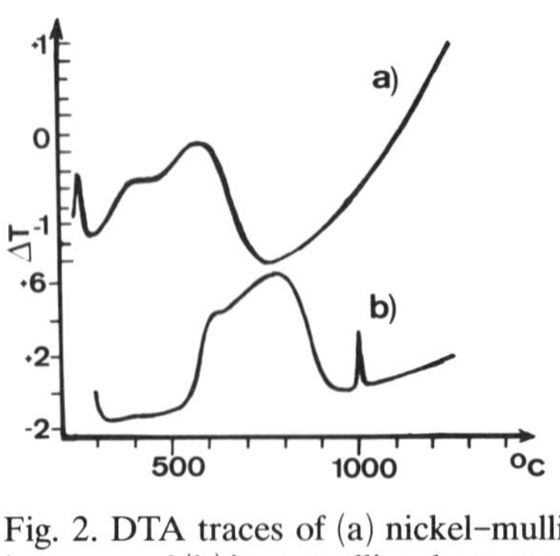

Fig. 2. DTA traces of (a) nickel–mullite composite prepared in hexane, and (b) iron–mullite glass composite prepared in acetone with addition of a non-ionic surfactant (batch 20;4).

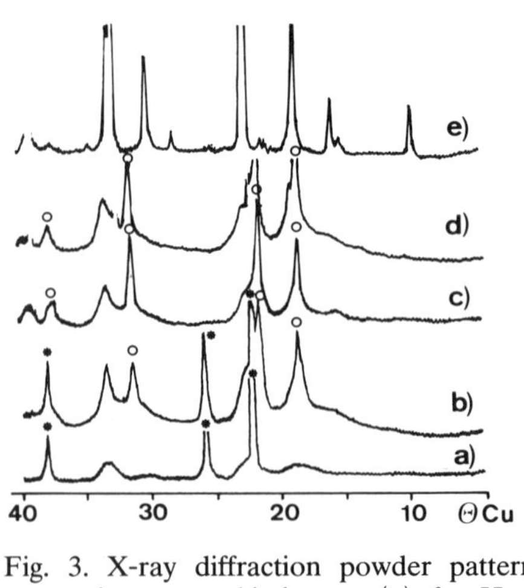

Fig. 3. X-ray diffraction powder patterns of nickel–mullite composite prepared in hexane: (a) after H_2 reduction at 1100 °C, and after heat treatment in air at (b) 450 °C, (c) 600 °C, (d) 800 °C, and (e) 1300 °C. * Ni, ○, NiO.

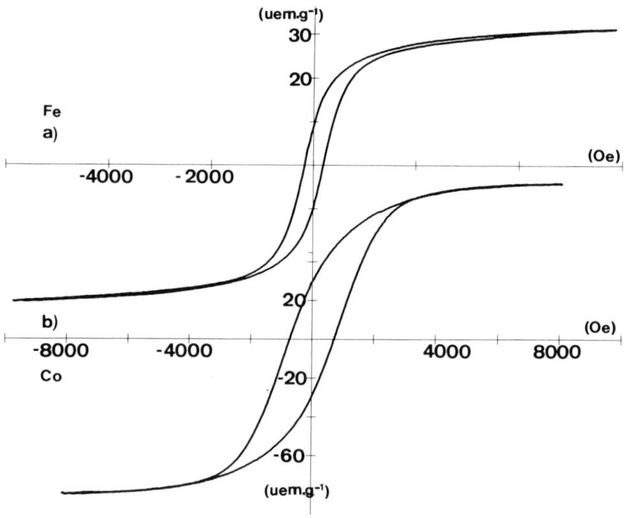

Fig. 4. Hysteresis of (a) iron–alumina glass composite prepared with acetone, and (b) cobalt–alumina glass composite. The temperature reduction was 600 °C.

with theoretical values for cobalt and iron (160 and 220 emu g^{-1} respectively).

Susceptibility measurements, as a function of temperature, were also taken on some samples, and showed that the susceptibility of alumina containing nickel remains constant up to its Curie temperature (354 °C), and then becomes zero in its paramagnetic domain.

4. Conclusion

The infiltration of mesoporous inorganic membranes from salt solutions allows the synthesis of magnetic nanocomposites. Such porous materials which also exhibit both Lewis centres (owing to the protonic traces) and transition metals may have catalytic properties. In the porous state, oxidation occurs in air above 500 °C (nickel) or 600 °C (iron, cobalt). The oxidation of dense nanocomposites by oxygen of the host matrix will be reported in a later paper.

References

1 L. C. Klein, *Sol-Gel Technologies* Noyes, Park Ridge, NJ, 1988.
2 Ph. Colomban, *Ceram. Int., 15* (1989) 23.
3 Ph. Colomban and L. Mazerolles, *J. Mater. Sci., 26* (1991) 3501.
4 R. Roy, S. Komarneni and D. M. Roy, *Mater. Res. Soc. Symp. Proc., 32* (1984) 347.
5 R. A. Roy and R. Roy, *Mater. Res. Bull., 19* (1984) 169.
6 J. Petrullat, S. Ray, U. Shubert, G. Guldner, Ch. Egger and B. Breitscheidel, *J. Non-Cryst. Solids, 147/148* (1992) 594.
7 E. Breval, Z. Deng, S. Chiou and C. G. Pantano, *J. Mater. Sci., 27* (1992) 1464.
8 A. Chatterjee and D. Chakravorty, *J. Phys. D, 22* (1989) 1386.
9 A. Chatterjee and D. Chakravorty, *J. Phys. D, 23* (1990) 1097.
10 R. D. Schull, J. J. Ritter, A. J. Shapiro, L. J. Swartzendruber and L. H. Bennett, *J. Appl. Phys., 67* (1990) 4490.
11 Ph. Colomban, *J. Mater. Sci., 24* (1989) 3002, 3011.
12 Ph. Colomban and V. Vendange, *J. Non-Cryst. Solids, 147/148* (1992) 245.
13 V. Vendange and Ph. Colomban, in C. A. C. Sequeira and M. J. Hudson (eds.), Multifunctional Mesopous Inorganic Solids, *Nato ASI Series*, Kluwer, Dordrecht, 1993.
14 V. Vendange and Ph. Colomban, to be published.
15 J. F. Gouyet, in A. L. Laskar *et al. (eds.), Diffusion in Materials*, Kluwer, Dordrecht, 1990, p. 155.
16 M. Brieu and M. Gillet, Thin Solid Films, 100 (1983) 53.
17 M. Brieu and M. Gillet, *Thin Solid Films, 167* (1988) 149.
18 D. Girardin and M. Maurer, *Mater. Res. Bull., 25* (1990) 119.

Materials Science and Engineering, A168 (1993) 205–208

Preparation and properties of sol-gel derived nanostructured thin ceramic layers

D. Sporn, F. Raether and S. Merklein

Fraunhofer-Institut für Silicatforschung, Neunerplatz 2, 8700 Würzburg (Germany)

Abstract

Thin ceramic layers of two different compositions were deposited onto substrates from sols with controlled viscosities. First, mullite ($3Al_2O_3 \times 2SiO_2$) layers were deposited onto SiC platelets with resulting thicknesses of up to 1 μm. By thermal treatment at up to 900 °C, dense layers consisting of crystals of about 20 nm were obtained. The measured hardness and Young modulus of the mullite layers were higher than those of mullite ceramics with grain sizes in the micrometre range. Uncoated SiC platelets were oxidized three times faster than mullite-coated platelets. Second, thin PZT layers with composition near the morphotropic phase boundary ($PbZr_{0.53}Ti_{0.47}O_3$) were deposited from sols onto platinum-coated silicon wafers exhibiting at the crystallization temperature of less than 500 °C, dense packaging of crystals 30 nm in diameter. Increasing the applied temperature up to 700 °C led to crystal sizes of up to 70 nm. After this thermal annealing the layers exhibited a remanent polarization of 14 μC cm^{-2}, a coercive force of 7 kV mm^{-1}, $\varepsilon_r > 1000$, and a resistance greater than 10 MΩ.

1. Introduction

Nanometre-sized ceramic structures have been produced using the sol-gel technique [1]. Both structural and compositional nanometre structures can be obtained by starting with highly dispersed precursors (metal alkoxides, soluble salts, metal complexes) on the molecular level, fixing in a gel state, and the formation of nucleii during thermal treatment generating crystallite sizes in the range 3–25 nm [2]. Significant advantages of the sol-gel technique in comparison with methods involving high temperatures and expensive equipment (*e.g.* evaporation) are the low processing temperatures of the method and the use of conventional technologies for material preparation.

Owing to the high surface energy of the small primary crystallites in the resulting powders, the preparation of bulk materials caused difficulties. Therefore, the most favourable application of sol-gel derived materials can be seen in layered systems [3]. Typical layer thickness after thermal treatment of gel layers are in the range 50–1000 nm. The crystallites are dense and well packed. Such layers are suitable for investigating the properties of nanostructured materials.

The aim of this work was to demonstrate the processing and resulting properties of two typical sol-gel derived ceramic layers. Owing to the similar thermal expansion coefficient and low oxygen diffusion at high temperatures, thin mullite layers are promising candidates for the thermal protection of non-oxidic ceramic materials, especially as coating material for SiC components. Another point of interest is measurement of the mechanical properties of supported mullite coatings, allowing comparison with data for mullite material with grain sizes in the micrometre range. Ferroelectric thin films are suitable for a wide range of applications in electronics [4]. Their electric properties are strongly influenced by the resulting microstructure. The influence of thermal annealing temperature and variation of the layer composition on permittivity, remanent polarization and coercive strength will be presented.

2. Experimental details

2.1. *Preparation of supported thin mullite layers on SiC platelets*

As supporting material SiC platelets (American Matrix, Inc.) were used. The mullite sol used for layer preparation was prepared in the following way: 0.09 mol $Al(OBu^s)_3$ were dissolved in 760 ml ethanol and chelated with 0.045 mol acetylacetone. After chelating, 0.03 mol tetraethoxysilane were added whilst stirring. The hydrolysis was carried out by adding 0.39 mol water containing 0.005 mol hydrogen chloride as catalyst. The resulting sol was refluxed for 3 h. After cooling down to room temperature, 1.2 g hydroxypropylcellulose was dissolved in the sol to improve the layer formation.

0921-5093/93/$6.00

The preparation of mullite coatings on SiC platelets was carried out by spray drying of platelet suspensions in mullite sols. The resulting coating thickness can be influenced by the solid content of the sol. The dried material was calcined at 500 °C in air and then suspended in a water–ethanol mixture containing 0.3 g polyacrylic acid per gram of calcined material in order to separate the coated platelets from pure coating material, *i.e.* sol droplets in which no platelet had been enclosed by sedimentation. After separation the mullite coated platelets were dried and heated up stepwise to 1000 °C in N_2 in order to crystallize and to densify the layers. The development of crystalline phases was investigated by X-ray diffraction (XRD) measurements. The layer morphology was examined by scanning electron microscopy (SEM).

2.2. Measurement of thermal and mechanical properties of mullite layers

Thermogravimetry was used to investigate the influence of mullite layers on the oxidation of SiC platelets at 1300 °C in air. The mechanical properties of the mullite coatings were measured by a microindentation method. A commercial device for microhardness testing (Fischer H 100) was mechanically coupled to an optical microscope to select the position of the indentation with the crosshair of the microscope. The lateral position of the indentation could be set with an accuracy of 5 μm. This was sufficient to keep it inside the area of the platelet which had a diameter of approximately 50 μm. The platelets were fixed on microscope slides by cellulose varnish. The varnish had to be predried before the platelets were sprinkled onto it in order to prevent dipping in of the platelets. The thickness of the varnish was 5 μm. The indenter was a Berkovich diamond. The indentation force was incremented stepwise from 0.4 mN to 5.3 mN in 22 steps and then decremented using the same procedure. For each step a time delay of 1 s was chosen and the indentation depth was measured. The indentation depth due to the highest force was 0.1 μm which is roughly 1/10 of the thickness of the mullite layer (see below).

The hardness H was calculated from the indentation depth d_{max} measured with the highest force F_{max}:

$$H = \frac{F_{max}}{26.75\ d_{max}^{2}}$$

The Young modulus E was calculated from the slope of the unloading curve by the method of Loubet *et al.* [5]. The elastic backbending of the material was determined by the ratio R of the indentation depth d_{min} measured with the lowest force in the unloading cycle and d_{max}:

$$R = \frac{d_{min}}{d_{max}}$$

2.3. Preparation of $PbZr_{0.53}Ti_{0.27}O_3$ (PZT) layers

The preparation of PZT layers was carried out according to ref. 6. The sol preparation started with a mixture of Zr- and Ti-*n*-propoxides in the desired ratio which was chelated with one equivalent of acetylacetone and diluted in 1,3-propanediol. Lead acetate trihydrate was dissolved in this solution by heating. Volatile reaction products were distilled off under normal pressure leading to a yellow and viscous precursor solution with a PZT content of more than 40 mass%. This corresponds to a molarity of 2 M. The viscosity of the sols was adjusted, allowing spin off coating of platinum-coated silicon wafers at 3000 rotations min^{-1}. The wet films were pyrolysed and densified quickly by putting them on a hot plate at 300 °C. With a heating rate of 1 K s^{-1} the temperature was increased to 450–470 °C in order to remove the organic groups. To obtain completely crystallized films, they were heated up to 600–800 °C with a heating rate of 100 K h^{-1} and kept for 2 h at the desired temperatures.

2.4. Characterization of PZT layers

The thicknesses of the prepared crystalline PZT layers were characterized using profilometer measurements and supplementary SEM. The crystalline phases were estimated by XRD and the crystal size by the method of Scherrer from XRD line broadening [7]. In order to investigate the dielectric and ferroelectric properties, circular gold electrodes (1 mm in diameter) were sputtered onto the layer surface to form capacitor-like structures. The electrical properties were measured at room temperature using standard equipment and methods: impedance bridge for permittivity and loss factor measurements and integrating series capacitor circuit for registration of the ferroelectric hysteresis loop with typically a 0.3 Hz cycle time.

3. Results

3.1. Properties of mullite layers

Using XRD measurements the crystallization behaviour of the mullitic coating material supported on SiC platelets was examined. It was found that in accordance with ref. 8, the crystallization of pure mullite phases starts at 900 °C. During the crystallization process the layers were densified. Typical morphology is shown in the high resolution SEM image of Fig. 1, indicating a

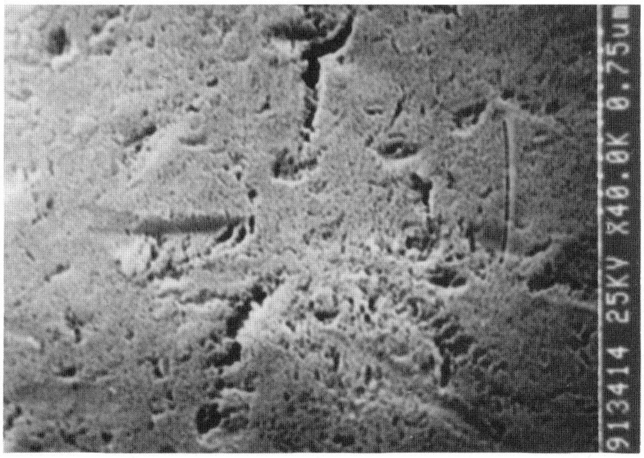

Fig. 1. SEM image of a mullite coating after thermal treatment at 900 °C, scale bar 0.75 μm.

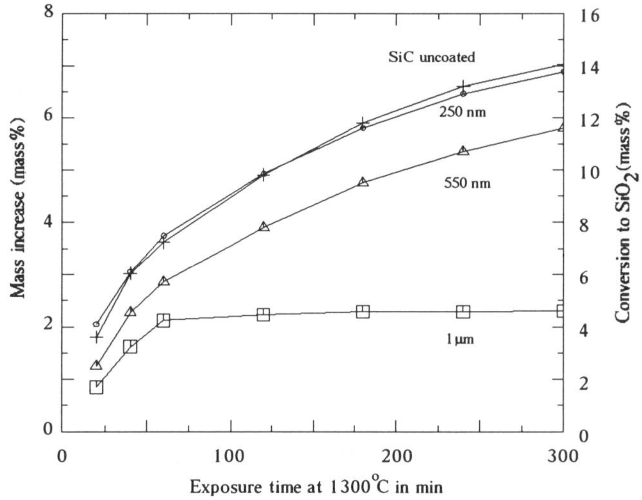

Fig. 2. Thermogravimetric of uncoated SiC platelets and SiC platelets coated with various thicknesses of mullite.

dense and crack-free layer of mullite with crystal sizes of about 20 nm.

To investigate the oxidation behaviour three samples of coated SiC platelets with different thicknesses of mullite coating were prepared. These and an uncoated sample were exposed to air for 5 h at 1300 °C. The expected mass increase caused by the decomposition reaction

$$SiC + 2O_2 \longrightarrow SiO_2 + CO_2$$

was monitored gravimetrically (Fig. 2).

In comparison with uncoated SiC platelets for which the decomposition is controlled by the diffusion of oxygen through the SiO_2 surface layer [8], the reaction rate can be significantly lowered with increasing coating thickness. Very thin mullite coatings are not so effective owing to insufficient coating of the edges of the platelets. Coatings with a thickness of approximately 1 μm acted as a very effective oxidation barrier. After an exposure time to air of 5 h, the uncoated platelets had an SiO_2 content which was higher by a factor of 3.1 than that of platelets coated with 1 μm of mullite. This effect is possibly caused by the lower oxygen diffusion coefficient in mullite compared with the amorphous SiO_2 layer if the coating is sufficiently thick and crack free.

The hardness of the mullite coating determined by the microindentation method from ten single measurements is $H = 15000 \pm 1000$ MPa and the Young modulus is $E = 370 \pm 30$ GPa. The reliability of the indentation method was examined by measuring bulky mullite with a grain size of approximately 1 μm using exactly the same procedure. This resulted in $H = 11500 \pm 300$ MPa and $E = 270 \pm 10$ GPa [9] which is in reasonable agreement with other measurements on mullite of similar grain size ($H = 12000$ MPa,

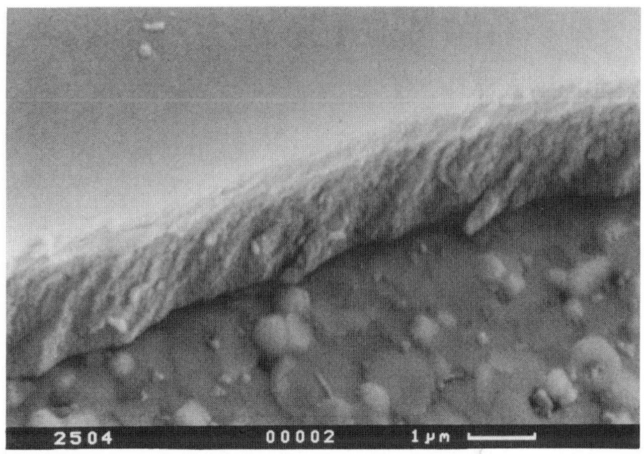

Fig. 3. SEM image of a crack-free PZT layer after thermal treatment at 600 °C, scale bar 1 μm.

$E = 250$ GPa [10]). The higher hardness and Young modulus of the nanosized mullite coating than of the bulky mullite demonstrated that a very dense layer could be produced on the SiC platelets. A high elastic backbonding of the coating of $R = 69 \pm 6\%$ leads to the expectation that the coating could enhance the mechanical performance of the platelets, in addition to providing chemical protection.

3.2. Properties of $PbZr_{0.53}Ti_{0.47}O_3$ layers

The described one-step process led to crack-free and smooth PZT layers with final thicknesses up to 1 μm (Fig. 3). The pyrolysed layers were still amorphous at a temperature of 450 °C. The crystallisation of PZT in the desired perovskite structure started typically at 500 °C, leading to crystal sizes of 30 nm. The crystallization could be promoted using compositions with an

TABLE 1. Typical dielectric and ferroelectric values for PZT thin films with composition $Pb_xZr_{0.53}Ti_{0.47}O_3$

x	Thickness (μm)	Temperature (°C)	ε_r	tan δ	P_R (μC cm^{-2})	E_C (kV mm^{-1})
1.0	0.5	600	425	0.0500	7	5
1.0	0.5	700	530	0.0800	11	7
1.1	0.67	600	950	0.0600	11	5
1.1	0.67	670	1150	0.0800	14	5

excess of 10 mol% PbO and led to crystal sizes of 50 nm at 500 °C. By increasing the applied temperature to 700 °C for 2 h, the crystals could be grown up to 55 nm (1.0 mol PbO in the composition) and 70 nm (1.1 mol PbO in the composition).

The typical electrical properties, which were found to be dependent on the lead content in the layer and the resulting annealing temperatures, are given in Table 1.

The measured data, which are in accordance with the reviewed data in ref. 11, show that the properties are influenced in a sensitive manner by the annealing temperature and composition. Therefore, the influence of the crystallite sizes in the PZT layers is evident. The tailoring of specific properties for different applications can be performed by variation of the composition and the annealing temperature.

4. Conclusions

Sol-gel preparations are suitable for the preparation of layered nanostructured materials. The mullite coated SiC platelets showed improved thermal protective as well as mechanical properties. They are promising candidates for technical applications. The preparation of PZT layers by sol-gel methods lead to dense and well crystallized material with ferroelectric properties, allowing the development of devices in the field of sensors and non-volatile memories.

References

1 H. Gleiter, *Progr. Mater. Sci., 33* (1989) 223.
2 S. L. Nail, J. L. White and S. L. Hem, *J. Pharm. Sci., 65* (1976) 1192.
3 C. J. Brinker and G. W. Scherer, *Sol-Gel Science*, Academic Press, Boston, MA, 1990, p. 787.
4 G. H. Haertling, *Ceram. Trans., 25* (1992) 1.
5 J. L. Loubet, J. M. Georges and G. Meile, *ASTM Committee, 889* (1985) 72.
6 S. Merklein, A. Schönecker, U. Schubert and D. Sporn, submitted to *Fortschrittsber. DKG*.
7 H. P. Klug and L. E. Alexander, *X-ray Diffraction Procedures for Polycrystalline and Amorphous Materials*, Wiley, New York, 1954, p. 491.
8 D. Sporn and H. Schmidt, *Euro-Ceramics 1*, Elsevier, London, 1989, p. 1.120.
9 G. Erwin, *J. Am. Ceram. Soc., 41* (1958) 347.
10 P. Krackhardt, *Diploma Thesis*, ISC Würzburg, FH Gießen Friedberg, 1991.
11 S. L. Swartz, *IEEE Trans., Topics in electronic ceramics, 25* (1990) 5.

Materials Science and Engineering, *A168* (1993) 209–215

Electrochromic materials: metal oxide nanocomposites with variable optical properties

C. G. Granqvist

Physics Department, Chalmers University of Technology and University of Gothenburg, S-412 96 Gothenburg (Sweden)

Abstract

Crystals of electrochromic metal (Me) oxides consist of MeO_6 octahedra in various corner-sharing and edge-sharing arrangements. This basic microstructure also exists in electrochromic films. The coordination of the ions leads to schematic electronic bandstructures which, at least for the (defect) perovskite and rutile lattices, are able to explain the presence or absence of cathodic and anodic electrochromism.

1. Introduction

Electrochromic devices are able to change their optical properties in a reversible and persistent manner under the action of a voltage pulse. The optical modulation is related to the amount of mobile ions in an electrochromic material, which here is taken to be a metal oxide. This paper points at some fundamental microstructural features that are of the decisive importance for electrochromism in oxides, and shows how the microstructure leads to schematic electronic bandstructure models capable of explaining, at least for most of the electrochromic materials, why certain materials become absorbing under ion insertion while others become transparent under ion insertion.

Electrochromic devices comprise several layers backed by a substrate (normally a glass plate) or layers positioned in between two substrates in a laminate configuration [1]. The substrate has a transparent conducting film and a film of the electrochromic oxide. Then follow a fast ion conductor or electrolyte, a layer serving as ion storage—which can be another electrochromic material—and a second (transparent) conductor. A specular or diffuse reflector may be incorporated in the design. When a voltage is applied between the (transparent) conductors, ions are inserted into or extracted from the electrochromic film, whose optical properties are thereby changed. The insertion–extraction process can be represented, highly schematically, by

$$MeO_n + xI^+ + xe^- \leftrightarrow I_xMeO_n \tag{1}$$

where Me is a metal atom, I^+ is a singly charged small ion such as H^+ or Li^+, e^- is an electron, and n depends on the particular type of oxide. For example, n is 3 for defect perovskites a 2 for rutiles.

Electrochromic devices have several important applications which are presently spurring on scientific and technical development. Among these, we note smart windows capable of regulating the inflow of radiant energy in buildings so that an optimum indoor climate is maintained at a minimum demand on paid energy [2, 3], antidazzling rear-view mirrors for automobiles, and non-emissive information displays with superior viewing properties.

This paper is organized as follows. Section 2 is devoted to the microstructure and surveys the bulk crystal structures for most of the metal oxides known to exhibit electrochromism. These oxides comprise octahedral building blocks of MeO_6 type that are organized in a plethora of different corner-sharing and edge-sharing configurations. The ubiquitous octahedral building blocks also prevail in thin films of electrochromic oxides. Section 3 then considers the electronic bandstructure, and ensuing optical properties, consistent with the octahedral MeO_6 units. Particular attention is given to perovskite-like and rutile-like materials. The schematic bandstructures for these classes of materials, following ideas originally put forward by Goodenough [4, 5], are used as a basis for discussion of electrochromism in several oxides. Heavily disordered oxides have absorption governed by polaron hopping, and crystalline tungsten oxide can have properties governed by free-electron effects. A

0921-5093/93/$6.00

summary and some concluding remarks are given in Section 4.

2. Microstructure

2.1. The ubiquitous octahedron

The electrochromic oxides can be divided into three main groups with regard to their bulk crystalline structures: perovskite-like, rutile-like, and layer and block structures. They are discussed in this order below, and the general occurrence of octahedral MeO_6 building blocks is emphasized. Reference 6 may serve as a standard source for much of the structural information. Along with the discussion of structure, we point at characteristic features of electrochromism.

The perovskite structure, with general composition $CMeO_3$, is illustrated in Fig. 1. As drawn, the Me ions occupy the corners of a primitive unit cell and O ions bisect the unit cell edges. The central atom, denoted C, is neglected at the moment since it is absent in the electrochromic oxides of most interest. The corresponding (defect) perovskite MeO_3 configuration is often referred to as the rhenium oxide structure. It can be visualized as an infinite array of corner-sharing octahedra each with a metal ion surrounded by six equidistant oxygen ions. In between these octahedra there are extended tunnels that can serve as conduits and intercalation sites for small ions.

Tungsten oxide, WO_3, was the first electrochromic material discovered, and it remains today the most viable option for devices. The electrochromism was first discussed in 1969 by Deb [7]. The crystal structure is perovskite type, but some atomic displacements and rotations of WO_6 octahedra normally occur so that different symmetries are found. Tungsten oxide also has a tendency to form substoichiometric (Magnéli) phases with edge-sharing octahedra and extended tunnels with large pentagonal or hexagonal cross-sections. At present it is very well established [1, 2] that disordered tungsten oxide films transform from an optically transparent to an absorbing state under ion insertion. This is referred to, in electrochemical terms, as cathodic colouration. The opposite case, with absorption associated with ion extraction, is called anodic colouration.

Molybdenum oxide, MoO_3, has been known as a cathodic electrochromic material at least since 1974 [8]. Bulk crystals at room temperature have a structure consisting of corner-sharing chains of MoO_6 octahedra which share edges with two similar chains; these layers are stacked and are held together by weak van der Waal's forces. This unique structure is referred to as the α-phase. There is also a β-phase [9], which can be viewed as a metastable analogue of WO_3, which is perovskite type. Available data on thin films of molybdenum oxide [10] strongly indicate that the structure is built from corner-sharing MoO_6 octahedra in much the same way as electrochromic tungsten oxide.

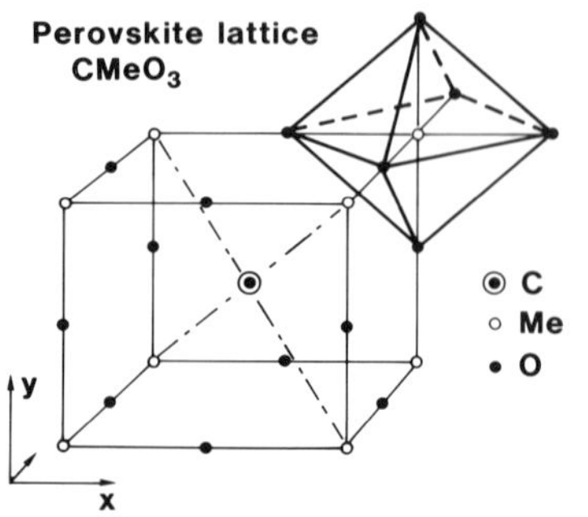

Fig. 1. Unit cell for the perovskite lattice. Octahedral symmetries are emphasized.

We also note that rhenium oxide itself, obviously a defect perovskite, is not known to exhibit electrochromism.

The rutile structure, or at least a rutile-like configuration, is present for a whole series of materials exhibiting cathodic or anodic electrochromism. The ideal structure can be thought of as built from almost octahedral MeO_6 units forming infinite edge-shared chains. These chains are arranged so that they form an equal number of ideal vacant tunnels.

Titanium oxide, TiO_2, can have different crystal structures, *i.e.* rutile, anatase and brookite. The anatase phase, which so far has attracted most interest in the context of electrochromism, consists of infinite planar double chains of TiO_6 octahedra; these chains are connected by corner-sharing. Empty sites also form double chains, or zig-zag tunnels. Cathodic electrochromism with a large degree of optical modulation was observed in some work [11].

Manganese oxide, MnO_2, has very complex crystal chemistry. However, most structures are built from edge-shared MnO_6 octahedra which share corners with other chains. Electrochromism was discovered recently, and ion insertion was found to lower the absorption [12].

Vanadium dioxide, VO_2, is rutile type (tetragonal) above 68 °C and is monoclinic—albeit with very slight atomic displacements from the ideal rutile structure—below this temperature. Vanadium dioxide films are electrochromic and show a decrease in absorptance under ion insertion irrespective of the temperature being below or above 68 °C [13].

Ruthenium oxide, RuO_2, has rutile structure. Ions can be inserted and extracted, but the material remains absorbing whatever the ionic content [14].

Iridium oxide, IrO_2, is another rutile material. It is electrochromic and, in contrast to the electrochromic oxides discussed so far, is clearly anodic and can go from an absorbing to a fully transparent state under insertion of one ion per formula unit [15].

Rhodium oxide, RhO_2, shows many similarities with iridium oxide, and ion insertion is able to lower the absorption significantly [16].

The layer and block structures form a somewhat undefined group of electrochromic materials with, generally speaking, not so well explored structural properties. Hydrated nickel oxide shows pronounced anodic electrochromism and goes from an absorbing to a transparent state under proton insertion [17]. The material is thought to contain layers of NiO_6 octahedra sharing edges, and the protonated material is probably akin to brucite, $Ni(OH)_2$.

Vanadium pentoxide, V_2O_5, can be described as consisting of distorted VO_6 octahedra or, alternatively and perhaps more properly, as consisting of square pyramidal VO_5 units. The latter representation highlights the layered structure. Vanadium pentoxide films are electrochromic and show increasing absorption upon ion insertion [18].

Niobium oxide, Nb_2O_5, has complex crystal chemistry. Various phases can be thought of as built from blocks (or columns) of NbO_6 octahedra arranged in an ReO_3-type structure; a structure derived from corner-sharing rutile blocks is also known. Niobium oxide is electrochromic and goes from a transparent to an absorbing state upon ion insertion [19].

2.2. Clusters and columns in thin films

Bulk crystals of, apparently, all the well known electrochromic oxides are built from highly regular arrangements of edge-sharing and corner-sharing MeO_6 octahedra. However, electrochromic thin films of these materials rarely exhibit long-range order. The local microstructure of the pertinent thin films has only been studied in detail for tungsten oxide, and the discussion below is focused entirely on this material.

When tungsten oxide is evaporated, the liberated species are of molecular type, with a preponderance of W_3O_9 [20]. Molecular ejection also prevails for sputtering [21]. Raman spectra gave clear evidence for vibrations in the framework of corner-sharing WO_6 octahedra and of terminal W=O bonds on internal surfaces in evaporated films [22]. IR spectroscopy indicated hydration and hydroxylation.

X-ray scattering is able to provide detailed structural information, and from Fourier transformations of the scattering data it is possible to extract radial distribution functions (RDFs) which yield density correlations in the analysed materials. The solid curves in the lower panels of Fig. 2 illustrate RDFs for tungsten oxide films evaporated onto substrates kept at three different

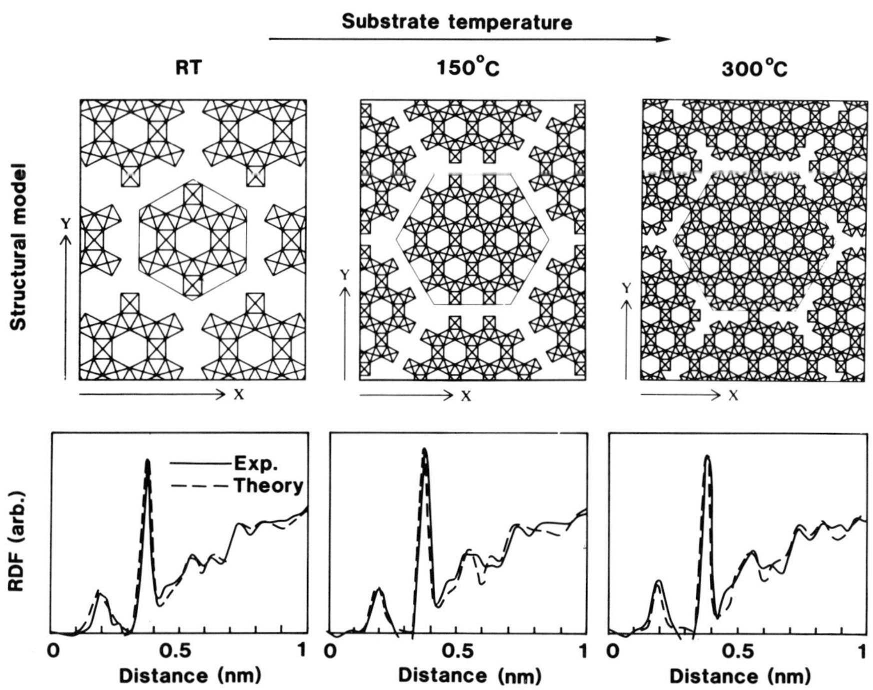

Fig. 2. Upper parts show structural models, based on corner-sharing WO_6 octahedra, for tungsten oxide films produced by evaporation onto substrates at room temperature (RT) and two elevated temperatures. The arrows in the X and Y-directions denote 2 nm. Lower parts show corresponding radial distribution functions (RDFs) as determined from X-ray scattering (solid curves) and as computed from the shown structural models. From Nanba and Yasui [23].

temperatures. A very ambitious structural model was used by Nanba and Yasui [23] to model the RDFs. Eight different crystallographic structures were employed as starting points, and corresponding RDFs were computed. All of the RDFs yielded peaks at less than 0.5 nm due to the octahedral WO_6 building blocks. The peaks at radii greater than 0.5 nm are of more decisive importance for the structural models and, in particular, a peak at approximately 0.73 nm can only be reconciled with a hexagonal structure. The upper left-hand panel in Fig. 2 shows a structural model based on hexagonal WO_3, in which three- and six-membered rings of octahedra are displayed in the projected X–Y plane and four-membered rings are parallel to the vertical Z direction. The clusters were arranged in space with consideration of their hexagonal plate configuration, so that they constituted a "film", and RDFs were computed. It is seen from the lower left-hand panel in Fig. 2 that theory and experiment are in excellent agreement, which gives strong credibility to the shown structural model. An extension of this model to films evaporated at elevated substrate temperatures is also shown in Fig. 2; cluster growth and cluster–cluster linking are apparent. The structural models give clear evidence that the tungsten oxide films should be viewed as nanocomposites—rather than truly amorphous materials—with grain sizes between approximately 1 and 5 nm depending on the deposition conditions.

Inhomogeneities on a larger scale are also possible and, in fact, microstructural features occur at all dimensions up to those comparable with the film thickness [24]. Columnar structures can be understood from models based on ballistic aggregation of structural units and ensuing reaccommodation of these, as well as from growth instabilities in continuum theories. For films prepared by normal evaporation—with the vapour flux incident perpendicular to the substrate—the columns are oriented normal to the substrate. The intercolumnar regions allow easy transport of ions across the films, which is of great importance for the colouration dynamics of electrochromic devices. The transport mechanism can be vehicular, Grotthuss type, or a mixture of these. It is possible to increase the columnar character by depositing tungsten oxide films at an oblique angle [25].

3. Optical properties

3.1. *Schematic bandstructures, and systematics for electrochromism*

Below we will show that highly simplified bandstructures, originally due to Goodenough [4] and presented in a particularly accessible way by Honig [5], allow us to understand why certain perovskite-type and rutile-type electrochromic oxides become absorbing under ion insertion, others become transparent under ion insertion, and still others show at least some absorptance irrespective of their ionic content. The layer and block structures can probably be handled in a similar way, but this remains to be done.

The perovskite structure was illustrated in Fig. 1. A corresponding energy level diagram for the defect perovskite, *i.e.* the rhenium oxide structure here denoted MeO_3, is illustrated in Fig. 3(a). The atomic s, p and d levels of Me are indicated, as well as the 2s and 2p levels of O. For WO_3 and ReO_3, the pertinent levels are 6s, 6p and 5d. Each Me ion is octahedrally surrounded by six oxygen ions, and each oxygen is linearly flanked by two Me ions. As a consequence of this arrangement, the d level is split up into e_g and t_{2g} levels, as shown in the left-hand part of Fig. 3(a). The splitting arises because the e_g orbitals point directly at the electronegative O whereas the t_{2g} orbitals point away from the nearest neighbours into empty space

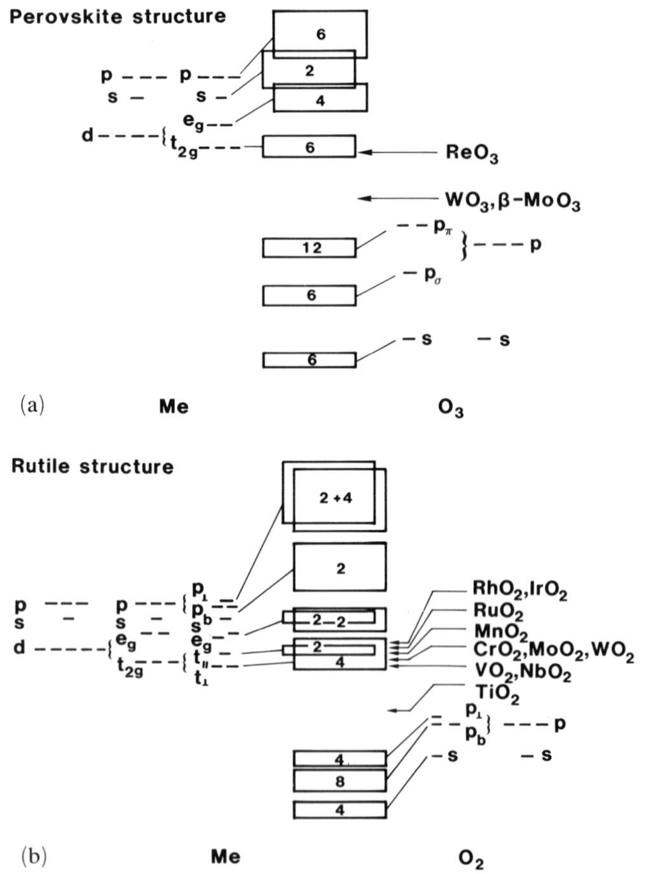

Fig. 3. Schematic bandstructures for the MeO_3 defect perovskite structure (a) and the MeO_2 rutile structure (b). The incipient atomic and molecular levels are shown, using standard notation. The numbers in the different bands denote electron capacities. Arrows indicate Fermi energies. After Honig [5].

and are hence lower in energy. Similarly, the O 2p orbitals are split as indicated in the right-hand part of Fig. 3(a); the $2p_\sigma$ orbitals point directly at the nearest electropositive Me ions whereas the $2p_\pi$ orbitals point into empty space. In the perovskite lattice, the incipient molecular energy levels broaden into bands, whose relative widths and repulsion can be inferred from general arguments [5].

The number of states available for electron occupancy is fixed for each band. For example, counted per MeO_3 formula unit, the t_{2g} band has a capacity for six electrons (allowing for spin degeneracy), and the p_π band has a capacity for twelve electrons (allowing for spin and electron degeneracy). The pertinent electron capacities are indicated by the numbers in the various bands in Fig. 3(a). WO_3 and the isostructural β-MoO_3 have 24 electrons in the shown bands, so that the Fermi energy lies in the gap between the t_{2g} and p_π bands. The band gap is wide enough to render the material transparent. ReO_3, however, has 25 electrons so that the lower part of the t_{2g} band is occupied and the material should be non-transparent.

When ions and electrons are inserted according to reaction (1), the Fermi level is moved upwards in the presumed rigid-band scheme. In the case of WO_3 and β-MoO_3, the excess electrons must enter the t_{2g} band and the material, in principle, transforms from a transparent to an absorbing or from a transparent to a reflecting state depending on whether the electrons occupy localized or extended states. When the ions and the accompanying electrons are extracted, the material returns to its original transparent state. In ReO_3, the Fermi energy also moves upwards upon ion insertion, but it remains well within the t_{2g} band irrespective of the content of mobile ions. Consequently the optical properties are not expected to be changed qualitatively. Clearly, the discussion above explains why tungsten oxide and molybdenum oxide are cathodic electrochromic materials [1, 2, 8].

Rutile and rutile-like structures are found among many MeO_2 oxides, as pointed out above. Several of these are electrochromic, either cathodic or anodic, and may or may not attain a transparent state. The electrochromism can be understood from the basic bandstructure shown in Fig. 3(b). It deviates from its counterpart for the perovskites because the MeO_6 building blocks are distorted and edge sharing. We start from the atomic s, p and d levels for the Me ion and the 2s and 2p levels for the oxygen ions. Every Me ion is almost octahedrally surrounded by oxygen ions, and every oxygen ion is roughly trigonally surrounded by Me ions. Hence the d levels are first split into e_g and t_{2g} levels, and the oxygen p levels are split into levels with orbitals p_b in the basal plane of the triangular array (where the oxygen ion is surrounded by three Me ions) and with orbitals $p_\perp$ perpendicular to that array. A further splitting occurs because the ideal octahedral symmetry is absent, and the ensuing Me levels, whose origin we refrain from discussing here, are designated $t_\parallel$, $t_\perp$, p_b and $p_\perp$. In addition, the degeneracy of the e_g level is lifted. When the ions are arranged in the rutile lattice, one expects overlap among the $t_\parallel$ and $t_\perp$ bands as well as among the two e_g bands. Hence the final schematic bandstructure, in fact, is quite similar to that of the perovskites. Electron band capacities can be deduced as for the perovskites.

The arrows in the right-hand part of Fig. 3(b) indicate the Fermi levels for several rutile-like oxides. For TiO_2, the Fermi level lies in the gap between the Ti $3t_{2g}$ and the O $2p_\perp$ bands, and ion insertion is expected to lead to transformation from a transparent to an absorbing state in much the same way as for WO_3 and β-MoO_3. This is consistent with experimental results on titanium oxide films [11]. For VO_2, MnO_2 and RuO_2, the initial absorbing state cannot be eliminated by ion insertion [12–14]. Again this is as expected from the bandstructure. RhO_2 and IrO_2 are especially interesting, since their Fermi levels lie close to the gap between the t_{2g} and e_g bands. Indeed, if ions are inserted up to about one formula unit in an iridium oxide film it is transformed from an absorbing to a transparent state [15]. In the bandstructure picture, the transparency is associated with the band gap referred to above. Rhodium oxide films are similar to iridium oxide films [16], but their optical properties have not been studied in great detail.

Figure 3(b) also lists a number of rutile-type oxides other than those mentioned above, specifically NbO_2, CrO_2, MoO_2 and WO_2. Whereas these materials undoubtedly can be prepared as thin films, the fabrication is fairly awkward since the rutiles are not the highest oxides.

3.2. Polaron absorption and free-electron effects in thin films

We first consider disordered tungsten oxide films. When ions are inserted, both X-ray photoelectron spectra [26] and electron spin resonance [27] show that W^{5+} as well as W^{6+} are present, and that the W^{5+} density scales with the intensity of the electrochromic absorption peak. The picture that emerges is hence that the inserted electrons are localized at tungsten sites, where they participate in processes responsible for the electrochromism.

The small polaron concept offers the possibility to formulate a quantitative theory for optical absorption. Small polarons are created when electrons polarize their surrounding, so that localization of the wavefunction takes place essentially to one lattice site. A small overlap between wavefunctions corresponding to adjacent sites, as well as strong disorder, are conducive to polaron formation. Theoretically, the small polaron absorption α_{pol} can be expressed by [28]

$$\alpha_{pol}(\omega) \propto \omega^{-1} \exp\left[\frac{(\hbar\omega - 4U_p)^2}{8U_p\hbar\omega_{ph}}\right] \quad (2)$$

where $\hbar\omega$ is the photon energy, U_p is the polaron binding energy, and $\hbar\omega_{ph}$ is the typical phonon energy.

Figure 4 shows a comparison between experiments on a proton inserted tungsten oxide film and $\omega^2\alpha_{pol}(\omega)$ [29]. The fitting used $U_p = 0.275$ eV and $\hbar\omega_{ph} = 0.098$ eV. It is found that theory and experiment are in good agreement at energies lower than that of the peak absorption. Between this peak and the semiconductor band gap at approximately 3.25 eV, the observed absorption is higher than that predicted from theory; this discrepancy is not necessarily inconsistent with the small polaron concept, though.

If one examines the available experimental data one may notice a strong similarity among the absorption features at least for oxides based on tungsten, molybdenum, titanium and W–Mo. This lends credence to the view that polaron absorption prevails generally for the cathodic electrochromic materials. The case is less clear for the anodically colouring films, and for electrochromic hydrated nickel oxide it was found that polaron absorption was not present but that instead charge transfer theory could be used [30].

We now consider crystalline materials. The perovskite structure is open enough to permit ion insertion, and in this way it is possible to render tungsten oxide conducting and IR reflecting. The basic mechanism for the optical and electrical properties appears to be akin to that of heavily doped semiconductors such as In_2O_3:Sn [31]. We note, in passing, that if films based on β-MoO_3 are crystallized they transform into the comparatively dense α-phase.

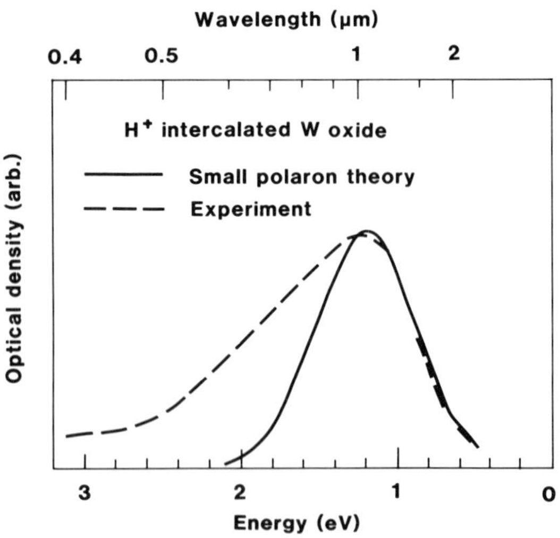

Fig. 4. Spectral optical density for an electrochromic evaporated tungsten oxide film in H_2SO_4 as measured and as computed from eqn. (2). From Schirmer *et al.* [29].

Free-electron theory can be employed to describe the optical properties of crystalline electrochromic tungsten oxide. Drude theory can be used for qualitative work but is unable to give quantitative predictions since electron scattering is treated as frequency independent and is represented only by a fitting parameter. Instead the Gerlach theory [32] is applicable, at least to first order, to give the performance for the limiting case when only ionized impurity scattering is present. Screening of the ions can be modelled within the random phase approximation or an extension thereof. An analysis based on this theoretical approach yielded qualitative agreement with experimental results [33].

4. Summary and remarks

All of the well known electrochromic oxides can be thought of as being built from MeO_6 octahedra in various corner-sharing and edge-sharing arrangements. The corresponding crystal structures are perovskite like, rutile like, and of layer or block character. The octahedral units are favourable for both ion and electron transport. The pertinent crystal structures allow long-range ion diffusion through tunnels or between layers. This capability exists, at least to some extent, in good crystals, and it is highly accentuated in thin films produced by standard techniques.

Once the ubiquity of the octahedra is realized, one may think of a canonical electron bandstructure capable of catching the essentials of the electrochromism among the oxides. Following Goodenough [4], it is indeed possible to represent the crystal-field-effects for the pertinent ion coordinations in terms of schematic bandstructures for the perovskites and rutiles. In this representation, an insertion of electrons has the sole effect of lifting the Fermi energy. The model is capable of explaining why oxides of tungsten, molybdenum and titanium show cathodic electrochromism, why oxides of iridium and rhodium show anodic electrochromism, and why several other oxides, although capable of ion insertion, are unable to be optically transparent.

Acknowledgment

Illuminating discussions with Dr. G. A. Niklasson and others are acknowledged with pleasure.

References

1 C. G. Granqvist, *Solid State Ionics, 53–56* (1992) 479.
C. G. Granqvist, in M. Francombe and J. Vossen (eds.), *Physics of Thin Films*, Vol. 17, Academic Press, San Diego, CA, 1993, in press.

2 C. M. Lampert and C. G. Granqvist (eds.), *Large-Area Chromogenics: Materials and Devices for Transmittance Control*, Vol. IS-4, SPIE Opt. Engr. Press, Bellingham, 1990.

3 C. G. Granqvist, in C. G. Granqvist (ed.), *Materials Science for Solar Energy Conversion Systems*, Pergamon, Oxford, 1991, p. 106.

4 J. B. Goodenough, in H. Reiss (ed.), *Progress in Solid State Chemistry*, Vol. 5, Pergamon, Oxford, 1971, p. 145.

5 J. M. Honig, in S. Trasatti (ed.), *Electrodes of Conductive Metallic Oxides*, Part A, Elsevier, Amsterdam, 1980, p. 1.

6 B. G. Hyde and S. Andersson, *Inorganic Crystal Structures*, Wiley, New York, 1989.

7 S. K. Deb, *Appl. Opt. Suppl., 3* (1969) 192.
S. K. Deb, *Philos. Mag., 27* (1973) 801.

8 M. R. Tubbs, *Phys. Status Solidi, A, 21* (1974) 253.

9 M. Figlarz, *Prog. Solid State Chem., 19* (1989) 1.

10 P. F. Carcia and E. M. McCarron III, *Thin Solid Films, 155* (1987) 53.

11 T. Ohtsuka, M. Masuda and N. Sato, *J. Electrochem. Soc., 134* (1987) 2406.

12 S. I. Córdoba de Torresi and A. Gorenstein, *Electrochim. Acta, 37* (1992) 2015.

13 M. S. R. Khan, K. A. Khan, W. Estrada and C. G. Granqvist, *J. Appl. Phys., 69* (1992) 3231.

14 J. Rishpon and S. Gottesfeld, *J. Electrochem. Soc., 131* (1984) 1960.

15 J. D. E. McIntyre, W. F. Peck, Jr., and S. Nakahara, *J. Electrochem. Soc., 127* (1980) 1264.

16 S. Gottesfeld, *J. Electrochem. Soc., 127* (1980) 272.

17 J. S. E. M. Svensson and C. G. Granqvist, *Appl. Phys. Lett., 49* (1986) 1566.
J. S. E. M. Svensson and C. G. Granqvist, *Appl. Opt., 26* (1987) 1554.

18 A. Talledo, A. M. Andersson and C. G. Granqvist, *J. Mater. Res., 5* (1990) 1253.
A. Talledo, A. M. Andersson and C. G. Granqvist, *J. Appl. Phys., 69* (1991) 3261.

19 R. Cabanel, J. Chaussy, J. Mazuer, G. Delabouglise, J. C. Joubert, G. Barral and C. Montella, *J. Electrochem. Soc., 137* (1990) 1444.

20 J. Berkowitz, W. A. Chupka and M. G. Inghram, *J. Chem. Phys., 27* (1957) 85.

21 A. Azens, G. Ramans, U. Kanders and A. Levrentyev, *Vacuum, 43* (1992) 943.

22 Y. Shigesato, Y. Hayashi, A. Masui and T. Haranou, *Jpn. J. Appl. Phys., 30* (1991) 814.

23 T. Nanba and I. Yasui, *J. Solid State Chem., 83* (1989) 304.

24 R. Messier, *J. Vac. Sci. Technol. A, 4* (1986) 490.

25 T. Motohiro and Y. Taga, *Appl. Opt., 28* (1989) 2466.

26 A. Temmink, O. Anderson, K. Bange, H. Hantsche and X. Yu, *Thin Solid Films, 192* (1990) 211.

27 J. V. Gabrusenoks, P. D. Chikmach, A. R. Lusis, J. J. Kleperis and G. M. Ramans, *Solid State Ionics, 14* (1984) 25.

28 O. F. Schirmer, *J. Phys. (Paris) Coll., 6* (1980) 479.

29 O. F. Schirmer, V. Wittwer, G. Baur and G. Brandt, *J. Electrochem. Soc., 124* (1977) 749.

30 T. M. J. Nilsson and G. A. Niklasson, *Proc. Soc. Photo-Opt. Instrum. Eng., 1272* (1990) 129.

31 I. Hamberg and C. G. Granqvist, *J. Appl. Phys., 60* (1986) R123.

32 E. Gerlach, *J. Phys. C, 19* (1986) 4585.

33 J. S. E. M. Svensson and C. G. Granqvist, *Appl. Phys. Lett., 45* (1984) 828.

Materials Science and Engineering, *A168* (1993) 217–224

Magnetic properties of granular compounds and other nanostructured materials

J. L. Dormann

Laboratoire de Magnétisme et Matériaux Magnétiques, CNRS, 92195 Meudon Cedex (France)

Abstract

The magnetic properties of some nanostructured compounds are reviewed. After considering the influence of size at micrometric and nanometric scales, the discussion is focused on the latter scale. The magnetic properties of fine particles are discussed from a fundamental point of view as well as with respect to applications. Magnetic properties of granular compounds consisting of fine particles embedded in a matrix are described and differences from the general properties of fine particles are underlined. Finally, the properties of other nanostructured materials, nanocrystalline compounds and multilayers, are discussed.

1. Introduction

The magnetic properties of materials are strongly affected when the size becomes very small. For micrometric and submicrometric scales, the main change concerns the structure of domains and domain walls. The existence of magnetic domains is due to the required minimization of the magnetostatic energy. The thickness of walls separating the domains depends on several factors, mainly the magnetocrystalline anisotropy. For strongly anisotropic materials, the order of magnitude of thickness is 20 nm, while this value is much higher for soft materials. Therefore, below a certain size, magnetic materials are monodomains. This size depends on the anisotropy constant K, the magnetization M and the shape of the particle. For example, for spheroidal cubic iron particles (low K), the critical diameter is around 20 nm, while for uniaxial MnBi particles (strong K), this diameter is around 800 nm [1]. For a monodomain, the coercive field H_c which measures the switching field is equal to about K/M, while for polydomain samples, H_c can be much smaller. So there are two methods of magnetic recording, for which the bit unit corresponds either to monodomain particle(s) or to domain(s) in the bulk sample. The former method has been in use for a long time, while materials research and improvements in technical characteristics are ongoing for the latter. *A priori*, the first method permits a higher bit density, but technical problems such as the coating of particles, and physical problems such as the interactions between particles, limit the bit density, in addition to numerous other problems such as noise, switching speed, and parameter distributions due to variations in size (and shape). Two media are used, a particulate medium which consists of coated magnetic monodomain particles [2–6] and thin films of segregated magnetic grains [6]. A new morphology formed from lithographic arrays of particles [7] will probably lead to the resolution of many of the problems cited above. The second method is now becoming attractive owing to the present possibility of producing almost perfect thin films, with a narrow distribution of shapes and properties of domains. The thin film morphology is also very useful because it leads to a bidimensional arrangement of domains (here the bit unit). In addition, the small thickness with regard to the other dimensions induces a demagnetizing field perpendicular to the film plane, which simplifies parameter adjustment. Perpendicular or longitudinal writing is possible, but at present it is difficult to know which is the better solution.

Two new features occur at the nanometric scale. One is related to the volume, the other to the surface. The fact that one half of the magnetic atoms are lacking, evidently changes the properties near and on the surface. This effect also exists for thick samples, but it is negligible with regard to the mean properties. Several theories and calculations have determined the magnetic state in the surface [8–10] and in monolayers [8–11], including magnetic anisotropy problems [12, 13]. It is a difficult task, particularly because a surface or a monolayer cannot exist alone. Numerous arrangements are possible where the surface plays a decisive role, such as monolayers, bilayers, sandwich films, all deposited onto substrates. Experiments on ultrathin structures [10, 12–19] are in agreement with the theoretical predictions which lead to an enhancement of the magnetic moment. This property could be useful for

0921-5093/93/$6.00

some applications. The same types of difficulty arise for multilayers with, in addition, the questions of exchange coupling across the interface, RKKY coupling, coupling through non-magnetic layers, and superlattice effects [8]. Since applications are now beginning to be devised for multilayers, we discuss briefly their properties in Section 4.

When the particle volume becomes small, the anisotropy energy becomes comparable with the thermal energy. In this case, the magnetic moment has a non-negligible probability to overcome the energy barrier and therefore it relaxes. The properties of this particular magnetic state, named superparamagnetism [20], are described and discussed in Section 2, as well as aspects of application. At low temperature, when the thermal energy is small, the properties are quasi-static, as for the bulk, but they are strongly influenced by the magnetic state of the particle surface, because the ratio of surface to volume becomes important. For example, the ratio is about 0.4 for particles of 4 nm.

Finally, we discuss in Section 3 the peculiar properties of small particles embedded in a matrix, named granular materials, and similar structures such as nanocrystalline materials. In fact, these properties are generally observed because the particles are almost never isolated and the questions have some similarity with those arising for multilayers.

Numerous papers devoted to fine particle and granular compound studies are collected in ref. 21.

2. General properties of fine particles

As mentioned above, the magnetic moment $\boldsymbol{m}$ of the particle relaxes between the easy directions when the anisotropy energy becomes comparable with the thermal energy. A rough approximation leads to a limiting volume for which relaxation does not occur, V (cm^3)$>10^{-12}/K$ at room temperature. For example, this corresponds to a diameter $\phi>10$ nm at room temperature for a medium K value equal to 2×10^6 erg cm^{-3}. Therefore, it is essential to know the anisotropies and their symmetry in order to calculate the relaxation time τ and to analyse parameters such as H_c. In addition to the usual magnetocrystalline and magnetostatic anisotropies, other terms often have a strong influence such as the anisotropies owing to strains, and surface and magnetic interactions due to other particles. We discuss the latter in Section 3. A detailed discussion of the anisotropy problem in the case of fine particles can be found in refs. 22 and 23.

After the pioneering work by Neel [24], Brown [25] solved the calculation of τ by assuming that the $\boldsymbol{m}$ motion obeys Gilbert's equation augmented by a random field term. An analytical solution is not possible and approximations for low and high values of E_a/kT, where E_a is the barrier energy, and for uniaxial and cubic anisotropies have been derived [26, 27], which for high values of E_a/kT and cubic symmetry leads to an Arrhenius law:

$$\tau=\tau_0 \exp(E_a/kT) \tag{1}$$

where τ_0 is the pre-exponential factor related to various parameters. References 22 and 28–30 include discussions on the validity of models and approximations. Recently a new resolution method [31] has provided a quasi-analytical solution and a unique (approximative) formula, valid for $0<E_a/kT<60$ (the upper limit corresponds to a value of τ larger than the age of the solar system), which is written in the form of eqn. (1) and gives a precise determination of τ_0. A synchronous rotation of spins is expected when $\boldsymbol{m}$ changes its direction. This is a rough approximation when the surface part becomes large because the magnetic disorders existing near and on the surface lead to incoherent rotation. Very small particles, ranging from a few atoms to roughly $\phi=2$ nm, for which the best name is aggregates, have a fully disorganized particle structure and particular electronic states of atoms, which lead to specific properties outside the superparamagnetism field [32]. However, the curling mode for the rotation of spins is more favourable for large particles under certain conditions [30]. This incoherent mode is also expected for particles used in magnetic recording. Moreover, in addition to $\boldsymbol{m}$ reversals, there are fluctuations of $\boldsymbol{m}$ in its potential well, as shown indirectly by Mössbauer spectroscopy for $T<T_B$ [33] and recently directly by neutron experiments in the whole range of temperature [34].

The time dependence is the prominent characteristic of the fine particle properties. When the measuring time τ_m of an experiment is much faster than τ, $\boldsymbol{m}$ appears motionless and measurements correspond to static conditions while a mean time is detected if τ_m is much slower than τ (superparamagnetic state). It is important to remember that the results under static conditions differ from those obtained for the bulk, owing to the influence of the surface. The blocking temperature T_B, which separates the two regimes, is the temperature for which $\tau=\tau_m$. The variation of the alternative susceptibility χ_{ac} *vs.* temperature (Fig. 1) well illustrates these phenomena. The χ'_{ac} maximum, obtained for $T\approx T_B$, corresponds to a temperature which increases with frequency $\nu=1/\tau_m$. The experimental variation of T_B *vs.* τ_m allows the calculations of τ to be checked. However, to our knowledge, the Arrhenius law eqn. (1) has never been truly checked (see Section 3). In the superparamagnetic state ($T>T_B$), a good approximation [35] of χ can be expressed by

$$\langle\chi\rangle\approx\langle|\boldsymbol{m}(T)|^2\rangle/3kT\langle V\rangle \tag{2}$$

Equation (2) is very similar to the formula for paramagnetic χ, except that $|\boldsymbol{m}|$ replaces the spin magnetic moment. Therefore giant "paramagnetic" χ are observed. For a hypothetical purely ferromagnetic particle, the superparamagnetic χ will be n times the paramagnetic χ, n being the number of magnetic atoms (for $\phi = 10$ nm, $n \approx 10^4$). However, it is important in eqn. (2) to take into account the thermal variation of $\boldsymbol{m}$ (Fig. 2) [36], even if the Curie temperature of the bulk is high, because on increasing the temperature the decrease in magnetization M is much more pronounced than in the bulk, owing to surface influence [37, 38].

The field cooled magnetization M_{FC} and zero field cooled magnetization M_{ZFC} under low applied field are very useful for characterizing fine particles. In fact, $M_{ZFC} = \chi$, but it is difficult in this case to define τ_m which depends on the cooling rate. For $T < T_{Bstatic}$, M_{ZFC} is outside equilibrium and varies with time while M_{FC} corresponds to a quasi-equilibrium state and therefore does not vary with time. Below $T_{Bstatic}$, M_{FC} remains constant but its value depends on the cooling rate. However, if there is a V distribution, M_{FC} will only be constant for temperatures lower than the smallest $T_{Bstatic}$. An example with a narrow V distribution is shown in Fig. 3 [39].

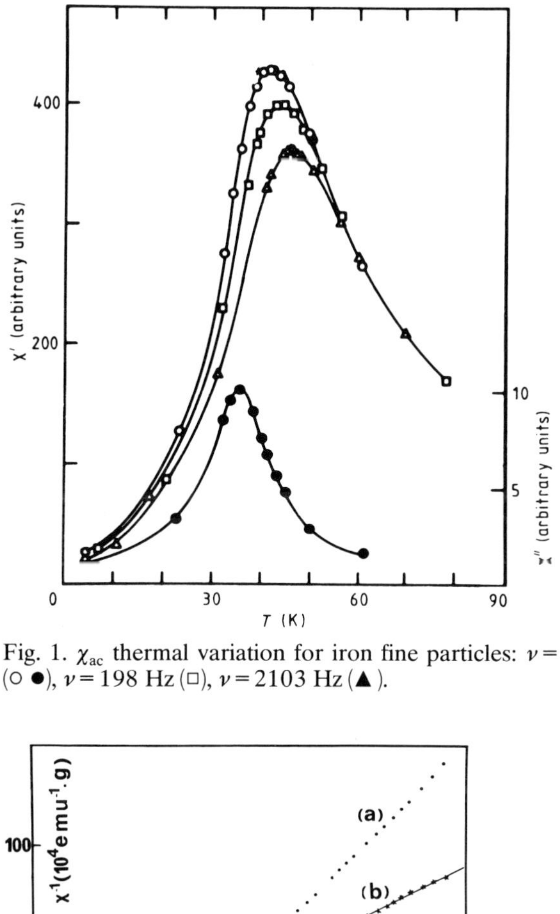

Fig. 1. χ_{ac} thermal variation for iron fine particles: $\nu = 17.3$ Hz (○ ●), $\nu = 198$ Hz (□), $\nu = 2103$ Hz (▲).

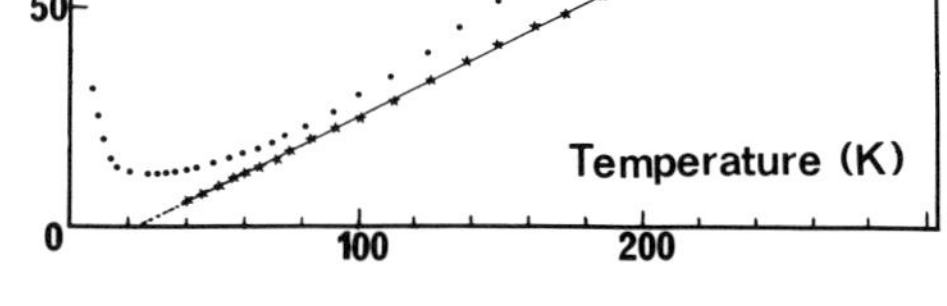

Fig. 2. $1/\chi$ thermal variation for iron fine particles, (a) as measured, and (b) corrected from the thermal dependence of relative magnetization.

The applied field H changes the anisotropy energy and causes asymmetric potential wells. On increasing H, $\boldsymbol{m}$ becomes more and more blocked in the lowest energy minimum and from $H > K/2M$ in uniaxial symmetry, relaxation of $\boldsymbol{m}$ is suppressed. However, it remains vibrating in the potential well and for H higher than the limit cited above, M can be expressed through a Langevin function [37, 40, 41]. Measurements of M under high H permit direct determination of the mean values of V and $|\boldsymbol{m}|$. Coupled with in-field Mössbauer spectroscopy, features are obtained on the magnetic state of the particle surface. In the case of iron fine particles embedded in an alumina matrix, we have shown that in the surface the spins are canted and the magnetic moment per spin is lowered with regard to the bulk value [41, 42]. This differs from the results obtained on ultrathin films [14–16, 19] but (i) the morphology is different and (ii) the iron surface layer is in contact with Al_2O_3, probably with bonds between them. References 43 and 44 describe in detail the results obtained by Mössbauer spectroscopy on particle surface magnetism.

Remanences obtained following various processes after the field cut-off are very interesting parameters because they are used to characterize particulate media and to evaluate the ability for recording. They have been discussed in detail by Chantrell and O'Grady [45, 46]. We will limit the discussion to the case of thermoremanent magnetization (TRM) and macroscopic quantum tunnelling (MQT). Recently, theoretical studies [47–49] have shown the possibility of MQT in

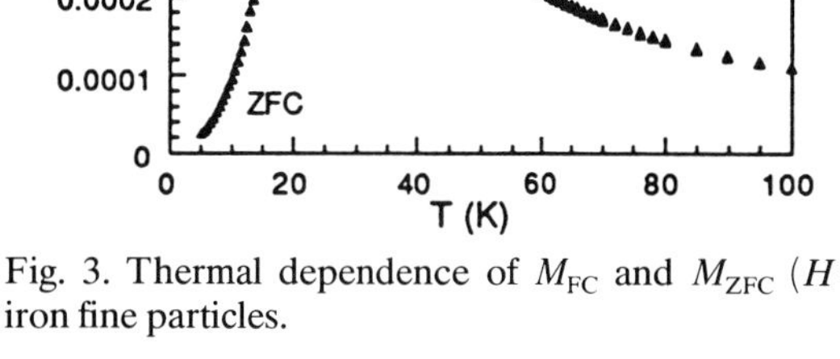

Fig. 3. Thermal dependence of M_{FC} and M_{ZFC} ($H = 5$ Oe) for iron fine particles.

small particles. Here, MQT corresponds to a relaxation of $\boldsymbol{m}$ across the barrier without paying E_a, while in the normal process the cost in energy is filled by the thermal energy. In fact, MQT concerns any system with two magnetic states separated by an energy barrier. It is an important problem for magnetic recording. If MQT occurs for a certain particle type, that will surely be an objection for its practical use. Analysis of the TRM variation *vs.* time is the direct way of detecting MQT. In the superparamagnetic model, this variation is supposed to be of TRM $= A \log t$ type, with A proportional to the temperature. However, the validity of this formula has not been demonstrated and it seems suspicious [30]. Our preliminary calculations show that the TRM variation is much more complex and, if TRM is simulated using the above formula, the variation in A is very far from a straight line. As long as we do not have a correct formulation for TRM in the superparamagnetic model, it will be very difficult to appreciate any MQT effect. MQT seems to exist in multilayers [50, 51], monodomain [52, 53] and bidomain [53, 54] particles, but the results need to be confirmed because very indirect methods or doubtful formulation has been used. In fact, the first results show that if MQT exists, it occurs at a very low temperature and its effect is very small. So, in our opinion, no disturbance will result for magnetic recording.

Several aspects have not been discussed in this paper, such as the variation in T_B with H [29, 55], and the hyperfine properties as seen by Mössbauer spectroscopy [23, 56]. They are discussed in review papers [22, 23, 29, 32, 39].

We turn now towards applications. The main point concerns magnetic recording. Better knowledge of the superparamagnetic properties will permit us to choose the smallest size for a material, for which the technical characteristics are adapted to magnetic recording. That will permit a smaller size than the size used at present and an increase in bit density. This raises the question of reading, but that is another problem. Other possible applications are related to the fact that T_B, which can be adjusted to room temperature, is a pseudo-transition temperature, with giant "paramagnetic" χ for $T > T_B$. That leads to χ_{ac} values at very high frequency which could be well adapted for certain uses. Enhanced magnetocalorific effects are also observed under certain conditions of operation (Fig. 4) [57, 58]. This property represents potential application to refrigeration.

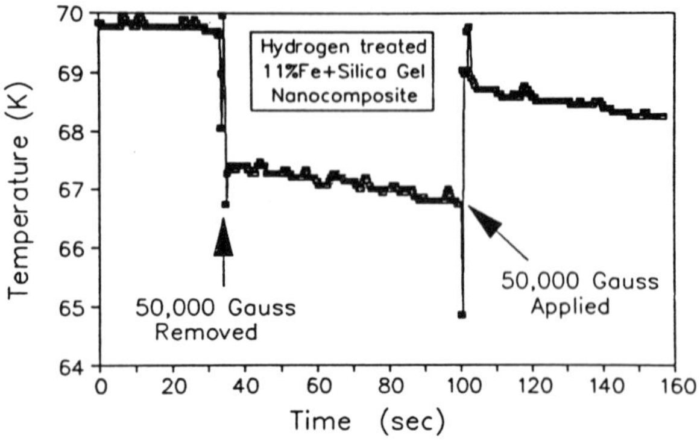

Fig. 4. Temperature *vs.* time for a gel nanocomposite when the magnetic field was first removed and then applied to the sample.

3. Peculiar properties of granular compounds

Granular materials consist of fine particles embedded in a matrix. Two main processes are used. Firstly, the particles are prepared alone using various techniques such as vapour deposition, aerosols, spray pyrolysis, and chemical methods [59], and then they are coated or placed in a polymer matrix. Aggregation due to van der Waals forces must be balanced by electrostatic strengths at the particle surface when the dispersed state is required [60]. Secondly, cosputtering of immiscible materials is utilized. This process enables a regular arrangement of particles to be obtained, the particle diameters being governed by the target composition and the temperature of the substrate during cosputtering. Various combinations have been prepared with magnetic particles: metallic particles in an insulating matrix, Ni/SiO_2 [61], Fe/SiO_2 [38, 62], Fe/Al_2O_3 [63], Fe_xCo_{1-x}/Al_2O_3 [64], amorphous $Fe_{80}B_{20}/Al_2O_3$ and $Fe_{64}Cr_{16}B_{20}/Al_2O_3$ [65]; insulating particles in an insulating matrix, Fe_2O_3/Al_2O_3 [66]; insulating particles in a metallic matrix, Fe_3O_4/Ag [67]; metallic particles in a metallic matrix, Co/Cu [68, 69].

The magnetic features of the fine particles discussed above are in general valid for granular compounds. However, the properties are more and less modified by the magnetic interactions. Dipolar magnetic interactions are always present and the interaction energy is proportional to $|\boldsymbol{m}|^2/d^3$, where d is the mean distance between the centres of two contiguous particles [70]. Then, the dipolar interactions are strong if the particles are near each other, and are negligible for antiferromagnetic particles owing to the small value of $[\boldsymbol{m}]$ [22]. The exchange interactions are appreciable if (i) the matrix and the particles are metallic, owing to RKKY coupling (the interaction energy is also proportional to $1/d^3$), and (ii) the particles are in contact or the surfaces are bonded with charges [59].

Several models have been proposed for which calculations are done on the one hand under static conditions [33, 46, 71], valid when $\boldsymbol{m}$ does not relax, for large particles or at very low temperatures, and on the other hand under dynamical conditions [29, 32, 70], valid for $\boldsymbol{m}$ relaxing. The latter model does not depend on the interaction type, while this is not true for the former. When $\boldsymbol{m}$ relaxes, the interactions produce an additional anisotropy energy, with uniaxial symmetry and an increasing anisotropy constant when the temperature decreases, which causes large deviations from the Arrhenius law, eqn. (1). The model has been checked from various experiments [34, 70] and excellent agreement is obtained. An interesting question concerns the low temperature magnetic state when the relaxation of $\boldsymbol{m}$ is very slow. The model leads to a magnetic transition at zero temperature but it is probable that there is a crossover of properties at a finite temperature. For a particle assembly, at zero temperature, the $\boldsymbol{m}$ magnetic order is driven by the interactions. When they are of dipolar type, a dipolar supermagnet is expected. Unfortunately, the order is unknown for a hexagonal compact arrangement. When the interactions are dominated by exchange, some frustrated magnetic states should occur. In our opinion, it is important to undertake works in this direction because particulate media are directly concerned, particularly if the particle is the unit bit and the packing fraction is high.

Finally, we note that the properties of interacting fine particles are similar to those of spin glasses, but with some essential differences [39], and are also similar to those of other magnetically frustrated systems [72].

With regard to magnetic properties, applicative aspects are the same for fine particles outside a matrix and granular compounds. However, magnetic interactions modify the parameters and this can be a restriction or a help. For example, an increase in H_c with ϕ was measured on Fe/SiO_2 particles, keeping the packing fraction constant. This leads to higher H_c values than expected for monodomains and therefore to high anisotropy values interesting for magnetic recording [38]. This feature could be due to interaction effects.

Magnetoresistance properties are specific to granular compounds. Two cases are to be considered. Firstly, for metallic particles embedded in an insulating matrix, the conduction is governed by tunnelling electrons [73, 74]. Without an applied field, the spin directions are decorrelated. For $T > T_B$, $\boldsymbol{m}$ relaxes, while for $T < T_B$, $\boldsymbol{m}$ is blocked in random directions. At low temperature, ferromagnetic correlations develop owing to increasing dipolar interactions. With high applied fields, all the spins are aligned and a large negative magnetoresistance is observed. For Ni/SiO_2 granular compounds, $\Delta\rho/\rho_0 \approx 1.6\%$ [61]. A low applied field aligns partially the spins only in the T_B range of temperature and a maximum is observed for $\Delta\rho/\rho_0$ for this temperature (Fig. 5) [61]. To our knowledge, only very few studies have been devoted to this feature which could be promising for magnetic heads.

Giant magnetoresistance was observed in Co/Cu granular compounds [68, 69, 75]. Here, the mechanisms are completely different from the above case. Roughly, without an applied field, RKKY coupling causes antiferromagnetic $\boldsymbol{m}$ order, the particle size being sufficiently large to avoid superparamagnetism. High magnetic fields align $\boldsymbol{m}$ and cause giant magnetoresistance (Fig. 6) [68]. In fact, the results obtained are very similar to those obtained for certain multilayers. We discuss this point briefly in the next section.

Nanocrystalline compounds are a new class of materials and are formed of single or multiphase polycrystals with crystallite sizes in the range of a few nanometres, typically 5–20 nm, so that about 30 vol.%

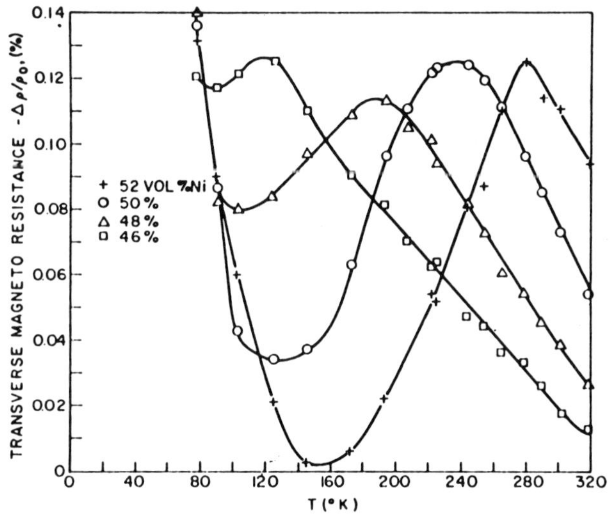

Fig. 5. Transverse magnetoresistance *vs.* temperature at 270 Oe for fine nickel particles embedded in SiO_2.

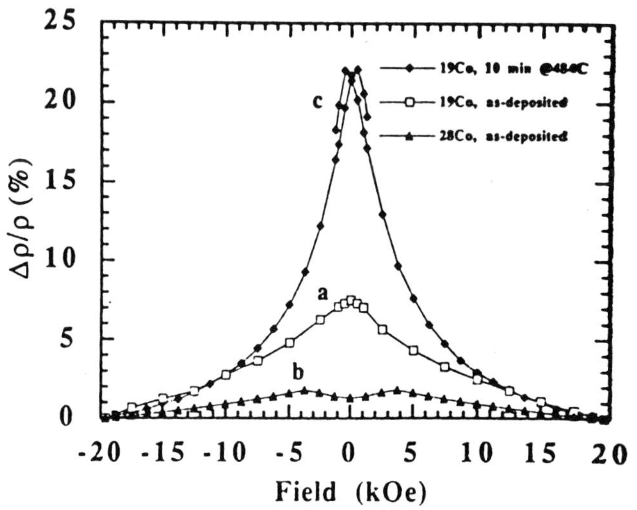

Fig. 6. Field dependence of $\Delta\rho/\rho = (\rho_H - \rho_{H=20\,\mathrm{kOe}})/\rho_{H=20\,\mathrm{kOe}}$ for different samples of granular cobalt in a copper matrix.

of the material consists of grain interface boundaries [76–78]. From a magnetic point of view, it is difficult at present to describe the properties because (i) only a few studies have been performed and (ii) the variety of materials probably leads to very different features. For example, one can expect that grains correspond to domains and boundaries to disordered walls, or that the properties are similar to those of granular metal/metal systems if the boundaries are not magnetic, or that superparamagnetic properties occur if the crystallite size is sufficiently small with non-magnetic boundaries. Work is necessary to sketch the magnetic properties of nanocrystalline materials.

Finally, fine magnetic particles are present in numerous compounds, of which the study is relevant to various fields such as mineralogy, catalysis, biological systems, fine arts. Some aspects are reviewed in ref. 21.

4. Multilayered materials

As we mentioned in the introduction, multilayered compounds show a very large variety of magnetic properties [8]. There are several questions from a fundamental point of view. The first is surface magnetism which is complicated by proximity and interfacial effects. Indeed it is very difficult to realize a perfect interface, requiring specific materials and techniques (as molecular beam epitaxy). Generally, interdiffusion occurs which confuses the analysis of properties. The second problem arises when the two layers are magnetic. Exchange coupling exists via RKKY interactions for metallic layers and short-range interactions in insulating layers. In this case, one can expect that only the monolayers near the surface will experience perturbed magnetism. However, when the bulk magnetic state is already perturbed, owing to magnetic frustrations, the change will probably be considerable. Only very few studies have been devoted to this problem. Another situation occurs when metallic magnetic layers are separated by non-magnetic metallic layers. Numerous experimental (see for example ref. 79) and theoretical studies [80] have discussed the properties of this combination. It has been shown that the exchange coupling oscillates as a function of the space thickness. For noble metals, RKKY formulation is applicable while for transition metals, modifications of the d electron band are decisive [80].

Interesting applications are derived from this property. When the space thickness is chosen such that antiferromagnetic coupling of the magnetic layers is effective, a sufficiently large applied field is able to align the whole spins and giant magnetoresistance is then observed (up to about 80%) [81]. This property is very interesting for possible use in reading heads although at present giant magnetoresistance is only effective for too large an applied field and at too low a temperature. Other main applications of multilayers are (i) magneto-optical recording or perpendicular recording media as Co/Pt which shows very large perpendicular anisotropy and coercivity when the cobalt layer thickness is only a few monolayers [82] and (ii) reading heads which require high saturation magnetization and high permeability in the megahertz range [83].

5. Conclusions

Nanostructured compounds show many new magnetic properties, some of which are promising for applications. Progress in fundamental research is necessary to understand the basic phenomena. This necessitates progress in material science to obtain reproducible compounds, well characterized and with improved qualities.

References

1 C. Kittel, *Phys. Rev., 70* (1946) 965.
C. Kittel, *Rev. Modern Phys., 21* (1949) 541.

2 G. Bate, *J. Magn. Magn. Mater., 100* (1991) 413.

3 W. M. Mularie and M. P. Sharrock, *J. Appl. Phys., 69* (1991) 4938.

4 D. E. Speliotis, *IEEE Trans. Magn., 24* (1988) 2850.

5 A. E. Berkowitz, F. T. Parker, F. E. Spada and D. Margulies, in J. L. Dormann and D. Fiorani (eds.), *Magnetic Properties of Fine Particles*, North-Holland, Amsterdam, 1992, p. 309.

6 K. E. Johnson, *J. Appl. Phys., 69* (1991) 4932.

7 J. F. Smyth, S. Schultz, R. D. Fredkin, D. P. Kern, S. A. Rishton, H. Schmid, M. Cali and T. R. Koehler, *J. Appl. Phys., 69* (1991) 5262.

8 A. J. Freeman and R.-Q. Wu, *J. Magn. Magn. Mater., 100* (1991) 497; *104–107* (1992) 1.

9 H. Hasegawa and F. Herman, *J. Phys. (Paris), 49* (1988) C8-1677.

10 U. Gradman, *J. Magn. Magn. Mater., 100* (1991) 481.

11 J. G. Gay and R. Richter, *J. Magn. Magn. Mater., 93* (1991) 315.

12 G. T. Rado, *J. Magn. Magn. Mater., 104–107* (1992) 1679.

13 A. S. Arrot and B. Heinrich, *J. Magn. Magn. Mater., 93* (1991) 571.

14 U. Gradman, H. J. Elmmers and M. Przybylski, *J. Phys. (Paris), 49* (1988) 1665.

15 C. Liu and S. D. Bader, *J. Magn. Magn. Mater., 93* (1991) 307.

16 C. Chappert, P. Beauvillain, P. Bruno, J. P. Chauyineau, M. Galtier, K. Le Dang, C. Marlière, R. Megy, D. Renard, J. P. Renard, J. Seiden, F. Trigui, P. Veillet and E. Velu, *J. Magn. Magn. Mater., 93* (1991) 319.

17 B. Heinrich, Z. Celinski, J. F. Cochran, A. S. Arrot and K. Myrtie, *J. Appl. Phys., 70* (1991) 5769.

18 C. Chappert and P. Bruno, *J. Appl. Phys., 64* (1988) 5736.

19 M. Stampanoni, A. Vaterlaus, M. Aeschlimann, F. Meier and D. Pescia, *J. Appl. Phys., 64* (1988) 5321.

20 C. P. Bean, *J. Appl. Phys., 26* (1955) 1381.
21 J. L. Dormann and D. Fiorani (eds.), *Magnetic Properties of Fine Particles*, North-Holland, Amsterdam, 1992.
22 J. L. Dormann, *Rev. Phys. Appl., 16* (1981) 275.
23 S. Mørup, J. A. Dumesic and H. Topsøe, in R. L. Cohen (ed.), *Applications of Mössbauer Spectroscopy*, Vol. 2, Academic Press, New York, 1980, p. 1.
24 L. Neel, *Ann. Geophys., 5* (1949) 99.
L. Neel, *C.R. Hebd. Acad. Sci. Paris, 228* (1949) 664.
25 W. F. Brown, Jr., *Phys. Rev., 130* (1963) 1677.
26 A. Aharoni, *Phys. Rev., 135* (1964) A447; *177* (1969) 793.
27 I. Eisenstein and A. Aharoni, *Phys. Rev. B, 16* (1977) 1278, 1285.
28 D. H. Jones, *Hyp. Int., 47* (1989) 289.
29 J. L. Dormann and D. Fiorani, in I. A. Campbell and C. Giovanella (eds.), *Relaxation in Complex Systems and Related Topics*, Plenum, New York, 1990, p. 39.
30 A. Aharoni, in J. L. Dormann and D. Fiorani (eds.), *Magnetic Properties of Fine Particles*, North-Holland, Amsterdam, 1992, p. 3.
31 L. Bessais, L. Ben Jaffel and J. L. Dormann, *Phys. Rev. B, 45* (1992) 7805.
32 J. L. Dormann, in J. L. Dormann and D. Fiorani (eds.), *Magnetic Properties of Fine Particles*, North-Holland, Amsterdam, 1992, p. 115.
33 S. Mørup, *J. Magn. Magn. Mater., 37* (1983) 39.
34 M. Hennion, C. Bellouard, I. Mirebeau, H. Blank, J. L. Dormann and C. Djega-Mariadassou, in J. L. Dormann and D. Fiorani (eds.), *Magnetic Properties of Fine Particles*, North-Holland, Amsterdam, 1992, p. 27.
35 R. W. Chantrell, N. Y. Ayoub and J. Popplewell, *J. Magn. Magn. Mater., 53* (1985) 199.
36 C. Djega-Mariadassou, S. Sayouri, J. L. Dormann and G. Villers, *J. Magn. Magn. Mater., 83* (1990) 453.
37 C. Djega-Mariadassou, J. L. Dormann, M. Noguès, G. Villers and S. Sayouri, *IEEE Trans. Magn., 26* (1990) 1819.
38 C. L. Chien, *J. Appl. Phys., 69* (1991) 5267.
39 D. Fiorani, in J. L. Dormann and D. Fiorani (eds.), *Magnetic Properties of Fine Particles*, North-Holland, Amsterdam, 1992, p. 135.
40 S. Mørup, B. S. Clausen and P. H. Christensen, *J. Magn. Magn. Mater., 67* (1987) 249.
41 C. Djega-Mariadassou and J. L. Dormann, in J. L. Dormann and D. Fiorani (eds.), *Magnetic Properties of Fine Particles*, North-Holland, Amsterdam, 1992, p. 191.
42 J. L. Dormann, C. Djega-Mariadassou and J. Jové, *J. Magn. Magn. Mater., 104–107* (1992) 1567.
43 K. Haneda, *Can. J. Phys., 65* (1987) 1233.
44 A. H. Morrish, in J. L. Dormann and D. Fiorani (eds.), *Magnetic Properties of Fine Particles*, North-Holland, Amsterdam, 1992, p. 181.
45 K. O'Grady and R. W. Chantrell, in J. L. Dormann and D. Fiorani (eds.), *Magnetic Properties of Fine Particles*, North-Holland, Amsterdam, 1992, p. 93.
46 R. W. Chantrell and K. O'Grady, in J. L. Dormann and D. Fiorani (eds.), *Magnetic Properties of Fine Particles*, North-Holland, Amsterdam, 1992, p. 103.
47 E. M. Chudnovsky and L. Gunther, *Phys. Rev. Lett., 60* (1988) 661.
48 L. Gunther, in J. L. Dormann and D. Fiorani (eds.), *Magnetic Properties of Fine Particles*, North-Holland, Amsterdam, 1992, p. 213.
49 *Symp. on Macroscopic Quantum Tunnelling, 3M Conf., Houston, TX, 1992*, in *J. Appl. Phys.*, in press.
50 X. X. Zhang, Ll. Balcells, J. M. Ruiz, O. Iglesias and J. Tejada, *Phys. Lett. A, 163* (1992) 130.
51 J. Tejada, X. X. Zhang, Ll. Balcells, C. Ferraté, J. M. Ruiz, F. Badia, O. Iglesias and B. Barbara, in J. L. Dormann and D. Fiorani (eds.), *Magnetic Properties of Fine Particles*, North-Holland, Amsterdam, 1992, p. 225.
52 D. D. Awschalom, J. F. Smyth, G. Grinstein, D. P. Di Vincenzo and D. Coss, *Phys. Rev. Lett., 68* (1992) 3092.
53 B. Barbara, C. Paulsen, L. C. Sampaio, M. Uehara, F. Fruchart, J. L. Tholence, A. Marchand, J. Tejada and S. Linderoth, in J. L. Dormann and D. Fiorani (eds.), *Magnetic Properties of Fine Particles*, North-Holland, Amsterdam, 1992, p. 235.
54 C. Paulsen, L. C. Sampaio, B. Barbara, D. Fruchard, A. Marchand, J. L. Tholence and M. Uehara, *Phys. Lett. A, 161* (1991) 319.
55 J. L. Dormann, D. Fiorani and M. El Yamani, *Phys. Lett. A, 120* (1987) 95.
56 S. Mørup, *Hyp. Int., 60* (1990) 959.
57 R. D. Shull, L. J. Swartzendruber and L. H. Bennet, in G. Green and M. Knox (eds.), *Proc. 6th Int. Cryocoolers Conf.*, David Taylor, Annapolis, 1991, p. 231.
58 R. D. Shull, R. D. McMichael, L. J. Swartzendruber and L. H. Bennet, in J. L. Dormann and D. Fiorani (eds.), *Magnetic Properties of Fine Particles*, North-Holland, Amsterdam, 1992, p. 161.
59 G. C. Hadjipanayis, Z. X. Tang, S. Gangopadhyay, L. Yiping, C. M. Sorensen, K. J. Klabunde, A. Kostikas and V. Papaefthymiou, in J. L. Dormann and D. Fiorani (eds.), *Magnetic Properties of Fine Particles*, North-Holland, Amsterdam, 1992, p. 35.
A. S. Edelstein, F. Kaatz, G. M. Chow and J. M. Perrit, in J. L. Dormann and D. Fiorani (eds.), *Magnetic Properties of Fine Particles*, North-Holland, Amsterdam, 1992, p. 47.
S. Linderoth, in J. L. Dormann and D. Fiorani (eds.), *Magnetic Properties of Fine Particles*, North-Holland, Amsterdam, 1992, p. 61.
60 E. Tronc and J. P. Jolivet, in J. L. Dormann and D. Fiorani (eds.), *Magnetic Properties of Fine Particles*, North-Holland, Amsterdam, 1992, p. 199.
61 J. I. Gittleman, Y. Goldstein and S. Bozowski, *Phys. Rev. B, 5* (1972) 3609.
62 J. L. Dormann, P. Gibart and P. Renaudin, *J. Phys. (Paris), 37* (1976) C6-281.
J. L. Dormann, C. Sella, P. Renaudin, A. Kaba and P. Gibart, *Thin Solid Films, 58* (1979) 265.
63 J. L. Dormann and D. Fiorani, *MRS Bull., 195* (1990) 429.
64 M. H. Annabi, J. L. Dormann, K. C. Sharma and C. Sella, *Proc. Int. Conf. on the Applications of Mossbauer Effect*, Indian National Science Academy, New Delhi, 1982, p. 441.
65 J. R. Cui, J. L. Dormann, C. Sella and P. Renaudin, *IEEE Trans. Magn., 18* (1982) 1589.
C. Djega-Mariadassou, L. Bessais, J. L. Dormann and G. Villers, *J. Phys. (Paris), 49* (1988) C8-1825.
C. Djega-Mariadassou, J. L. Dormann, L. Bessais, P. Renaudin, E. Agostinelli and D. Fiorani, *Hyp. Int., 55* (1990) 933.
66 J. L. Dormann, J. R. Cui and C. Sella, *J. Appl. Phys., 57* (1985) 4283.
67 R. D. Shull, U. Atzmony, A. J. Shapiro, L. J. Swartzendruber and L. H. Bennet, *J. Appl. Phys., 63* (1988) 4261.
68 A. E. Berkowitz, J. R. Mitchell, M. J. Carey, A. P. Young, S. Zhang, F. E. Spada, F. T. Parker, A. Hutten and G. Thomas, *Phys. Rev. Lett., 68* (1992) 3745.
69 J. Q. Xiao, J. S. Jiang and C. L. Chien, *Phys. Rev. Lett., 68* (1992) 3749.
70 J. L. Dormann, L. Bessais and D. Fiorani, *J. Phys. C, 21* (1988) 2015 and cited references therein.

71 S. Mørup, in J. L. Dormann and D. Fiorani (eds.), *Magnetic Properties of Fine Particles*, North-Holland, Amsterdam, 1992, p. 125.

72 J. L. Dormann and M. Noguès, *Phase Transitions, 33* (1991) 159.

73 J. Lafait, in J. L. Dormann and D. Fiorani (eds.), *Magnetic Properties of Fine Particles*, North-Holland, Amsterdam, 1992, p. 245.

74 C. G. Granqvist, *Mater. Sci. Eng., A*, in press.

75 *Symp. on New Giant Magnetoresistance Systems, 3M Conf., Houston, TX, 1992*, in *J. Appl. Phys.*, in press.

76 H. Gleiter, *Prog. Mater. Sci., 33* (1989) 223.

77 H. J. Flecht, in J. L. Dormann and D. Fiorani (eds.), *Magnetic Properties of Fine Particles*, North-Holland, Amsterdam, 1992, p. 55.
T. Haubold, in J. L. Dormann and D. Fiorani (eds.), *Magnetic Properties of Fine Particles*, North-Holland, Amsterdam, 1992, p. 67.
J. E. Epperson and R. W. Siegel, in J. L. Dormann and D. Fiorani (eds.), *Magnetic Properties of Fine Particles*, North-Holland, Amsterdam, 1992, p. 83.

78 R. W. Siegel, *Mater. Sci. Eng., A*, in press.

79 J. Kwo, M. Hong, D. B. McWhan, Y. Yafet, R. M. Fleming, F. J. Di Salvo, J. V. Waszczak, C. F. Majkrzak, D. Gibbs, A. I. Goldmann, P. Boni, J. Bohr, H. Grimm, C. L. Chien and J. W. Cable, *J. Phys. (Paris), 49* (1988) C8-1651.

80 J. Mathon, *J. Magn. Magn. Mater., 100* (1991) 527.
J. Mathon, D. M. Edwards, R. B. Muniz and M. S. Phan, *J. Magn. Magn. Mater., 104–107* (1992) 1721.

81 P. Grunberg, J. Barnas, F. Saurenbach, J. A. Fuss, A. Wolf and M. Vohl, *J. Magn. Magn. Mater., 93* (1991) 58.
A. Fert, A. Barthelemy, P. Etienne, S. Lequien, R. Loloce, D. K. Lottis, D. H. Mosca, F. Petroff, W. P. Pratt and P. A. Schroeder, *J. Magn. Magn. Mater., 104–107* (1992) 1712.

82 C. J. Lin, G. L. Gorman, C. H. Lee, R. F. C. Farrow, E. E. Marinero, H. V. Do, H. Notarys and G. J. Chien, *J. Magn. Magn. Mater., 93* (1991) 194.
F. J. A. Den Broeder, W. Hoving and P. J. H. Bloemen, *J. Magn. Magn. Mater., 93* (1991) 562.

83 H. J. de Witt, *Rep. Progr. Phys., 55* (1992) 113.

An examination of the implications of void growth in submicrometer and nanocrystalline structures

Yan Ma and Terence G. Langdon

Departments of Materials Science and Mechanical Engineering, University of Southern California, Los Angeles, CA 90089-1453 (USA)

Abstract

Tensile experiments were conducted on fully dense tetragonal zirconia stabilized with approximately 3 mol% yttria and with a submicrometer grain size. Examination of the specimens after superplastic-like flow revealed the development of internal cavitation and the extent of the cavitation increased with increasing strain rate. Nanocrystalline materials have an apparent potential for superplastic forming operations but this potential may be limited because of the presence of residual porosity from the fabrication process or because of the nucleation and growth of cavities during superplastic flow. It is shown by calculation that diffusion will play a dominant role in the growth of internal voids in nanocrystalline structures because each grain boundary impinging on a void provides a path for rapid diffusion.

1. Introduction

Structural superplasticity is of current scientific interest because of the potential for industrial applications. Although the word *superplasticity* was introduced into the English language in 1947 as a direct translation of the Russian word *sverkhplastichnost'* meaning ultra-high plasticity [1], it is surprising to note that a formal definition of superplasticity was put forward only in 1991 [2]: "Superplasticity is the ability of a polycrystalline material to exhibit, in a generally isotropic manner, very high tensile elongations prior to failure."

Superplasticity requires a very small and stable grain size. In metals, the grain sizes are typically of the order of 1–10 μm. Since superplasticity is a diffusion-controlled process, it also requires a high testing temperature of the order of $0.5T_m$, where T_m is the absolute melting temperature. In metallic alloys, the retention of a very small grain size at elevated temperatures is achieved by using a two-phase alloy or by incorporating a fine dispersion of a second phase to act as a grain refiner [3].

The first demonstration of superplasticity in a crystalline ceramic tested in tension was presented by Wakai *et al.* [4] in 1986. Subsequently, there were numerous reports of superplasticity in ceramics and these results have been summarized in several reviews [5–8]. An important difference between superplastic metals and ceramics is that the grain sizes in the ceramics are usually less than 1 μm and often of the order of 0.3–0.5 μm [9].

Very recently, interest has developed in the possibility of attaining superplasticity in materials with nanocrystalline grain sizes. However, although there has been speculation on the possibility of extreme ductilities in nanocrystalline structures [10, 11], and there are limited experimental data from the compression testing of nanocrystalline TiO_2 [12, 13] and ZrO_2 [14, 15] which tend to confirm these possibilities, no tensile experiments have been conducted to date on fully dense nanocrystalline materials to confirm the superplastic capabilities. Nevertheless, there are implications from tests on nanocrystalline TiO_2 using a gas pressure forming technique [16] that the tensile properties, and therefore the potential for superplastic forming, may be enhanced by reducing the grain size into the nanocrystalline range.

A potential limitation on superplastic forming operations is the development and growth of internal cavities. It is now well established that internal cavities develop in most superplastic metals during superplastic deformation [17–19] and there is also evidence for cavitation in superplastic ceramics [8]. This problem, when combined with the difficulty of attaining a fully dense nanocrystalline structure, may tend to limit the utility of these materials in forming operations.

This paper addresses two aspects of the problem. First, it demonstrates by experiment that the nucleation and growth of cavities is important in the deformation of a typical ceramic, essentially fully dense, having a grain size within the submicrometer range. Second, it shows by calculation that internal cavities in a nanocrystalline structure will grow extremely rapidly since

0921-5093/93/$6.00

each grain boundary impinging on a void will provide a path for rapid vacancy diffusion.

2. Experimental material and procedures

Tests were conducted on tensile specimens, with a gage length of 15 mm, prepared from tetragonal ZrO_2 stabilized with 5.2 wt.% Y_2O_3, equivalent to approximately 3 mol% yttria; this material is termed 3Y-TZP. The material was obtained from the Nippon Kagaku Togyo Co. in Sakai, Japan, and it was nominally of high purity containing, in weight per cent, 0.005 Al_2O_3, 0.002 Fe_2O_3, 0.003 Na_2O and 0.002 SiO_2. No porosity was visible in this material using optical microscopy prior to testing and, therefore, as in the earlier work of Wakai and coworkers [4, 20] using material from the same source, the specimens appeared to be fully dense. Prior to testing, the average grain size d was determined as approximately 0.5 μm.

Tests were carried out in tension using an Instron testing machine operating at a constant rate of cross-head displacement. The specimens were pulled to failure in air at a temperature of 1723 ± 2 K. After failure, specimens were sectioned, carefully polished on SiC papers and diamond paste, and examined optically for internal cavitation.

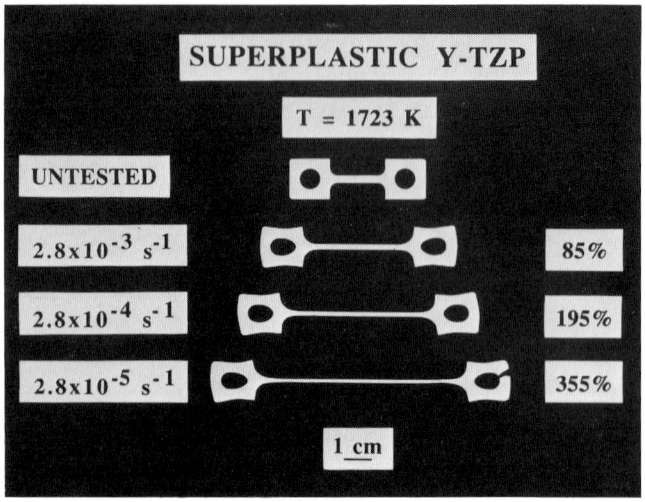

Fig. 1. Tensile specimens of 3Y-TZP after pulling to failure at 1723 K at strain rates from 2.8×10^{-3} to 2.8×10^{-5} s^{-1}; the upper specimen is untested.

3. Experimental results

Figure 1 shows three fractured tensile specimens tested over a range of strain rates from 2.8×10^{-3} to 2.8×10^{-5} s^{-1}; the upper specimen is untested and the elongations listed to the right of each specimen refer to deformation occurring within the original gage length. Inspection of Fig. 1 shows that there is superplastic-like flow with elongations of up to over 300% and with the specimens pulling out fairly uniformly prior to failure. There is also some deformation within the gripping region.

Using the experimental data for the three specimens shown in Fig. 1, the testing strain rate $\dot{\varepsilon}$ was plotted logarithmically against the flow stress σ. The result is shown in Fig. 2, together with experimental points from earlier experiments conducted at 1723 K on 3Y-TZP with a grain size of approximately 0.3 μm by Wakai *et al.* [4] and Nieh *et al.* [21]. The present results are very close to those of Nieh *et al.* [21] and the stress exponent n in both experiments is approximately 3.0. In the earlier investigation by Wakai *et al.* [4], the stress exponent was $n \approx 1.9$. The difference in the values of n in experiments conducted under nominally similar conditions has been attributed to differences in the impurity contents of the test samples [6]. Thus, a nominally "low" purity material, as in the experiments by Wakai *et al.* [4], tends to give a low value of n whereas a nominally "high" purity material, as in the present experiments and in the work of Nieh *et al.* [21], tends to give a high value of n. As discussed elsewhere [6], these

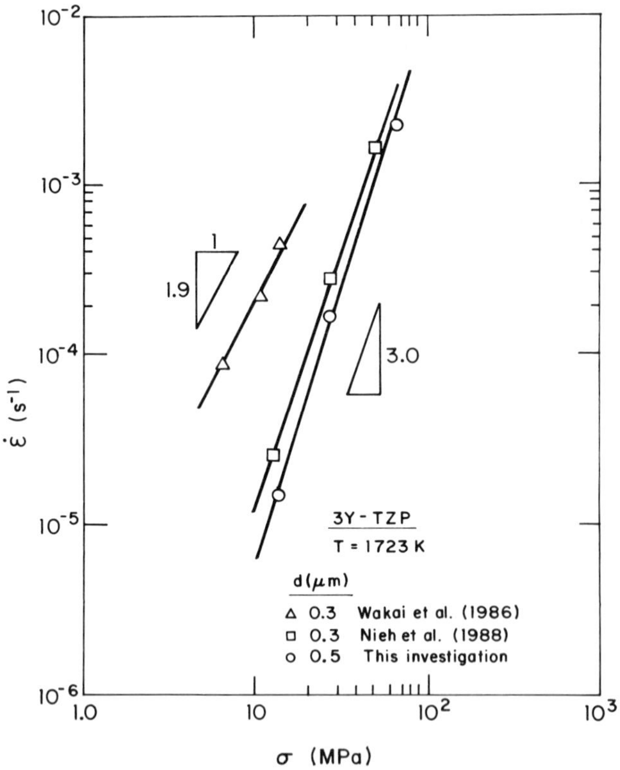

Fig. 2. Strain rate *vs.* stress for 3Y-TZP tested at 1723 K in the present investigation and in the experiments of Wakai *et al.* [4] and Nieh *et al.* [21].

differences are associated with the increasing presence of an intergranular glassy phase along the grain boundaries in materials with a high impurity content.

Inspection of the three fractured specimens after testing revealed the presence of extensive internal cavitation within the gage lengths. Examples are shown in Fig. 3 for the three initial strain rates of (a) 2.8×10^{-3} s^{-1}, (b) 2.8×10^{-4} s^{-1}, and (c) 2.8×10^{-5} s^{-1}; the tensile axis is horizontal in each photomicrograph. It is clear from Fig. 3 that the extent of cavitation increases with increasing strain rate, and there is some evidence at the fastest strain rate for the interlinkage of cavities perpendicular to the tensile axis. In general, however, the cavities are large, essentially rounded, and they appear to be reasonably stable.

4. Discussion

4.1. Significance of the experimental results

The experiments show that fully dense 3Y-TZP, with a submicrometer grain size, exhibits superplastic-like flow at 1723 K with tensile elongations of up to over 300%. However, extensive internal cavitation develops during the deformation process.

Three features of the cavitation appear to contrast with observations on typical superplastic metals. First, both qualitative observations [22] and quantitative measurements [23] on metals show that the extent of cavitation increases with decreasing strain rate whereas in 3Y-TZP, in both this investigation and in the experiments of Wakai and coworkers [4, 24], the extent of cavitation increases with increasing strain rate. A similar increase in cavitation with increasing strain rate was also reported in a 3Y-TZP/20%Al_2O_3 composite by Chokshi *et al.* [25] and in 6Y-TZP in Bravo-León *et al.* [26] using a material containing some initial residual porosity. Second, there is no evidence in these experiments on 3Y-TZP for the formation of stringers of cavities parallel to the tensile axis, although this is a common feature of superplastic metals [27, 28] and it was also reported in 3Y-TZP in a specimen deformed to an elongation of about 700% at 1823 K [29]. Third, the evidence for cavity interlinkage perpendicular to the tensile axis at a point remote from the fracture tip, as in Fig. 3(a), contrasts with metals where this type of interlinkage generally only occurs in the immediate vicinity of the fracture tip [30].

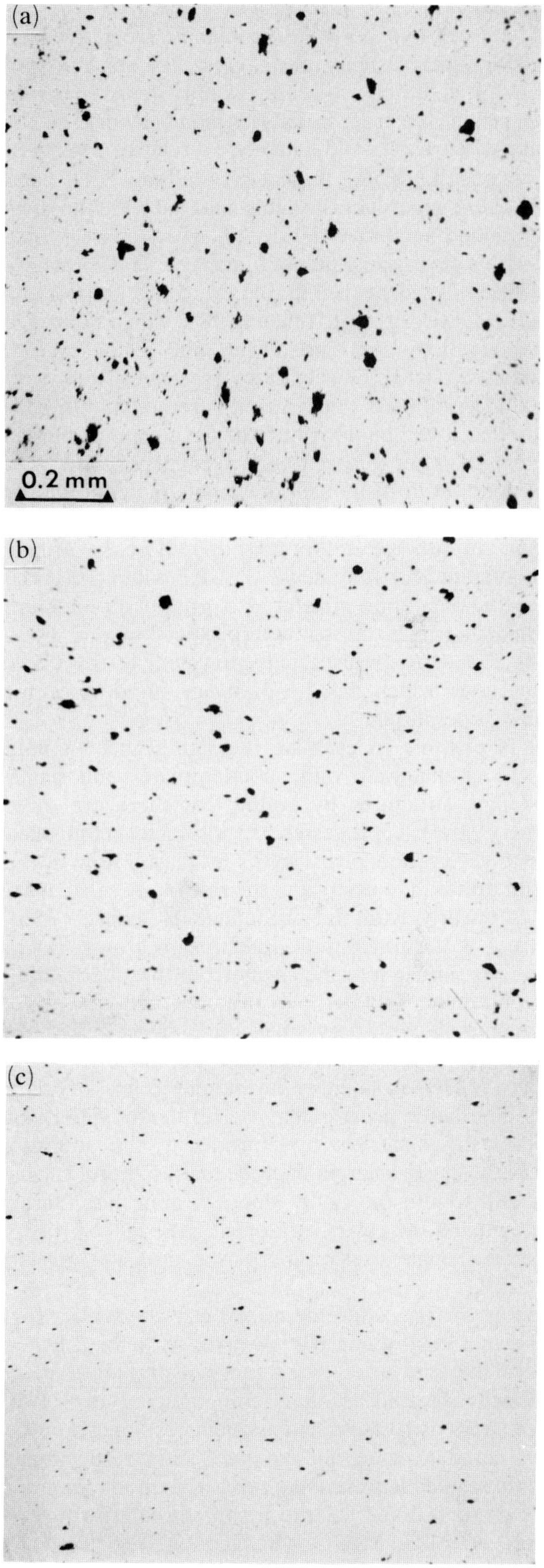

Fig. 3. Internal cavitation in 3Y-TZP tested to failure at 1723 K at strain rates of (a) 2.8×10^{-3} s^{-1}, (b) 2.8×10^{-4} s^{-1} and (c) 2.8×10^{-5} s^{-1}; the tensile axis is horizontal.

4.2. Implications for nanocrystalline structures

There are two potential problems associated with attaining superplasticity in nanocrystalline structures.

First, it is difficult to achieve a fully dense structure while retaining a grain size of less than 0.1 μm [31]. For example, Karch *et al.* [32] attained a relative density of approximately 75% in nanocrystalline TiO_2 and subsequent experiments on this material by Hahn *et al.* [33] yielded a relative density of approximately 96% but with a grain size of about 1 μm. For 3Y-TZP where a compressive stress is imposed, Mayo [15] reported a relative density of approximately 98% with a grain size of about 0.16 μm and Owen and Chokshi [34] obtained a relative density of approximately 99% with a grain size of 0.15 μm. Thus, it is reasonable to conclude that it will be difficult to achieve nanocrystalline ceramics without at least a small residual porosity.

Second, even if the nanocrystalline material is fully dense, internal cavities are likely to develop and grow during any superplastic forming operation. As already demonstrated for fully dense 3Y-TZP when the initial grain size was approximately 0.5 μm, large voids are formed internally during superplastic flow and these voids appear to be especially prevalent at the faster strain rates which will be of major importance for forming operations.

It is possible to estimate the significance of void growth in materials with submicrometer and nanocrystalline structures by noting that there are three primary growth processes. In plasticity controlled growth (PCG), the cavity grows by plastic flow in the region around the cavity and the rate of growth, $dr/d\varepsilon$, varies linearly with the instantaneous radius r [35], where ε is the total strain. In diffusion growth (DG), vacancies diffuse into the cavity from the surrounding grain boundary and $dr/d\varepsilon$ is proportional to $1/r^2$ [36]. For materials with very small grain sizes, it is more realistic to consider the diffusion growth of a single cavity (SCDG) isolated on an infinite grain boundary [37]. The latter process may be further modified for superplastic materials by considering the situation where the cavity is larger than the grain size and vacancies diffuse into the cavity along all grain boundaries impinging on the cavity surface [38]; the latter process is termed single cavity superplastic diffusion growth (SCSPDG).

Using the procedure described elsewhere [38], Fig. 4 shows a logarithmic plot of $dr/d\varepsilon$ *vs.* r for a hypothetical material tested over a range of strain rates at an arbitrarily selected absolute temperature T of 800 K and a stress of 10 MPa, and assuming exponents of the stress n and inverse grain size p of 3 and 2 respectively. The three lines represent the predicted trends for grain sizes of 10, 1.0 and 0.1 μm for the three growth processes of PCG, SCDG and SCSPDG respectively. Inspection shows that, because of the large numbers of boundaries impinging on the cavity when the grain size is very small, the SCSPDG process becomes of increasing importance at large cavity radii when the grain size is reduced. This type of plot may be further extended to a grain size of 0.03 μm (30 nm), as shown in Fig. 5 for $T = 800$ K and a strain rate of 10^{-4} s^{-1}. This plot serves to demonstrate the dominant role of SCSPDG when grain sizes are in the nanometer range.

Figure 6 shows examples of the effect of testing strain rate, from 10^{-2} to 10^{-6} s^{-1}, for grain sizes of (a) 0.1 μm and (b) 0.03 μm. A comparison of these two plots confirms again the important role of the SCSPDG process for cavity radii greater than about 10 μm. It is therefore concluded that the presence of a large number of grain boundary paths in nanocrystalline structures will lead to the rapid growth of internal voids, and this growth process will be enhanced when the experimental strain rate is very low.

Finally, it should be noted that the 3Y-TZP specimens tested in this work exhibit maximum cavitation at the fastest experimental strain rate. This suggests that, at least under the present experimental conditions, diffusion growth is not sufficiently rapid to account fully for cavity development and instead, as suggested also by Bravo-León *et al.* [26] for 6Y-TZP, cavities develop more readily at the higher strain rates owing to a failure to accommodate fully grain boundary sliding during superplastic deformation.

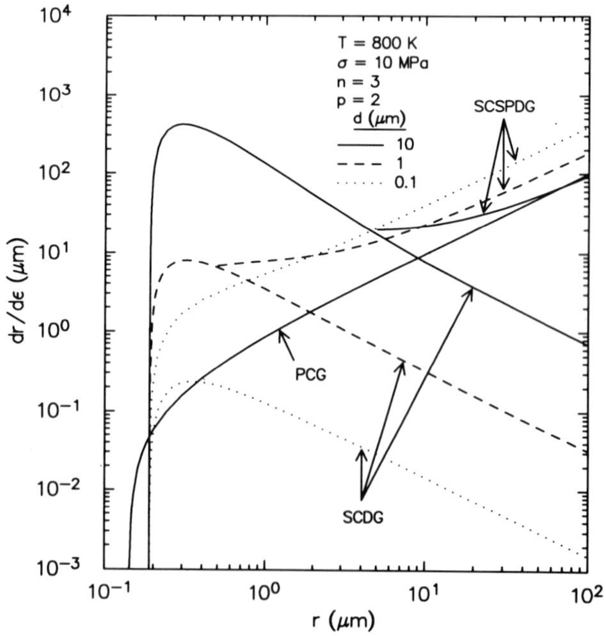

Fig. 4. Logarithmic plot of cavity growth rate *vs.* instantaneous radius for grain sizes from 10 to 0.1 μm and a stress of 10 MPa; PCG, SCDG and SCSPDG refer to three different growth processes.

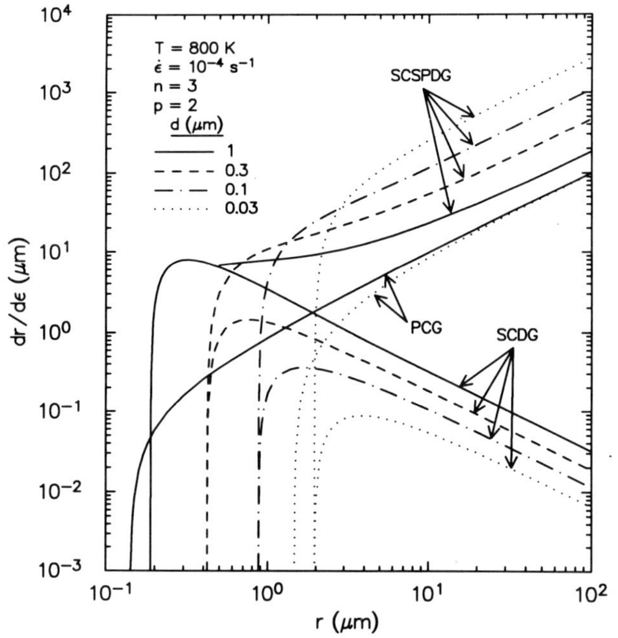

Fig. 5. As in Fig. 4 for grain sizes from 1 μm to 30 nm and a strain rate of 10^{-4} s^{-1}.

5. Summary and conclusions

(1) Tensile experiments were conducted at 1723 K on fully dense tetragonal zirconia stabilized with approximately 3 mol% yttria (termed 3Y-TZP) and with a submicrometer grain size. The results show that the material exhibits superplastic-like flow with elongations of up to over 300% but with the concomitant development of internal cavitation which increases in extent with increasing strain rate.

(2) It is concluded that the superplastic properties of nanocrystalline structures may be limited by the presence of internal cavitation resulting from either residual porosity after the fabrication process or the nucleation of cavities during superplastic flow. It is shown by calculation that the diffusion growth of cavities will be significantly enhanced in nanocrystalline structures because of the large number of grain boundaries intersecting each void.

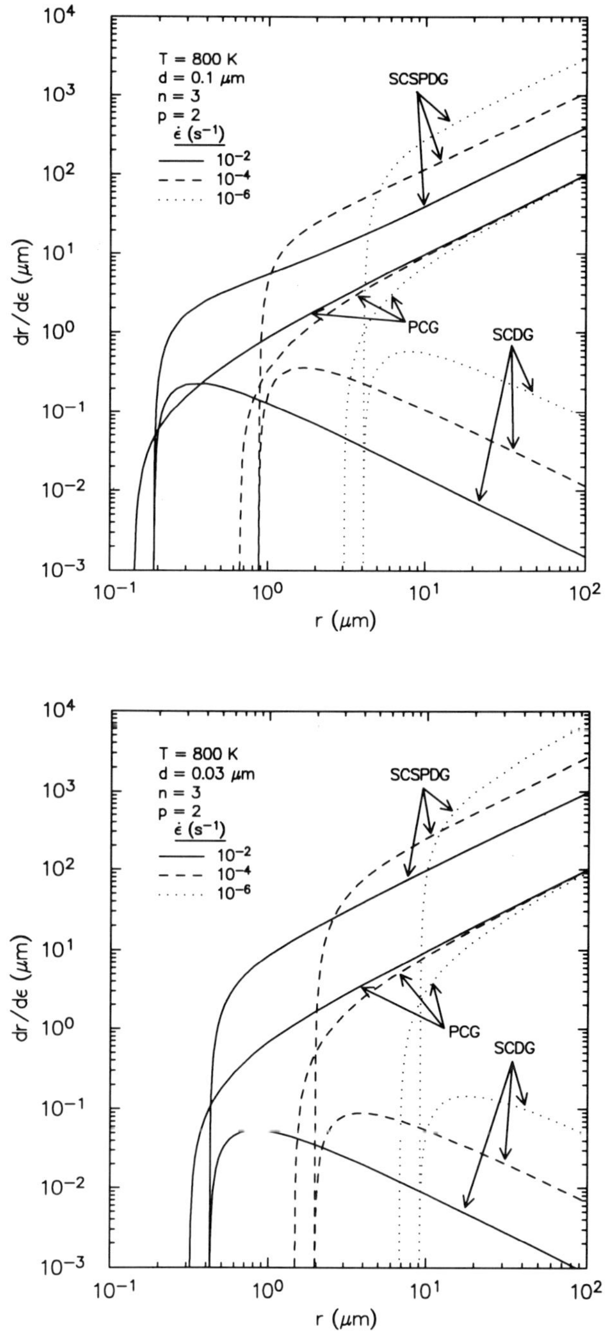

Fig. 6. As in Fig. 4 for strain rates from 10^{-2} to 10^{-6} s^{-1} and grain sizes of (a) 0.1 μm and (b) 30 nm.

Acknowledgment

This work was supported by the United States Department of Energy under Grant No. DE-FG03-92ER45472.

References

1 T. G. Langdon, in S. Hori, M. Tokizane and N. Furushiro (eds.), *Superplasticity in Advanced Materials*, The Japan Society for Research on Superplasticity, Osaka, 1991, p. 3.
2 T. G. Langdon and J. Wadsworth, in S. Hori, M. Tokizane and N. Furishiro (eds.), *Superplasticity in Advanced Materials*, The Japan Society for Research on Superplasticity, Osaka, 1991, p. 847.
3 T. G. Langdon, *Metall. Trans. A, 13* (1982) 689.
4 F. Wakai, S. Sakaguchi and Y. Matsuno, *Adv. Ceram. Mater., 1* (1986) 259.
5 Y. Maehara and T. G. Langdon, *J. Mater. Sci., 25* (1990) 2275.
6 T. G. Langdon, *J. Met., 42* (7) (1990) 8.

7 T. G. Nieh, J. Wadsworth and F. Wakai, *Int. Metall. Rev., 36* (1991) 146.
8 A. H. Chokshi, A. K. Mukherjee and T. G. Langdon, *Mater. Sci. Eng. Rep., R10* (1993) 237–274.
9 T. G. Langdon, *Bull. Jpn. Inst. Met., 30* (1991) 1001.
10 R. W. Siegel, *Mater. Res. Soc. Symp. Proc., 196* (1990) 59.
11 R. W. Siegel, *Ann. Rev. Mater. Sci., 21* (1991) 559.
12 H. Hahn, J. Logas, H. J. Höfler, P. Kurath and R. S. Averback, *Mater. Res. Soc. Symp. Proc., 196* (1990) 71.
13 H. Hahn and R. S. Averback, *J. Am. Ceram. Soc., 74* (1991) 2918.
14 M. Ciftcioglu and M. J. Mayo, *Mater. Res. Soc. Symp. Proc., 196* (1990) 77.
15 M. J. Mayo, in S. Hori, M. Tokizane and N. Furishiro (eds.), *Superplasticity in Advanced Materials*, The Japan Society for Research on Superplasticity, Osaka, 1991, p. 541.
16 Z. Cui and H. Hahn, *Nanostruct. Mater., 1* (1992) 419.
17 T. G. Langdon, *Met. Sci., 16* (1982) 175.
18 B. P. Kashyap and A. K. Mukherjee, *Res Mech., 17* (1986) 293.
19 J. Pilling and N. Ridley, *Res Mech., 23* (1988) 31.
20 F. Wakai, N. Murayama and S. Sakaguchi, in S. Somiya, N. Yamamoto and H. Yanagida (eds.), *Advances in Ceramics: Science and Technology of Zirconia III*, Vol. 24B, The American Ceramic Society, Westerville, OH, 1988, p. 583.
21 T. G. Nieh, C. M. McNally and J. Wadsworth, *Scripta Metall.. 22* (1988) 1297.
22 S.-A. Shei and T. G. Langdon, *J. Mater. Sci., 13* (1978) 1084.
23 Y. Ma, X. Zhao and T. G. Langdon, in B. Wilshire and R. W. Evans (eds.), *Proc. Fourth Int. Conf. on Creep and Fracture of Engineering Materials and Structures*, The Institute of Metals, London, 1990, p. 199.
24 F. Wakai, Y. Kodama and T. Nagano, in *Lattice Defects in Ceramics, Jpn. J. Appl. Phys., Ser., 2* (1989) 57.
25 A. H. Chokshi, T. G. Nieh and J. Wadsworth, *J. Am. Ceram. Soc., 74* (1991) 869.
26 A. Bravo-León, M. Jiménez-Melendo and A. Domínguez-Rodríguez, *Acta Metall. Mater., 40* (1992) 2717.
27 H. Ishikawa, D. G. Bhat, F. A. Mohamed and T. G. Langdon, *Metall. Trans. A, 8* (1977) 523.
28 A. H. Chokshi and T. G. Langdon, *Acta Metall., 37* (1989) 715.
29 D. J. Schissler, A. H. Chokshi, T. G. Nieh and J. Wadsworth, *Acta Metall. Mater., 39* (1991) 3227.
30 A. H. Chokshi and T. G. Langdon, *Acta Metall. Mater., 38* (1990) 867.
31 D. M. Owen, A. H. Chokshi, Y. Ma and T. G. Langdon, *Proc. Br. Ceram. Soc.*, in press.
32 H. Karch, R. Birringer and H. Gleiter, *Nature, 330* (1987) 556.
33 H. Hahn, J. Logas and R. S. Averback, *J. Mater. Res., 5* (1990) 609.
34 D. M. Owen and A. H. Chokshi, *Proc. 1st Int. Conf. on Nanostructured Materials*, in press.
35 J. W. Hancock, *Met. Sci., 10* (1976) 319.
36 M. V. Speight and W. Beeré, *Met. Sci., 9* (1975) 190.
37 Y. Ma and T. G. Langdon, *Scripta Metall. Mater., 26* (1992) 1239.
38 Y. Ma and T. G. Langdon, *Mater. Res. Soc. Symp. Proc., 196* (1990) 39.

Materials Science and Engineering, A168 (1993) 231–234

The electrical characterization of grain boundaries in ultra-fine grained Y-TZP

C. S. Chen, M. M. R. Boutz, B. A. Boukamp, A. J. A. Winnubst, K. J. de Vries and A. J. Burggraaf
Laboratory of Inorganic Chemistry, Catalysis and Inorganic Materials Science, Faculty of Chemical Technology, University of Twente, P.O. Box 217, 7500AE, Enschede (Netherlands)

Abstract

Starting from a sinter reactive powder prepared by a gel precipitation technique, dense, ultra-fine grained (100–200 nm) yttria stabilized tetragonal zirconia ceramics were obtained by sinter forging at a temperature of 1100 °C or by pressureless sintering at 1150 °C. The pressureless sintered compacts were subjected to further heat treatments at temperatures of 1250–1450 °C or compressive deformation at 1250 °C under uniaxial stresses of 20–100 MPa. The obtained samples were characterized mainly by impedance spectroscopy. After compressive deformation a decrease in grain boundary resistivity was found which increased with applied stress. This can be interpreted in terms of a decrease in impurity segregation and a partial removal by compressive deformation of a poorly conducting amorphous film around the grains. It was also found that the grain boundary resistivity of samples sintered at 1150 °C could be considerably reduced by further pressureless heat treatments at temperatures above 1250 °C. This effect is probably owing to dewetting of the grain boundary and dissolution of grain boundary impurities into the bulk of the grains.

1. Introduction

At present there is increasing interest in nanoscale ceramics [1]. As indicated by Burggraaf [1] and Burggraaf *et al.* [2], the characteristics of ceramics with grain sizes in the nanometer range are (i) a large fraction of atoms is contained in the grain boundary region or in a region with characteristic physical length (Debye length, domain size), and (ii) transport properties are changed, *e.g.* enhanced diffusivity and deformability. One of these nanoscale ceramics under investigation in our laboratory is yttria stabilized tetragonal zirconia polycrystal (Y-TZP). Boutz *et al.* [3] have reported on the preparation of 2.6mol%Y_2O_3–ZrO_2 (2.6Y-TZP) with a grain size of 100 nm and a relative density of 94% using sinter-forging. The particle sizes in the starting green compacts were around 7–10 nm, the pore sizes were around 6 nm. It is expected that these ultra-fine grained ceramics may possess special grain boundary related properties.

It is recognized that the presence of an amorphous silicate phase at grain boundaries in Y-TZP has a strong influence on both the electrical [4] and creep [5] properties. The distribution of this amorphous phase strongly depends on its composition and the thermo-mechanical history. Several ceramic processing parameters such as heat treatment temperature, cooling rate, pressure [6] and ambient [7] all affect the nature and location of the grain boundary phase.

In the present investigation the influence of processing parameters (temperature, pressure) on the grain boundary properties of Y-TZP (initial grain size 100–200 nm) was studied. Impedance spectroscopy (IS), in combination with transmission electron microscopy (TEM) and X-ray photoelectron spectroscopy (XPS), was used to gain a better insight into the grain boundary properties of ultra-fine grained Y-TZP.

2. Experimental details

2.1. Preparation and processing

The ultra-fine grained starting powders, 2.6mol%Y_2O_3–ZrO_2 with a crystallite size of 7–10 nm, were prepared by a gel precipitation technique using metal chlorides as precursor chemicals. Details of the synthesis have been presented elsewhere [8]. Of these powders cylindrical compacts were prepared by cold isostatic compaction (400 MPa).

In order to improve the strength of the green compact samples for the sinter forging experiments a pre-sintering step was carried out by heating to 950 °C at a rate of 120 °C h^{-1}, immediately followed by cooling. Dimensions after the pre-sintering step were typically

0921-5093/93/$6.00

15.7 mm (height) × 7.3 mm (diameter). Sinter forging tests were performed in air at 1100 °C under constant load corresponding to initial stresses of 28, 58 and 84 MPa respectively. Further details are given elsewhere [3].

Samples for hot forging experiments were prepared by sintering for 10 h in air at 1150 °C, followed by machining to identical dimensions (6.5 mm diameter and 8.5 mm height). Compression was carried out at 1250 °C under constant loads corresponding to initial stresses of 20, 60 and 100 MPa respectively. When the sample height was reduced to half the original size the pressure was removed and the sample cooled to room temperature at a rate of 300 °C h^{-1}.

The samples that were sintered at 1150 °C for 10 h without applying pressure were subjected to further heat treatments in air at temperatures between 1250 and 1450 °C for 3 h, followed by cooling at a rate of 300 °C h^{-1}.

2.2. Characterization of the Y-TZP

The grain diameter was determined by the linear intercept technique [9] from SEM images of polished and subsequently thermally etched samples.

For the impedance measurements disk-shaped samples were cut with a diameter of 5.1 mm and a thickness of 3.0 mm. After polishing, a 200 nm thick platinum layer was deposited onto both sides by d.c. sputtering. The electrodes were cured and stabilized by heat treatment at 750–950 °C for 2 h and cooling at a rate of 300 °C h^{-1}. The impedance was measured using a Solartron 1255 frequency response analyzer over the frequency range 1–10^6 Hz and in the temperature range 300–590 °C. Only results obtained at 341 °C are presented here.

The impedance spectra were analysed using the non-linear least squares impedance analysis package Equivalent Circuit [10]. As model function the equivalent circuit proposed by Bauerle [11] was used. By simultaneously fitting all equivalent circuit parameters in the NLLSF procedure, the bulk and grain boundary resistances can be separated [10]. The grain boundary resistivity is affected by the impurity segregation at the grain boundaries as well as by the size of the grains. Hence, in order to account for both effects, the grain boundary resistivity was corrected for grain size. This grain boundary resistivity per unit surface area R_{sp} is calculated by [12]

$$R_{sp} = R_{gb}^{mac} D_g$$

where R_{gb}^{mac} is the macroscopic grain boundary resistivity normalized for specimen dimensions, and D_g is the grain size.

3. Results and discussion

3.1. Grain boundary properties of the deformed Y-TZP

In Fig. 1 the impedance spectra are shown for 2.6Y-TZP samples prepared by sinter forging at 1100 °C under respectively 28, 58 and 84 MPa initial stress. The grain sizes for these samples were in the range 100–130 nm. In Fig. 2 the impedance spectra are shown for three 2.6Y-TZP samples treated by hot forging at 1250 °C at initial stresses of respectively 20, 60 and 100 MPa. The grain sizes were in the range 210–240 nm. Samples that were deformed either by sinter forging or by hot forging showed, with increasing initial stress, a decrease in the corrected grain boundary resistivity R_{sp}. This effect is most pronounced for the highest stress applied, as can be seen in Fig. 3.

Although samples prepared by sinter forging at 1100 °C and samples treated by hot forging at 1250 °C showed similar trends of a decrease in R_{sp} with increas-

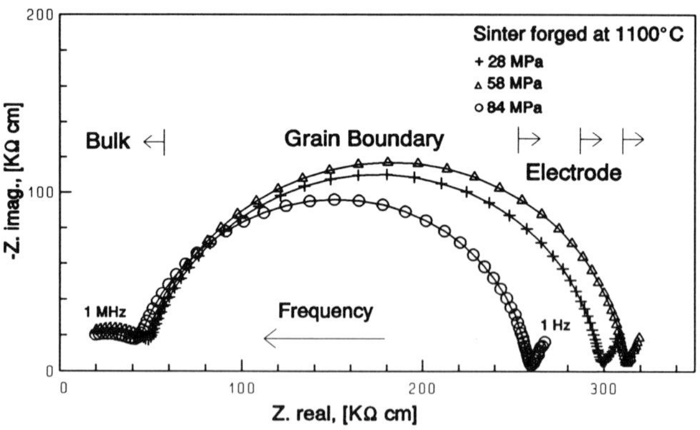

Fig. 1. Impedance spectra for 2.6mol%Y_2O_3–ZrO_2 ceramics prepared by sinter forging at 1100 °C under various initial stresses, measured at 341 °C in air.

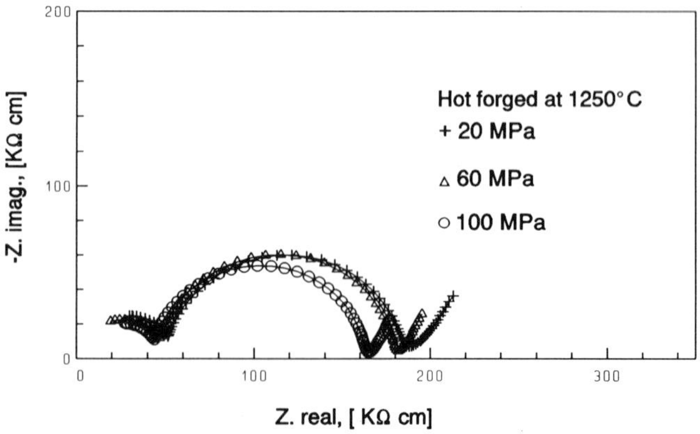

Fig. 2. Impedance spectra for 2.6mol%Y_2O_3–ZrO_2 ceramics deformed at 1250 °C under various initial uniaxial stresses, measured at 341 °C in air.

ing stress, the mechanism of each trend may be quite different. In the case of sinter forging, the decrease in R_{sp} should be attributed to a decrease in the segregation of impurities. As demonstrated by Boutz *et al.* [3], sinter forging allows a reduction in both the sintering temperature and sintering time required to obtain dense ceramics. The important advantages of these reductions are (i) grain growth is markedly suppressed, and (ii) there is less time available for impurities to segregate to the grain boundaries, forming amorphous films around the grains. By increasing the stress imposed on the sintering compact the sintering temperature can be lowered and the sintering time required to obtain a dense ceramic, can be reduced. Consequently the extent to which segregation can take place is reduced, hence R_{sp} will be decreased.

Samples obtained by pressureless sintering at 1150 °C, which were to be used in the hot forging experiments, contained a continuous film around the grains which impedes the transport of oxygen ions across the grain boundaries. After compressive deformation at 1250 °C, this film became discontinuous. Impurities in the films were partially removed from the grain boundaries, resulting in a reduction in R_{sp}. In a separate investigation by Boutz [13], using TEM and XPS, some direct evidence was provided to support the above explanation. The TEM observations showed that a continuous film (thickness less than or equal to 1 nm) of an amorphous intergranular phase existed in the as-sintered materials, while this amorphous phase was absent at certain grain boundaries after deformation. XPS analysis revealed that the average grain boundary concentrations of yttrium, aluminium and silicon were decreased after compressive deformation. This effect is most pronounced in the samples deformed at the higher stresses.

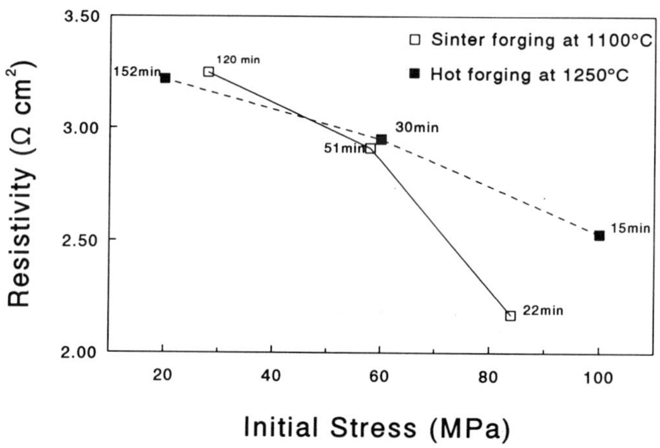

Fig. 3. Corrected grain boundary resistivity for the deformed 2.6mol%Y_2O_3–ZrO_2 ceramics as a function of initial stress and processing time. Temperature is 341 °C.

3.2. *Grain boundary properties of Y-TZP heat treated at various temperatures*

2.6Y-TZP ceramics sintered at 1150 °C for 10 h were further pressureless heat treated at temperatures of 1250–1450 °C for 3 h. Figure 4 shows the impedance spectra of the heat treated Y-TZP ceramics. As shown in Fig. 5, the corrected grain boundary resistivity R_{sp} decreases sharply after heat treatment at 1250 °C while the bulk resistivity remains almost unchanged. This possibly indicates that dewetting of the grain boundaries takes place, resulting in cleaning of the grain boundaries [14]. Furthermore, impurities such as iron and aluminium, initially present at the grain boundaries, will dissolve into the bulk of the grains at higher temperatures, as has been demonstrated by Hughes and Badwal [15].

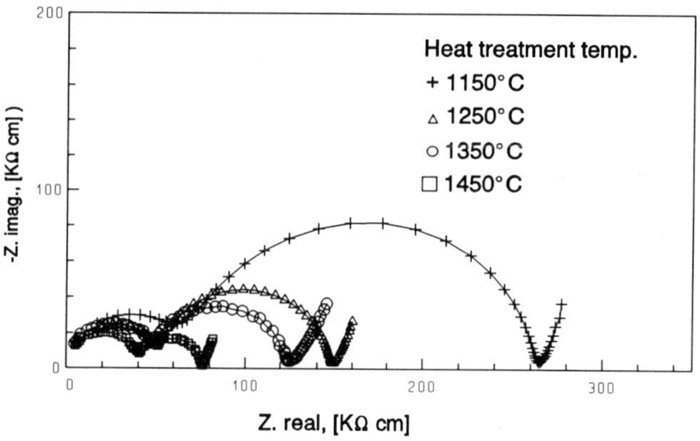

Fig. 4. Impedance spectra for 2.6mol%Y_2O_3–ZrO_2 ceramics pressureless heat treated at different temperatures for 3 h, measured at 341 °C in air.

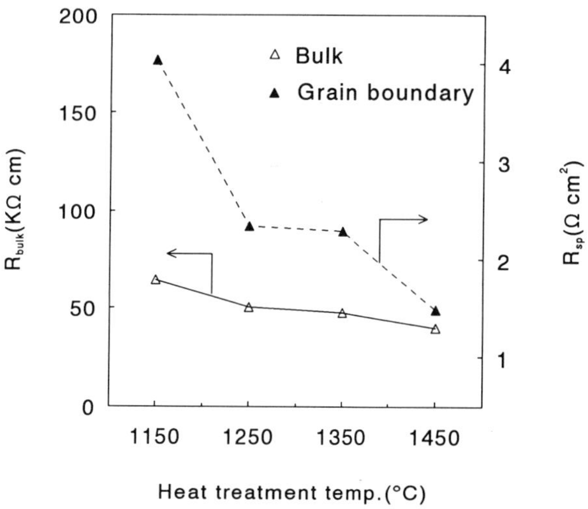

Fig. 5. Bulk resistivity and corrected grain boundary resistivity of 2.6mol%Y_2O_3–ZrO_2 pressureless heat treated at different temperatures, measured at 341 °C in air.

4. Conclusions

Dense, ultra-fine grained (100–130 nm) Y-TZP ceramics were prepared by sinter forging at a temperature of 1100 °C. Examination of the grain boundaries by impedance spectroscopy indicated that segregation of the impurities was suppressed owing to a lowering of the sintering temperature and a shortening of the required sintering time, which is a consequence of the exertion of uniaxial stress on the sintering body.

After compressive deformation at 1250 °C the pressureless sintered samples showed a decrease in the grain boundary resistivity, revealing that a continuous amorphous film which is formed during sintering at 1150 °C and which is rich in impurities can be partially removed from the grain boundaries, becoming discontinuous. This improves the transport of oxygen ions across the grain boundaries.

The grain boundary resistivity of the sample pressureless sintered at 1150 °C could also be considerably reduced by further heat treatment at temperatures above around 1250 °C. This effect is probably due to dewetting of the grain boundaries and dissolution of grain boundary impurities into the bulk of the grains.

Acknowledgement

The investigations were supported by the Netherlands Foundation for Chemical Research (SON) with financial aid from the Netherlands Organization for Scientific Research (NWO).

References

1 A. J. Burggraaf, *J. Eur. Ceram. Soc., 9* (1992) 245.
2 A. J. Burggraaf, K. Keizer and B. A. van Hassel, Nanophase ceramics, membranes and ion implanted layers. In L. C. Dufour, C. Marty and G. Petot-Ervas (eds.), *Surfaces and Interfaces of Ceramic Materials*, Kluwer, Dordrecht, 1989, pp. 705–725.
3 M. M. R. Boutz, A. J. A. Winnubst, A. J. Burggraaf, M. Nauer and C. Carry, in S. P. S. Badwal, M. J. Bannister and R. H. J. Hannink (eds.), Low temperature sinter-forging of Y-TZP ceramics, *Proc. Zirconia V, Melbourne, August 1991*, Australian Ceramic Society, 1993, in press.
4 S. P. S. Badwal, J. Drennan and A. E. Hughes, Segregation in oxygen-ion conducting solid electrolytes and its influence on electrical properties. In J. Nowotny (ed.), *Science of Ceramic Interfaces*, Elsevier, Amsterdam, 1991, pp. 227–285.
5 C. Carry, Microstructure, grain boundaries and superplasticity in fine grained ceramics. In M. Mayo, M. Kobayashi and J. Wadsworth (eds.), *Superplasticity in Metals, Ceramics and Intermetallics, Mater. Res. Soc. Symp. Proc., 196* (1990) 313.
6 S. P. S. Badwal, F. T. Ciacchi, M. V. Swain and V. Zelizko, *J. Am. Ceram. Soc., 73* (1990) 2505.
7 S. P. S. Badwal and A. E. Hughes, *J. Eur. Ceram. Soc., 9* (1992) 115.
8 W. Groot Zevert, A. J. A. Winnubst, G. S. A. M. Theunissen and A. J. Burggraaf, *J. Mater. Sci., 25* (1990) 3449.
9 M. I. Mendelson, *J. Am. Ceram. Soc., 52* (1969) 443.
10 B. A. Boukamp, *Solid State Ionics, 18–19* (1986) 136.
11 J. E. Bauerle, *J. Phys. Chem. Solids, 30* (1969) 2657.
12 M. Miyayama and H. Yanagida, *J. Am. Ceram. Soc., 67* (1984) C194.
13 M. M. R. Boutz, *Internal Rep. CT92/182/128*, 1992 (University of Twente, 7500AE Enschede).
14 T. Stoto, M. Nauer and C. Carry, *J. Am. Ceram. Soc., 74* (1991) 2615.
15 A. E. Hughes and S. P. S. Badwal, *Solid State Ionics, 46* (1991) 265.

Ultrasonic evaluation of elastic properties of silica aerogels

J. Groß, T. Schlief and J. Fricke

Physikalisches Institut der Universität Würzburg, Am Hubland, D-97074 Würzburg (Germany)

Abstract

The velocity and attenuation of ultrasonic waves were measured in nanostructured silica aerogels as a function of frequency, density, gas pressure within the pores and elastic strain. The sound velocity was found to be independent of frequency. For the density dependence of the elastic modulus a scaling law was observed for each class of specimens with exponents ranging from 2.0 to 3.6. The gas pressure strongly influences the sound velocity only for aerogels with densities below 30 kg m^{-3}. Aerogels exhibit elastic non-linearity of an unexpected kind: the sound velocity decreases with increasing static compression. This is attributed to structural properties of these materials. The ultrasonic attenuation of silica aerogels is found to be lower than that of other highly porous materials. The main contributions under ambient conditions are the interaction between the pore gas and skeleton, and the water content, which is typically about 5 wt.%. Evacuated and heat treated specimens show elastic hysteresis with a loss tangent of about 10^{-3}. This value is compared with preliminary data derived from creep measurements.

1. Introduction

Silica aerogels are produced via hypercritical drying of nanostructured alcogels made from TMOS (tetramethoxysilane), TEOS (tetraethoxysilane) or waterglass (sodium silicate) [1–3]. Densities between 3 kg m^{-3} and 500 kg m^{-3} can be achieved; this corresponds to porosities in the range 99.8%–75% [4]. Higher densities (lower porosities) are obtained by sintering at temperatures around 1000 °C, which is considered low for a pure silica system [5, 6]. Typical pore diameters of aerogels are in the range 5–100 nm. The inner surface is typically between 300 and 1000 m^2 g^{-1}. Onc of the key applications of aerogels is for thermal insulation. Aerogels are also well suited for acoustic applications, *i.e.* quarterwave matching layers for ultrasonic transducers used as sensors in gases [7].

Some results on the sound velocity and elastic moduli of silica aerogels, especially on the scaling of these quantities with density, have already been published by other groups (*e.g.* refs. 8, 9). None of these studies examined the influence of gas pressure. The frequencies for which data were taken were either very high (Brillouin scattering [8]) or very low (quasistatic measurements [9]). One study examined the elastic non-linearity of silica aerogels [10]. The ultrasonic attenuation of aerogels has to our knowledge only been measured once in the frequency range above 50 MHz without evacuating or heat treating the specimens [11].

This paper reports the influence of density, gas pressure and elastic pre-strain on the sound velocity of silica aerogels as well as the frequency dependence of ultrasonic attenuation. The basic attenuation mechanisms are identified by measurements on evacuated and baked specimens.

2. Specimens and experimental procedures

All silica aerogels used for this study were prepared by hydrolysation and polycondensation of TMOS in methanol at DESY (Hamburg, Germany) [12], Airglass (Staffanstorp, Sweden) [13] and LLNL (Lawrence Livermore National Laboratory, Livermore, CA). The ultra low density aerogels were made at LLNL in a two-step process [14]. The organic aerogels used for comparison of elastic non-linearity were also made at LLNL [15]. Sintering of silica aerogels was performed at temperatures of 800–1000 °C for 2–200 h [5].

All sound velocity measurements presented here were taken using the pulse echo method. Two ultrasonic transducers were used as transmitter and receiver respectively. The elastic modulus c_{11} was calculated according to

$$c_{11} = c_l^2 \rho \quad (1)$$

where c_l is the longitudinal sound velocity and ρ the density of the specimen. To vary the pre-strain, the sample was compressed using the transducers as pistons. In order to vary the gas pressure within the (open) skeleton, the whole sample assembly was placed in a vacuum chamber. As the sound velocity of aerogels is independent of frequency at least up to 10 MHz, the

0921-5093/93/$6.00

measuring frequency was chosen arbitrarily in the range 33 KHz–5 MHz to match precision and intensity requirements.

The ultrasonic attenuation was measured using Debye–Sears diffraction of a He–Ne-laser beam by the sound wave. The specimen with a transducer glued to it was placed in a small vacuum chamber in order to vary the gas pressure; it could be heated to 100 °C to evaporate physisorbed water from the inner surface of the aerogel.

3. Results and discussion

3.1. Dependence of sound velocity and elastic modulus on aerogel density

The density ρ is the most important parameter that influences the sound velocity and thus the elastic modulus of aerogels. In Fig. 1, the elastic modulus c_{11} is plotted *vs.* density. The straight lines correspond to scaling relationships of the form

$$c_{11} \sim \rho^{\alpha} \quad (2)$$

where α is the scaling exponent. In Table 1, the values of α are given for different groups of specimens. These values may be compared with the theoretically expected exponent $\alpha_t = 2.0$ for bending deformation of the microstructural beams (or "string of pearls" [5]). As no "softer" deformation mechanism exists than the bending of beams, the high values of α found for aerogels are only possible if the density of the mechanically relevant part of the structure decreases faster than the macroscopic density. In other words, with decreasing density an increasing fraction of the total mass does not contribute to the stiffness of the skeleton, for example because it is located in "dead ends" (beams that are attached to the skeleton only at one end) [16].

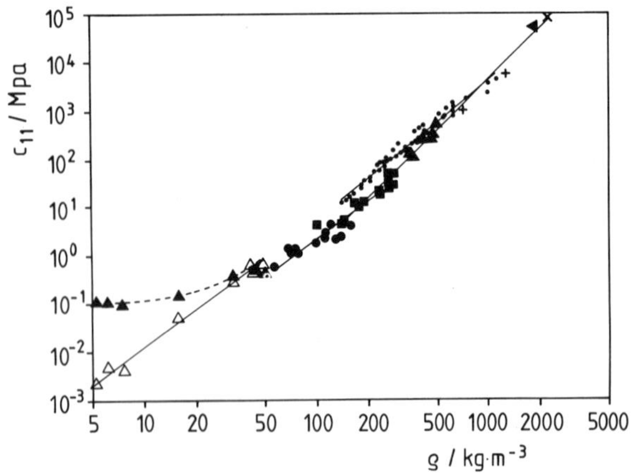

Fig. 1. Elastic modulus c_{11} as a function of density ρ for SiO_2 materials: vitreous silica (×), opal (◄), xerogels (+), aerogels from DESY (●), Airglass (■), LLNL (▲) and sintered aerogels (·); evacuated specimens are denoted by open symbols.

For sintered aerogels, the elastic modulus increases very fast at the beginning of the heat treatment, while the density remains almost unchanged. This is caused by increasing contact radii between the gel particles and "capturing" of dead ends by the skeleton. Thus, heat treatment allows considerable strengthening of the aerogels with little loss of porosity.

3.2. Dependence of sound velocity and elastic modulus on gas pressure

In Fig. 2, the sound velocity c_l is plotted *vs.* the pressure of gas inside the pores p_g for an ultralow density aerogel with $\rho = 5$ kg m^{-3}. In vacuum, a value of only 20 m s^{-1} is measured, while in air under ambient conditions the result is 140 m s^{-1}. The reason for this dramatic influence on gas pressure is that the elastic moduli and densities of the aerogel skeleton and of the gas have to be added in order to obtain the properties of the "composite" material:

$$c_l = \left(\frac{c_{11} + \Pi k p_g}{\rho + \Pi \rho_g} \right)^{1/2} \quad (3)$$

where all the properties of the gas are weighted with the porosity Π. This is true for homogeneous deformation of the skeleton and the gas. The effective adiabatic constant k is very close to unity owing to effective heat transfer between the gas and the skeleton during deformation by the sound wave [17]. The solid line in Fig. 2 is calculated according to the above equation. A very good agreement between measured and calculated data is observed. This confirms the fact that the aerogel can be considered as a homogeneous composite material even for ultrasonic frequencies. The skeleton and the gas inside the pores are strained in a uniform manner by the sound wave.

3.3. Elastic non-linearity

Almost all solid materials become stiffer when compressed. Vitreous silica is one of the (rare) exceptions to this rule. In most materials, however, this elastic

TABLE 1. Fitted scaling exponents α for different density ranges and classes of silica aerogels

α	Density range (kg m^{-3})	Samples
3.53 ± 0.08	150–2200	Vitreous silica, opal, xerogels, aerogels from LLNL
2.89 ± 0.06	150–1200	Sintered aerogels
2.57 ± 0.09	60–280	Aerogels from DESY, Airglass
2.59 ± 0.10	5–50	Low density aerogels (LLNL)

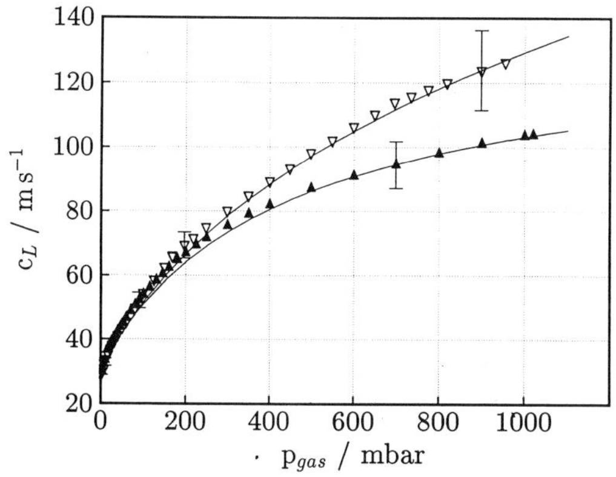

Fig. 2. Sound velocity c_l as a function of gas pressure in a low density aerogel ($\rho = 5.3$ kg m^{-3}) in air (V) and SF_6 (▲).

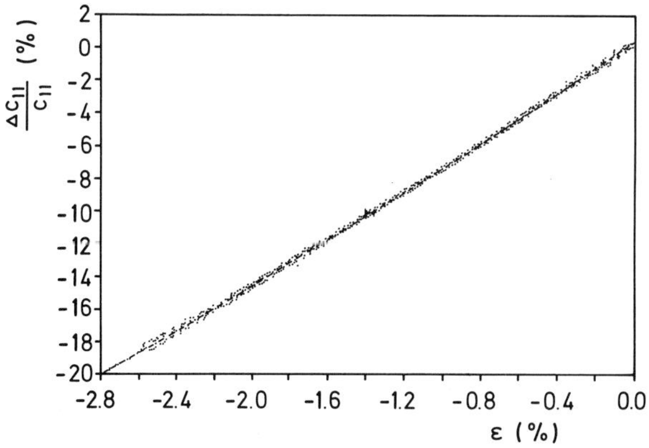

Fig. 3. Relative change in elastic constant c_{11} as a function of strain ε for an SiO_2 aerogel ($\rho = 48$ kg m^{-3}, $S_c = 8.0 \pm 0.4$). The solid curves are fits according to eqn. (3).

non-linearity is not easily observed because very high pressures are required to achieve non-infinitesimal strains. Aerogels are outstanding for two reasons. Firstly, they are highly compressible and thus high strains are easily obtained; secondly, they belong to the small class of materials that become "softer" when compressed. In Fig. 3, the relative change in the elastic modulus c_{11} is depicted as a function of uniaxial strain ε (negative strain means compression). The slope of the resulting curve near $\varepsilon = 0$, S_c, is a measure of the elastic non-linearity of the material. This quantity was determined for aerogels of different densities and compositions and calculated from the higher order elastic constants [18] of vitreous silica [16]. In Fig. 4, these values are plotted *vs.* density. The following is now evident.

(i) Aerogels of different chemical composition exhibit the same unusual positive non-linearity.

(ii) The non-linearity depends on the density of the material; it is generally more pronounced for very low densities.

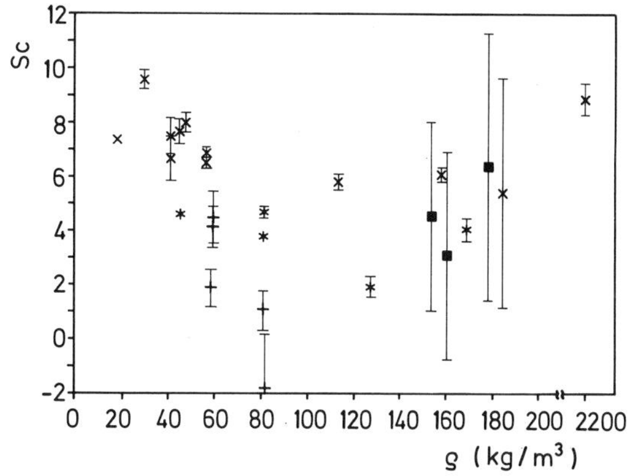

Fig. 4. Slopes S_c of low density aerogels as a function of density ρ: SiO_2 (×), resorcinol formaldehyde (∗), melamine formaldehyde (■) and carbon (+). The value of vitreous silica ($\rho = 2250$ kg m^{-3}) is calculated from higher order elastic constants.

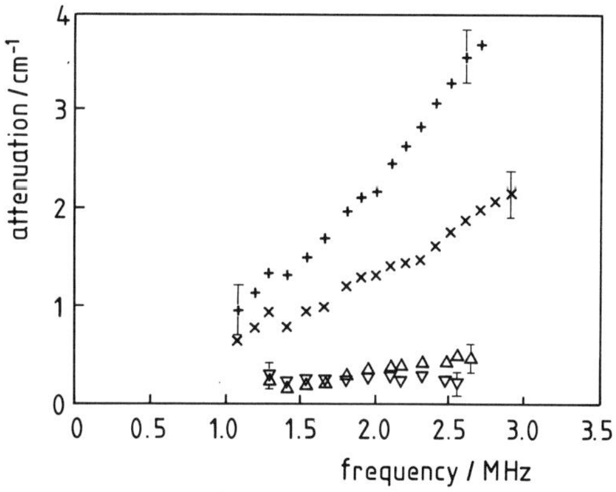

Fig. 5. Ultrasonic attenuation for an aerogel of $\rho = 71$ kg m^{-3} under ambient conditions in air (+), evacuated to 10 mbar (×) and after heat treatment (△, ▽). The slight difference between △ and ▽ is due to diffusion of water vapour into the vacuum vessel during the experiment.

We conclude from this that the material the aerogel is made of is only partially responsible for the elastic non-linearity. The main influence, especially at low densities, results from the bar-like nanostructure, which is common to all the examined aerogels. Bending deformation of these bars causes the dependence of the effect on the sign and density, according to the proposed knee model [19].

3.4. Attenuation

In Fig. 5, the ultrasonic attenuation μ of a silica aerogel is plotted as a function of frequency. The attenuation is highest for ambient conditions and rises as a parabolic function of frequency. Upon evacuation, the quadratic contribution disappears. It is obviously related to interaction between the gas and the aerogel skeleton. Indeed, calculations according to the model of the non-ridgid absorber give a satisfactory (within a factor of 2) agreement for both the pressure- and frequency-dependence of the attenuation [20]. After heat treatment at 100 °C for 2 h in vacuum, the attenuation is again drastically reduced. The main effect of the baking is desorption of water from the inner surface of the aerogel. Thus, most of the linearly frequency dependent attenuation is caused by water. Two models exist that can explain ultrasonic attenuation caused by the presence of water in a silica system. One relies on stress-induced diffusion of adsorbed water [21], the other on water-induced bond cracking of SiO_2 bonds [22].

The attenuation during one cycle, $\lambda\mu$, (λ is the wavelength) is related to the loss tangent $\tan\delta$ and the quality factor Q by

$$\tan\delta = Q^{-1} = \frac{\lambda\mu}{2\pi} \tag{4}$$

According to Kjartansson [23], this quantity can also be determined by creep measurements. Although only a few experiments on aerogels are currently available (and none for heat treated and evacuated specimens), an estimate of the loss tangent may be derived. In Fig. 6, $\tan\delta$ from creep measurements and from ultrasonic attenuation measurements are depicted *vs.* frequency, together with literature data from Bragg reflection by sound waves [11]. The frequency of creep measurements is deduced from the time scale (0.3–1000 s) of the experiment. The loss tangent is remarkably constant over a frequency range of more than 11 orders of magnitude. Only at very high frequencies do Brillouin scattering experiments reveal a steep increase in attenuation of elastic waves [8].

4. Conclusions

The propagation of sound waves in highly porous silica aerogels is described successfully by simple models. However, the attenuation data in particular suffer from large gaps in the frequency coverage of the experiments. Further creep measurements and sound attenuation measurements at kilohertz frequencies are necessary to understand fully the attenuation mechanisms.

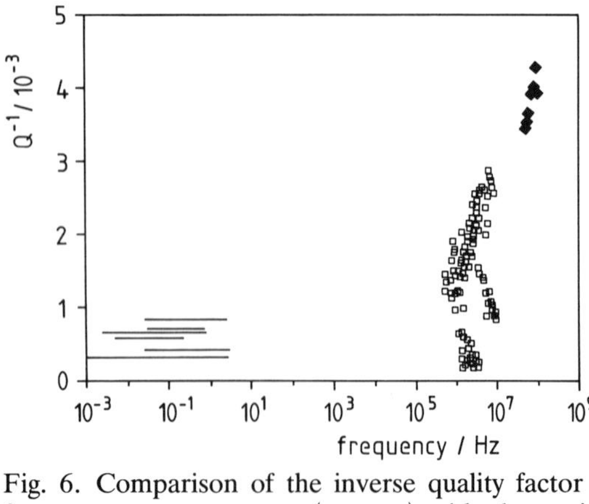

Fig. 6. Comparison of the inverse quality factor $Q^{-1} = \tan\delta$ from creep measurements (———), with ultrasonic attenuation measurements (□ our data, ◆ from ref. 11).

References

1 J. Fricke (ed.), *Aerogels, Proc. ISA 1*, Springer Proceedings in Physics, Vol. 6, Springer, Berlin, 1985.
2 R. Vacher, J. Phalippou, J. Pelous and T. Woignier (eds.), *Proc. 2nd Int. Symp. on Aerogels, ISA 2*, in *Rev. Phys. Appl., 24, Coll. C4* (1989).
3 J. Fricke (ed.), *Aerogels 3, Proc. 3rd Int. Symp. on Aerogels, ISA 3*, in *J. Non-Cryst. Solids, 145* (1992).
4 T. M. Tillotson, L. W. Hrubesh and I. M. Thomas, in C. J. Brinker, D. E. Clark and D. R. Ulrich (eds.), *Better Ceramics Through Chemistry III, Mater. Res. Soc. Symp. Proc., 121* (1988) 685–689.
5 A. Emmerling, J. Groß, R. Gerlach, R. Goswin, G. Reichenauer, J. Fricke and H. G. Haubold, *J. Non-Cryst. Solids, 125* (1990) 230–243.
6 T. Woignier, J. Phalippou and M. Prassas, *J. Mater. Sci., 25* (1990) 3118–3126.
7 R. Gerlach, O. Krauß and J. Fricke, *Proc. Ultrasonics Int. 91*, Butterworth Heinemann, Oxford, 1991, pp. 323–326.
8 E. Courtens, J. Pelous, J. Phalippou, R. Vacher and T. Woignier, *Phys. Rev. Lett., 58* (1987) 128–131.
9 J. D. LeMay, *Mater. Res. Soc. Symp. Proc., 207* (1991) 21–25.
10 P. Xhonneux, E. Courtens, J. Pelous and R. Vacher, *Europhys. Lett., 10* (1989) 733–738.
11 B. Nouailhas, F. Michard, R. Gohier and A. Zarembowitch, *11' ICA Rev. Acoust. Proc.*, 1983, pp. 179–182.
12 G. Poelz and R. Riethmüller, *Nucl. Instrum. Methods Phys. Res., 195* (1982) 491–503.
13 S. Henning and L. Svensson, *Phys. Scripta, 23* (1981) 697–702.
14 L. W. Hrubesh, T. M. Thomas and J. F. Poco, in C. J. Brinker, D. E. Clark, D. R. Ulrich and B. J. Zelinski (eds.), *Better Ceramics Through Chemistry IV, Mater. Res. Soc. Symp. Proc., 180* (1990) 315–319.
15 R. W. Pekala and F. M. Kong, in R. Vacher, J. Phalippou, J. Pelous and T. Woignier (eds.), *Proc. 2nd Int. Symp. on Aerogels, ISA 2*, in *Rev. Phys. Appl., 24, Coll. C4* (1989) 33–40.
16 J. Groß, *PhD Thesis*, Universität Würzburg, Germany, 1992 (Report E21-0692-2, 1992).
17 J. Groß, J. Fricke and L. W. Hrubesh, *J. Acoust. Soc. Am., 91* (1992) 2004–2006.
18 E. H. Bogardus, *J. Appl. Phys., 36* (1965) 2504–2513.
19 J. Groß, J. Fricke, R. W. Pekala and L. W. Hrubesh, *Phys. Rev. B, 45* (1992) 12774–12777.
20 T. Schlief, J. Groß and J. Fricke, in J. Fricke (ed.), *Aerogels 3, Proc. 3rd Int. Symp. on Aerogels, ISA 3*, in *J. Non-Cryst. Solids, 145* (1992).
21 W. F. Murphy III, *J. Acoust. Soc. Am., 71* (1982) 1458.
22 S. Hyodo and Y. Togami, *J. Appl. Phys., 44* (1973) 2237.
23 E. Kjartansson, *J. Geophys. Res., 84* (1979) 4737–4748.

Materials Science and Engineering, A168 (1993) 239–243

Production of β'-SiAlONs through carbon/oxide nanocomposites obtained from montmorillonite/aromatic-ammonium complexes

A. Seron, I. Ben Maimoun, M. Crespin and F. Beguin

C.R.M.D.-C.N.R.S. Université, 1 B, rue de la Férollerie, 45071 Orleans Cedex 2 (France)

Abstract

Carbon/oxide nanocomposites were formed by thermal treatment of montmorillonite/ammonium complexes, in the range 500–1000 °C. Although montmorillonite is decomposed above 700 °C, the lamellar structure is maintained up to 1000 °C by the presence of "intercalated" carbon. The amount of carbon in the composites can be adjusted by varying the nature of the organic ammonium cations so as to reach the computed value given by the theoretical equation for the formation of β'-SiAlON (19%). Starting from the montmorillonite/safraninium complex, the carbon content in the nanocomposite was 17.3%. β'-SiAlON was synthesized by heat treatment of this composite to 1350 °C under nitrogen. Since the oxide and the carbon source are intimately mixed on the nanoscale, β'-SiAlON is even obtained at temperatures as low as 1100 °C. Compared with other processes, the temperature for β'-SiAlON formation can be lower by approximately 300 °C.

1. Introduction

β'-SiAlONs have usually been synthesized either by sintering mixtures of Si_3N_4, SiO_2, AlN and Al_2O_3 at 1700 °C [1], or by carbothermal reduction of clay minerals in nitrogen flow [2]. Until recently, nearly all the processes based on clay minerals involved mixtures of small particles of clay and carbon obtained by mechanical grinding of powders. Since it seemed beneficial to produce better homogeneity, especially for kinetic reasons, some workers tried to obtain mixtures of clay and carbon on the "nanoscale", from organic molecules intercalated in clay.

Organic molecules can be intercalated by two different processes (Fig. 1). The first is solvation, where organic polar molecules solvate the exchangeable cations. The other is cationic exchange: the intercalated cations (Na^+ in purified Wyoming montmorillonite) can be exchanged by organic cations, such as quaternary ammonium derivatives. At first, Sugahara *et al.* [3] tried to use polar molecules such as acrylonitrile for intercalation in montmorillonite. The complex was then pyrolysed to obtain a "clay"/carbon composite. Owing to the weak ion–dipole interactions, much acrylonitrile was evolved during the heat treatment and the most important part of it probably polymerized outside the oxide lattice. This loss resulted in a carbon content of less than the 20% required to obtain the highest yield of β'-SiAlON [4].

Recently, Oya and coworkers reported the synthesis of "clay"/carbon composites using cationic exchange [5, 6]. In the present study, we tried to synthesize β'-SiAlON from carbon/oxide composites obtained by the pyrolysis of montmorillonite/arylammonium cation complexes. This method, starting from organic ammonium cations, seemed more favourable than the method developed by Sugahara owing to the strong electrostatic interactions between the cations and the negative charges of the aluminosilicate sheets.

We also characterized the intermediate clay–ammonium complexes and carbon/oxide composites. We show that the intermediate complex is a nanocomposite which favours the formation of pure β'-SiAlON at a lower temperature than that used in other carboreduction processes. We only present here the results obtained after cationic exchange with safranine.

2. Experimental details

The first step of the process consists in intercalation of the safraninium cation (Fig. 2) in sodium Wyoming montmorillonite $Na_{0.65}Si_8Al_3Mg_{0.65}Fe_{0.35}O_{20}(OH)_4H_2O$ (Fig. 1). This clay mineral is a layered structure with two tetrahedral layers on either side of an octahedral layer. Owing to cationic substitutions in the octahedral layer, the sheet is negatively charged. This charge is compensated by exchangeable cations in the interlayer space (here sodium cations) which can be replaced by organic cations. This 2/1 clay material has a cationic exchange capacity of 87 meq per 100 g. The exchange was obtained by stirring 3 g of sodium montmorillonite

0921-5093/93/$6.00

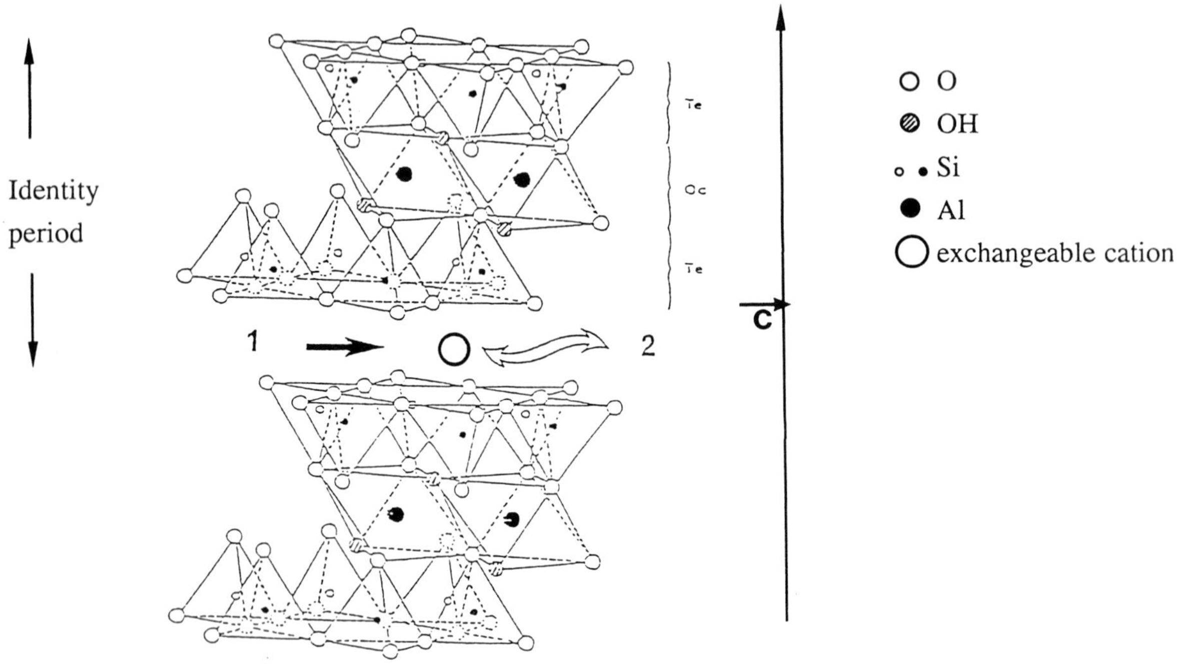

Fig. 1. Structure of montmorillonite and processes for the intercalation of organic molecules (1 solvation, 2 cationic exchange).

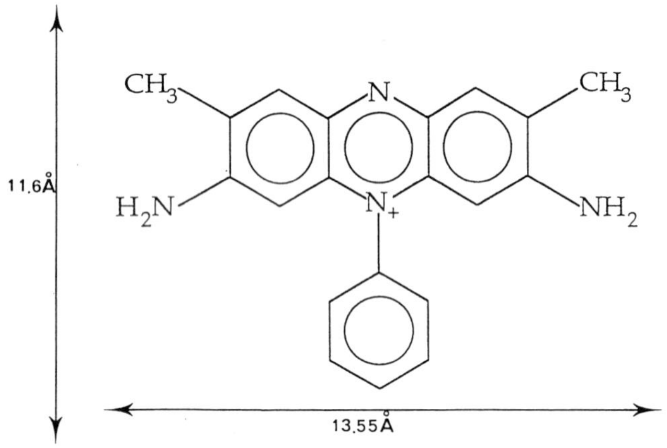

Fig. 2. Structure and dimensions of the safraninium cation. The maximum thickness is 4.08 Å.

in an aqueous solution (pH 1 or 5) of the aromatic ammonium in small excess (1.5 NR_4^+ per Na^+ equivalent) at 353 K for 4 days. The so-called "complexes" thus obtained were characterized by elemental analysis, X-ray diffraction (XRD, $\lambda = 1.5418$ Å, Siemens D 500 goniometer) and Fourier transform IR (FTIR) spectroscopy (Nicolet 710).

These complexes were pyrolysed under argon at various temperatures (500 °C $\leqslant T \leqslant$ 1000 °C) to obtain the carbon/oxide nanocomposite, then characterized by XRD and thermogravimetric analysis (TGA). The intermediate carbon present in the interlayer space could also be identified after dissolution of the oxide moiety in HF and HCl [7]. For β′-SiAlON synthesis, the composites were heated in nitrogen at temperatures higher than 1000 °C to permit nitriding carboreduction of the oxide by carbon. The products obtained were then analysed by XRD.

3. Results and discussion

The ammonium intercalation was confirmed by studying the 001 reflection of montmorillonite. The identity period increased from 12.9 Å for sodium montmorillonite to 20.1 Å after intercalation of the safraninium cation (Fig. 3). The dimensions of the safraninium cation were computed from the following data (Fig. 2): $d_{C\text{-}C} = 1.395$ Å; $d_{C\text{-}H} = 1.08$ Å; $d_{N\text{-}H} = 1.03$ Å; $d_{C\text{-}N} = 1.235$ Å; HNH angle 109.5°; van der Waals radius of hydrogen $R_H = 1.2$ Å. As for linear amines either in silicates or aluminosilicates [8], the aromatic ammonium is intercalated with its plane perpendicular to the layers of the host. Indeed, the difference (10.5 Å) between the identity period of collapsed sodium montmorillonite (9.6 Å) and the identity period of the montmorillonite/safraninium complex (20.1 Å) is comparable with the dimension of the molecule along its symmetry plane, *i.e.* 11.6 Å (Fig. 2).

The FTIR spectra of montmorillonite, safranine and the montmorillonite/safraninium complex confirmed the existence of an intercalation compound. For the complex, two strong bands at 3470 cm^{-1} and 3380

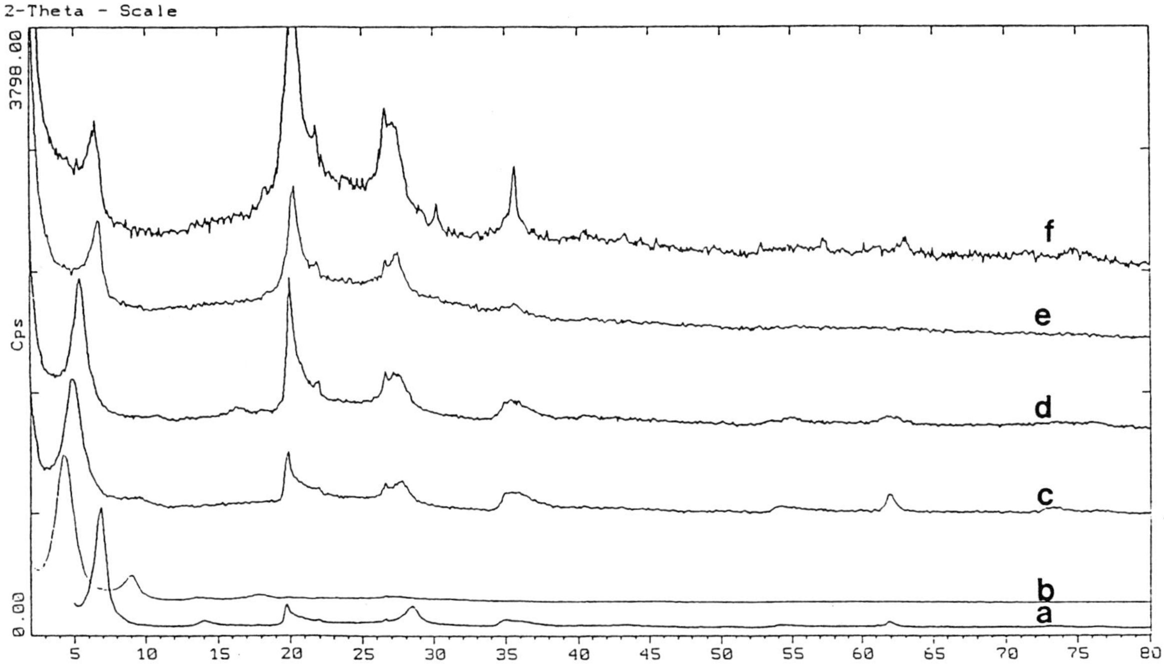

Fig. 3. XRD spectra ($\lambda = 1.5418$ Å): (a) Na-montmorillonite, and (b) Na-montmorillonite exchanged with safraninium; montmorillonite/safraninium complex heat treated at (c) 500 °C, (d) 700 °C, (e) 900 °C and (f) 1000 °C.

TABLE 1. Variation in identity period and thickness of the intercalated layer with pyrolysis temperature of the montmorillonite/safraninium complex

Temperature (°C)	d_{001} (Å)	$d_{001} - 9.60$ (Å)
20	20.1	10.5
500	17.80	8.20
700	16.48	6.88
900	13.98	4.38
1000	13.92	4.32

cm^{-1} are assigned to the N–H vibrations of NH_2. The band at 1635 cm^{-1} could be assigned either to the C=N or C=C elongations or to NH_2 deformation.

After heat treatment of montmorillonite/safraninium under argon, the lamellar structure is still present up to 1000 °C (Fig. 3). However, the structure of sodium montmorillonite was completely destroyed by heating above 700 °C. This point was confirmed by TGA which showed that all the hydroxides were destroyed at this temperature, and by XRD indicating the presence of spinel at 900 °C. The lamellar structure is apparently stabilized by the presence of "intercalated" carbon. Table 1 shows that the organic molecules tilt progressively as the heat treatment temperature increases. At 1000 °C, the thickness of the intercalated layers is not far from the value of a single carbon layer (approximately 3.4 Å), indicating an almost regular sequence of oxide planes and graphene planes, hence the existence of a nanocomposite.

The carbon content of the composites prepared from the montmorillonite/safraninium complex was determined by measuring the weight loss after oxidation by air at 800 °C in the TGA apparatus. The amount of carbon strongly depends on the pH value, 17.3% carbon at pH 1, 13.3% carbon at pH 5. There is good agreement between the observed carbon content at pH 1 and the value computed taking into account the cationic exchange capacity. Indeed, chemical analyses have demonstrated that nearly all the sodium cations were replaced by safraninium (sodium content 2.1% in sodium montmorillonite, 0.14% in montmorillonite/safraninium complex).

We have already shown that the synthesis of β'-SiAlON is possible after 5 h treatment at 1350 °C using the montmorillonite/acriflavinium complex [9], even if the amount of carbon is lower than that predicted theoretically (19%) from eqn. (1) corresponding to β'-SiAlON formation from dehydrated montmorillonite:

$$(Al_2O_3, 4SiO_2) + 3N_2 + 9C \longrightarrow Si_4Al_2O_2N_6 + 9CO \quad (1)$$

This is a limiting case since SiO can be evolved during heat treatment [10]. We also demonstrated that the nature of the phases synthesized depends on the heating time and nitrogen flow rate. However, the amount of carbon predicted from eqn. (1) can be reached by using molecules with a higher carbon content than acriflavine. We therefore decided to use the montmorillonite/safraninium complex prepared at pH 1 for the synthesis of the carbon/oxide nanocomposite. In

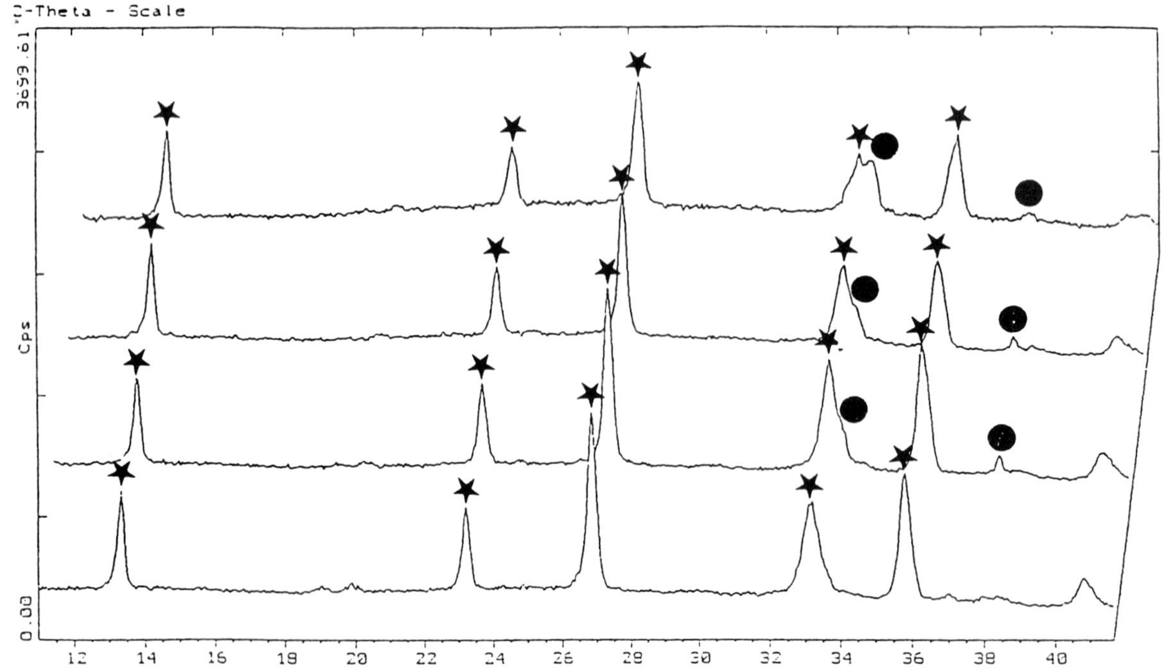

Fig. 4. XRD spectra ($\lambda = 1.5418$ Å) after nitriding heat treatment of the nanocomposites at 200 °C h^{-1} up to 1350 °C, plateau at 1350 °C held for various times: from bottom to top, 0 min, 30 min, 1 h, 2 h (★ β'-SiAlON, ● AlN).

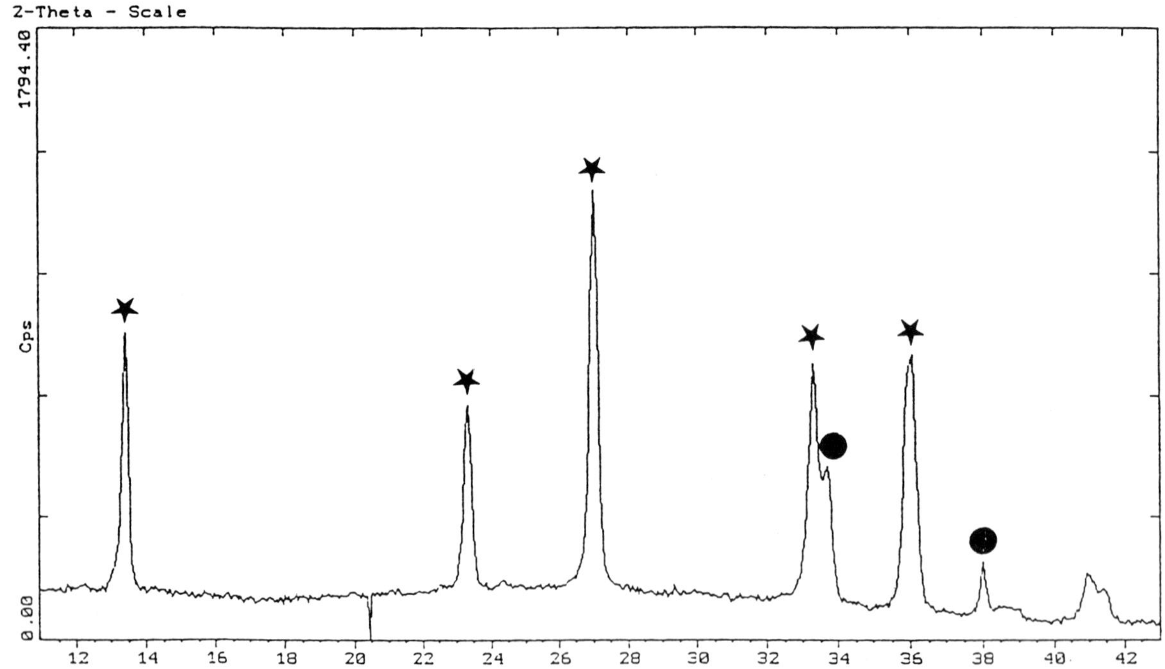

Fig. 5. XRD spectrum ($\lambda = 1.5418$ Å) after nitriding heat treatment of the nanocomposite at 1100 °C for 48 h (★ β'-SiAlON, ● AlN).

this composite, the amount of carbon is very close (17.3%) to the value calculated from eqn. (1) (19%). The synthesis of β'-SiAlON from this composite was quite easy (Fig. 4). Pure β'-SiAlON was obtained by heating at 200 °C h^{-1} up to 1350 °C. Because of the strong homogeneity of the mixture (on the "nano-scale"), the reaction rate was very important and a plateau at 1350 °C was not necessary. When the heating time at the highest temperature (1350 °C) was increased, a small amount of AlN appeared (Fig. 4). Safranine therefore provided enough carbon for the synthesis of β'-SiAlON but not enough for the formation of pure AlN. Again, the evolution of SiO can explain why such a mixture of β'-SiAlON and AlN could be obtained even if the amount of carbon is smaller than predicted from eqn. (1).

Some experiments were performed at lower temperatures. After 48 h of heat treatment at 1100 °C, we obtained a mixture of β'-SiAlON and AlN (Fig. 5). It is likely that pure β'-SiAlON could be obtained by using a shorter heat treatment time at 1100 °C.

4. Conclusion

Montmorillonite/safraninium complexes can easily be obtained by cationic exchange. The exchanged cation is oriented perpendicular to the montmorillonite layers. Owing to the strong electrostatic attraction between the ammonium cations and the negatively charged layers, the carbon source is tightly maintained in the host even during thermal treatment. After heating at 700–900 °C a nanocomposite is formed with a quite regular sequence of oxide layers and carbon layers. Depending on the exchanged cation and on the pH value, the carbon content of the composite can reach the theoretical value (19%) for the nitriding carbothermal reduction of the oxide moiety. Pure β'-SiAlON is obtained after heating at 200 °C h^{-1} up to 1350 °C without a plateau. Owing to the high homogeneity of the composite, we were able to obtain β'-SiAlON even at a very low temperature (1100 °C). This process is an improvement allowing the preparation of β'-SiAlON under very mild conditions.

References

1 L. J. Gauckler, H. L. Lukas and G. Petzow, *J. Am. Ceram. Soc., 58* (1975) 346.
2 J. G. Lee and I. B. Cutler, *Ceram. Bull., 58* (1979) 869.
3 Y. Sugahara, K. Kuroda and C. Kato, *Clay Sci., 7* (1987) 17.
4 J. B. Baldo, V. C. Pandofelli and J. R. Casarini, *Ceramic Powders*, Vincenzini, Amsterdam, 1983, p. 437.
5 A. Oya, M. Saito and S. Otani, *Appl. Clay Sci., 3* (1988) 291.
6 A. Oya, J. Sakano and S. Otani, *Yogyo Kyokai-Shi, 12* (1987) 1164.
7 N. Sonobe, T. Kyotani and A. Tomita, *Carbon, 26* (1988) 573.
8 G. Lagaly and K. Beneke, *Am. Miner., 60* (1975) 642.
9 A. Seron, F. Beguin and F. Bergaya, *Mater. Sci. Forum, 91–92* (1991) 265.
10 P. D. Miller, J. G. Lee and I. B. Cutler, *J. Am. Ceram. Soc., 62* (1979) 147.

Materials Science and Engineering, A168 (1993) 245–247

Spinel ferrite–metal alloy nanocomposites arising from disproportionation in a liquid media

A. Malats i Riera, G. Pourroy* and P. Poix

IPCMS, Groupe des Matériaux Inorganiques, EHICS, 1 rue Blaise Pascal, BP 296, 67008 Strasbourg Cedex (France)

Abstract

A new method for synthesizing a composite made up of a spinel ferrite and an iron–cobalt alloy based on a disproportionation reaction is described. In addition to a spinel phase, two metallic phases were encountered, one isomorphous to α-Fe (b.c.c. structure), the other isomorphous to cubic cobalt (f.c.c. structure). The chemical formula was determined by chemical analysis and X-ray measurements. It can be schematized by $Fe_{0.17}Co_{0.65}[Fe_{2.43}Co_{0.57}O_4]$. No oxidation of the composite is observed up to 300 °C, showing that the iron–cobalt alloy and spinel phase are surrounded and protected against oxidation. The small grain sizes, 0.1–0.5 μm, are in agreement with nanocomposite characteristics, giving outstanding magnetic properties: the saturation magnetization and coercive field are high, 90 emu g^{-1} and 2300 Oe respectively.

1. Introduction

The most widely known compounds among the ferrites of spinel structure are without doubt magnetite Fe_3O_4, maghemite γ-Fe_2O_3 and cobalt-doped γ-Fe_2O_3. The latter are widely used as recording materials when coercive fields of about 250–1250 Oe are required [1]. The most common synthesis methods always involve precipitation of hydroxides or oxohydroxides [2–8]. Fine metal particles based on iron, cobalt or nickel, are also very attractive because they offer the possibility to combine higher magnetic moments and higher coercivities than can be achieved with any known oxide particles. The principal methods of obtaining metal particles are reduction in hydrogen of oxides, hydroxides and oxalates, reduction in aqueous solution, decomposition of metallo-organic compounds, vacuum deposition and sputtering, and electrodeposition [1]. However, metallic particles are highly reactive. The most effective methods for protecting metal particles against atmospheric corrosion are deliberate oxidation of their surface, encapsulating the particles in a polymer film, or coating the particles with an inorganic compound [9–12]. To our knowledge, nobody has yet proposed the synthesis of metallic particles protected against oxidation, directly by simple precipitation. We report here a method based on a disproportionation reaction in a liquid media. The precipitate was analyzed. Its crystallographic properties, thermal stability and magnetic properties are reported.

*Author to whom correspondence should be addressed.

2. Experimental details

Cobaltous and ferrous chlorides, $CoCl_2 \cdot 6H_2O$ and $FeCl_2 \cdot 4H_2O$, in ratio 1/2, were dissolved in water and mixed. This mixture was then added to boiling potassium hydroxide and a black powder precipitated. The precipitate was filtered and washed with boiling water. It was first analyzed by energy-dispersive X-ray analysis using a JEOL 840 scanning electron microscope and a KEVEX analyzer. This indicated [Fe]/[Co] = 2.12. The amount of iron was determined using the potassium dichromate method. This led to the global formula $Co_{0.94}Fe_2O_{3.08}$ for the precipitate. X-ray diffraction data were collected at room temperature with a D500 Siemens diffractometer equipped with a primary beam quartz monochromator (Co Kα_1 1.78897 Å).

Thermal gravimetric and thermal differential analyses were carried out in platinum crucibles under air using SETARAM 92 apparatus. The rate of temperature increase was 1 °C min^{-1}. The variation in weight was calculated after removing the contribution of the empty crucible.

Magnetization was measured using FONER apparatus at room temperature.

0921-5093/93/$6.00

3. Results

Examinations of the precipitate with the electron beam microscope show that the grains have sizes 0.1–0.5 μm (Fig. 1). They exhibit geometrical shapes, most of them having an octahedral form. The precipitate appears to be a well crystallized phase. The X-ray diffraction pattern, displayed in Fig. 2, is characterized by very fine diffraction lines and a very low background signal. Three phases are detected: a spinel phase, a phase isomorphous to α-Fe of b.c.c. structure, and a phase isomorphous to cobalt of f.c.c. structure. The crystallographic parameters of the three phases were refined by a least square method. They have the following values: spinel $a = 8.396(1)$ Å, b.c.c. phase $a = 2.838(1)$ Å, and f.c.c. phase $a = 3.561(7)$ Å. The spinel crystallographic parameter, close to that of Fe_3O_4 ($a = 8.397$ Å), suggests a strong Fe^{2+} concentration [13]. The composition of the f.c.c. and b.c.c. phases estimated from crystallographic parameters is $Fe_{0.2}Co_{0.8}$ [14]. Therefore, a complete chemical formula can be proposed. As the chemical analysis gives $Co_{0.94}Fe_2O_{3.08}$, *i.e.* $Co_{1.22}Fe_{2.6}O_4$, the composite chemical formula can be expressed by $Fe_{0.17}Co_{0.65}$-$[Fe_{2.43}Co_{0.57}O_4]$.

Fig. 1. Scanning electron micrograph of the precipitate.

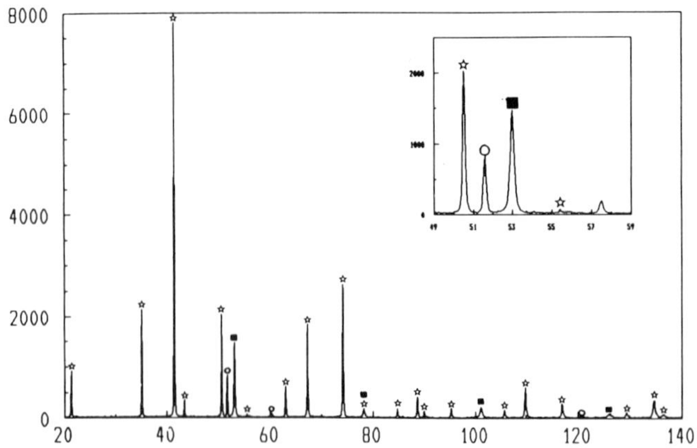

Fig. 2. Observed X-ray diffraction pattern of the composite: * diffraction lines of the spinel phase, ■ the phase isomorphous to α-Fe of b.c.c. structure, and ○ the phase isomorphous to cubic cobalt of f.c.c. structure. Details of the 2θ range 45°–60° are displayed in the inset.

The evolution of the composite during heating under air was studied by thermal gravimetric and thermal differential analyses (Fig. 3). No weight loss is observed below 300 °C, showing that the precipitate is completely dehydrated, in agreement with the global chemical formula. Despite the presence of metallic iron or cobalt, this phase only oxidizes above 300 °C. The weight increase of 5.9% between room temperature and 800 °C is consistent with transformation of the precipitate into $Co_{0.94}Fe_2O_{3.94}$. Furthermore, X-ray diffraction measurements at room temperature of the powder heated at 800 °C show that the powder has a spinel structure with $a = 8.3755$ Å. This parameter is close to that of $CoFe_2O_4$, in agreement with the given formula.

The magnetization *vs.* magnetic field measured at room temperature and its derivative are displayed in Fig. 4. We note a saturation magnetization of 90 emu g^{-1} and a coercive field of 2300 Oe.

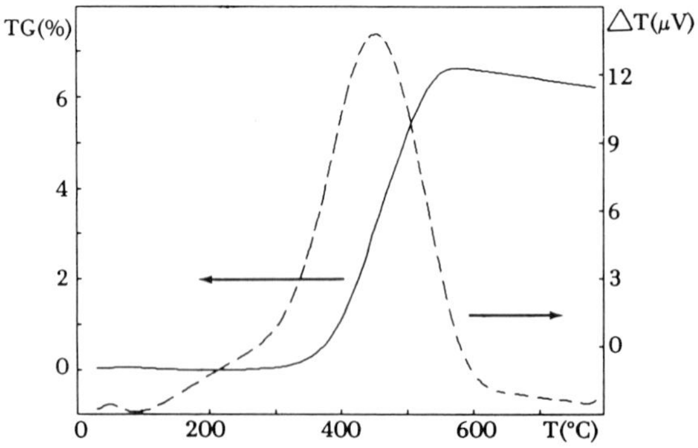

Fig. 3. Thermal gravimetric and thermal differential analyses of the precipitate under air.

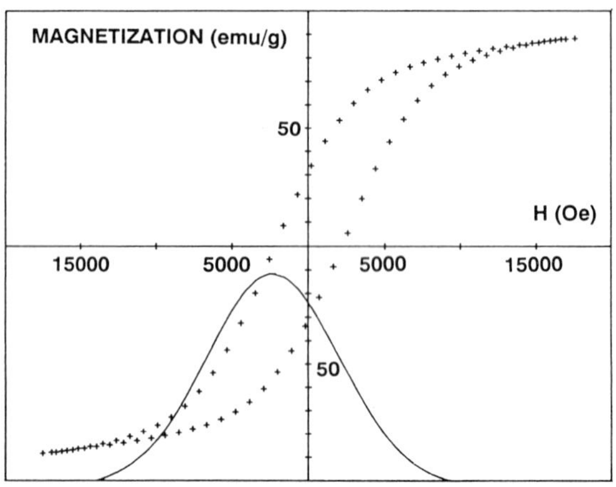

Fig. 4. Hysteresis cycle observed at room temperature for the precipitate and derivative.

4. Discussion and conclusions

The precipitation of cobaltous and ferrous chlorides in a strongly basic medium leads to an unexpected result: the formation of a composite phase made up of a spinel oxide and two metallic phases which are iron–cobalt alloys. Chemical analysis and crystallographic measurements give the chemical formula $Fe_{0.17}Co_{0.65}[Fe_{2.43}Co_{0.57}O_4]$. The formation of such a composite can be explained by considering the following successive steps: $CoFe_2(OH)_6$ precipitates; immediately, both dehydration to $CoFe_2O_3$ and disproportionation into Fe^0, Fe^{3+} and Co^0 occur; then the precipitate oxidizes slightly and loses small amounts of cobalt in solution.

A similar disproportionation is encountered for FeO at 400–500 °C following the reaction $4FeO \rightarrow Fe_3O_4 + Fe$ [15]. It has also been reported in aqueous suspensions of $Fe(OH)_2$ between 100 and 300 °C following the reaction $4Fe(OH)_2 \rightarrow Fe + Fe_3O_4 + 4H_2O$ [16–18]. However, metallic iron was not revealed by X-ray diffraction measurements, while in our case the metallic alloys were easily detected in the X-ray diffraction pattern. Although the composite grains are small, 0.1–0.5 μm, as shown by microscopic observations, it is remarkably stable in air, compared with magnetite, iron or cobalt, all of which oxidize below 300 °C. No oxidation is observed up to 300 °C. That means that the metallic phases and even spinel are protected. The only explanation we have found is the formation of a coating thin film of Co_3O_4 around the grains which hinders the diffusion of oxygen.

The magnetic properties are quite interesting. The saturation magnetization $M_s = 90$ emu g^{-1} is consistent with the presence of metallic iron and cobalt. The coercive field is high and is probably due to a small grain effect. Indeed, $CoFe_2O_4$ grains of 700 Å give a coercive field of 1000–2000 Oe [19]. The magnetization derivative is characterized by a unique broad peak at $H_c = 2300$ Oe, showing that the coercive fields of all the phases have values of the same order of magnitude.

The new method for synthesizing nanocomposites, based on disproportionation in the liquid medium, has never been reported before. It allows the preparation of a nanocomposite consisting of metallic alloys and spinel oxide, thus combining good stability against oxidation and magnetic properties at least as good as those of metal particles. This composite has without doubt a promising future. This synthesis method itself may allow the development of new phases with interesting properties for magnetic recording as well as for catalysis.

Acknowledgments

We thank P. Panissod IPCMS, Groupe d'Etude des Matériaux Métalliques, 4 rue Blaise Pascal, 67000 Strasbourg, France, for fruitful discussions and the European Commission for the grant BRITE/EURAM (ref. B/BREU-913010) awarded to A. Malats i Riera.

References

1 E. P. Wohlfarth (ed.), *Ferromagnetic Materials*, North-Holland, Amsterdam, 1980.
2 W. J. Schuele and V. D. Deetscreek, *J. Appl. Phys., 32* (1961) 235S.
3 M. Kimaya, *Bull. Chem. Soc. Jpn., 51* (1978) 134.
4 K. Matsuda and I. Kayama, *Proc. Mater. Sci. Tokai Univ., 17* (1982) 101.
5 T. Sato, T. Iijima, M. Seki and N. Inagaki, *J. Magn. Magn. Mater., 65* (1987) 252.
6 H. Tamura and E. Matijevic, *J. Colloid Interface Sci., 90* (1982) 100.
7 R. E. Vandenberghe, R. Valeerberghe, E. Degrave and G. Robbrecht, *J. Magn. Magn. Mater., 15* (1980) 1117.
8 Y. Makino, I. Kamiya and G. Suguira, *Japanese Patent 7415757*, 1974.
9 R. J. Deffeyes, C. E. Johnson, Jr. and J. H. Judy, *IEEE Trans. Magn., 8* (1972) 465.
10 A. A. van der Giesen, *IEEE Trans. Magn., 9* (1973) 191.
11 J. S. Roden, *US Patent 3 932 293*, 1976.
12 M. Robbins, J. H. Swisher and R. C. Sherwood, *J. Appl. Phys., 42* (1971) 1352.
13 V. N. Sharma, *Phys. Lett., 19* (1965) 463.
14 W. B. Pearson, *Handbook of Lattice Spacings and Structures of Metals*, Pergamon, Oxford, 1964.
15 R. Collongues and G. Chaudron, *C. R. Acad. Sci. Paris, 234* (1952) 728.
16 G. Schikorr, *Z. Elektrochem., 36* (1929) 65.
17 F. J. Shipko and D. L. Douglas, *J. Phys. Chem., 60* (1956) 1519.
18 U. R. Evans and J. N. Wanklyn, *Nature (London), 162* (1948) 27.
19 A. E. Berkowitz and W. J. Schuele, *J. Appl. Phys., 30* (1959) 134S.

Materials Science and Engineering, *A168* (1993) 249–252

Synthesis of $BaTiO_3$ and ZnO varistor precursor powders by reaction spray pyrolysis

O. Milošević and D. Uskoković*

Institute of Technical Sciences of Serbian Academy of Sciences and Arts, Belgrade (Serbia)

Abstract

High-purity, fine, homogeneous ceramic powders with complex composition which are commonly used for electronic materials (ZnO-based non-linear resistors–varistors and piezoelectric $BaTiO_3$) were synthesized by the reaction spray process. The particles obtained were hollow spherical shells and shell fragments with average particle size ranging from 3 to 4 μm, representing the aggregates of small individual crystallites each less than 1 μm. Owing to the extremely high heating rate of a salt/droplet particle, mixing on an atomic or nanometre scale in the feeding solution should persist in the resulting powder. It was shown that the chlorides $BaCl_2$ and $TiCl_4$ are suitable precursors for $BaTiO_3$ ceramics. Varistor ceramics with a non-linearity coefficient $\alpha = 30$ and breakdown voltage $K = 700$ V mm^{-1} were produced by sintering the reaction spray-derived multicomponent ZnO-based powder. The compositional homogeneity of the resulting powders and the phase content were determined by the processing and solution parameters which affect the dehydration, precipitation and decomposition processes.

1. Introduction

The methods of wet chemistry offer many advantages in material processing compared with conventional synthesis [1]. These include better homogeneity, high purity, lower calcination temperatures, the ability to prepare powders with a narrow distribution of particle size and a small mean size, the ability to prepare ceramic materials with novel microstructures and distribution of phases [2].

Investigations into producing stoichiometric high purity submicrometre precursor powders for electronic materials are now performed to an increasing extent by applying chemical methods (*e.g.* coprecipitation [3–5], sol-gel [6, 7], hydrothermal synthesis [8] etc.). However, it is often difficult to obtain adequate chemical homogeneity, even by the solution techniques, owing to some compositional segregation. Mixing on an atomic (or a nanometre) scale [2], which is provided by wet chemical preparation techniques, is hardly maintained in the resulting powders during calcination.

The reaction spray process is one of the dispersion phase techniques and is recognized as an important method for making fine powders and materials [9–12]. It involves atomizing the solution into a heated reaction column. The mist of the solution is dried and subsequently decomposed at elevated temperatures. This permits higher surface reaction and prevents any compositional segregation. Therefore, this method is considered to produce powders maintaining compositional homogeneity and to have the advantage of ease of close compositional control for multicomponent powders.

We consider here electronic materials where controlled and/or fine grain size and grain boundary chemistry are important or critical to performance. Additionally, the chemistry of the grain boundary and the depletion layer, the defect structure and the dopant concentration are phenomena relating to the lattice structure in electronic materials and are controlled on a nanometer scale. The difficulty of controlling the properties and attaining compositional homogeneity from micrometre size powders is easily appreciated. The goal of this work was to investigate the potential of the reaction spray pyrolysis method for processing of electronic materials (especially multicomponent ZnO-based non-linear resistors (varistors) and piezoelectric $BaTiO_3$ ceramics) when the above mentioned demands are presumed.

2. Experimental procedure

2.1. *Laboratory equipment for the reaction spray process*

The main parts of the equipment were as follows:

(i) an atomizing system, including a twin fluid atomizer, which was used in this work; an ultrasonic atomizer, Gapusol 9001 type, 2.5 MHz, RBI was also available;

*Correspondence address: Pro Ignis Consultancy, Mr. Miša Ignis, Postbus 82430, NL2508EK Denhaag, Netherlands.

0921-5093/93/$6.00

(ii) a vertical tube furnace (Electron-Ignis Combo Lab), height 1750 mm with plasma generated alumina tube 150 mm in diameter with three independently controllable temperature zones, each of 1473 K maximum; a microcomputer based temperature indicated controller MCM-100 series (Shinko) and AR-201 six-point Toshiba recorder.

The general flow diagram of the reaction spray process for the production of ceramic powders is given elsewhere [13]. The solution (suspension) of metal salts was atomized into a hot reaction column, where the droplets were dried and decomposed in dispersed phase. The reaction spray temperature was 1173 K. To cause the gas–liquid (droplets) dispersed system to travel through the furnace a pressure drop was maintained at the exhaust by a vacuum pump. Separation of the gas-borne particles from the gas stream (mixture of gases and water vapour) was achieved by deposition onto a collecting surface in gravity settling chambers. After spraying and decomposition of all the suspension, the resulting powder was exposed to flowing hot gases at the exhaust for about 30 min. The process parameters, the gas and liquid flow rates and the temperatures in the reaction zones were adjusted preliminarily.

2.2. Preparation of multicomponent ZnO-based varistor precursor powder

In order to obtain powder with complex initial composition $(100-X)$ZnO + Y(additives) the following starting components were used: $Zn(NO_3)_2 \cdot 6H_2O$ (Merck, extra pure), $Bi(NO_3)_2 \cdot 9H_2O$ (Kemika, p.a.), $SbCl_3$ (Kemika, 99%), $Co(NO_3)_2 \cdot 6H_2O$ (Merck, p.a.), $Mn(CH_3COO)_2 \cdot 4H_2O$ (Merck, extra pure), $Cr(NO_3)_2 \cdot 9H_2O$ (Merck, p.a.) and $Ni(NO_3)_2 \cdot 6H_2O$ (Merck, p.a.).

The starting components were dry mixed and heated to form a homogeneous common grey solution [5]. The solution obtained was diluted with distilled water in the proportion of desired composition and mixed thoroughly. The suspension was then fed to an atomizing system and sprayed into the reaction column. The gas pressure (N_2) was 0.5 bar and the liquid flow rate was 0.5 l h^{-1}. The powder was compacted into samples having a diameter of 8 mm and a height of 1–2 mm at a uniaxial pressing pressure of 80 MPa. The samples were sintered in air from 1373 to 1573 K, held for 60 min; the heating and cooling rates were controlled at 5 K min^{-1}.

2.3. Preparation of $BaTiO_3$ precursor powders

$BaTiO_3$ precursor powders were prepared according to the following initial solutions and suspensions:

(i) solution of chlorides $TiCl_4$ and $BaCl_2$ in ethanol–water mixture as solvent, with concentration 0.2 M; based on this solution, two experiments were performed, after 1 day (powder B-1) and after 4 days (powder B-2);

(ii) suspension of TiO_2–$X H_2O$ + $Ba(OH)_2$ in nitric acid, with concentration 0.2 M (powder B-3);

(iii) solution of nitrates $TiO(NO_3)_3$ and $Ba(NO_3)_2$ (powder B-4).

The solutions–suspensions were fed to an atomizing system and sprayed after stationary conditions had been established in the system with the parameters denoted in Table 1.

2.4. Characterization

Development of the crystal phases in the powders was studied with a Philips PW 1710 diffractometer using a graphite-monochromatized Cu Kα radiation. Differential thermal analysis (DTA) and thermogravimetric analysis (TGA) were performed in air at heating rates of 10 K min^{-1} (Netzch 409 EP). The distribution of powder size was measured using a Coulter Multisizer. The powder morphology was examined by scanning electron microscopy (SEM, AMR-1600 T). The chemical homogeneity of the microstructures was determined by energy-dispersive spectroscopy (EDS, PGT system IV from Princeton Gamma Tech.).

Electrical measurements were registered on sintered varistor ceramics in the interval 1–100 A m^{-2} by a d.c. power supply. The non-linearity coefficients were determined within the range 1–10 A m^{-2} (α_1) and 10–100 A m^{-2} (α_2); the breakdown field K_c was determined at 10 A m^{-2} and the leakage current J_L was measured at a voltage of 0.8 K_c.

3. Results and discussion

A SEM image of ZnO based varistor precursor powder particles is shown in Fig. 1. The particle shape is typical of nitrate-derived powders, consisting of hollow, spherical shells and shell fragments, with the mean particle size measuring approximately 3 μm in diameter. Their morphology is a remnant of the original dried salt structure, which forms as the droplet dries and the salt precipitates onto the surface to form

TABLE 1. Feed solutions and process parameters during reaction spray pyrolysis

Powder	Initial composition	Liquid flow rate ($l\ h^{-1}$)	Gas pressure (bar)
B-1	$TiCl_4 + BaCl_2$	0.3	0.6
B-2	$TiCl_4 + BaCl_2$	0.2	0.6
B-3	TiO_2–$X H_2O + Ba(OH)_2$	0.3	0.6
B-4	$TiO(NO_3)_3 + Ba(NO_3)_2$	—	0.5

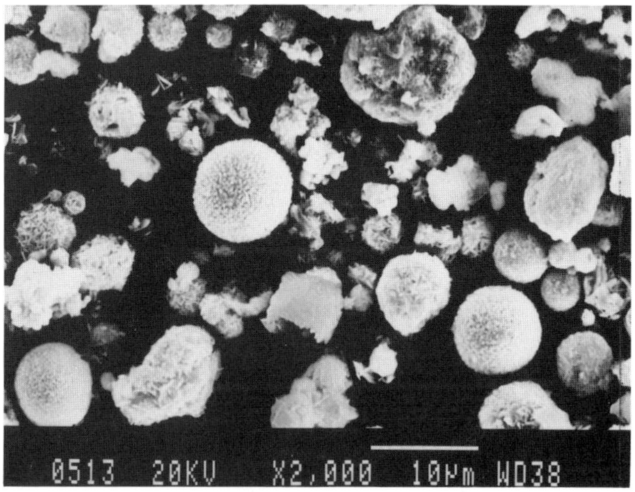

Fig. 1. SEM image of ZnO based varistor precursor powders obtained by the reaction spray process.

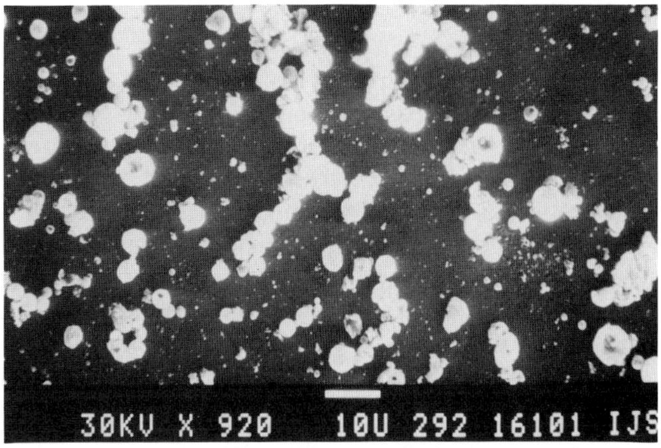

Fig. 2. SEM image of chloride-derived $BaTiO_3$ precursor powder.

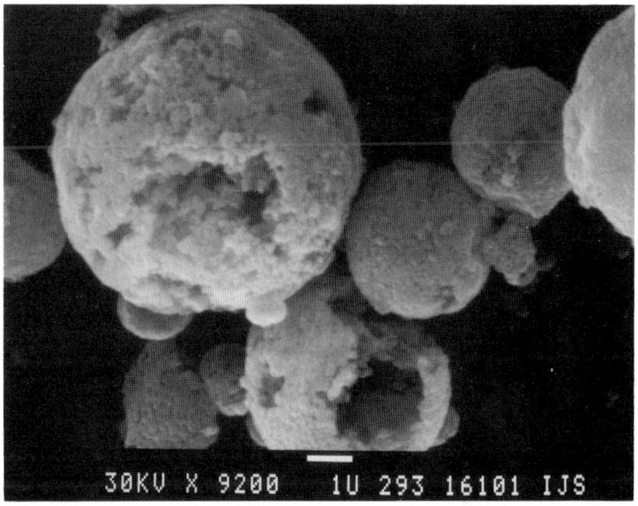

Fig. 3. SEM image of the inner structure of one chloride-derived $BaTiO_3$ particle.

a crust [10]. With the rapid temperature increase, the precipitate salt crystals then decompose to form the oxide crystallites, while the original spherical shape of the particle persist. So the particle surfaces are not smooth and each particle seems to be an aggregate of small individual particles which are less than or equal to 1 μm in diameter. The decomposition curves indicate completed reaction and X-ray diffraction (XRD) implies the presence of a well crytallized ZnO phase, a poorly crystallized spinel phase, as well as a small amount of an unidentified phase [13].

From the SEM image of chloride-derived $BaTiO_3$ precursor powder (powder B-1), spherical particles with diameters ranging from 1 to 7 μm are visible, although there are some irregularly shaped particles (Fig. 2). As in the case of the previous powder, the particles obtained are hollow spherical shells and shell fragments consisting of aggregates of primary particles (crystallites) (Fig. 3). The particle size distribution indicates a mean particle size $D_{50}=4$ μm (powder B-1) and $D_{50}=4.5$ μm (powder B-2). XRD patterns for the chloride-derived powders show the presence of a well crystallized c-$BaTiO_3$ phase as well as about 10% $BaCl_2 \cdot 2H_2O$ phase, when the thermal analysis results are considered. The appearance of c-$BaTiO_3$ instead of tetragonal $BaTiO_3$ may be determined by the size of the crystallites and may occur for crystallite sizes less than 250 nm [14]. EDS analysis of chloride-derived $BaTiO_3$ powder indicates extremely homogeneous composition when the particle surface and the bulk are analysed. This implies that mixing in the feeding solution has persisted in the resulting powder.

SEM analysis of powder obtained from the suspension TiO_2–$X H_2O + Ba(OH)_2$ in nitric acid (powder B-3) indicates highly agglomerated powder with mostly irregularly shaped particles and a mean particle size $D_{50}=3.5$ μm. XRD powder analysis reveals that decomposition of the salt is not complete for powders B-3 and B-4.

The non-linearity coefficient observed for varistor ceramics in the range 10–100 A m^{-2} realized its maximum value ($\alpha_2 \approx 30$) for sintering at 1473 K, while the non-linearity coefficient in the low current region (α_1) is relatively low; the leakage current is about 40 μA cm^{-2} and the breakdown field is 700 V mm^{-1} for these sintering conditions. The relatively higher values of breakdown fields observed in this work, compared with similar systems [13], may be the result of refinement of the microstructure due to the very uniform and spherical particles during preparation of the powder in the dispersion phase.

4. Conclusion

High purity, fine precursor powders for multicomponent ZnO-based non-linear varistor and $BaTiO_3$ ceramics were prepared by the reaction spray process at 1173 K. The particles obtained are hollow spherical shells and shell fragments and seem to be an aggregate of small individual particles less than 1 μm in size. Compositional homogeneity and phase content are determined by the process parameters, by the uniformity of the feed solution and by the characteristics of the starting compounds, which affect the dehydration, precipitation and decomposition processes. The chlorides $TiCl_4$ and $BaCl_2$ are convenient salts for $BaTiO_3$ precursor powder synthesis. Varistor ceramics with non-linearity coefficient $\alpha = 30$ and breakdown voltage $K \approx 700$ V mm^{-1} were produced by sintering of the reaction spray pyrolysis derived multicomponent powder with composition $(100 - X)$ZnO + Y additives.

Acknowledgments

This research was financially supported by the Republic Science Foundation through the project "Physicochemical Processes in Homogeneous and Heterogeneous Systems", and "Metal and Ceramic Matrix Composites" and partly by the National Institute for Standards and Technology, Gaithersburg, MD, through funds made available to the US–Yugoslav Joint Board on Scientific and Technological Cooperation for the project Synthesis and Microanalysis of Ceramics. In addition, the authors gratefully acknowledge Professor M. Trontelj, Dr. Lj. Karanović, Dr. D. Poleti and Ms. D. Vasović whose contributions were significant in making the work successful.

References

1 D. R. Uhlmann, B. J. J. Zelinski and G. E. Wnek, in C. J. Brinker, D. E. Clark and D. R. Ulrich (eds.), *Better Ceramics Through Chemistry*, Elsevier, Amsterdam, 1984, p. 59.
2 D. Hennings, in G. de With, R. A. Terpstra and R. Metselaar (eds.), *Euro-Ceramics*, Vol. 2, Elsevier, Amsterdam, 1989, p. 2273.
3 R. G. Dosch, in L. L. Hench and D. R. Ulrich (eds.), *Science of Ceramic Chemical Processing*, Wiley, New York, 1986, p. 311.
4 T. T. Fang, H. B. Lin and J. B. Hwang, *J. Am. Ceram. Soc., 73* (11) (1990) 3363.
5 O. Milošević, D. Vasović, D. Poleti, Lj. Karanović, V. Petrović and D. Uskoković, in L. M. Levinson (ed.), *Ceramic Transactions: Advances in Varistor Technology*, Vol. 3, The American Ceramic Society, Westerville, OH, 1989, p. 395.
6 J. Lauf and W. D. Bond, *Ceram. Bull., 63* (1984) 278.
7 Ph. Colomban, *Ind. Ceram., 792* (3) (1985) 186.
8 A. K. Maurice and R. C. Buchanan, *Ferroelectrics, 74* (1987) 61.
9 T. Q. Liu, O. Sakurai, N. Mizutani and M. Kato, *J. Mater. Sci., 21* (1986) 3698.
10 D. W. Sproson, G. L. Messing and T. J. Gardner, *Ceram. Int., 12* (1986) 3.
11 E. Ivers-Tiffee and K. Seitz, *Am. Ceram. Soc. Bull., 66* (1987) 1384.
12 E. Ivers-Tiffee, *Ferroelectrics, 68* (1986) 99.
13 O. Milošević, Lj. Karnović, M. Tomasevic-Canovic, Lj. Trontelj and D. Uskoković, *J. Mater. Sci.*, in press.
14 H. Yamamura, A. Watanabe, S. Shirasaki, Y. Moriyoshi and M. Tanada, *Ceram. Int., 11* (1) (1985) 17.

Materials Science and Engineering, *A168* (1993) 253–256

A novel approach to control of ZnO-based varistor microstructures at the nanometre level

Z. Milosavljević, O. Milošević and D. Usković*
Institute of Technical Sciences of the Serbian Academy of Sciences and Arts, Belgrade (Serbia)
D. Vasović and D. Poleti
Faculty of Technology and Metallurgy, Belgrade (Serbia)
Lj. Karanović
Faculty of Mining and Geology, Belgrade (Serbia)

Abstract

ZnO based multiphase ceramics with non-linearity coefficients exceeding 40 and breakdown fields ranging from 200 to 1600 V mm^{-1}, were obtained by a method referred to as "directed synthesis of the constituent phases". This method is based on reaction sintering of a mixture of ZnO, spinel and γ-Bi_2O_3 crystal phases, which are the main constituents of the ZnO varistor ceramics. Each phase was synthesized separately as a single phase. Sintering was performed in air, in the temperature range 1073–1573 K. This method enables us to achieve an optimum and precisely defined phase content in the varistors. Directed synthesis of the constituent phases represents a novel approach to control of the uniformity of varistor microstructures at the nanometre level.

1. Introduction

ZnO based varistors are ceramic semiconductors obtained by sintering ZnO with a small amount of additives, such as oxides of bismuth, antimony, cobalt, manganese, nickel, chromium, aluminium and some other metals. Their characteristic feature is high non-linearity of the current–voltage characteristics [1]. The electrical properties of ZnO varistors are determined by their functional microstructure, *i.e.* those microstructural features which have a direct influence on the electrical performance. ZnO grains with controlled doping level, grain boundaries and intergranular network of bismuth-rich phases constitute the varistor functional microstructure [2]. It is generally assumed that the grain boundaries and the intergranular layer, the thickness of which is measured in nanometres, are responsible for the varistor effects [3, 4]. For this reason, we must be able to control the structure at the nanometre level in order to obtain optimum varistor characteristics.

Since the discovery of varistors, the fabrication technology has proceeded from research into new additives and optimization of the processing conditions [5–10] towards synthesis research. Directed synthesis of the constituent phases is a new method, differing from all other methods in its approach to the problem of producing varistors. This approach is based on the well known phase structure of ZnO varistors.

This new method of directed synthesis of the constituent phases has been applied to obtain varistors with precisely defined and optimum composite structure and to control the phase composition and the distribution of additives. From the point of view of the functional microstructure, the purpose of this method is to control the chemistry of ZnO grain boundaries (concentration of lattice defects, dopants and oxygen) by introducing previously prepared phases into the system: ZnO phase, uniformly doped with Co^{2+} and Mn^{2+} ions; spinel phase, acting as an inhibitor of grain growth and thus providing a homogeneous microstructure; γ-Bi_2O_3 phase, with surplus of oxygen (Bi_2O_{3+x}) [11]. The role of γ-Bi_2O_3 is directly related to the chemistry of ZnO grain boundaries, especially control of the concentration of oxygen at the grain boundaries.

2. Experiment

All chemicals used were of reagent grade quality.

The ZnO phase of composition 99.8 mol% ZnO + 0.2 mol% (Co^{2+} + Mn^{2+}) was prepared by a chemical method to obtain better homogeneity of composition. ZnO was suspended in an aqueous solution of

*Correspondence address: Pro Ignis Consultancy, Mr. Miša Ignis, Postbus 82430, NL2508 EK, Denhaag, Netherlands.

0921-5093/93/$6.00

$Mn(CH_3COO)_2$ and $Co(NO_3)_2$. After evaporation of water, the material was dried at 408 K, calcined at 1473 K for 2 h and homogenized in an agate planetary ball mill (Retsch, type PM4) for 4 h. The ZnO phase was examined by X-ray powder diffraction and the lattice constants were $a = 3.2507(4)$ Å and $c = 5.207(3)$ Å.

Two mixtures of the Bi_2O_3 phase were prepared: A $6Bi_2O_3{\cdot}MnO_2$, B $6Bi_2O_3{\cdot}ZnO$.

Both mixtures were homogenized by dry milling in an agate mortar and heated at 800 K in a platinum crucible until the presence of the pure γ-modification was established by X-ray powder diffraction. The lattice constants for the obtained γ-Bi_2O_3 were $a = 10.221(1)$ Å for mixture A and $a = 10.196(3)$ Å for mixture B.

The mixture of components of the spinel phase (ZnO, NiO, Co_3O_4, Cr_2O_3, MnO_2, Sb_2O_3) corresponding to the composition $Zn_{1.971}Ni_{0.090}Co_{0.030}Cr_{0.247}$-$Mn_{0.090}Sb_{0.545}O_4$, was homogenized in an agate mortar, and heated at 1373 K for 2 h. The experimental value of the lattice constant was $a = 8.546(1)$ Å.

The phases obtained were used for preparation of the varistor mixtures. Two varistor mixtures were prepared, differing only in the additive used for γ-Bi_2O_3 stabilization. The phase composition of both mixtures was 90.9 mol% ZnO (Co^{2+}, Mn^{2+}) + 0.4 mol% γ-Bi_2O_3 + 8.7 mol% $Zn_{1.971}Ni_{0.090}Co_{0.030}Cr_{0.247}$-$Mn_{0.090}Sb_{0.545}O_4$. Type A γ-Bi_2O_3 was used to prepare mixture Z1, and type B γ-Bi_2O_3 was used for the mixture Z2. The mixtures were homogenized in an agate planetary ball mill for 2 h. The resulting powder was pressed into pills 1 mm in height and 8 mm in diameter. Samples were sintered in air, at temperatures in the range 973–1573 K for 1 h. Heating and cooling rates were controlled (5 K min^{-1}). Heat treatment of as-sintered samples was performed at 873 K for 1 h.

Differential thermal analysis (DTA) of γ-Bi_2O_3 phases was performed in air, at the heating rate of 10 K min^{-1} (Netzch 409 EP). Identification of the crystal phases in the initial powders and resulting ceramics was carried out using a Philips PW 1710 diffractometer, with graphite-monochromatized Cu Kα radiation. Optical and scanning electron microscopy examinations were carried out on polished and chemically etched (HNO_3 solution for 5 s) sintered samples.

Electrical properties were registered within the interval 0.1–10 mA cm^{-2} using the direct current method. The non-linearity coefficients were determined within the ranges 0.1–1.0 mA cm^{-2} (α_1) and 1–10 mA cm^{-2} (α_2), the breakdown field K_c was measured at 1 mA cm^{-2} and the leakage current was determined at a voltage of 0.8 K_c. Calculations of α_1, α_2, K_c and U_b values were based on the well known relations [1].

3. Results and discussion

Prominent endothermal peaks were registered in the temperature interval 1005–1059 K on the DTA diagram for type A γ-Bi_2O_3. These peaks presumably result from the melting process, which occurred after transition from the γ phase to another modification of Bi_2O_3. From comparison of these results with the phase diagram of the bismuth oxide–manganese oxide system [12] it is supposed that this was a $\gamma \rightarrow \delta$ transition. The DTA diagram for sample B showed only one endothermal peak at temperature $T = 1026.6$ K which is in congruence with the melting process. This is in good agreement with data given in the phase diagram of the bismuth oxide–zinc oxide system [12].

X-ray powder analysis of the sintered samples Z1 and Z2 showed the same diffraction patterns with small differences in peak intensities. The lattice constants of ZnO and spinel in Z1 samples are shown in Table 1. The ZnO phase, which has the highest degree of crystallinity in the samples sintered at 1573 K, was the dominant phase in all samples. Differences in ZnO lattice parameters can be explained by diffusion of Co^{2+}, Mn^{2+}, Ni^{2+} ions to the ZnO phase during sintering. The γ-Bi_2O_3 phase is identified at all temperatures except 1573 K. The diffraction maxima of this phase decreased with increasing temperature and finally disappeared at 1573 K. Weak peaks of β-Bi_2O_3 were registered at that temperature. The differences in spinel lattice parameters were relatively small and they could be due to cation redistribution between different phases.

The mean particle sizes D were measured by optical microscopy. Larger ZnO grains and smaller spinel grains were found in Z2 varistor samples.

Characteristic parameters which define the electrical properties of the varistors are shown in Table 2. The values of non-linearity coefficients α_1 and α_2 for samples sintered for 1 h in the temperature interval

TABLE 1. Lattice constants of ZnO and spinel phases in Z1 varistor samples

Temperature (K)	ZnO		Spinel, a (Å)
	a (Å)	c (Å)	
1073	3.2518(7)	5.211(4)	8.551(3)
1173	3.2530(6)	5.208(4)	8.549(5)
1273	3.2530(9)	5.210(6)	8.552(3)
1373	3.259(2)	5.22(1)	8.561(6)
1473	3.2520(5)	5.211(3)	8.550(3)
1573	3.2556(8)	5.217(5)	8.559(2)
JCPDS card	3.2496(1)	5.2070(9)	
Starting sample	3.2507(4)	5.207(3)	8.550(3)

TABLE 2. Characteristic parameters for Z1 and Z2 varistor samples

Temperature (K)	Z1 D (μm)	Z1 α_1	Z1 α_2	Z1 K_c (V mm^{-1})	Z1 J_L (μA cm^{-2})	Z2 D (μm)	Z2 α_1	Z2 α_2	Z2 K_c (V mm^{-1})	Z2 J_L (μA cm^{-2})
1073	1.9	29	26	1586	12.8	2.1	28	33	1393	12.5
1173	2.6	18	42	1141	56.6	2.7	20	35	1103	53.8
1273	3.7	45	36	811	10.0	4.4	30	40	688	15.0
1373	5.5	37	31	541	7.4	6.7	33	28	445	6.8
1473	6.8	32	28	443	5.9	7.8	37	26	385	10.0
1573	10.4	31	25	288	6.4	11.4	34	26	263	9.6

α_1 non-linearity coefficients in the range 0.1–1.0 mA cm^{-2}; α_2 non-linearity coefficients in the range 1.0–10.0 mA cm^{-2}; K_c breakdown field measured at 1 mA cm^{-2}; J_L leakage current determined at a voltage of $0.8K_c$; D grain size.

1073–1573 K are in the range 18–45. Z1 varistor samples have slightly better electrical characteristics. Samples sintered below 1048 K do not show varistor effects. The lowest temperature at which the varistor effect is detectable is in agreement with the DTA results, *i.e.* the varistor microstructure resulted from liquid phase sintering, so it would not be expected to obtain the varistor effect at temperatures below the melting point of Bi_2O_3. The leakage current depends on temperature and is in the range 5–15 μA cm^{-2}, except for the samples sintered at 1173 K. We could not find a reasonable explanation for the high values of leakage current in samples sintered at 1173 K. The best values of α_1, α_2 and J_L were obtained at 1273 K, for both varistor mixtures. Raising the sintering temperature resulted in an increase in ZnO-grain growth and a decrease in K_c. K_c values were relatively high (200–1600 V mm^{-1}). Slightly lower values were obtained for the Z2 varistor samples. The difference between the Z1 and Z2 values is in agreement with results obtained by optical microscopy, which reveal the presence of larger ZnO grains in the Z2 varistor samples. The voltage per barrier was approximately 3 V for both varistor mixtures.

Heat treatment of the as-sintered samples at 873 K for 1 h did not improve the varistor characteristics. It is known that any additional heat treatment can result in δ- or $\beta \rightarrow \gamma$-Bi_2O_3 transition [13]. This transition improves the electrical properties. In this experiment, the γ-Bi_2O_3 modification persisted in the sintering ceramics, and it is supposed that further improvement of properties by heat treatment could not be expected.

4. Conclusions

ZnO varistor ceramics with non-linearity coefficients exceeding 40 and breakdown fields in the range 200–1600 V mm^{-1} were obtained by the directed synthesis of the constituent phases at sintering temperatures of 1073–1573 K. It was shown that the varistor properties already appear at 1073 K. This is in accordance with the increase in functional microstructure in the varistor composite during liquid phase sintering. Structural examination revealed the presence of a well crystallized ZnO phase uniformly doped with Mn^{2+} and Co^{2+} ions, a spinel phase doped with NiO, Co_3O_4, Cr_2O_3 and MnO_2, and a γ-Bi_2O_3 phase along the ZnO grain boundaries. These microstructural features are crucial for varistor performance from the viewpoint of the ZnO grain defect structure and microstructural homogeneity. From the viewpoint of the conductivity mechanisms in varistor ceramics, this method enables full control of the properties, especially if one considers chemistry of the depletion layer and γ-Bi_2O_3 as a supplier of oxygen ions.

Acknowledgments

This research was financially supported by the Republic Science Foundation through the project “Physical Chemistry Processes in Homogeneous and Heterogeneous Systems”, and “Synthesis, Characterization and Properties of Metal and Ceramics Matrix Composites”.

References

1 M. Matsuoka, *Jpn. J. Appl. Phys., 10* (1971) 736.
2 E. Olsson and G. L. Dunlop, in *Advances in Varistor Technology*, in *Ceram. Trans., 3* (1989) 57.
3 U. Schwing and B. Hoffmann, *J. Appl. Phys., 51* (1980) 4558.
4 P. F. Bongers and P. E. C. Franken, *Grain Boundary Phenomena in Electronic Ceramics*, in *Adv. Ceram., 1* (1981) 38.
5 K. Eda, *Advances in Varistors Technology*, in *Ceram. Trans., 3* (1989) 10.
6 M. V. Vlasova, N. G. Kakazey, P. Kostić, O. Milošević and D. Uskoković, *J. Mater. Sci., 16* (1984) 127.
7 P. Kostić, O. Milošević and D. Uskoković, in G. C.

Kuczynski, D. P. Uskoković, H. Palmour III and M. M. Ristić (eds.), *Sintering '85*, Plenum, New York, 1987, p. 301.

8 P. Kostić, O. Milošević, M. M. Ristić and D. P. Uskoković, *Physica B, 150* (1988) 175.

9 O. Milošević, D. Vasović, D. Poleti, Lj. Karanović, V. Petrović and D. Uskoković, *Ceram. Trans., 3* (1989) 395.

10 O. Milošević, D. Vasović, D. Poleti, Lj. Karanović, V. Petrović and D. Uskoković, in D. P. Uskoković, H. Palmour III and R. M. Spriggs (eds.), *Science of Sintering: New Direction for Materials Processing and Microstructural Control*, Plenum, New York, 1990, p. 117.

11 T. K. Gupta and W. G. Carlson, *J. Mater. Sci., 20* (1985) 3487.

12 E. M. Levin and R. S. Roth, *J. Res. NBS, 68A* (1964) 197.

13 A. Iga, M. Matsuoka and T. Masuyama, *Jpn. J. Appl. Phys., 15* (1976) 1161.

Materials Science and Engineering, A168 (1993) 257–261

Influence of the deposition conditions on the optical, electrical and magnetic properties of sputtered $Fe–Al_2O_3$ and $Ni–Al_2O_3$ nanocermet films

M. Gadenne and P. Gadenne

Laboratoire d'Optique des Solides, Université P. et M. Curie, 4 Place Jussieu, 75252 Paris Cedex 05 (France)

M. T. Ramdou and J. P. Segaud

Direction des Matériaux, ONERA, 91120 Palaiseau (France)

H. Lassri and R. Krishnan

Laboratoire du Magnétisme et des Matériaux Magnétiques, CNRS, 1 place A. Briand, 91195 Meudon Cedex (France)

C. Sella

Laboratoire de Physique des Matériaux, CNRS, 1 place A. Briand, 92195 Meudon Cedex (France)

Abstract

We reveal the influence of the deposition conditions on the physical properties of cermet films made of nanometer sized magnetic metal ($\mu \neq \mu_0$) and alumina grains. The films were prepared by r.f. cosputtering under two typical base pressures (10^{-6} and 10^{-7} Torr), using either an argon plasma or an argon + 10% hydrogen plasma under 6×10^{-3} Torr. They present a wide range of compositions and thicknesses. Electron microprobe analysis, transmission electron microscopy and high energy electron diffraction were used to study their crystallographic and granular structure. We report the electrical d.c. conductivity, optical transmittivity with respect to wavelength, magnetization and ferromagnetic resonance as a function of the deposition conditions. These physical properties are governed to a large extent by oxidation of the metal.

1. Introduction

It is now well known that, for a given cermet type, the percolation threshold mainly depends on the concentration of metal [1]. We have shown that the threshold occurs for a critical thickness of film for a high enough metal concentration [2, 3]. Moreover, when comparing the physical properties in the literature, one notices that they can be very different depending on the elaboration technique. For example, the d.c. conductivity and the optical properties vary over a wide range, even if the thickness of the films and the apparent concentration of metal are fixed. Studying a mixture of gold and chromium, it has been shown [4] that the deposition conditions play a prominent part in the oxidation of this compound. The magnetic properties of cermet made of magnetic metal inclusions in alumina may also be modified by partial oxidation of the magnetic metal [5]. Taking into account these preliminary remarks, we investigate here in a systematic way the influence of the deposition conditions (base pressure, composition of the sputtering gas, etc.).

2. Preparation of the samples

Our samples were thin cermet films ($Fe–Al_2O_3$ and $Ni–Al_2O_3$), deposited by r.f. cosputtering onto several kinds of substrate. The target was an alumina disc (110 mm diameter) on which small nickel or iron plates (5 mm diameter) were regularly placed in hexagonal lattice. The number of small discs varied between 55 and 223, to prepare samples with very different metal concentrations. A rotating substrate holder, allowing *in situ* measurement of the d.c. resistance, provided samples with good homogeneity. Experiments were performed under various base pressures (from a few 10^{-6} Torr to 10^{-7} Torr) using either an argon plasma or an argon + 10% hydrogen plasma, under a working pressure of 6×10^{-3} Torr. The thicknesses of the films

0921-5093/93/$6.00

varied from a few tenths of nanometers to several micrometers for all concentrations.

3. Deposition rate

The global deposition rate is mainly dependent on the target composition, as iron, nickel and alumina do not have the same sputtering rate. In Table 1 we compare the deposition rate measured for two target configurations under vacuum and plasma conditions. The total sputtering time was the same (100 min) for all measurements. Obviously the global rate is higher when the base pressure is lower. The addition of hydrogen slows down this rate, probably by modifying the quantity and nature of iron oxides.

4. D.c. conductivity

The variations of the d.c. resistance was measured *in situ* during sputtering. As an example we show these variations in Fig. 1 for one target configuration (169 nickel plates on the alumina disc) and for four deposition conditions. It is evident that the better the base pressure, the sooner the critical percolation thickness is reached. The same kind of dependence is found for iron cermets, and will influence strongly their IR optical properties (see Section 6). The correct explanation is suggested by observation of the electron diffraction diagrams (Fig. 2). In fact the characteristic diffraction rings of NiO appear much clearer for the sample obtained under a base vacuum of approximately 10^{-6} Torr than for that obtained at approximately 10^{-7} Torr. We note here that there is no ring corresponding to metallic aluminum. From this observation we infer that oxidation is the main factor delaying the percolation threshold by lowering the conductivity.

TABLE 1. Comparison of the global deposition rate *vs.* the plasma composition and base pressure, for two different target configurations

Plasma composition	Base pressure (Torr)	Rate for 180 metal plates (Å min^{-1})	Rate for 79 metal plates (Å min^{-1})
Pure argon	4×10^{-6}	137.7	93.3
	2.6×10^{-6}	131.5	94.0
Pure argon	2×10^{-7}	160.9	127.5
	2×10^{-7}	180	
Argon + 10%H_2	2.8×10^{-6}	96.2	71.4

5. Atomic composition of the samples

The composition was determined by electron microprobe analysis (Table 2). The results show the number of atoms of metal (nickel or iron), aluminum and oxygen. The argon included cannot be detected. On the one hand, the number of oxygen atoms is always greater than the alumina stoichiometry suggests. This means that there is significant reorganization of atoms during sputtering. On the other hand, when present in the plasma, hydrogen increases the amount of oxygen atoms in a very remarkable way. These remarks do not confirm the commonly adopted explanation that there are molecules of H_2O trapped in the amorphous alumina and the samples. However, it can reasonably be stated that there is a lack of oxygen in the deposited alumina.

6. Optical properties

We took optical measurements of the reflectance R and transmittance T in the range 0.35–2.7 μm, using a Varian V spectrophotometer, of $Fe-Al_2O_3$ and $Ni-Al_2O_3$ cermet thin films. In order to reveal the influence of the deposition conditions, we focus on $Fe-Al_2O_3$ cermet films, comparing the transmittance of two sets of three films. All these films had very similar thicknesses (100 nm $< d <$ 120 nm). Each set of three films was produced with the same target configuration. Throughout the wavelength spectrum, the optical properties depend essentially on the deposition conditions, *i.e.* the base vacuum pressure and the composition of the plasma (pure argon or argon + 10% hydrogen). We divide the study into two ranges, above and below about 1.5 μm.

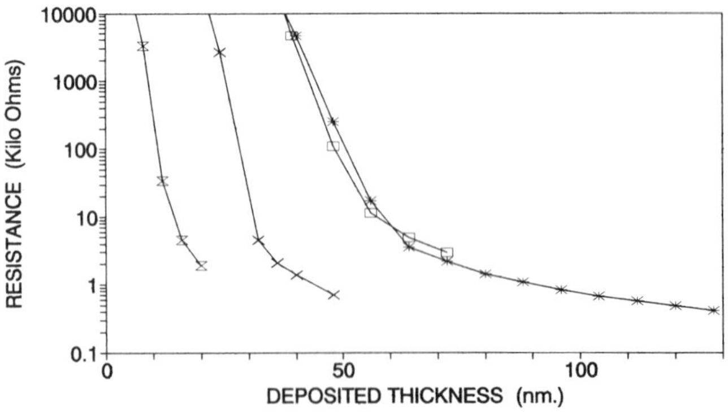

Fig. 1. Variation of the d.c. resistance *vs.* the deposited thickness for four deposition conditions with the same target configuration (169 nickel plates on alumina): $*$ base pressure 4.6×10^{-6} Torr, pure argon plasma; □ base pressure 2.6×10^{-6} Torr, pure argon plasma; × base pressure 8×10^{-7} Torr, pure argon plasma; ⧖ base pressure 2×10^{-6} Torr, argon + 10% hydrogen plasma.

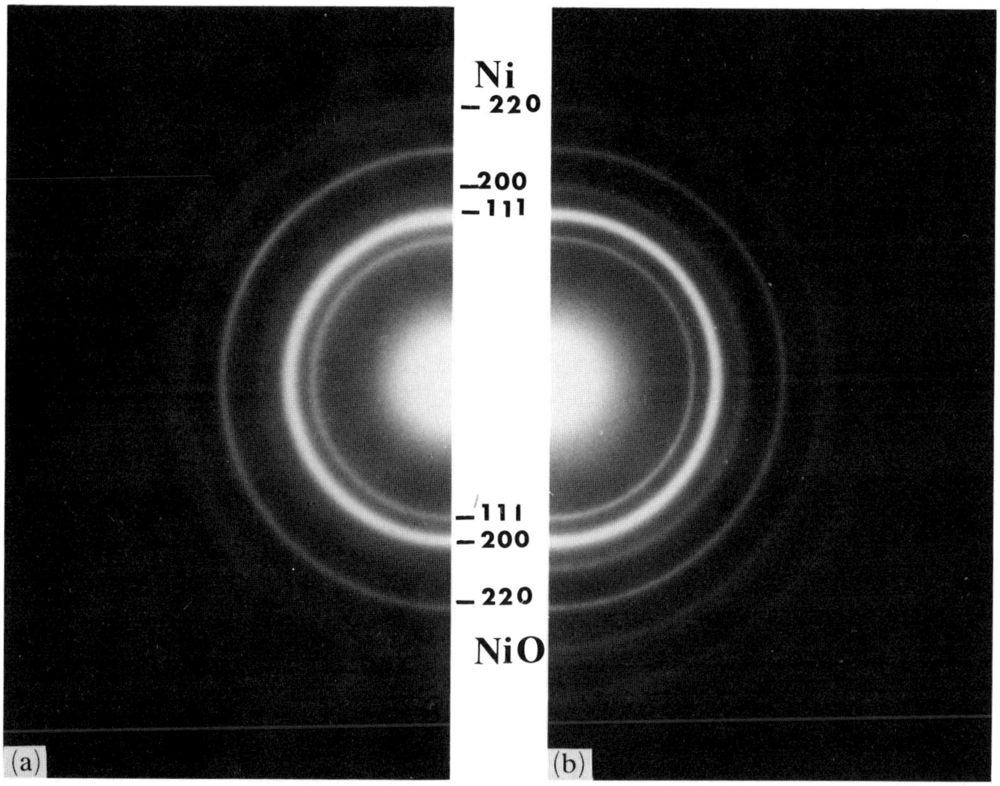

Fig. 2. Electron diffraction pattern of two nickel–alumina thin films (223 nickel plates), elaborated under different base pressures (a) 10^{-7} Torr, (b) 10^{-6} Torr.

Above 1.5 μm, we recognize the characteristic optical behavior of percolative samples. As already pointed out for the conductivity measurements (Section 4), although films of the same set are obtained from the same target configuration and have the same thickness, they behave differently, with regard to both electrical d.c. conductivity and optical properties (Fig. 3). As the d.c. resistance of the samples increases, the transmittance increases for a given wavelength and its slope *vs.* wavelength increases too. As already shown for granular gold [6] and gold alumina cermets [7], the transmittance of the samples close to the percolation threshold (sample d in Fig. 3) remains almost constant in that range of wavelength.

Below 1.5 μm, the transmittance of the samples obtained under the best base pressure (approximately 10^{-7} Torr) behaves monotonically *vs.* wavelength. In contrast, the transmittance of samples obtained either under a poor base pressure (a few 10^{-6} Torr) or with a plasma containing 10% hydrogen shows a dip. The wavelength at which this dip occurs increases with metal concentration (1.1 → 1.5 μm). We also took reflectance measurements and can deduce that the dip in transmittance is an absorption peak. The same kind of absorptance appears in the same range of wavelength in Ni–Al_2O_3 cermets deposited under the same vacuum conditions, possibly owing to the well known grain resonance [8]. However, we need more experimental results (at least a study of spectra of the oxides) to confirm the explanation of this phenomena.

7. Magnetic properties

We measured the magnetization with a vibrating sample magnetometer (VSM) and observed the ferromagnetic resonance (FMR) in the samples. The measurements were taken at room temperature. The magnetization M (emu cm^{-3}) was expressed in terms of the volume of the sample and not of the metal content. These results are quite complex to understand in detail although one can reveal some trends.

7.1. Fe–Al_2O_3 films

VSM measurements show that, as expected, M increases with the iron content and, for a given concentration of iron, M is slightly dependent on the deposition conditions and particularly on the sputter gas composition. For instance, in the 180 series, M increases from 530 emu cm^{-3} to 617 emu cm^{-3} when the sputtering gas is changed from argon to argon + 10% H_2. The contributions to the magnetic

TABLE 2. Composition in percentage of atoms (microprobe analysis) of cermets

Fe–Al_2O_3 cermets

Number of iron disks	Deposition conditions	Atomic concentration				Σ
		Fe	Al	O	Ar	
180	Ar	49	13	36	4	102
	Ar + LN_2	49	13	35	4	101
	Ar + 10%H_2	46	13	38	4	101
144	Ar	46	17	35	3	101
	Ar + LN_2	47	17	34	4	102
	Ar + 10%H_2	44	13	40	3	100
79	Ar	30	22	45	4	101
	Ar + LN_2	31	24	38	4	98
	Ar + 10%H_2	33	18	44	4	99
60	Ar	24	25	46	4	99
	Ar + LN_2	26	27	41	4	98
	Ar + 10%H_2	24	18	54	5	98

Ar, pure argon plasma, base pressure approximately 10^{-6} Torr; Ar + LN_2 pure argon plasma, base pressure approximately 10^{-7} Torr; Ar + 10%H_2 argon + 10% hydrogen plasma, base pressure approximately 10^{-6} Torr.

Ni–Al_2O_3 cermets

Number of nickel disks	Sample	Atomic concentration			Precision
		Ni	Al	O	
180	1	43.8	22.5	33.7	±0.5
	2	43.8	22.5	33.7	±0.5
144	1	40.5	23.8	35.7	±0.5
	2	40.5	23.8	35.7	±0.5
114	1	33.5	26.6	39.9	±0.5
	2	33.2	26.7	40.1	±0.5
79	1	24.8	30.1	45.1	±0.5
	2	24.8	30.1	45.1	±0.5
42	1	13.4	34.6	52	±0.5
	2	13.5	34.6	51.9	±0.5

moment in the samples arise from various sources, such as metallic iron or the formation of a spinel type ferrite, either simply the ordinary magnetite or magnetite with Al^{3+} ions substituted in place of Fe^{3+} ions. The more oxidation occurs, the lower will be the magnetization, because iron has the highest value of M.

We can obtain more information from FMR studies. For example, inducing resonance with the fields applied in the plane of the film and perpendicular to it, it is possible to obtain the effective magnetization $4\pi M_{eff}$ which is equal to $4\pi M - H_u$, where H_u is the anisotropy field. Knowing $4\pi M$, one can calculate H_u. However, if it is assumed that the anisotropy is negligible, because the hysteresis loops show easy saturation in the plane, then the FMR measurements yield directly the value of magnetization M. The FMR spectra are quite complex and show a greater sensitiv-

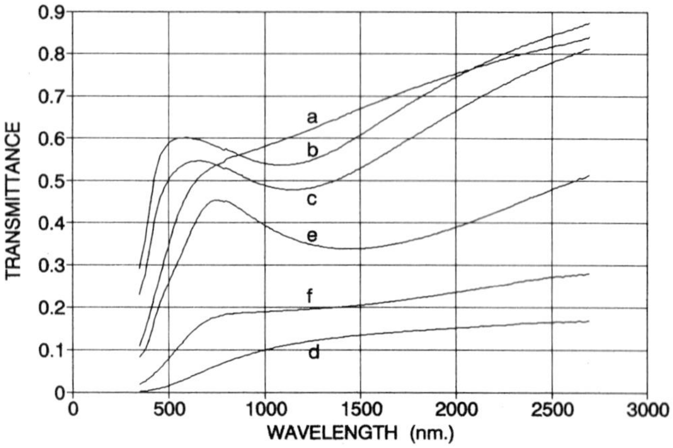

Fig. 3. Transmittance of iron–alumina films *vs.* wavelength, for two sets of films. 73 iron plates on the alumina target: (a) thickness 114.2 nm, resistance 10.7 MΩ, 2×10^{-7} Torr, pure argon plasma; (b) thickness 100 nm, resistance 2.1 MΩ, 2×10^{-6} Torr, argon + 10% hydrogen plasma; (c) thickness 103.8 nm, resistance 0.68 MΩ, 2.6×10^{-6} Torr, pure argon plasma. 180 iron plates on the alumina target: (d) thickness 119.2 nm, resistance 750 Ω, 2×10^{-7} Torr, pure argon plasma; (e) thickness 105.7 nm, resistance 104 kΩ, argon + 10% hydrogen plasma; (f) thickness 115.2 nm, resistance 2.9 kΩ, 2.6×10^{-6} Torr, pure argon plasma.

ity to the preparation conditions. Also, the presence of more than one absorption mode of comparable intensity is attributed to the signal from regions of different magnetization. However, the calculations were performed for all modes, but only those which also give a g factor close to 2.2, which is the signature of iron, were retained for discussion. Using just argon as sputter gas leads to unreasonable values of g. However for argon + 10% H_2 the spectrum gives correct g values. The value of M from FMR is higher, indicating that the VSM values were underestimated as a result of uncertainty in the volume of the magnetic species. Also, the line width is very low for the second case, which indicates better quality and homogeneity of the sample.

7.2. Ni–Al_2O_3 films

In the case of nickel, one could expect a simpler situation compared with iron, because in Ni–Al_2O_3 the only magnetic species is metallic nickel. Neither its oxide nor nickel aluminate spinel is magnetic.

The magnetization obtained by VSM increases with the nickel content from 48 emu cm^{-3} for 79 discs on the target to 125 emu cm^{-3} for 144 discs. Compared with the magnetization of 450 emu cm^{-3} for bulk nickel, these values are considerably lower, indicating some oxidation and/or mixing with Al_2O_3. Again the exact volume of magnetic nickel being not known, the M values may be an underestimate. Examination of the results from FMR will give more information. Figure 4

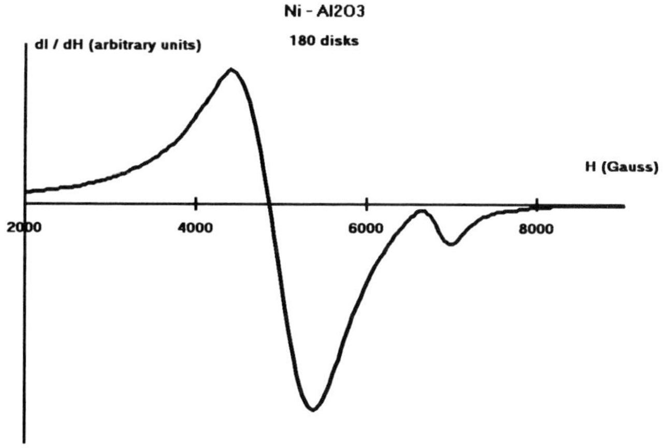

Fig. 4. FerroMagnetic resonance (FMR) spectrum of a nickel–alumina film (180 nickel plates).

shows the FMR spectrum for the sample of 180 nickel plates on the alumina target prepared under the best base pressure. The external field was applied normally to the film plane. The spectrum is relatively simple consisting of an intense mode, and another mode about one tenth as intense, at a higher field. This rules out the possibility of considering them as spin wave modes. A simple interpretation would be for the two modes to arise from two regions with different magnetic properties. The volume corresponding to the stronger mode is about ten times that for the weaker mode. Once more, the calculations were done here assuming the absence of any anisotropy. The g factors (2.00 and 2.30) are reasonably close to that of metallic nickel (2.18). However, it is seen that the magnetization (260 and 150 emu cm^{-3}) is still lower than that of bulk nickel. The two values of M indicate that there are two regions with different magnetic properties. The region corresponding to mode 1 shows a smaller line width, indicating that this region is more homogeneous. It is not possible to draw any more conclusions from these measurements.

The magnetic properties show a strong dependence on the deposition conditions but are not well understood at present. We need to carry out other studies such as the temperature dependence of M to determine the Curie temperature and Mössbauer studies to obtain a more complete picture. In the case of nickel, the reduction in magnetization can be clearly attributed to the oxidation or mixing of nickel with alumina.

8. Conclusion

When studying cermets, the dispersion of the experimental results corresponds mainly to the multiplicity of the deposition conditions. The influence of metal oxidation is of prime importance. Curiously the presence of hydrogen in the argon plasma does not prevent oxidation, but the good quality of the base pressure is very effective. The composition, the size of the clusters, and consequently the electrical d.c. conductivity, the critical percolation thickness, the optical transmission in the visible and near IR are very sensitive to the deposition conditions of the films. It is thus necessary to be very cautious before planning a volumic metal concentration and a value of conductivity or transmittance.

However, with carefully characterized samples, one can confirm the strong correlation between the composition in conducting inclusions and the electrical and optical properties above and below the percolation threshold. As the metallic inclusions are magnetic, they induce a magnetization in the samples more strongly if they are less oxidized. The nickel and iron cermets show very similar behavior, but the explanation of the mechanisms is much more complicated for iron cermet films, because there are three different iron oxides with different magnetic behaviors.

References

1 B. Abeles, P. Sheng, M. D. Couts and Y. Arie, *Adv. Phys., 24* (1975) 407.
B. Abeles, in R. Wolfe (ed.), *Applied Solid State Science*, Vol. 6, Academic Press, New York, 1976, p. 2.
2 M. Gadenne and P. Gadenne, *Physica A, 157* (1989) 344.
3 M. Gadenne, P. Gadenne, C. Sella and J. C. Martin, *Thin Solid Films, 221* (1992) 183.
4 P. Gadenne, C. Sella, M. Gasgnier and A. Benhamou, *Thin Solid Films, 165* (1988) 29.
5 J. L. Dormann, P. Gibbart, G. Suran and C. Sella, *Physica B, 86–88* (1977) 1431.
C. Sella, A. Kaba, S. Berthier and J. Lafait, *Sol. Energy Mater., 16* (1987) 143.
6 P. Gadenne, *Thesis*, Universite Pierre et Marie Curie, Paris, 1986.
P. Gadenne, A. Beghdadi and J. Lafait, *Opt. Commun., 65* (1) (1988) 17.
7 M. Gadenne, *Thesis*, Universite Pierre et Marie Curie, Paris, 1987.
M. Gadenne, J. Lafait and P. Gadenne, *Opt. Commun., 71* (5) (1989) 273.
8 S. Berthier, *Ann. Phys. Fr., 13* (1988) 503.
C. G. Granqvist and O. Hunderi, *Phys. Rev. B, 16* (8) (1977) 3513.

CdS semiconductor doped silica xerogels

A. Othmani, C. Bovier, J. C. Plenet and J. Dumas
Département de Physique des Matériaux, URA 172, CNRS, Lyon-1, 69622 Villeurbanne cedex (France)

B. Champagnon
Lab. de Phys. Chim. des Materiaux Luminescents, URA 422, CNRS, Lyon-1 69622, Villeurbanne cedex (France)

C. Mai
GEMPPM, Insa-Lyon, 69621 Villeurbanne cedex (France)

Abstract

CdS doped amorphous silica was prepared by the sol-gel method. The CdS concentration in weight varied from 5% to 20%. The samples were characterized by four different and complementary techniques: X-ray diffraction (XRD), high resolution transmission electron microscopy (HRTEM), low frequency inelastic Raman scattering (LOFIRS) and small angle X-ray scattering (SAXS). These observations show that the size of the nanocrystals does not depend on the CdS concentration.

1. Introduction

Ultrasmall semiconductor particles have gained considerable attention in recent years because of their enhanced non-linear optical properties and photocatalytic properties [1, 2]. The observation of photoinduced blue shifts in the absorption spectra and exciton peak [3] is evidence of quantum size effects due to confinement of the exciton hydrogen-like function in nanocrystals. Since Jain and Lind reported high non-linear optical properties of $CdS_xSe_{(1-x)}$ doped glass [4], much work has been performed on this type of material.

The conventional melting to produce semiconductor-doped glass uses multicomponent oxides to reduce the glass transition temperature. In fact, it does not allow a cadmium concentration higher than around 1%, owing to the volatility and instability of CdO species which occur around 900 °C in the bulk phase.

The sol-gel process is a good technique for obtaining a well chemically defined matrix of pure silica at a temperature lower than 500 °C [5, 6], in which the CdO concentration can reach 20 wt.%. In this work, we report on the preparation of silica glass doped with CdS nanocrystals by the sol-gel route. The nanocrystalline structure was determined by X-ray diffraction (XRD) and high resolution transmission electron microscopy (HRTEM). The size distribution of the crystallites was determined by low frequency inelastic Raman scattering (LOFIRS) and small angle X-ray scattering (SAXS).

2. Preparation

The CdS doped glass samples were obtained by a two-step method. In the first stage, a pure silica glass was produced by a sol-gel technique involving two solutions at different pH values; this is one of the best ways of obtaining hard monolithic glass if the use of chemical additives for drying control is not desired. The fabrication involves three solutions: the acid solution A contains tetraethylorthosilicate (TEOS), ethanol, water and chlorhydric acid in the molar proportion 1:1:1:0.27; the basic solution B consists of water, ethanol and ammoniac in molar proportion 4:1.25:0.005; the doping mixture C is a solution of 0.05 mol hydrated cadmium acetate in 1 mol methanol.

Mixing of the different components of the initial solution has to be done in the ratio (one part of solution A+ n parts of solution C) + one part of solution B, with $n = 0.1$ for 1 mol TEOS to obtain roughly 1% cadmium by weight doped glass. The amount of cadmium is conveniently adjusted by adjusting the quantity n.

The following precautions have to be taken to ensure good homogeneity of the gelling solution: the doping solution C is mixed with solution A, stirred

0921-5093/93/$6.00

carefully and left for 2 h to hydrolyse; the basic solution B is then added and mixed; the resulting liquid is left in a closed container for 10 days and then the temperature is raised by 5 °C h^{-1} to 500 °C and left for 4 h at this temperature.

The second step is thermal treatment in H_2S atmosphere, at 200 °C for 2 h, to precipitate the CdS microcrystals in the silica matrix. The sample then becomes a characteristic light yellow colour.

The glass obtained has an apparent density of 1.75 when no cadmium is added. Samples are typically 15×15 mm^2 with a thickness ranging from 0.1 mm to 5 mm, depending on the experiments to be carried out.

3. X-ray diffraction results

XRD measurements were taken using a Siemens D500 diffractometer. Figure 1 shows XRD patterns of glass heated for 4 h at 500 °C before and after treatment with H_2S at 200 °C. The spectra show clearly that crystallization occurs after exposing the sample to H_2S. The glass heated at 500 °C only has an amorphous structure Fig. 1(a). Curves (b), (c) and (d) are spectra for CdO concentrations of 5, 10 and 15 wt.% respectively. These curves show that CdS nanocrystals have the hexagonal wurtzite structure [7] and that the line intensities increase regularly with cadmium concentration. No evidence of cubic blend structure is observed.

4. Electron microscopy

HRTEM was used to confirm the presence of CdS nanocrystals in the silica amorphous matrix. Figure 2 shows a micrograph of a CdS cluster. The particles observed are not agglomerated and have a spherical-like shape.

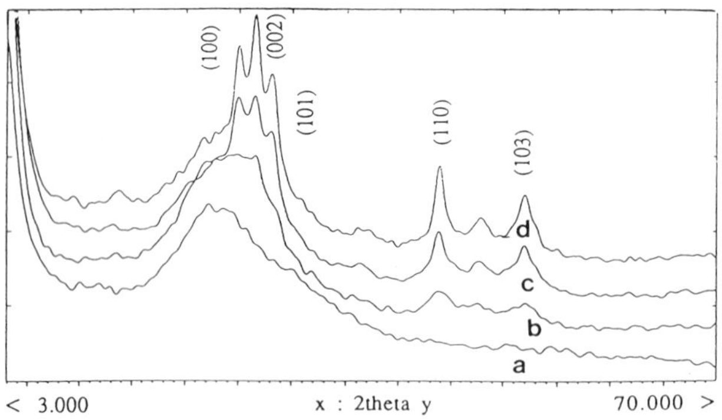

Fig. 1. XRD patterns for CdO doped silica xerogels: (a) before treatment with H_2S; (b), (c) and (d), 5, 10 and 15 wt.% CdO concentration respectively, after treatment with H_2S.

5. Raman scattering

Raman scattering on doped silica xerogels was used to ascertain the formation of CdS and to determine the size of the nanocrystals by LOFIRS. The theory has been discussed in several papers dealing mainly with glass ceramic and polymers [8–10].

The LOFIRS line corresponding to the interaction of light with the surface vibration modes of the nanocrystals. The breathing mode, which is observed, does not depolarize the laser light. The diameter of the particles is deduced from the measured low frequency ω_0 peak (see Fig. 3(b)) using the relation:

$$d = 0.9 v_l / \omega_0 c$$

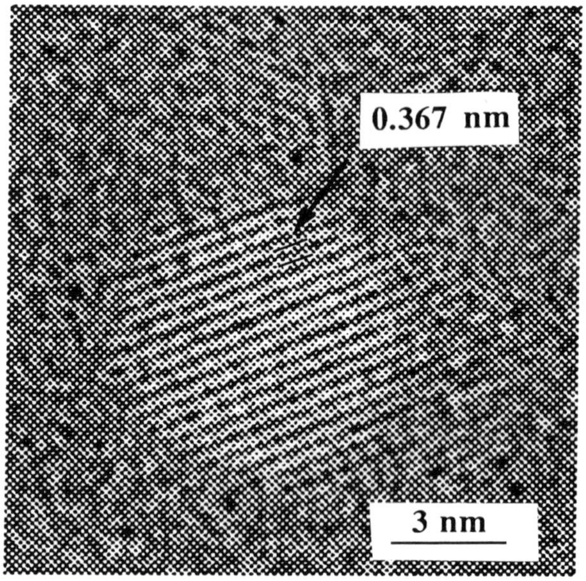

Fig. 2. CdS cluster observed by HRTEM.

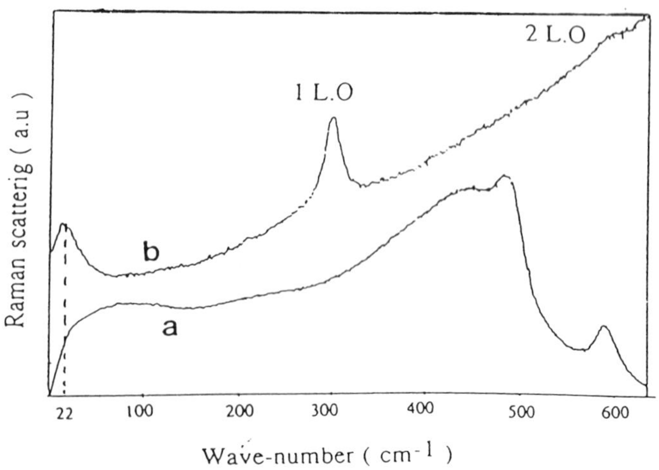

Fig. 3. Raman spectra for 10 wt.% doped silica xerogels: (a) not treated with H_2S, (b) treated for 2 h at 200 °C with H_2S.

v_l is the longitudinal sound velocity (4277 m s^{-1}) and c is the light velocity.

The Raman measurements were obtained using a Z40 Dilor monochromator with the 5145 Å line of an argon ion laser as excitation source and a cooled photomultiplier.

Figure 3 shows Raman spectra of xerogels heated for 4 h at 500 °C (a) and then reacted with H_2S for 2 h at 200 °C (b). The glass heated at 500 °C, without the H_2S treatment, shows a spectrum typical of porous silica (a) with no boson peak and no low frequency features. The 300 cm^{-1} longitudinal optic line (1LO) (b) which is characteristic of CdS indicates that crystallization only occurs after treatment with H_2S. It is a single phonon forbidden line which is nevertheless observed under resonant Raman conditions in small particles. The Frölich interaction, different for electrons and holes, is considered to be responsible for this effect [11]. The peak at 22 cm^{-1}, which corresponds to the particle modes, is evidence of a fairly narrow size distribution. In our samples the 2LO mode at 600 cm^{-1} is very weak.

Figure 4 shows Raman spectra of glasses containing 5, 10, 15 and 20 wt.% CdO, heated for 4 h at 500 °C followed by reaction with H_2S for 2 h at 200 °C.

The peak at 22 cm^{-1} corresponds to the particle breathing modes. The maximum of both the LOFIRS peak at 22 cm^{-1} and the 300 cm^{-1} longitudinal optic line increases roughly linearly with concentration and, for all concentrations, correspond to $d = 55$ Å [11]. For the same preparation conditions and heat treatment, the nanocrystal size does not change with concentration. This important result is confirmed by the following SAXS measurements.

6. Small angle X-ray scattering

The instrument was a Rigaku X-ray diffractometer. The diffusion spectra show a correlation feature for the four concentrations of xerogel treated by H_2S, which was analysed by a method described elsewhere [12] to obtain the particle size distribution and the average distance between particles. Figure 5 shows this distribution for the four concentrations studied. According to these measurements, the same conditions of preparation and H_2S treatment give a size distribution which does not depend on CdS concentration. This confirms the LOFIRS results which implied the same conclusion.

The average distance between CdS clusters decreases when the CdO concentration increases (Table 1). The number of clusters increases with CdO concentration without a corresponding increase in their size. Quantitative studies are detailed in a forthcoming paper.

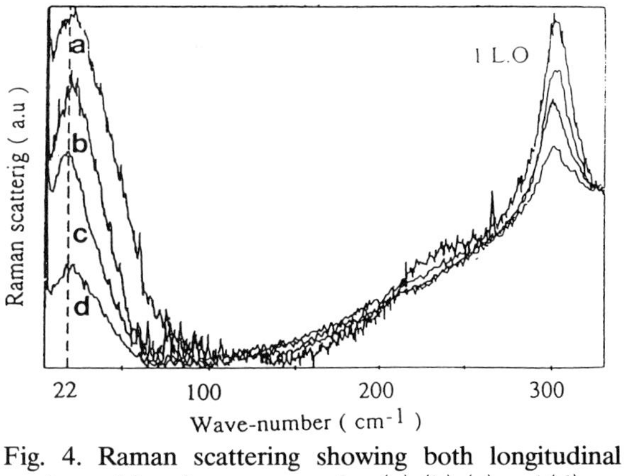

Fig. 4. Raman scattering showing both longitudinal optical modes and low frequency modes: (a), (b), (c) and (d) correspond to 5, 10, 15 and 20 wt.% CdO concentration respectively.

7. Conclusion

The sol-gel method is a simple technique for obtaining CdS semiconducting nanocrystals dispersed at high concentrations (in the range 2–20 wt.% CdO) in a silica glass. The doped silica glass was characterized by small angle X-ray scattering, high resolution electron micros-

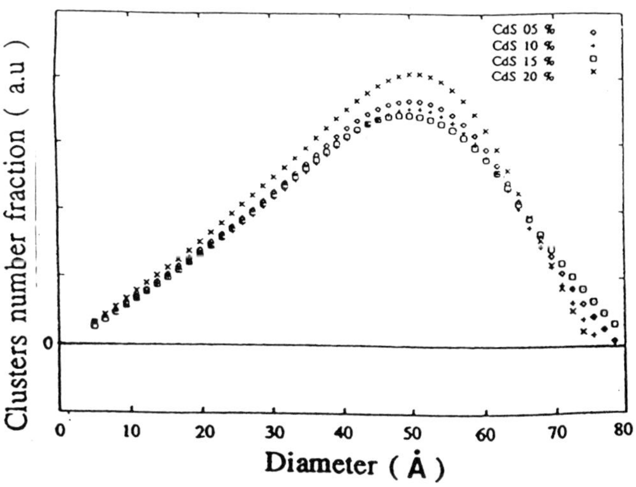

Fig. 5. The CdS size distribution determined by SAXS measurements. The CdS concentrations corresponding to the curves are indicated on the figure.

TABLE 1. The average distances between CdS clusters determined by small angle X-ray scattering

CdO concentration (wt.%)	Cluster distance (Å)
5	342
10	308
15	264
20	215

copy and low frequency Raman scattering. The results are coherent and show mainly that in the samples studied, and for all these concentrations, the size distribution is the same. This characteristic seems only to depend on the structure of the glass matrix: measurements are being carried out to confirm this conclusion in more detail.

References

1 Y. Wang, *J. Phys. Chem., 95* (1991) 1119.
2 E. Vogel, M. J. Weber and D. M. Krol, *Phys. Chem. Glass, 32* (1991) 231.
3 A. I. Ekimov and A. A. Onushchnko, *Sov. Phys. Semicond., 16* (1982) 757.
4 R. K. Jain and R. C. Lind, *J. Opt. Soc. Am., 73* (1983) 647.
5 M. Nogami, K. Nagasaka and E. Kato, *J. Am. Ceram. Soc., 73* (1990) 2097.
6 M. Nogami, M. Watabe and K. Nagasaka, *Sol-Gel Optics, 1328* (1990) 119.
7 A. Othmani, B. Champagnon and J. Dumas, to be published.
8 B. Champagnon, A. Boukenter and E. Duval, *J. Non-Cryst. Solids, 94* (1987) 216.
9 A. E. Love, *The Mathematical Theory of Elasticity*, Cambridge University Press, Cambridge, 1927.
10 A. Tamura, K. Higeta and T. Ichninokawa, *J. Phys. C, 15* (1982) 4957.
11 E. Duval, A. Boukenter and B. Champagnon, *Phys. Rev. Lett., 56* (1986) 2052.
12 C. Mai, I. Livet and G. Vigier, *Scripta Metall., 15* (1981) 1179.

Author Index of Volume 168 Issue 2

Subject Index of Volume 168 Issue 2